T0189443

Advances in Intelligent Systems and Computing

Volume 1056

The series "Advances in Intelligent Systems and Computing" contains publications on theory, applications, and design methods of Intelligent Systems and Intelligent Computing. Virtually all disciplines such as engineering, natural sciences, computer and information science, ICT, economics, business, e-commerce, environment, healthcare, life science are covered. The list of topics spans all the areas of modern intelligent systems and computing such as: computational intelligence, soft computing including neural networks, fuzzy systems, evolutionary computing and the fusion of these paradigms, social intelligence, ambient intelligence, computational neuroscience, artificial life, virtual worlds and society, cognitive science and systems, Perception and Vision, DNA and immune based systems, self-organizing and adaptive systems, e-Learning and teaching, human-centered and human-centric computing, recommender systems, intelligent control, robotics and mechatronics including human-machine teaming, knowledge-based paradigms, learning paradigms, machine ethics, intelligent data analysis, knowledge management, intelligent agents, intelligent decision making and support, intelligent network security, trust management, interactive entertainment, Web intelligence and multimedia.

The publications within "Advances in Intelligent Systems and Computing" are primarily proceedings of important conferences, symposia and congresses. They cover significant recent developments in the field, both of a foundational and applicable character. An important characteristic feature of the series is the short publication time and world-wide distribution. This permits a rapid and broad dissemination of research results.

** Indexing: The books of this series are submitted to ISI Proceedings, EI-Compendex, DBLP, SCOPUS, Google Scholar and Springerlink **

More information about this series at http://www.springer.com/series/11156

Subhransu Sekhar Dash · C. Lakshmi ·
Swagatam Das · Bijaya Ketan Panigrahi
Editors

Artificial Intelligence and Evolutionary Computations in Engineering Systems

 Springer

Editors
Subhransu Sekhar Dash
Department of Electrical Engineering
Government College of Engineering
Keonjhar, India

C. Lakshmi
Department of Software Engineering
SRM Institute of Science and Technology
Kattankulathur, Chennai, Tamil Nadu, India

Swagatam Das ⓘ
Electronics and Communication
Sciences Unit
Indian Statistical Institute
Kolkata, West Bengal, India

Bijaya Ketan Panigrahi
Department of Electrical Engineering
IIT Delhi
New Delhi, Delhi, India

ISSN 2194-5357   ISSN 2194-5365 (electronic)
Advances in Intelligent Systems and Computing
ISBN 978-981-15-0198-2   ISBN 978-981-15-0199-9 (eBook)
https://doi.org/10.1007/978-981-15-0199-9

This Springer imprint is published by the registered company Springer Nature Singapore Pte Ltd.
The registered company address is: 152 Beach Road, #21-01/04 Gateway East, Singapore 189721, Singapore

4th International Conference on Artificial Intelligence and Evolutionary Computations in Engineering Systems

April 11–13, 2019

Organized by Department of Software Engineering, School of Computing, SRMIST—Kattankulathur

Chief Patrons

Dr. T. R. Paarivendhar, Chancellor, SRMIST, India
Mr. Ravi Pachamoothoo, Chairman, SRM Group, India
Mr. P. Sathyanarayanan, President, SRMIST, India
Dr. R. Shivakumar, Vice President, SRMIST, India

Patrons

Dr. Sandeep Sancheti, Vice Chancellor, SRMIST, India
Dr. T. P. Ganesan, Pro-Vice Chancellor, SRMIST, India
Dr. C. Muthamizhchelvan, Director (Engg & Tech), SRMIST, India
Dr. D. Narayana Rao, Director (Research), SRMIST, India

General Chair

Dr. Bijaya Ketan Panigrahi, IIT Delhi, India
Dr. Subhransu Sekhar Dash, GCE, Keonjhar, India
Dr. Swagatam Das, ISI, Kolkata

General Co-chair

Dr. P. N. Suganthan, NUS, Singapore
Dr. A. A. Jimo, TUT, South Africa

Organizing Chair

Dr. C. Lakshmi, SRMIST, India
Prof. V. Sivakumar, SRMIST, India

Organizing Co-chair

Prof. S. Karthik, SRMIST, India
Prof. G. Senthil Kumar, SRMIST, India
Dr. T. S. Shiny Angel, SRMIST, India

Programme Chair

Dr. C. Lakshmi

Programme Co-chair

Dr. Ilhami Colak, Gazi University, Turkey
Dr. Ramazan Bayindir, Gazi University, Turkey
Dr. J. S. Femilda Josephin, SRMIST, India

Publicity Chair

Prof. S. Krishnaveni, SRMIST, India
Prof. S. Aruna, SRMIST, India

Public Relation Chair

Prof. C. G. Anupama, SRMIST, India
Prof. V. Haribaabu, SRMIST, India

Publication Chair

Dr. G. Usha, SRMIST, India
Dr. M. Ferni Ukrit, SRMIST, India

Finance Chair

Prof. Sasi Rekha Shankar, SRMIST, India
Prof. S. Selvakumarasamy, SRMIST, India

Organizing Committee

Prof. M. Uma, SRMIST, India
Prof. M. S. Abirami, SRMIST, India
Dr. N. Snehalatha, SRMIST, India
Prof. K. Vijayakumar, SRMIST, India
Prof. D. Anitha, SRMIST, India
Prof. J. Jeyasudha, SRMIST, India
Prof. S. Amudha, SRMIST, India
Prof. B. Jothi, SRMIST, India
Prof. M. Maheswari, SRMIST, India
Prof. S. Ramraj, SRMIST, India
Prof. C. Arun, SRMIST, India
Prof. S. Joseph James, SRMIST, India
Prof. A. Saranya, SRMIST, India
Prof. H. Karthikeyan, SRMIST, India

International/National Advisory Committee

Dr. S. R. S. Prabaharan, Joint Director Research, SRMIST, India
Dr. B. Amutha, SRMIST, India
Dr. S. S. Sridhar, SRMIST, India
Dr. G. Vadivu, SRMIST, India
Dr. S. Rajendran, SRMIST, India
Dr. R. Subburaj, SRMIST, India
Dr. V. Ganapathy, SRMIST, India
Dr. Akhtar Kalam, VU, Australia
Dr. Alfredo Vaccaro, USA

Preface

This AISC volume contains the papers presented at the 4th International Conference on Artificial Intelligence and Evolutionary Computations in Engineering Systems (ICAIECES-2019) held during 11–13 April 2019 at SRM Institute of Science and Technology, Kattankulathur, Chennai, India. ICAIECES-2019 is the fourth international conference aiming at bringing together the researchers from academia and industry to report and review the latest progresses in the cutting-edge research on various research areas of artificial intelligence, evolutionary computing, image processing, computer vision and pattern recognition, machine learning, data mining and computational life sciences, management of data including big data and analytics, distributed and mobile systems including grid and cloud infrastructure, information security and privacy, VLSI, antenna, computational fluid dynamics and heat transfer, intelligent manufacturing, signal processing, intelligent computing, soft computing, web security, privacy and e-commerce, e-governance, optimization, communications, smart wireless and sensor networks, networking and information security, mobile computing and applications, industrial automation and MES, cloud computing, electronic circuits, power systems, renewable energy applications, green and finally to create awareness of these domains to a wider audience of practitioners.

ICAIECES-2019 received 196 paper submissions, including seven foreign countries across the globe. All the papers were peer-reviewed by the experts in the area in India and abroad, and comments have been sent to the authors of accepted papers. Finally, 67 papers were accepted for oral presentation at the conference. This corresponds to an acceptance rate of 34% and is intended to maintain the high standards of the conference proceedings. The papers included in this AISC volume cover a wide range of topics in intelligent computing and algorithms and their real-time applications in problems from diverse domains of science and engineering.

The conference was inaugurated by Dr. Ivan Zelinka, Technical University of Ostrava, on 11 April 2019. The conference featured distinguished keynote speakers as follows: Dr. P. N. Suganthan, NTU, Singapore, Dr. Ramazan Bayindir, Gazi University, Turkey, Ho Chin Kuan, Multimedia University, Malaysia,

Dr. B. K. Panigrahi, IIT Delhi, India, Dr. Swagatam Das, ISI, Kolkata, India, and Dr. S. S. Dash, GCE, Keonjhar, Odisha, India.

We take this opportunity to thank the authors of the submitted papers for their hard work, adherence to the deadlines, and patience with the review process. The quality of a referred volume depends mainly on the expertise and dedication of the reviewers. We are indebted to the Technical Committee members, who produced excellent reviews in short time frames. First, we are indebted to the honourable Chancellor Thiru. T. R. Paarivendhar, President Thiru. P. Sathyanarayanan, and Vice Chancellor Professor Sandeep Sancheti wholeheartedly for the confidence they entrusted on us for organizing this ICAIECES-2019. We sincerely thank our beloved Pro-Vice Chancellor for their continuous support and guidance in organizing the conference. Our heartfelt thanks to the Registrar, Director (E&T), HODs, Professors, and the staff members of SRMIST, Kattankulathur, for their valuable support for the success of this programme. We thank the International Advisory Committee Members for providing valuable guidelines and inspiration to overcome various difficulties in the process of organizing this conference. We would also like to thank the participants of this conference and LDRA, India, for sponsoring. The members of faculty and students of SRM Institute of Science and Technology, Chennai, deserve special thanks because, without their involvement, we would not have been able to face the challenges of our responsibilities. Finally, we thank all the volunteers who made great efforts in meeting the deadlines and arranging every detail to make sure that the conference could run smoothly. We hope the readers of these proceedings find the papers inspiring and enjoyable.

Keonjhar, India	Dr. Subhransu Sekhar Dash
Kattankulathur, Chennai, India	Dr. C. Lakshmi
Kolkata, India	Dr. Swagatam Das
New Delhi, India	Dr. Bijaya Ketan Panigrahi
April 2019	

Contents

About the Editors

Dr. Subhransu Sekhar Dash is presently a Professor and HOD in the Department of Electrical Engineering, Government College of Engineering, Keonjhar, Odisha, India. He received his Ph.D. degree from College of Engineering, Guindy, Anna University, Chennai, India. He has more than 22 years of research and teaching experience. His research areas are AI techniques application to power system, modeling of FACTS controller, power quality and smart grid. He is a Visiting Professor at Francois Rabelais University, POLYTECH, France. He has published more than 220 research articles in peer-reviewed international journals and conferences.

Dr. C. Lakshmi is working as a Professor and Head in the Department of Software Engineering, SRM Institute of Science & Technology, India. She received her B.E., M.E., and Ph.D., degree in Computer Science and Engineering in 1990, 2001 and 2010 respectively. Her research interests include Pattern Recognition, Image Processing, Machine learning and Software Engineering. She has published many papers in International Journals and Conferences. She has served as guest editor in various international journals.

Dr. Swagatam Das received the B. E. Tel. E., M. E. Tel. E (Control Engineering specialization), and Ph.D. degrees, all from Jadavpur University, India, in 2003, 2005, and 2009, respectively. Currently, he is serving as an Assistant Professor at the Electronics and Communication Sciences Unit of Indian Statistical Institute, Kolkata. His research interests include evolutionary computing, pattern recognition, multi-agent systems, and wireless communication. Dr. Das has published one research monograph, one edited volume, and more than 150 research articles in peer-reviewed journals and international conferences. He is the founding co-editor-in-chief of "Swarm and Evolutionary Computation", an international journal from Elsevier. He serves as associate editor of the IEEE Transactions on Systems, Man, and Cybernetics: Systems and Information Sciences (Elsevier). He is an editorial board member of Progress in Artificial Intelligence (Springer), Mathematical Problems in Engineering, International Journal of Artificial

Intelligence and Soft Computing, and International Journal of Adaptive and Autonomous Communication Systems. He is the recipient of the 2012 Young Engineer Award from the Indian National Academy of Engineering (INAE).

Dr. Bijaya Ketan Panigrahi is working as a Professor in the Electrical Engineering Department, IIT Delhi, India. Prior to joining IIT Delhi in 2005, he has served as a faculty in Electrical Engineering Department, UCE Burla, Odisha, India, from 1992 to 2005. Dr. Panigrahi is a senior member of IEEE and Fellow of INAE, India. His research interest includes the application of soft computing and evolutionary computing techniques to power system planning, operation, and control. He has also worked in the fields of bio-medical signal processing and image processing. He has served as the editorial board member, associate editor, and special issue guest editor of different international journals. He is also associated with various international conferences in various capacities. Dr. Panigrahi has published more than 100 research papers in various international and national journals.

Automatic Generation Control Using Novel PD Plus FOPI Controller Tuned by Salp Swarm Algorithm

Nimai Charan Patel, Manoj Kumar Debnath, Binod Kumar Sahu and Subhransu Sekhar Dash

Abstract To reduce the frequency variation problem, here a novel controller is projected named as proportional derivative plus fractional order proportional integral (PD plus FOPI) controller. The newly designed controller is realised in a dual control area-based multi-unit model. The desired and suitable value of the parameter of the projected controller is obtained by salp swarm algorithm (SSA) concerning integral time absolute error (ITAE) as fitness or cost function. The designed multi-unit model is inspected with proportional integral derivative (PID), proportional derivative plus proportional integral (PD plus PI) and PD plus FOPI controller, and the dominance of the projected PD plus FOPI controller is established over others concerning different performance indices such as least undershoot, settling time and maximum overshoot. The system response is surveyed under two circumstances, namely (i) an abrupt load deviation of magnitude 0.01 per unit in control area 1 and (ii) abrupt load deviation of magnitude 0.06 per unit in control area 1. The later investigation also reveals that projected controller is robust as it can successfully manage the frequency abnormalities without further tuning the controller gains.

Keywords PD plus FOPI controller · Automatic generation control · Controller optimisation · Salp swarm algorithm · Multi-area multi-unit system

---

N. C. Patel (✉) · S. S. Dash
Government College of Engineering, Keonjhar, Odisha 758002, India
e-mail: ncpatel.iter@gmail.com

S. S. Dash
e-mail: munu_dash_2k@yahoo.com

M. K. Debnath · B. K. Sahu
Institute of Technical Education & Research, Siksha 'O' Anusandhan University,
Bhubaneswar, Odisha 751030, India
e-mail: mkd.odisha@gmail.com

B. K. Sahu
e-mail: binoditer@gmail.com

© Springer Nature Singapore Pte Ltd. 2020
S. S. Dash et al. (eds.), *Artificial Intelligence and Evolutionary Computations in Engineering Systems*, Advances in Intelligent Systems and Computing 1056,
https://doi.org/10.1007/978-981-15-0199-9_1

1 Introduction

The configuration of the modern power system is very complex in nature. It comprises a large network with different control areas interlinked through tie lines for power exchange. Each of these control areas consists of multiple generating units to meet the power demand of the consumers. Providing secured, reliable and uninterrupted power supply to the consumers and at the same time maintaining the power quality of the delivered power are stringent requirements of the present-day power system. Due to the large complexity of the power system, it is not so easy to meet this requirement. Two important parameters measuring the power quality of the power delivered by a power system are the voltage and the frequency which must be maintained at a nominal value irrespective of any changes in the power system. The voltage profile of a power system is affected by the reactive power demand, whereas the frequency profile of a power system depends upon the active load on the power system. The voltage fluctuation is addressed by implementing automatic voltage regulator (AVR) while the frequency fluctuation is addressed by implementing automatic generation control (AGC) or load frequency control (LFC). The load on the power system is random in nature. An increase in load tends to reduce the frequency and vice versa. AGC helps to uphold the system frequency and interline power flow at a nominal value within the tolerable limits during any load changes on the power system by maintaining the balance between the generation and the load [1–3].

The initial work on the AGC was undertaken by Cohn in the year 1956 [4]. In the subsequent years, many research works on the topic were carried out by various researchers as evident from the literature review [5–7]. Elgerd and Fosha in their work presented an optimal control method to claim that lower bias settings help to give a better response for LFC of power system [8]. Lyapunov technique was used to select the various control parameters for LFC of nonlinear power system [9]. It was established that the use of battery energy storage system (BES) in LFC significantly enhances the system performance [10]. LFC under deregulated environment was studied by Christie and Bose [11]. In recent years, extensive study on LFC of power systems has been carried out by implementing various controllers using different optimisation algorithms to optimise the controller parameters. Multi-stage fuzzy controller for LFC of deregulation-based power system was optimally designed using particle swarm optimisation (PSO) [12]. Teaching–learning-based optimisation (TLBO) designed fuzzy PID controller was employed to investigate the LFC issues of power system [13]. Performance of hybrid local unimodal sampling teaching–learning-based optimisation (LUS-TLBO) algorithm optimised fuzzy PID controller with derivative filter (fuzzy PIDF) was studied, and its dominance over PID, PIDF and fuzzy PID controller was established [14]. Differential evolution and particle swarm optimisation algorithm were hybridised to optimally design fuzzy PID controller for LFC study [15]. Differential evolution and grey wolf optimisation algorithm have been hybridised to set the parameters of fuzzy PID controller for LFC of power system with different sources of energy [16]. Performance comparison of PID, fractional order PID (FOPID), fuzzy PID (FPID) and fractional order fuzzy PID (FOFPID) controllers optimally designed with ant lion optimiser (ALO) algorithm has been carried out for power system with different energy sources [17]. Multi-staged PID controller optimised by invasive weed optimisation (IWO) has been recommended for LFC issues of solar energy-based power system, and its supremacy over the classical PID controller has been established [18]. Whale optimisation algorithm (WOA) was used to tune the gains of fuzzy integrated proportional integral (FIPI) controller to solve the LFC problems of multi-area power system [19]. Cuckoo search algorithm (CSO)-based 2DOF-PID controller has been employed for LFC of a nonlinear power system including governor dead band [20].

In the present work, a two-area multi-unit steam power plant with reheat turbine is considered. Each area consists of three equal thermal units with equal participation factor of 1/3. A novel PD plus FOPI controller is proposed for load frequency control of the considered power system. The performance of the power system is studied with three different controllers, i.e. (i) PID, (ii) PD plus PI and (iii) PD plus FOPI controller, when a step load of 0.01 p.u. is applied in area 1, and it is established that the PD plus FOPI controller outperforms the other controllers. Lastly, the robustness of the proposed controller is validated by increasing the load in area 1 by 5%.

2 System Modelling

The power system under study for this work is shown in Fig. 1. It consists of two equal-area multi-unit reheat steam power plants. Various parameters of the system under consideration are depicted in Appendix 1. The performance of the power system is investigated with three different controllers such as (i) PID, (ii) PD plus PI and (iii) PD plus FOPI controller. The parameters of these controllers are optimally selected by using SSA.

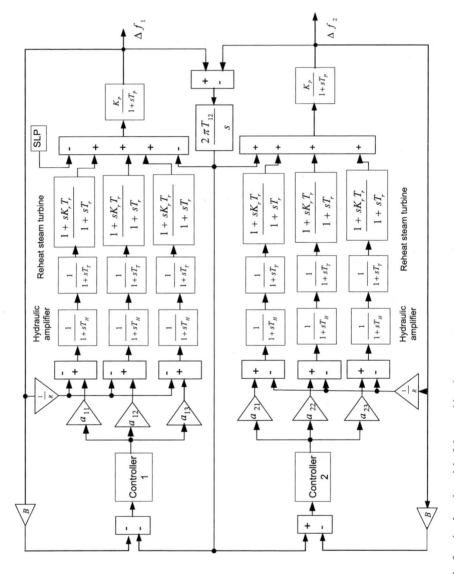

Fig. 1 Transfer function-based model of the considered power system

3 Structure and Optimal Design of Controllers

3.1 Structure of Controllers

PID controller delivers the proportional, integral and derivative control actions embedded in parallel. PD plus PI controller is a combination of PD and PI controller where PD controller includes the proportional and derivative actions and PI controller includes the proportional and integral actions. The structure of PD plus FOPI controller is shown in Fig. 2 where PD and FOPI controllers are cascaded together. FOPI controller is different than the traditional PI controller in the sense that the former comprises an integrator of fractional order λ, whereas the later has integer order integrator. Thus, the extra knob λ present in the FOPI integrator has the advantage of offering more flexibility in terms of design and adjustment of system dynamics. In order to obtain superior dynamic performance of the system, parameters of the different controllers such as the gains and the fractional order of the integrator are optimised by employing the SSA.

3.2 Objective Function

Integral time absolute error (ITAE) has been taken as the objective function in the proposed optimisation algorithm for optimal selection of the various parameters of the controllers. The mathematical expression for the objective function is given by Eq. (1).

$$\text{ITAE} = \int_{0}^{t} (|\Delta f_1| + |\Delta f_2| + |\Delta P_{\text{tie}}|) \cdot t dt \qquad (1)$$

3.3 Salp Swarm Algorithm

SSA is a stochastic optimisation technique inspired by the swarming nature of salps during their navigation in search of foods in ocean. It is a metaheuristic algorithm based on swarm intelligence technique which simulates the foraging behaviour of salps. Stochastic optimisation technique can be either individualist or collective. SSA is a collective technique which produces random solution in the search space. These solutions are evolved during the optimisation process and collaborate to compute the global optimum in the search area. The main advantage of SSA is that it does not require the gradient information because it considers the optimisation problem as a black box and monitors the inputs, outputs and constraints of the problem. Other advantages of the SSA are its simplicity and its ability to avoid local

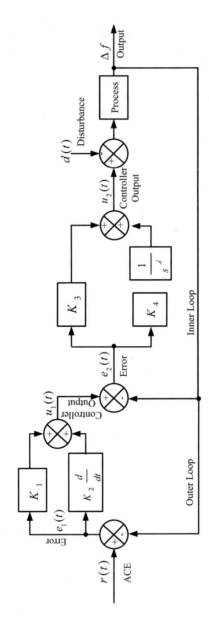

Fig. 2 Structure of PD plus FOPI controller

optima. SSA undergoes through both the exploration and exploitation phase. The algorithm finds the promising regions in the search space and avoids the local optima during exploration phase, whereas the exploitation phase helps to enhance the precision of the best solutions achieved in the exploration phase.

Salps move in a similar manner as jellyfishes forming a swarm called salp chain. The salp chain can be mathematically modelled by dividing the population into two clusters as (i) leader and (ii) followers. The salp in front of the chain guiding other salps in the chain is the leader. All other salps following each other thus directly or indirectly following the leader in the chain are the followers. A two-dimensional matrix $x = [NS \times D]$ stores the position of all the salps where NS is the numbers of solutions and D is the dimensions of search space. Considering a food source F in the search area as the target of the swarm, the location of the salp leader is updated using Eq. (2).

$$x_n^1 = F_n + k_1((ub_n - lb_n)k_2 + lb_n), \quad k_3 \geq 0$$

and

$$x_n^1 = F_n - k_1((ub_n - lb_n)k_2 + lb_n), \quad k_3 < 0 \tag{2}$$

where x_n^1 indicates the position of the leader salp in the nth dimension, F_n indicates the food source position in the nth dimension, ub_n denotes the upper bound of the nth dimension, lb_n denotes the lower bound of the nth dimension, and k_1, k_2, k_3 are the random numbers. Coefficient k_1 is used to maintain balance between the exploration and exploitation and is given by the following equation.

$$k_1 = 2e^{-(4i/I)^2} \tag{3}$$

where i is the current iteration and I is the maximum number of iterations. Parameters k_2 and k_3 are uniformly produced random numbers in the interval $[0, 1]$ where k_2 and k_3 are used to find the subsequent position in the nth dimension and the step size.

The position of the followers is updated with the help of the following equation which represents Newton's law of motion.

$$x_n^m = ut + \frac{1}{2}at^2 \tag{4}$$

where $m \geq 2$, x_n^m denotes the position of the mth follower salp in nth dimension, t is the time, u is the initial velocity, and a is the acceleration.

In optimisation, time means iteration, and therefore, the time between the subsequent iteration is equal to 1. With this and considering the initial velocity $u = 0$, the above equation can be rewritten as follows.

$$x_n^m = \frac{1}{2}\left(x_n^m + x_n^{m-1}\right) \tag{5}$$

In the algorithm, the food source F in the search space is the global optimum towards which the salp leader and hence the salp chain moves and it is assumed that the finest solution achieved so far is the global optimum towards which the salp chain chases. The various steps of the algorithm are explained below.

Step 1: Initialisation: In this stage, the initial solution with random positions of salps approaching the global optimum is generated and stored in a two-dimensional matrix $x = [\text{NS} \times D]$, where NS is the number of population and D is the dimension of the search space.

Step 2: Evaluation: In this stage, the fitness function $f(x_i)$ of each salp is evaluated and the salp with the best fitness is determined. The position of the best salp is assigned to the variable F towards which the salp chain moves.

Step 3: Updation: In this stage, the coefficient k_1 is updated using Eq. (3) and also the position of the leader and the follower salps for each dimension is updated using Eqs. (2) and (5), respectively.

Steps (2) and (3) are repeated until the ending criterion are fulfilled, and finally, the best fitness solution is selected.

4 Result Analysis

The power system under study is subjected to an abrupt load of 1% in area 1, and the dynamic performance of the system is studied separately in the presence of the PID, PD plus PI and PD plus FOPI controllers. The study is carried out by simulating the power system model in MATLAB–Simulink environment where the optimisation process SSA is coded in .m file and the Simulink model of the considered system is called through the SSA program for optimal design of the controllers.

The optimised parameters of the controllers are depicted in Table 1, whereas the transient performance specifications of the power system are depicted in Table 2. The frequency oscillations in area 1 and area 2 are shown in Figs. 3 and 4, respectively, whereas the inter-line power oscillation is shown in Fig. 5. These figures along with Table 2 reveal that the power system exhibits superior transient

Table 1 SSA tuned values of the controller parameters

Controllers	Area 1					Area 2				
	K_1	K_2	K_3	K_4	λ	K_1	K_2	K_3	K_4	λ
PD-FOPI	3.8343	3.1896	2.6564	3.8059	0.95	2.8665	2.3825	0.6822	2.5026	0.89
PD-PI	3.9008	1.2132	3.6134	3.9043	NA	2.0125	2.0013	2.7215	0.1000	NA
PID	K_p		K_i	K_d		K_p		K_i	K_d	
	2.966		6.5902	0.9320		0.710		0.6827	0.7419	

Table 2 Transient response specifications like least undershoot, settling time and maximum overshoot

Frequency/ inter-line power	Transient response specifications	PD-FOPI	PD-PI	PID
Δf_1	Undershoots (Hz)	−0.0031	−0.0043	−0.0107
	Overshoots (Hz)	0.00009725	0.0008830	0.0031
	T_s in sec (0.05% band)	0.6500	0.6700	3.0700
Δf_2	Undershoots (Hz)	−0.0006	−0.0009	−0.0060
	Overshoots (Hz)	0.00002948	0.0000366	0.0014
	T_s in sec (0.05% band)	0.6500	0.7900	3.4900
ΔP_{tie}	Undershoots (PU)	−0.0002	−0.0004	−0.0020
	Overshoots (PU)	0.00002093	0.0000176	0.0005
	T_s in sec (0.005% band)	2.5100	2.8600	2.8655

Fig. 3 Frequency fluctuation in area 1

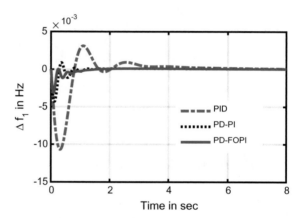

Fig. 4 Frequency fluctuation in area 2

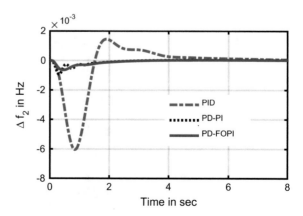

Fig. 5 Inter-line power
fluctuation

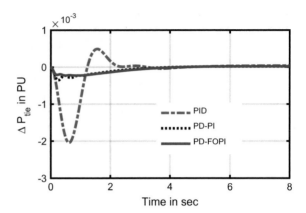

Fig. 5 Inter-line power
fluctuation

response in the presence of PD plus FOPI controllers as compared to the other
controllers. Further, it can be seen that the performance of PD plus PI controller is
much better than the conventional PID controller in terms of the peak magnitude of
undershoot as well as overshoot and settling time of the frequency and inter-line
power oscillations. Observation also reveals that the performance of PD plus PI
controller can further be improved by replacing the integer order integrator of the
controller by a fractional order integrator which leads to the development of PD
plus FOPI controller.

4.1 Robustness Analysis: Area

The robustness of the proposed PD plus FOPI controller is verified by applying an
SLP of 0.06 p.u. in area 1 with the existing controller parameters. The frequency
oscillations in area 1 and area 2 are depicted in Figs. 6 and 7, respectively, whereas

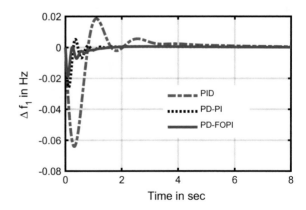

Fig. 6 Frequency fluctuation
in area 1

Fig. 7 Frequency fluctuation in area 2

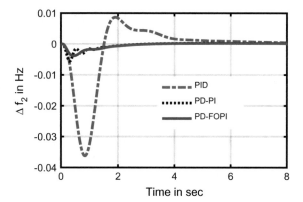

Fig. 8 Inter-line power fluctuation

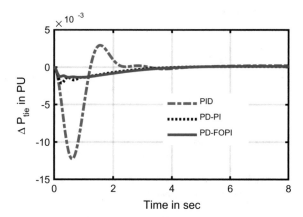

the inter-line power oscillation is depicted in Fig. 8. It is seen from these figures that even if the SLP in area 1 is increased by 5%, the stability of the power system remains intact and there is no significant deviations in the transient response of frequencies and interline power in case of the PD plus FOPI controller.

5 Conclusion

In this work, a dual area, multi-unit thermal system with reheat turbine was considered to study the dynamic performance of power system separately in the presence of PID, PD plus PI and PD plus FOPI controllers optimally designed by SSA. An SLP of 0.01 p.u. was applied in area 1, and it was observed that the PD plus FOPI outperforms the other two controllers as far as the transient responses of frequencies and inter-line power are concerned. It was also found that the PD plus

PI controller delivers significantly improved performance over the conventional PID controller. Lastly, the sturdiness of the proposed PD plus FOPI controller was established by increasing the SLP in area 1 from 1 to 6%.

Appendix 1

$T_H = 0.08$ s, $T_T = 0.3$ s, $T_r = 10$ s, $K_r = 0.5$, $T_P = 20$ s, $K_P = 120$, $R = 2.5$ Hz/MW, $B = 0.425$ MW/Hz, $T_{21} = 0.086$, $\alpha_{11} = \alpha_{12} = \alpha_{13} = \alpha_{21} = \alpha_{22} = \alpha_{23} = 1/3$.

References

1. Kundur, P., Balu, N.J., Lauby, M.G.: Power System Stability and Control. McGraw-hill, New York (1994)
2. Elgard, O.I.: Electric Energy Systems Theory, pp. 299–362. McGraw-Hill, New York (1982)
3. Saadat, H.: Power System Analysis. In: McGraw-Hill Series in Electrical Computer Engineering (1999)
4. Cohn, N.: Some aspects of tie-line bias control on interconnected power systems. Trans. American Inst. Electrical Eng. Part III Power Apparatus Syst. **75**(3)1415–1436 (1956)
5. Shankar, R., Pradhan, S.R., Chatterjee, K., Mandal, R.: A comprehensive state of the art literature survey on LFC mechanism for power system. Renew. Sustain. Energy Rev. **1**(76), 1185–1207 (2017)
6. Pandey, S.K., Mohanty, S.R., Kishor, N.: A literature survey on load–frequency control for conventional and distribution generation power systems. Renew. Sustain. Energy Rev. **1**(25), 318–334 (2013)
7. Kumar, P., Kothari, D.P.: Recent philosophies of automatic generation control strategies in power systems. IEEE Trans. Power Syst. **20**(1), 346–357 (2005)
8. Elgerd, O.I., Fosha, C.E.: Optimum megawatt-frequency control of multiarea electric energy systems. IEEE Trans. Power Appar. Syst. **4**, 556–563 (1970)
9. Tripathy, S.C., Hope, G.S., Malik, O.P.: Optimisation of load-frequency control parameters for power systems with reheat steam turbines and governor dead band nonlinearity. In: IEEE Proceedings C (Generation, Transmission and Distribution), vol. 129, no. 1, pp. 10–16. IET Digital Library (1982)
10. Aditya, S.K., Das, D.: Application of battery energy storage system to load frequency control of an isolated power system. Int. J. Energy Res. **23**(3), 247–258 (1999)
11. Christie, R.D., Bose, A.: Load frequency control issues in power system operations after deregulation. IEEE Trans. Power Syst. **11**(3), 1191–1200 (1996)
12. Shayeghi, H., Jalili, A., Shayanfar, H.A.: Multi-stage fuzzy load frequency control using PSO. Energy Convers. Manag. **49**(10), 2570–2580 (2008)
13. Sahu, B.K., Pati, S., Mohanty, P.K., Panda, S.: Teaching–learning based optimization algorithm based fuzzy-PID controller for automatic generation control of multi-area power system. Appl. Soft Comput. **1**(27), 240–249 (2015)
14. Mohanty, P.K., Sahu, B.K., Pati, T.K., Panda, S., Kar, S.K.: Design and analysis of fuzzy PID controller with derivative filter for AGC in multi-area interconnected power system. IET Gener. Transm. Distrib. **10**(15), 3764–3776 (2016)

15. Sahu, B.K., Pati, S., Panda, S.: Hybrid differential evolution particle swarm optimisation optimised fuzzy proportional–integral derivative controller for automatic generation control of interconnected power system. IET Gener. Transm. Distrib. **8**(11), 1789–1800 (2014)
16. Debnath, M.K., Mallick, R.K., Sahu, B.K.: Application of hybrid differential evolution–grey wolf optimization algorithm for automatic generation control of a multi-source interconnected power system using optimal fuzzy–PID controller. Electr. Power Compon. Syst. **45**(19), 2104–2117 (2017)
17. Patel, N,C., Sahu, B.K., Bagarty, D.P., Das, P., Debnath, M.K.: A novel application of ALO-based fractional order fuzzy PID controller for AGC of power system with diverse sources of generation. Int. J. Electrical Eng. Educ. 0020720919829710 (2019). https://doi.org/10.1177/0020720919829710
18. Patel, N.C., Debnath, M.K., Sahu, B.K., Dash, S.S., Bayindir, R.: Application of invasive weed optimization algorithm to optimally design multi-staged PID controller for LFC analysis. Int. J. Renew. Energy Res. **9**(1), 470–479 (2019)
19. Patel, N.C., Debnath, M.K.: Whale optimization algorithm tuned fuzzy integrated PI controller for LFC problem in thermal-hydro-wind interconnected system. In: Applications of Computing, Automation and Wireless Systems in Electrical Engineering, pp. 67–77. Springer, Singapore (2019)
20. Patel, N.C., Debnath, M.K., Sahu, B.K., Das, P.: 2DOF-PID controller-based load frequency control of linear/nonlinear unified power system. In: International Conference on Intelligent Computing and Applications, pp. 227–236. Springer, Singapore (2019)

Plant Disease Identification and Detection Using Support Vector Machines and Artificial Neural Networks

S. Iniyan, R. Jebakumar, P. Mangalraj, Mayank Mohit and Aroop Nanda

Abstract In growing nations like India, agriculture plays a vital role in the economy. Increase in agro-products affects the GDP of the nation to a good extent. To increase the productivity in agriculture, early detection of diseases needs to be identified and addressed. In the research work, we have concise our discussion with detection of crop diseases using machine learning techniques, especially with support vector machine (SVM) and artificial neural network (ANN). We have concluded our survey with the pros and cons of every method in context with input parameters (Crop type).

Keywords Agriculture · Disease detection · Machine learning · Artificial neural network · Support vector machine

1 Introduction

Technological innovations in the field of agriculture have drastically improved in the last few decades. Still, some of the pertinent issues like detecting diseases in crop during its early stage are a challenging one. Due to the large plantation area of crops, it is very difficult to identify the affected plants that have been infected with diseases [1, 2]. In general, most of the agricultural areas in India are supervised

S. Iniyan · R. Jebakumar · P. Mangalraj (✉) · M. Mohit · A. Nanda
SRMIST, Kattankulathur, Chennai 603203, India
e-mail: mangal86@gmail.com

S. Iniyan
e-mail: iniyan.sv@gmail.com

R. Jebakumar
e-mail: jebakumar.r@ktr.srmuniv.ac.in

M. Mohit
e-mail: mayank.mohit1122@gmail.com

A. Nanda
e-mail: aroopnanda@gmail.com

© Springer Nature Singapore Pte Ltd. 2020   15
S. S. Dash et al. (eds.), *Artificial Intelligence and Evolutionary Computations in Engineering Systems*, Advances in Intelligent Systems and Computing 1056,
https://doi.org/10.1007/978-981-15-0199-9_2

manually. Even if some plantations are inspected by experts, it is very expensive and not efficient. To identify the severity of diseases in crops/plants, the automation becomes a necessary part. Recent developments in machine learning and artificial intelligence have helped researchers to address these pertinent issues to a good extent. In the machine learning techniques, the widely used techniques are SVM and ANN due to their advantages. For example, SVM needs a fewer sample to identify the classes precisely, wherein ANN the error rate used to improve the input parameters for prescriptive analytics [3]. In the short survey, we have concise our discussions only with these two methods. The paper has organized as follows: In Sect. 1.1, we have discussed the few commonly occurring diseases in crop, in Sect. 2, we have discussed basic methodology widely in identifying diseases, in Sect. 2.3, we have discussed the primary feature extraction methods, in Sect. 3, we have provided few backgrounds about SVM and different ANN methods, in Sect. 4, we have provided the survey of different methods in context with the different disease and crop, and in Sect. 5, we have concluded our survey with the merits and demerits of every method in context with input parameters (Crop type).

1.1 Diseases Commonly Found in Crops

In this section, we will discuss few commonly occurring diseases in Indian crops.

Blight Blight disease in crops usually occurred due to the presence of a pathogenic organism such as bacteria or fungus. Blight disease is of many types such as Late Blight, Fire Blight, Early Blight, or Bacterial Blight. The major symptoms are the presence of spotting, yellowing, or discoloring of leaves.

Cankers Cankers disease is found in crops and shrubs and the main cause of Cankers disease is biotic fungi, Geosmithia fungus, and Fusarium solanium. The major symptoms are yellowing and thinning of leaves in the upper crown.

Rust Rusts is fungal disease found in crops like wheat and rice. Rust is majorly caused by Puccinia graminis. The minor symptoms are pale-leaf spots, a formation of pustules. The major symptoms of rusts lead to premature shredding of leaves.

Wilt Wilts is a water stress disease that usually occurs in pine trees and scrubs. Wilt disease basically caused by pathogens called Ceratocystis and Bursaphelenchus xylophilus. The plant affected by Wilt leads to death of trees/shrubs.

2 Basic Methodology

In general, the identification of disease in crops/plants/shrubs carried out in four steps. The basic methodology used for the identification of disease in crops/plants/shrubs has been provided in Fig. 1.

Fig. 1 Basic methodology

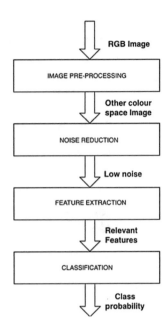

RGB Image

IMAGE PRE-PROCESSING

Other colour
space Image

NOISE REDUCTION

Low noise

FEATURE EXTRACTION

Relevant
Features

CLASSIFICATION

Class
probability

2.1 Input Preprocessing

In our research work, we have considered the images as input. During the image preprocessing, the given image will be converted into appropriate color space. Several color space models are available such as

- Hue, saturation, and intensity (HSI),
- Lightness, green-red, and blue-yellow (CIELAB),
- Hue, saturation, and luminescence (HSL),
- Hue, saturation, and value (HSV), and so on.

Usually, researchers use appropriate color models for their research work. The HSI model is one of the widely used color space models in this research area.

2.2 Filtering

Image filtering is an important task in the image preprocessing, unlike color space conversion image filtering plays a crucial role in the research. We have provided list of few noises and their corresponding filters used widely in the current state of the art:

- Gaussian noise—Gaussian, inverse Gaussian, mean/average filters, and so on,
- Salt and pepper noise—Median, adaptive filters, and so on,

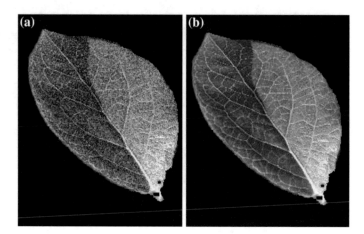

Fig. 2 a Original image; **b** After applying adaptive median filter

- Impulse filter—Bilateral, Gaussian Markov based, and so on,
- Speckle—Lee, frost, and so on.

In order to apply any filter, identifying the noise type is more important. For example, in case of salt and pepper noise, if we apply Gaussian filter, it results in over smoothening of features like textures, spots, etc. Identifying the type of noise becomes a crucial step because of its impacts in feature extraction. Figure 2 shows the effect of median filter in an image having salt and pepper noise while applying adaptive median filter (Fig. 2).

2.3   Feature Extraction

Feature extraction plays an important role in any machine learning algorithms. The precision of any machine learning algorithms totally depends on the feature selected for the analytical procedure. Extracting the appropriate features is called as feature engineering, and usually, a domain expert is needed to perform feature engineering for better results [3]. In image processing, there are few common features that are widely used in the analytics and the same are provided below.

Edges Edge detection uses various mathematical models to find points of discontinuities in an image. A discontinuity might represent a sudden change in image brightness or color. This helps us divide the image into various parts [3]. Unnecessary parts of the image can be removed to reduce the complexity of the image, for example. The blight disease usually has a drastic change in the discoloring of leaves. The edges extracted from blight-affected leaves have been depicted in Fig. 3. Usually for edge detection, several methods have been used such as gradient based, sobel, canny, and so on. Selection of method usually depends on the type of feature needed for the analytical procedure.

Corner Points and Blob Detection Corner points or interest points can be said as the intersection of multiple edges at a point. These interest points generally give the main points of focus in the image thus sparing the trouble of focusing on each and every pixel in an image. However, corner detection algorithms are usually not very robust, and hence, it is required to be carried out several times in order to prevent errors which might cost the whole recognition task, for example. If canker disease affects any plants, it usually thins the leaves. The corner or interest point features can be easily used to classify the canker affected and not affected plants. The corner point detection in leaves has been depicted in Fig. 4a, b. Usually to detect the corner or interest points, the researchers prefer scale-invariant feature transform (SIFT) or speeded up robust features (SURF).

Blob detection uses various methods to detect regions in the image that differ in properties with their surrounding regions. For example, points inside a blot are similar in properties such as dots or contours. This can be used to discard irrelevant pixels in an image. The blob detection to detect blots has been depicted in Fig. 4c. Usually, the blobs can be detected for blot-affected leaves which can be carried out with the help of morphological operations such as erosion and dilation.

2.4 Classification

Classification is the final step in identifying the plant disease using images. There are several machine learning algorithms to perform classification. Each model has its own advantages and disadvantages. In our short survey, we have discussed the widely used two models: 1. Support Vector Machine (SVM) and 2. Artificial Neural Network [ANN]. Support vector machine is one of the most widely used methods

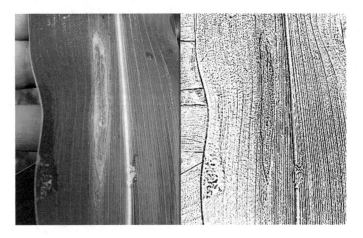

Fig. 3 Edge detection for blight disease identification

Fig. 4 **a** Original image; **b** Corner point detection; **c** Blob detection

Fig. 5 Binary classification using SVM

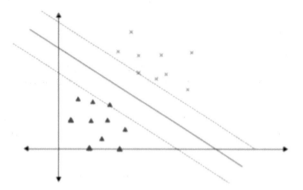

for classification purposes due to its advantages like converting features to the high-dimensional space and creating a hyperplane to classify according to the feature set. Another major advantage of SVM is that even with fewer samples, the machine will be able to create a stable boundary to classify the inputs. Figure 5 depicts how a hyperplane classifies the input into binary classes.

There are several types such as multi-class SVM, where multiple hyperplanes used to categorize the inputs into different classes.

Artificial Neural Networks: Artificial neural network works in similar way like our brain, how a network of neurons transfers different signals and makes us to take decisions. ANN is inspired by our nervous system and contains an interconnected group of nodes known as artificial neuron. Figure 6 depicts how an artificial network classifies the inputs into binary classes.

The major advantage of ANN is to retune the input according to the desired output. So vastly ANN is used for prescriptive analytics. There are different types of ANN present in the state of the art such as

– Back-propagation model,
– Convolutional neural network,

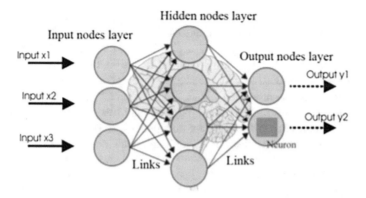

Fig. 6 Binary classification using ANN

- Pulse coded neural network,
- Recurrent neural network, and so on.

3   Discussion on SVM-Based Methods

In this section, we will discuss few SVM-based methods which used images as input and SVM as classifier. We have scrutinized few methods which have been widely used in the field of disease identification in plants.

In [3], the authors have used K-means algorithm for image segmentation and linear support vector machine (LSVM) for classification. Features like shape, color, and textures were extracted and combined to generate feature vector before the training of LSVM. 88.89% of average accuracy was achieved using LSVM classifier.

In [4], the authors explore the detection algorithm on leaf images on VGG16 and transfer learning. VGG16 is used to extract features, and SVM is used to detect the plant disease or pests. Initially, the input image (224 × 224 RGB image) is passed on to a stack of convolution layers and then through three fully connected (FC) layers. The output is employed to train a SVM model. Next, fine-tuning is used to construct an end-to-end classification model based on the original VGG16 model by replacing the three FC layers with two FC layers with 2048 and 11 nodes respectively corresponding to the ten categories of pests and diseases and healthy. The overall accuracy achieved was 89%, given that high-quality test images were used.

In [5], the authors used K-means clustering for image segmentation. Texture, color, and shape of the leaf are used as features, and gray level co-occurrence matrix (GLCM) is used for feature extension. Neural network and support vector machine are used for classification. Neural network used texture, color, and shape features achieving an overall accuracy of 80.21%, and on the other hand, SVM used texture and shape features achieving an overall accuracy of 89.23%.

In [6], the authors have suggested a pipeline for plant disease detection from RGB image captured using a digital camera. In the first step, the RBG image is converted into its respective HIS translation in order to make the image device independent. Next the unwanted pixels, i.e., the green pixels in the image, are masked and removed from the image. The resultant image is then segmented into components, and useful components are obtained. Color co-occurrence and SGDM are used to find the texture features in the image that can be fed to a classifier for disease identification. They have used two classifiers, minimum distance criterion, and SVM to identify the disease. The SVM classifier gave a better accuracy of 94.74% than that of minimum distance criterion (86.77%).

In [7], the authors have acquired images of the leaves captured using a high-resolution digital camera resized them to 256×256. The image is then sent for preprocessing that aims at enhancing the required portions of the image and suppress the distortions. This step also includes image enhancement using SFCES and color space conversion to YCbCr and L*a*b*. Then segmentation is done using K-means, and the features extracted using GLCM. These extracted features are fed to SVM-RBF (radial basis kernel) and SVMPOLY (polynomial kernel) which give an accuracy rate of 96% and 95%, respectively.

In [8], the authors have proposed a new algorithm with the help of simple linear iterative cluster (SLIC) combined with support vector machine (SVM) SLICSVM for identification of disease in Tea plant leaf. SLIC is used for extracting the boundary region of disease-infected areas in the leaf using super-pixel block and convex hull method. Classification of disease over the boundary region is done by the SVM classifier achieving 98.5% accuracy from 261 diseased images.

In [9], the authors have used Haar wavelet and gray level co-occurrence matrix (GLCM) for feature extraction of fungal disease detection in Maize Leaf, SVM, and KNN for disease classification. Nearly 200 images consist of three categories of diseased leaf images combined with images of healthy leaves producing 85% of accuracy in terms of detecting and classification of leaf disease in maize.

In [10], they have proposed a novel neutrosophic logic-based segmentation technique, an intuitionistic fuzzy-based approached which segments the GLCM into true, false, and intermediate sets. Different classifiers such as decision tree, Naive Bayes, KNN, SVM, random forest, Adaboost, ANN, discriminant analysis, generalized linear model (GLM) along with neutrosophic sets compared for classification of plant disease. From 400 cases of leaf images both healthy and diseased, the proposed system produced accuracy of 98.4% classification accuracy is achieved with the combination of GLM and neutrosophic sets.

4 Discussion on Artificial Neural Network-Based Methods

In this section, we have confined our survey specifically with two types of artificial neural networks, namely convolutional and recurrent neural networks.

4.1 Convolutional Neural Network (CNN)

CNN or convolutional neural network is made up of neurons that take inputs (usually multi-channeled image) and then take the weighted sum over them and respond with an output by passing through the activation function. CNN is very powerful image processing technique that uses deep learning to perform both generative and descriptive tasks.

ImageNet Large Scale Visual Recognition Challenge (ILSVRC) evaluates the algorithm for object detection and image classification. List of various CNN architecture along with the year in which they won ILSVRC competition has been summarized in Table 1.

4.2 Recurrent Neural Network (RNN)

Recurrent neural network is an artificial neural network in which a directed graph is formed by the connections between the nodes. RNN has memory (internal state) to process the sequence of inputs. So, RNN saves the output of a layer and feeds this back to the input to help in predicting the output of the layer. Figure 7 depicts the working of RNN. Among all the artificial neural networks, convolution neural networks are considered the best algorithms for image classification. Performance of CNN in ImageNet is close to that of humans. However, they are still prone to misclassification if the objects are very small or thin or have been distorted by filters.

Table 1 Various CNN architectures	CNN architecture	Year	Top-5 error rate (%)
	LeNet	ILSVRC—1998	——
	AlexNet	ILSVRC—2012	15.3
	ZFNet	ILSVRC—2013	14.8
	GoogLeNet	ILSVRC—2014	6.67
	VGGNet	ILSVRC—2014	6.67
	ResNet	ILSVRC—2015	3.6

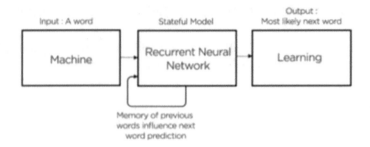

Fig. 7 Recurrent neural network (RNN)

4.3 Discussions

In this section, we will discuss few ANN-based methods which used images as input and used either CNN or RNN as classifier. We have confined our discussion with few methods which have been commonly used in the field of disease identification in plants.

In [11], the authors have implemented two architectures, i.e., AlexNet and GoogLeNet on PlantVillage dataset with three different visual representations of image data, i.e., segmented image, color image, and grayscale image. They have implemented two training techniques, i.e., transfer learning and training from scratch with the various training–testing set distributions such as 80–20 or 60–40. As a result, they were able to achieve the accuracy ranging from 85.53% (in case of AlexNet, training from scratch, grayscale, 8020) to 99.34% (in case of GoogLeNet, transfer learning, color, 8020). On a comparison between GoogLeNet and AlexNet architectures, GoogLeNet performed better.

In [12], the authors used a high-resolution camera to capture pictures of the leaf so as to get better results and greater efficiency. At first, they use K-means to find and mask the green parts of the leaves as they do not provide any useful information. In the next steps, the masked pixels are completely removed and the infected parts of the leaf are converted to HSI color format. Using the H and S values of each pixel, a graylevel co-occurrence matrix (GLCM) is generated. Next SGDM is used to calculate statistics for texture analysis. The features extracted are fed to a back-propagation neural network (BPNN) where the training weights are updated in every iteration until the defined iteration number or until they reach an acceptable error.

In [13], the authors have used convolutional neural network (CNN) with various layers such as convolutional layers, pooling layer, and activation layer. Softmax function is used in the final layer of CNN to assign probability values of each class label. Training of Inception V3 [14] model was done on ImageNet database with learning rate set to 0.01, which achieved an accuracy of 88.6%. With Inception model, high accuracy was achieved but the time taken was quite high. To tackle this problem, 28 layer MobileNet architecture is trained with training, testing, and validation ratio of 75:5:20 and learning rate as 0.01, which achieved the final test accuracy of 92.12%.

In [15], the authors proposed the use of fuzzy entropy and a probabilistic neural network model for paddy disease identification. The paper classifies the given image of paddy leaf into four diseases. After preprocessing the image, the image is fed to a fuzzy entropy function that extracts the features using two membership functions based on the bright and dark parts of an image. The feature set is fed to a PNN with four layers (input, pattern, summation, and output lays). The experiment results show an accuracy of 91.46%.

In [16], the authors used unsupervised learning technique self-organizing map (SOM) neural network using 100×80 input nodes, 2 output nodes, 50 epochs for training with 300 training patterns. 92.34% accuracy was achieved in classification using RGB of spot, and 84.97% accuracy was achieved by using Fourier transform.

In [17], the authors implemented a convolutional neural network using a deep learning framework 'Caffee' which is an open source and developed by Berkley Vision and Learning Centre (BVLC). A python script is used to resize the training image sample to reduce the time of training using the OpenCV framework. The CaffeNet architecture used in this paper has five convolutional layers, three fully connected layers, and eight learning layers. The final overall accuracy achieved by this model was 96.3.

In [18], the authors have worked on the classification of fungal disease in fruits crops, cereal crops, commercial crops, and vegetable crops. In fruits, they have considered mango, pomegranate, and grape. Features were extracted from the image samples using gray level run length matrix (GLRLM) and gray level co-occurrence matrix (GLCM) and block-wise. Images were classified into three categories, i.e., severely affected, moderately affected, and partially affected using nearest neighbor (NN) classifier. The accuracy achieved using GLCM features was 91.37% while the accuracy achieved using GLRLM features was 86.715%. The overall accuracy achieved using block-wise feature was 94.085%. For vegetable crops, artificial neural network (ANN) was used which achieved the accuracy of 84.11%.

In [19], the authors used two deep learning architectures, namely VGG16 and AlexNet for classification of diseases in tomato. The AlexNet architecture used in this model consists of three fully connected layers and five convolutional layers while the VGG16Net architecture consists of 13 convolutional layers and every convolutional layer was followed by ReLU layer. Six different disease samples of tomato crop were taken from the Plant Village dataset, and transfer learning approach was used. Using 13,262 images, the classification accuracy obtained by AlexNet was 97.49% and that of VGG16 was 97.29%.

In [20], the authors have focused on the identification of diseases in grape plants. They have used a specific camera to capture all the images for the dataset in order to reduce the chances of false classification that may be caused due to device dependency of the images. After acquiring the images, the images are resized to 300×300 images and have their background removed. Then the green pixels are masked in order to increase the accuracy as well as decreasing the overall processing time. The image is then enhanced using five iterations of anisotropic diffusion to preserve the information of the affected portions. The image is then segmented using K-means into six clusters. After obtaining the six clusters, lesions are extracted from the clusters. GLCM and SGDM are used to extract features from the affected portions of the leaf which are fed to a BPNN of three layers to identify the disease in the plant.

5   Concluding Remarks

In the short survey, we have discussed the methods utilizing SVM and ANN for identifying disease in plants. The SVM-based methods used few samples and features to identify the disease-affected plants at ease with lesser time complexity and space complexity. When it comes to accuracy, the SVM-based methods have lesser accuracy when compared to ANN-based methods. The ANN-based methods worked well in identifying the disease; at the same time, the advanced models like CNN and RNN can be used for prescriptive analytics. The ANN-based methods provide better accuracy compared to other SVM-based methods, but the complexity has increased while features as well as hidden layers get increased to get a desired output. This short survey helps the researchers to identify the pros and cons of SVM and ANN as well as the different types of identifying diseases in plants using machine learning techniques.

References

1. Agarwal, A., Sarkar, A., Dubey, A.K.: Computer vision-based fruit disease detection and classification. In: Smart Innovations in Communication and Computational Sciences, pp. 105–115. Springer (2019)
2. Al-Khaffaf, H.S., Talib, A.Z., Abdul, R.: Salt and pepper noise removal from document images. In: International Visual Informatics Conference, pp. 607–618. Springer (2009)
3. Arivazhagan, S., Shebiah, R.N., Ananthi, S., Varthini, S.V.: Detection of unhealthy region of plant leaves and classification of plant leaf diseases using texture features. Agricu. Eng. Int. CIGR J. **15**(1), 211–217 (2013)
4. Badnakhe, M.R., Deshmukh, P.R.: Infected leaf analysis and comparison by otsu threshold and k-means clustering. Int. J. Adv. Res. Comput. Sci. Softw. Eng. **2**(3) (2012)
5. Bankar, S., Dube, A., Kadam, P., Deokule, S.: Plant disease detection techniques using canny edge detection & color histogram in image processing. Int. J. Comput. Sci. Inf. Technol. **52**(2), 1165–1168 (2014)
6. Bashir, K., Rehman, M., Bari, M.: Detection and classification of rice diseases: an automated approach using textural features. Mehran University Res. J. Eng. Technol. **38**(1), 239–250 (2019)
7. Blum, A.L., Langley, P.: Selection of relevant features and examples in machine learning. Artif. Intell. **97**(1–2), 245–271 (1997)
8. Cortes, C., Vapnik, V.: Support-vector networks. Mach. Learn. **20**(3), 273–297 (1995)

9. Deokar, A., Pophale, A., Patil, S., Nazarkar, P., Mungase, S.: Plant disease identification using content based image retrieval techniques based on android system. Int. Adv. Res. J. Sci. Eng. Technol **3**(2) (2016)
10. Deshapande, A.S., Giraddi, S.G., Karibasappa, K., Desai, S.D.: Fungal disease detection in maize leaves using haar wavelet features. In: Information and Communication Technology for Intelligent Systems, pp. 275–286. Springer (2019)
11. Dey, A., Bhoumik, D., Dey, K.N.: Automatic multi-class classification of beetle pest using statistical feature extraction and support vector machine. In: Emerging Technologies in Data Mining and Information Security, pp. 533–544. Springer (2019)
12. Dhingra, G., Kumar, V., Joshi, H.D.: A novel computer vision based neutrosophic approach for leaf disease identification and classification. Measurement **135**, 782–794 (2019)
13. Gaikwad, V.P., Musande, V.: Wheat disease detection using image processing. In: 2017 1st International Conference on Intelligent Systems and Information Management (ICISIM), pp. 110–112. IEEE (2017)
14. Gandhi, R., Nimbalkar, S., Yelamanchili, N., Ponkshe, S.: Plant disease detection using CNNS and GANS as an augmentative approach. In: 2018 IEEE International Conference on Innovative Research and Development (ICIRD), pp. 1–5. IEEE (2018)
15. Gavhale, K.R., Gawande, U., Hajari, K.O.: Unhealthy region of citrus leaf detection using image processing techniques. In: 2014 International Conference for Convergence of Technology (I2CT), pp. 1–6. IEEE (2014)
16. Ghaiwat, S.N., Arora, P.: Detection and classification of plant leaf diseases using image processing techniques: a review. Int. J. Recent Adv. Eng. Technol. **2**(3), 1–7 (2014)
17. Husin, Z.B., Shakaff, A.Y.B.M., Aziz, A.H.B.A., Farook, R.B.S.M.: Feasibility study on plant chili disease detection using image processing techniques. In: 2012 Third International Conference on Intelligent Systems, Modelling and Simulation (ISMS), pp. 291–296. IEEE (2012)
18. Krizhevsky, A., Sutskever, I., Hinton, G.E.: Imagenet classification with deep convolutional neural networks. In: Advances in Neural Information Processing Systems, pp. 1097–1105 (2012)
19. KumarPatidar, P., Lalit, L., Singh, B., Bagaria, G.: Image filtering using linear and non linear filter for Gaussian noise. Int. J. Comput. Appl. **93**(8), 29–34 (2014)
20. Lippmann, R.: An introduction to computing with neural nets. IEEE Assp Mag. **4**(2), 4–22 (1987)

Efficiency Comparison and Analysis of Pseudo-random Generators in Network Security

Asis Kumar Tripathy, Satyabrata Swain and Alekha Kumar Mishra

Abstract Random data encryption algorithm (RDEA) is an advanced security model as compared to the Data Encryption Standard. It makes use of a random key generator to get the key from the cipher key database. The RDEA seems to be a great option for attaining greater security when we compare with 64-bit key of DES algorithm. Most of the random generators offered today are software-based random generators, which are not capable of generating actual random information. Because software random generators generate random information by using a fixed data set, their output can be predictable. This predictability weakens software encryption capability. Therefore, we would develop a modification to this algorithm by using pseudo-random key generation, which reduces the computation at both the encryption and decryption side without affecting the encryption capability. The cipher keys generated by means of pseudo-random methods will enhance the diffusion rate, which will attract higher security as well as throughput.

Keywords RDEA · DES · Random key · Pseudo-random generators · Security

1 Introduction

In the 1970s, DES was one of the most used encryption algorithm in the history of cryptography. Despite the vulnerabilities, the use of DES was quite enormous until it is publicly broke down by distributed.net and Electronic Frontier Foundation in the year 1999. DES is presently considered as insecure in some cases due to its less

A. K. Tripathy (✉) · S. Swain
School of Information Technology and Engineering, VIT, Vellore, India
e-mail: asistripathy@gmail.com

S. Swain
e-mail: satya.swai10@gmail.com

A. K. Mishra
Department of Computer Application, NIT, Jamshedpur, India
e-mail: alekha.ca@nitjsr.ac.in

© Springer Nature Singapore Pte Ltd. 2020
S. S. Dash et al. (eds.), *Artificial Intelligence and Evolutionary Computations in Engineering Systems*, Advances in Intelligent Systems and Computing 1056, https://doi.org/10.1007/978-981-15-0199-9_3

key size. As the DES algorithm failed due to possible brute force attack, certain other encryption algorithms have come into pictures like the Asymmetric Encryption Standard (AES), International Data Encryption Algorithm (IDEA), Secure Hash Algorithm (SHA), and certain other versions of DES like triple DES to improve its encryption capability. Some of them are quite good in the efficiency but the computation in those is quite high. To solve this, a randomized data encryption algorithm (RDEA) was developed which proved to be of great use. In that, a random block was attached to the DES mechanism based on which the key from the cipher key database was selected but again the computation was more.

In this paper, we aim to compare and analyze several methods of pseudo-random number generators. We implement these as an advancement to the existing Data Encryption Standard (DES) Algorithm by selecting the key from the cipher key database based on a pseudo-random number generated from these methods. The major parameter we would focus on the basis of which analysis would be performed is the diffusion rate. It is actually the change in the ciphertext caused by a small change in the plain text. Greater the diffusion rate, better is the encryption process. Study of random number generators: random number generation refers to the generation of a sequence of numbers that is nearly impossible to predict without the help of hardware. Random number generation has a great advantage for cryptographic applications. Despite the fact that a computer is totally unfit to make a random number, producing genuinely random numbers in programming is in reality impractical; however, it is possible with hardware to build a device which can generate truly random numbers.

Recently, random number generators have started being used in the DES encryption algorithm to boost its encryption capability. A separate random generator block is attached to the cipher key database, and based on the output from that block, cipher key is selected for each round. This surely is secure but at the same time increases the overall complexity of the algorithm.

Pseudo-random numbers are just like normal random numbers but they require a seed value by which they can be easily generated using several methods. These methods differ in their efficiency and complexity and can be used in various applications. Again, these have been implemented alongside DES algorithm. This has reduced the complexity of the random number generators and has maintained the encryption capability.

Some of the methods we have implemented and analyzed are:

1. Multiply-with-Carry (MWC) Method

The MWC strategy is a variation of the add-with-carry generator presented by Marsaglia and Zaman [1]. In its least complex form, MWC utilizes a comparable formula to the linear congruential generator, yet with c fluctuating on every emphasis: $x = (a * x + c) \bmod m$.

Properties of the MWC algorithm include the following:

- With appropriate parameters, it passes statistical tests that LCG generators fail.
- If b is made to be a power of 2 half the size of a register, then the division and mod operations become trivial.

- The period is not exactly a power of 2, an interesting property for use in combined random number generators.

2. Park–Miller Method

Park–Miller random number generator is also known as Lehmer random number generator. It is a type of linear congruential generator (LCG) that works on multiplicative group of integers modulo n. The general formula is:

$$X_{k+1} = g * X_k \bmod n$$

where the modulus n should be a prime number or a power of a prime number, the multiplier g is an element of high multiplicative order modulo n, with particular parameters $n = 2^{31} - 1 = 2,147,483,647$ and $g = 7^5 = 16,807$. Using a modulus n is of power of two makes for a particularly convenient computer implementation but comes at a cost. The period is at most $n/4$, and the low bits have periods shorter than that. This is because the low k bits form a modulo-2 k generator all by themselves.

3. Middle Square Method

The middle square technique is a method for producing pseudo-random numbers. In this to create a sequence of n digit pseudo-random numbers, a n digit beginning quality is made and squared, delivering a $2n$ digit number. Also, there is a chance that the outcome has less than $2n$ digits, then zeroes are added to adjust. The center n digits of the outcome would be the following number in the arrangement and returned as the outcome. This procedure is then rehashed to produce more numbers. In this, if the center n digits are all zeroes, the generator at that point yields zeroes for eternity. If the first half of a number in the grouping is zeroes, the consequent numbers will decline to zero.

2 Literature Review

Tuncer et al. introduced a new modified method for improving the efficiency of the existing DES model. Their algorithm was developed using data signal processing for audio signals or text images. They added a randomized number generator block to the existing DES mechanism stating that this would help in selecting the cipher key form the cipher key database. They ensured that this was fully compatible with the existing DES algorithm to ensure the usability of the modified mechanism. The major advantages of their proposed system were that they tried to provide solutions to the vulnerabilities of the existing DES algorithm by minimal modifications but their model did not effectively generate the random numbers; hence, the computations were still more [2].

Ranjan Kumar proposed in his paper that data traffic over both public and private networks increases overtime making it critical to protect the privacy of all the information that is stored in storage devices or is exchange between several computing machines. Therefore, they proposed a random number generator (RNG) in the encryption process. Hardware random number generators are non-deterministic according to them meaning that by nature there is no algorithm which can be used to determine the sequence bits that are generated from the process hence ensuring the non-vulnerability to existing algorithm. Thus, hardware RNGs are not susceptible to the disclosure or algorithm disassembly attack. They have not tried to implement this with the DES encryption algorithm which limits the use of their proposed method in our application of encryption mechanisms [3].

Topaloglu et al. proposed that Random numbers have an extensive variety of usage in areas such as simulation, gaming, sampling and IT sector of cryptography, game programming, and data transmission. In order to use random numbers in programming, there are three basic necessities. First, the random numbers that are being generated have to be unpredictable. Second, these numbers should have great measurable statistical properties. Third, the numbers that are produced once must not be generated again. Random number generators (RNGs) have been created to acquire randomize numbers with the properties mentioned above. PRNGs utilized for producing random numbers is stream encryption algorithms [4].

Couture et al. proposed that with the increase in cryptanalysis, securing multimedia information against various sorts of attacks has become a challenging task. The examination demonstrates that this method is resistant to many measurable statistical and differential assaults and attacks, and also since the size of data encrypted is small, it became more beneficial on large amount of data [5].

Park et al. proposed that the current pattern of computer tech is toward securing computational power and communication between two or more parties which require solid and strong cryptographic methods, particularly with agent-based issues. Since random number generators are the main principle constituents of such method and autonomous activities, they are required to be quick and enough secure. Despite the fact that there are some great quality and quick approaches, the vast majority of them either utilize large primes which are difficult to handle or the cost of getting information from the source is very high [6].

Kelly et al. proposed the multiply-with-carry family of generators proposed by Marsaglia as a speculation of previous add-with-carry families. He also characterizes for them a vast state space and concentrates on the limited subset of repetitive states. This subset will, thus, split into conceivably a few subgenerators. We talk about the consistency of the d-dimensional distribution of the yield of these subgenerators over their full period [7].

Marsaglia et al. presented the rationale for choice of a minimal standard generator. He also proposed that these generators should always be used unless one has to access to a random number generator known to be better. He also implemented this generator of variety of systems to check its accuracy. He also discussed the theoretical aspect of it. Also concluded that with properly chosen parameters, minimal standard generator can give very satisfactory results [1].

3 Proposed System

Most of the newly emerged computer programs use software-generated pseudo-random number rather than the true random numbers. This is because of the ease of generation of pseudo-random numbers rather than generating a true big random number which can be used. The concept of pseudo-random number generators (PRNGs) requires a seed value which is used as a mathematical operand to create pseudo-random numbers. Pseudo-random numbers are increasingly used because of the major properties like:

- A seed value to initialize the equation;
- Generated sequence repeats eventually.

There are certain algorithms with which pseudo-random numbers can be generated. So we would implement the three above-mentioned algorithms for pseudo-random number generation which are Park–Miller method, middle square method, and multiply-with-carry method. The major aim would be to implement these alongside the DES algorithm for key selection process for each of the 16 rounds and then to compare them with the help of the diffusion parameter. With this parameter, we would get an idea about the best method that can be used to further improve the encryption capability. Figure 1 explains the working model of the proposed system. For the comparison, a set of plain texts are used which differ by some bits only and the diffusion rate for all the outputs are generated for all the three methods. The differences in the diffusion rate would depict the efficiency in generating different cipher texts even for a small change in the plain text and in turn would tell us about the effectiveness of the method of pseudo-random number generator in this particular DES application.

Fig. 1 Working model of the proposed system

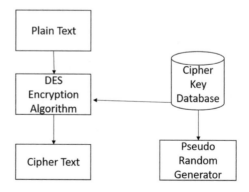

4   Result Analysis

The inputs that we use for the experimental analysis consist of a 64-bit or 8-byte words which differ by only a single bit. These are passed to 64-bit implementation of the DES algorithm, and the resulted ciphertext is analyzed for the diffusion rate with respect to the original ciphertext obtained for the original dataset. We have considered two experimental sets of strings, and for each of them, 5 different permutations are passed which only differ by a single bit to compare about the change in the resultant cipher key for each of them. The method with the highest diffusion rate in both the experimental sets is the best method for DES application.

Random numbers have a wide range of applications such as game of chance, simulation, game programming, and cryptography. But cryptography has strongly accepted the use of these in the encryption and decryption process. Pseudo-random number generation algorithms help in improving the encryption capability of DES algorithms.

With our proposed comparison in Table 1, we could prove that we can easily make the concept of random numbers compatible with DES algorithm. Also, we

Table 1 Diffusion rate comparison

Sl. no.	Plain text	64-bit representation	Diffusion		
			MS	MWC	PM
1. *Bit-level change*					
1	abcdefgh	01100001 01100010 01100011 01100110 01100101 01100110 01100111 01101000	0	0	0
	abcddfgh	01100001 01100010 01100011 01100100 01100100 01100110 01100111 01101000	100	100	100
	abcdefgi	01100001 01100010 01100011 01100100 01100101 01100110 01100111 01101100	87.5	87.5	87.5
	ibcdefgh	01101001 01100010 01100011 01100100 01100101 01100110 01100111 01101000	100	87.5	100
	aBcdefgh	01100001 01000010 01100011 01100100 01100101 01100110 01100111 01101000	100	87.5	100
2	zxcvbnml	01111010 01111000 01100011 01110110 01100010 01101110 01101101 01101100	0	0	0
	zpcvbnml	01111010 01110000 01100011 01110111 01100010 01101110 01101101 01101100	100	100	100
	zxcwbnml	01111010 01111000 01100011 01110111 01100010 01101110 01101101 01101100	100	100	100
	zxcvbNml	01111010 01111000 01100011 01110110 01100010 01001110 01101101 01101100	87.5	100	87.5
	zxcvbnol	01111010 01111000 01100011 01110110 0110010 01101110 01101111 011011000	87.5	100	100

(continued)

Table 1 (continued)

Sl. no.	Plain text	64-bit representation	Diffusion		
			MS	MWC	PM
2. *Byte-level change*					
1	abcdefgh	01100001 01100010 01100011 01100110 01100101 01100110 01100111 01101000	0	0	0
	abcddfgh	01100001 01100010 01100011 01100100 01100100 01100110 01100111 01101000	100	100	100
	abcdefgi	01100001 01100010 01100011 01100100 01100101 01100110 01100111 01101100	87.5	100	100
	ibcdefgh	01101001 01100010 01100011 01100100 01100101 01100110 01100111 01101000	100	100	100
	aBcdefgh	01100001 01000010 01100011 01100100 01100101 01100110 01100111 01101000	100	100	100
2	zxcvbnml	01111010 01111000 01100011 01110110 01100010 01101110 01101101 01101100	0	0	0
	zpcvbnml	01111010 01110000 01100011 01110111 01100010 01101110 01101101 01101100	100	100	100
	zxcwbnml	01111010 01111000 01100011 01110111 01100010 01101110 01101101 01101100	100	100	87.5
	zxcvbNml	01111010 01111000 01100011 01110110 01100010 01001110 01101101 01101100	87.5	100	100
	zxcvbnol	01111010 01111000 01100011 01110110 0110010 01101110 01101111 011011000	100	100	100
3. *Significant change*					
1	abcdefgh	01100001 01100010 01100011 01100110 01100101 01100110 01100111 01101000	0	0	0
	abcddfgh	01100001 01100010 01100011 01100100 01100100 01100110 01100111 01101000	100	100	100
	abcdefgi	01100001 01100010 01100011 01100100 01100101 01100110 01100111 01101100	100	100	87.5
	ibcdefgh	01101001 01100010 01100011 01100100 01100101 01100110 01100111 01101000	100	100	87.5
	aBcdefgh	01100001 01000010 01100011 01100100 01100101 01100110 01100111 01101000	100	100	100
2	zxcvbnml	01111010 01111000 01100011 01110110 01100010 01101110 01101101 01101100	0	0	0
	zpcvbnml	01111010 01110000 01100011 01110111 01100010 01101110 01101101 01101100	100	100	100
	zxcwbnml	01111010 01111000 01100011 01110111 01100010 01101110 01101101 01101100	100	100	100
	zxcvbNml	01111010 01111000 01100011 01110110 01100010 01001110 01101101 01101100	100	87.5	100
	zxcvbnol	01111010 01111000 01100011 01110110 0110010 01101110 01101111 011011000	100	100	100

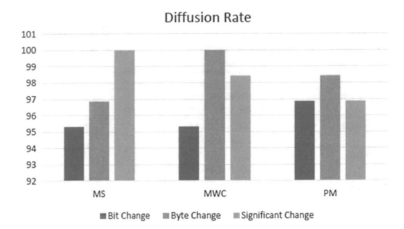

Fig. 2 Diffusion rate comparison

can state which of the method of pseudo-random number generation is suitable according to the diffusion level we obtain from above experimental results.

From the experimental results shown in Fig. 2, it is obtained that Park–Miller method is the best among the rest, whereas for byte level change multiply-with-carry suits to be the best. For any significant or big change, again multiply-with-carry is the best. Hence, it can be concluded that greater the change in the inputs, the better is the performance of multiply–with-carry. Middle square methods prove to be the most inefficient. Park–Miller is highly efficient for bit-level changes thus it helps in better encryption capability. Hence, among the various methods available for pseudo-random number generation, we think the multiply-with-carry is the best and can be used in most of the cases and is nicely compatible with the Data Encryption Standard algorithm as an advancement to the existing algorithm.

5   Conclusion

It can be summarized that pseudo-random number generators are an advantageous modification to the existing DES algorithm rather than simple random number generators which add to the complexity of the encryption and decryption process. PRNG block is a simple alternative to counter the brute force technique which could somehow crack the DES algorithm. Among the most methods available for pseudo-random number generation, multiply-with-carry can be said as the best. It is observed that by using pseudo-random generators makes the security schemes more fruitful as compared to the normal random numbers.

References

1. Marsaglia, G., Zaman, A.: A new class of random number generators. Ann. Appl. Probab. 462–480 (1991)
2. Tuncer, T., Avaroğlu, E.: Random number generation with LFSR based stream cipher algorithms. In: 2017 40th International Convention on Information and Communication Technology, Electronics and Microelectronics (MIPRO), pp. 171–175. IEEE (2017)
3. Ranjan, K.H.S, SP Fathimath, S.S.P., Shetty, S., Aithal, G.: Image encryption based on pixel transposition and Lehmer Pseudo random number generation. In: 2017 2nd IEEE International Conference on Recent Trends in Electronics, Information & Communication Technology (RTEICT), pp. 1188–1193. IEEE (2017)
4. Topaloglu, U., Bayrak, C., Iqbal, K.: A pseudo random number generator in mobile agent interactions. In: 2006 IEEE International Conference on Engineering of Intelligent Systems, pp. 1–5. IEEE (2006)
5. Couture, R., L'Ecuyer, P.: Distribution properties of multiply-with-carry random number generators. Math. Comput. American Math. Soc. **66**(218), 591–607 (1997)
6. Park, S.K., Miller, K.W.: Random number generators: good ones are hard to find. Commun. ACM **31**(10), 1192–1201 (1988)
7. Kelly, J.R.: Cryptographically secure pseudo random number generator. US Patent 6,275,586, issued 14 Aug 2001

A Detailed Analysis of Intruders' Activities in the Network Through the Real-Time Virtual Honeynet Experimentation

Rajarajan Ganesarathinam, M. Amutha Prabakar, Muthukumaran Singaravelu and A. Leo Fernandez

Abstract The menace of attackers over the network is unstoppable for the past two decades. The security practitioners and researchers are devising mechanisms to safeguard the network and its components, but still attackers emerge with cutting edge technologies to disturb the intention of legitimate users in the network. Thus, before devising proper defensive mechanisms against a specific attack, it is essential to understand the motive and strategies of the attackers with the proper clarity. This paper presents a virtual honeynet framework to record all the attackers' activities and analyzes the strategies, tools, and mechanisms followed by the attacker, in a real-time manner. We analyzed the recorded attacks in our framework with respect to different parameters like protocol, ports, honeypots, and IDPS tools to understand the motive behind the attacks. This novel virtual honeynet architecture will give insight to the readers and security practitioners to understand the strategies followed by the attackers as well as the way of designing different traps to secretly follow the attackers in the road toward foolproof safeguarding mechanisms.

Keywords Honeypots · Network intrusion · Attackers · Virtual honeynet

R. Ganesarathinam (✉)
Department of Computational Intelligence, Vellore Institute of Technology, Vellore, India
e-mail: rajarajan.g@vit.ac.in

M. A. Prabakar
Department of Analytics, Vellore Institute of Technology, Vellore, India
e-mail: amuthaprabakar.m@vit.ac.in

M. Singaravelu
Department of Computer Science and Engineering, Anna University, Chennai, India
e-mail: smuthukumaran97@gmail.com

A. L. Fernandez
Department of Software Engineering, Kirirom Institute of Technology, Traeng Trayueng, Cambodia
e-mail: leo.fernandez@kit.edu.kh

© Springer Nature Singapore Pte Ltd. 2020
S. S. Dash et al. (eds.), *Artificial Intelligence and Evolutionary Computations in Engineering Systems*, Advances in Intelligent Systems and Computing 1056, https://doi.org/10.1007/978-981-15-0199-9_4

1 Introduction

Intrusion detection and prevention systems (IDPS) are allowing organizations to identify attacks and provide reaction mechanism against those attackers. But these mechanisms lack some functionality like detecting new kinds of threats, collecting needed data about activities of the attacker, strategies, and skills. For example, signature-based intrusion detection and prevention systems are incapable of finding out new attacks due to lack of new attacks signatures. Thus, they are solely capable of detecting the known attacks.

In these days, many organizations and research institutions investigate and interpret the tactics of intruders' community, which harms their networks. These institutions mostly make use of honeypots [1, 2] to study and understand attack loopholes in their network and acquire details of the attackers. Also, the usage of honeypots permits us to introspect how the intruders act for manipulation of vulnerabilities of system [3]. This is what our work in this paper focuses on. This investigation gives relevant information to learn the prowess of attackers. The real-time experimentation analysis of attacker's competence could be helpful in designing appropriate defensive model against the network intruders in the network. This paper contributes to the following in terms of discerning the attacker's potential.

(i) Implementation of virtual honeynet framework.
(ii) Classification and analysis of recorded vulnerabilities and attacks.

2 Proposed System

In our experiment, both low- and high- interaction honeypots [4] are used. It allows us to provide comprehensive details about threats and monitor the activities performed by the attackers (human beings, automated tools).

For the low-interaction honeypot, honeyd [5] is used and virtual honeynet architecture is implemented, i.e., based on VirtualBox vitualization, software is used for high-interaction honeypot [6]. The implemented architecture is shown in Fig. 1. The high- and low-interaction honeypots are placed apart and a common database is maintained on a remote machine in order to store the collected information from the honeypots. For this work, two physical machines are used which have the virtual honeypots as well as another remote machine to monitor the details related to attacks and to control the honeypots' activities. Here, the entire honeypot system is configured and deployed over the virtual machines.

Using virtualization [7], it is possible to configure and deploy many virtual machines over a single physical device. An OS running on a virtual and physical device is referred as guest and host system. The term virtual means, all OS have the

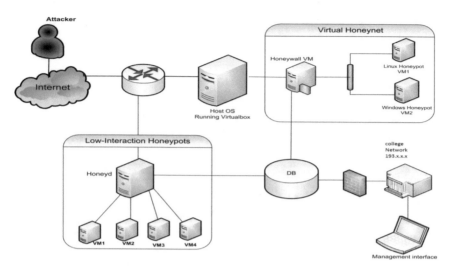

Fig. 1 Proposed honeynet architecture

appearance of being run over separate machine. Using virtualization, it is possible to configure many virtual machines on an individual physical device.

In order to perform virtualization, the host system must allot the resources like files, memory, I/O, CPU, etc. with guest system. In this work, virtual honeynet [8] has been deployed and built as high-interaction system of our proposed architecture. Within an individual physical device, the entire honeynet is installed. The detailed experimentation of our virtual honeynet is represented in next section and the Fig. 2 depicts the virtual honeynet architecture.

In our implementation, honeywall [9] is used as a gateway to high-interaction honeypots. In general, honeywall can be deployed to protect honeypots present in the network from attackers. It has three important goals.

(a) Data capture: It captures all the attacker's activity inside the honeynet as well as information enters and leaves the honeynet without the knowledge of attackers.
(b) Data control: It controls the inflow and outflow of traffic in the honeynet.
(c) Data analysis: It helps the administrator to simplify the analysis of capture details that helps in network forensics.

3 Implementation of Honeynet Architecture

Here, we describe the procedure of setting up the honeypots which are implemented in our experiment and also emphasize the issues involved during the deployment of honeypots in the network. The following Table 1 lists out variety of software and hardware resources used in our experimentation.

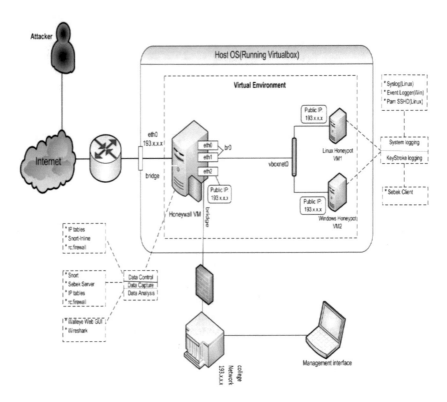

Fig. 2 Virtual honeynet architecture

Table 1 List of software tools used in this work

Software/OS	Description
Snort	Intrusion detection system
Sebek	Client–server-based data capture tool
Psacct	Monitoring tool on linux honeypot
Linux, Ubuntu 9. 04	Host machine (virtual honeynet)
Linux, Ubuntu 9.10	Honeyd host (physical device)
Linux, Honeywall Roo, CENTOS 5	Virtual host (honeywall gateway)
Linux ubuntu 7.10	Virtual honeypot
VirtualBox	Virtualization software
Windows 2003 server standard service pack 1	Virtual honeypot
Windows 7 professional	Remote management machine (physical machine)
Snort	Intrusion detection system

Three physical machines are used here: One for Honeyd, (low-interaction honeypot), another one is used as virtual honeynet, and last one as remote management machine.

3.1 Honeyd

Honeyd is a framework for virtual honeypots, created and maintained by Niels provos [10], which simulates virtual systems at the network level. It permits us to set up and run multiple virtual machines as well as network services on a single physical machine. It is a low-interaction honeypot that simulates UDP, TCP, and ICMP services and binds a script to specific port in order to emulate a particular service.

In our implementation, a virtual server that appears to run windows is set up and listens to port 110. The following configuration template of honeyd shows the presence of windows virtual honeypot running on IP address 193.10.X...X. Thus, whenever attacker fingerprints the honeypot using Xprobe or Nmap, it seems as server standard Edition of Windows 2003. When the attacker connects to TCP port 110 of virtual windows system (windows 2003 server), honeyd begins to execute the script "./scripts/pop3.sh".

"create windows
set windows personality „Windows 2003 Server Standard Edition"
add windows tcp port 110 „sh scripts/pop3.sh"
bind windows 193.10.X.X"

Using honeyd, it is possible to develop virtual routing topologies that have many virtual networks with hundreds of devices on a single System. The following Fig. 3 depicts the overview of honeyd. It consists of a single physical honeyd machine which hosts four virtual machines. Honeyd uses a text-based configuration file in which low-interaction honeypots are specified. It specifies the open ports, IP addresses, and available services of each virtual machine.

The following command shows part of the configuration file.

create linux
set linux personality "linux 2.4.16"
set linux default udp action reset
set linux default tcp action reset
set linux uptime 5184002
add linux tcp port 25 "sh/scripts/unix/general/smtp.sh"
add linux tcp port 110 "sh/scripts/unix/general/pop3.sh"
add linux tcp port 21 "sh/scripts/unix/linux/ftp.sh"
add linux tcp port 22 open bind x.x.67.128 linux.

In order to run our honeyd with configuration file, the following rule is used.

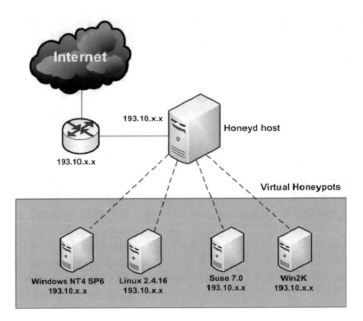

Fig. 3 Honeyd framework

*Honeyd −d −i −etho −f honeyd.conf −p nmap.prints −x xprobe2.conf −l
/var/log/honeyd/honeyd-packet.log −s/var/log/honeyd/honeyd-
service.log < IP1,IP2, IP3, IP4>*

Here, IP1 to IP4 refers the IP addresses of configured virtual honeypots in the network. In order to check the correct installation and working of honeyd, Nmap, traceroute tools are used to ensure the correctness of our installed honeynet.

The firewall in our institution [11] is configured to allow all incoming traffic to virtual honeypots, but restricts the outgoing traffic from honeypots. The services like http, www, and ssh access are allowed from remote system to physical honeyd to monitor and backup logs. IPTables are purposefully used to secure our virtual honeypots from external worms and other threats.

*iptables −A INPUT −p tcp −s x.x.66.230 −d X.X.66.129 −dport 80 −
mstate −state RELATED, ESTABLISHED, -j ACCEPT.*

In order to log the details of packet, honeyd provides two types of logging. 1. Packet-level logging. 2. Service-level logging. The command line option (−1) of packet-level logging provides a complete timestamp of when a packet was received, destination IP and port, source IP and port, protocols, utilities, etc. Likewise, the service-level logging with (−s) command line gives us detailed information about inbound traffic. These commands are specified at the time of running honeyd in order to capture logs. The following shows the sample log file of packet-level logging.

2018-11-10-16:45:12.1612 tcp(6) S 207.32.16.14 53002 193.10.67.83 80
[windows XP sp1]
2018-11-10-10:12:41.2611 tcp(6) E 207.32.16.14 53002 193.10.67.83 80: 207
528
2018-11-28-15:03:25.1165 tcp(6) – 192.168.3.12 53256 193.10.67.111 22: 60 S
[Linux 2.6 0.1-7]
2018-11-23-17:12:36.8965icmp(1)–64.33.72.112 193.10.67.111:80 61

The first field indicates the exact date and time in which packet was received. The next field shows the information related to protocols like tcp, udp, or icmp, etc. The third field indicates the type of connection. It can be either *S* (starting of fresh connection), *E* (ending of existing connection) or—Neither (*E* or *S*). Next fields represent source IP, source port, destination IP, and destination port. The last field shows the type of operating systems from which packet has arrived. This following is the part of service-level log.

/var/log/honeyd/honeyd.log
December 15 10:15:46 armakedon honeyd [31256]: Connection request:tcp
(207.32.16.55:50124 -193.10.67.81:25)
December 15 10:19:21 armakedon honeyd [31256]: Connection established:tcp
(207.32.16.55:50124 –193.10.67.81:25)
December 15 10:42:29 armakedon honeyd [31256]: Killing attempted connection:
tcp (207.32.16.55:50124 –
193.10.67.81:25)

3.2 Virtual Honeynet

In order to lure the attackers to interact with real OS, programs, and services, virtual honeynet framework is chosen. The important reason for selecting the virtual honeynet was devoid of hardware services and resources, and hence, it is decided to make use of VirtualBox as virtualization software to deploy honeynet. Moreover, we benefitted a lot from the chapters of honeynet within the honeynet project which also made use of various virtualization tools [12] like VMware, Qemu, etc. In this work, VirtualBox is used on single physical machine running ubuntu 9.04 version. In our work, VirtualBox is made to allow running of three virtual machines simultaneously with available resources. The details of those three virtual machines are as follows.

1. Honeywall
2. Honeypot 1 (Ubuntu)
3. Honeypot 2 (Windows).

The honeywall roo 1.4 iso-image and its operating system based on CentOS 5 is used. This honeywall is configured with 1024 MB RAM, 40 GB HDD, and three interfaces: The first one for host-only and other two for bridged interfaces. The second machine, Ubuntu honeypot, was configured with 360 MB RAMS, 20 GB HDD, and one host-only interface. The third machine windows honeypot has 256 MB RAM, 20 GB storage, and single interface for host-only network. Using this VirtualBox, it is possible to configure up to four network cards in each virtual machine. For our implementation, PCNet FAST III virtual network card is selected and enabled network address translation (NAT), bridged networking, and host-only networking mode. The details of our installed machines on VirtualBox are shown in Fig. 4.

The installation of honeywall is fully automated. It provides the significant features like data analysis, data control, and data capture. The honeynet project provides a bootable CD-ROM named Honeywall-Roo 1.4 through which we installed in our VirtualBox. The Honeywall Roo [13] contains security tools like sebek, snort, snort_inline, TCPdump, Hflow, and Walleye Web interface [12, 14].

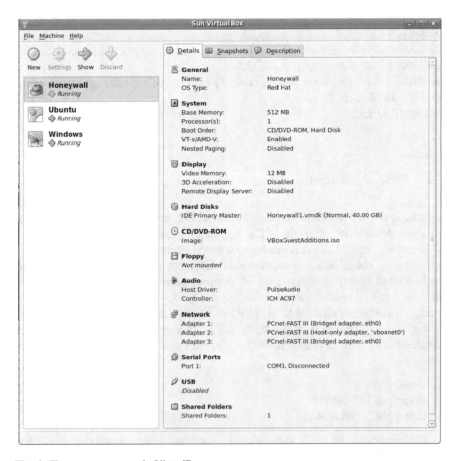

Fig. 4 Three guest systems in VirtualBox

In order to configure this honeywall, Dialog menu interface is used, through which DNS server addresses, IP addresses, inbound and outbound connection restriction, accessible ports on honeywall, snort-inline, and sebek parameters are provided.

3.3   Virtual High-Interaction Honeypots—Configuration

The virtual honeypots are set up in two steps. As a first step, OS is installed on virtual machines similar to OS installation of physical machine. In the second step, vulnerable applications and softwares are deliberately installed for attracting attackers toward the honeynet. Linux and Windows are specifically chosen as our virtual honeypot operating systems, because it is more common OS in Internet than MacOS or some other mobile operating systems.

(a) Honeypot 1 (Ubuntu)

The lower version of Ubuntu is purposefully set up and installed vulnerable Web services, applications with older versions of FTP, Apache, pHp, SSH, and MySQL. This is useful in attracting attackers toward this honeypot. Also, honeypot is configured in such a way that all incoming connections for services like SSH and FTP are open without any restriction, but outgoing connections are restricted for preventing attack traffic to other machines in our network. As attackers preferring to get connected through SSH service is expected, the OpenSSH logs contain details of every SSH connections and termination. But, by default, this OpenSSH does not log password attempted by user. In order to find the attempted password by attacker, a file called auth_password.c is modified from openSSH library. The auth_passwd.c file is modified as below for finding attempted password.

(b) Honeypot 2 (Windows)

In our second guest machine, windows 2003 service pack 1 is used. In order to lure attackers, older versions of some vulnerable software are installed. Here, XAMPP 1.6.6 version Web server package is installed that has Apache 2.2.7, MySQL 4.0.51, PHP 4.4.8, OpenSSL 0.9.8 g, FileZilla FTP server 0.9.25, and PhpMyadmin 2.11.4. The default user/password or blank password is made for all the tools in XAMPP for inviting attackers to exploit.

4   Obseration and Analysis

According to [15, 16], the process of collecting, manipulating, storing, and analysis of data recorded from machine, network, or computing resource to predict the attack source from the evidences is called computer forensics. Our proposed honeynet architecture is set up and all the installed honeypot systems were put online for the duration of almost180 days (6 months) from June 2018 to January 2019, at

Information Security laboratory in VIT, Vellore campus [11]. During this period, more than 50,000 attack connections were received. These results give insight about attackers' skill and their activities, which helps further research.

In our experiment, after observing multiple connections to our honeynet around three months, we try to find out the starting point of all the intrusion. From the observation, it is being found that most of the attackers use scanners [17, 18] like nmap, nicto, nessus, etc., to find open ports. The following Fig. 5 shows one of the snort alerts which display the attacker usage of nessus tool to probe the http (80), before launching an attack.1

4.1 Witnessed Attack Cases

Among the low- and high-interaction honeypots in our honeynet setup, more than half of the connections occurred over honeyd (low-interaction honeypot). The frequent traces that we detected were UDP, TCP, and ICMP port scans by attackers to find out vulnerabilities present in our system. However, few of the connections to honeypot intruded into our network and carried out brute-force, spamming, cross site-scripting, and remote file inclusion [11, 19–21].

4.2 Scanning of Ports

In general, port scanning is done by both system administrator and attacker. The scanning of port by authorized system administrators is to check the security level of network. But the attacker's intention of port scanning is all about illegal intrusion in the network. From the snort log, it is found that common port scans were TCP, UDP, and ICMP by variety of scanner tools like Portsweep, Nessus, and Nmap. The attacker mostly sends empty UDP datagram to the destination port (target port). When the port is closed, the attacker will receive "ICMP port Unreachable" reply and when it is open, it replies with some datagram or error messages. From that, attacker learns about status of port. The usage of Dfind and cyberkit 2.2 software is observed in order to check the status of the honeypot. The Dfind is the port scanner which scans only the predefined ports and scan IIS servers and web banners. The cyberkit software is used to generate anonymous ICMP echo requests continuously

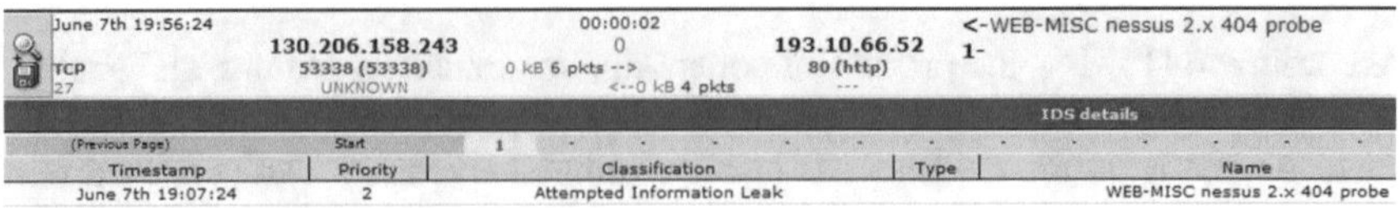

Fig. 5 Nessus scanning http

toward target. It issues ping command to check whether the machine is active or not. The following log shows the part of request carried out by Dfind scanner.

GET /w00tw00t.at.ISC.SANS.DFind :) HTTP/1.1 GET/HTTP/1.1
Accept: application /x-ms-application, application/x-ms, -xbap,
image/jpg,image/gif,image/pjpeg,*/* Accept-Language; fr-FR
User-Agent: Mozilla/4.0 Accept-Encoding: default Connection: Keep-Alive

4.3 Brute-Force SSH Attack

In this experimentation, almost 28,735 SSH login attempts are observed against the installed honeypots. Out of such huge attempts, only three attempts successfully connected to our honeypots. The first SSH brute-force attack was received from 92.16.59.213 on December 21 13:35:06. After numerous failed attempts, this attacker gains user shell access on the honeypot and tried to unset bash history configuration variable. The authentication log (/var/log/auth.log) gives information about attackers' activity which is shown below.

December 20 14:19:36 cselab04 sshd[834]: Failed password for invalid user
anonymous from 92.16.59.213 port 44798 ssh2
December 20 16:20:15 cselab04 sshd[834]: Failed password for invalid user
anonymous from 92.16.59.213 port 44798 ssh2
December 21 12:30:12 cselab04 CRON[8347]: Failed password for invalid
user anonymous from 92.16.59.213 port 44798 ssh2
December 20 12:30:14 cselab04 CRON[837]: Failed password for invalid user
anonymous from 92.16.59.213 port 44798 ssh2
December 21 13:35:06 cselab04 sshd[849]: Accepted password for test from
92.16.59.213 port 37082 ssh2

   The following list of commands executed by attacker is recorded by Psacct tool.
   linux@ cselab04: /usr/src/linux-header $ sudolastcomm test

uname -aDecember 21 16:12
cat /proc/cpuinfo December 21 16:12
sshd December 21 16:13
id December 21 16:14
ls-alDecember 21 16:14
ifconfig December 21 16:15
ps December 21 16:15
su December 21 16:16
wget www.vacam-steaua.com/eggdrop-aa.tgz December 21 16:18
history −cDecember 21 16:20

As vulnerable XAMPP server is already on honeypot 2 (Windows machine), the host with IP address 211.110.122.98 probes the open ftp port (21) on honeypot 2 and intruded into ftp server after many failed attempts. Meanwhile, another host with IP address 83.105.122.16 (might be the same host) also tried different passwords to log into honeypot. By using the sebek log shown below, it is learnt that different rootkits are installed on our honeypot by 211.110.22.98, and finally, attacker cleared all the access records. It is explicitly clear from the sebek log that attacker installed many backdoor tools explained below:

1. *wget.exe*—This is the very first file uploaded by the attacker. This utility is used to download files from remote host or Web server via command prompt.
2. *goonshell.php*—This is a common backdoor file such that the attacker uploaded into Web server to get remote control of honeypot.
3. *nc.exe*—This netcat utility which reads and writes data across network connections. By using this tool, the attackers can remotely control the rootkits and other tools on compromised machine.
4. *Kill.exe*—This is used to end (terminate) the processes on our honeypot.
5. *y.php*—This file helps attacker to clear all the traces and logs on the honeypot.
6. *Winhelp.exe*—It is a self-extractable archive contains files like Trojan horse, rootkits, install.bat, etc., and it can be included in system directory.

Our assumption is, this attacker (211.110.22.98) is an experienced script kiddie, because this attacker has the set of tools to control the remote anonymous machines very easily.

4.4 Directory Traversal

Before initiating an attack, most of the attackers utilizing this path traversal attack to obtain the URL tree of Web site. The intention of this search is to find hidden file configuration, set up files, etc. This is more common attack on old versions of phpMyAdmin, Web mail, and Web forums. The following snort log shows the attacker intention of finding main.php file, through many HTTP GET requests.

GET/admin/phpMyAdmin/main.php HTTP/1.0
GET/admin/sqladmin/main.php HTTP/1.0
GET/admin/sqladmin/main.php
GET/admin/sqlmanager/main.php HTTP/1.0
GET/admin/sysadmin/main.php HTTP/1.0
GET/admin/db/main.php HTTP/1.0
GET/admin/web/main.php HTTP/1.0

4.5   Statistical Analysis

In this part, the overall statistics about attacks happened over our installed honeypots from June 2018 to January, 2019 (around 180 days) is presented. From our observation, it is found that TCP is the commonly used protocol by the attackers. The following Table 2 shows the details of different protocol connections by attacker.

Out of 31,554 connection attempts, around 90% of them are TCP connections. Since multiple vulnerable services have been installed on our honeypots, many attempts were tried against these honeypots. The following graph in Fig. 6 shows the details of such blind attack over honeypots. All the login attempts made by the attackers are to get into the administrator privileges as well as for many vulnerable services and applications. The following Table 3 shows the part of login attempts made up by attackers with different user name and password.

Table 2 Connections attempted using different protocols

Protocol	Connections
ICMP	2588
UDP	332
TCP	28,634

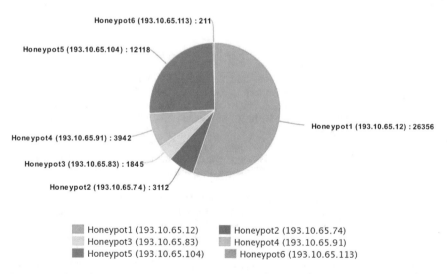

No. of connections attempted over honeypots

Honeypot6 (193.10.65.113) : 211
Honeypot5 (193.10.65.104) : 12118
Honeypot1 (193.10.65.12) : 26356
Honeypot4 (193.10.65.91) : 3942
Honeypot3 (193.10.65.83) : 1845
Honeypot2 (193.10.65.74) : 3112

- Honeypot1 (193.10.65.12)   - Honeypot2 (193.10.65.74)
- Honeypot3 (193.10.65.83)   - Honeypot4 (193.10.65.91)
- Honeypot5 (193.10.65.104)   - Honeypot6 (193.10.65.113)

Fig. 6 Number of connections attempted over honeypots

Table 3 Most tested
username and password

Login attempt	User name	Password
102	Root	1qaz2wsx
110	Mysql	mysql
216	Root	root
208	Root	123456
219	Root	password
92	Apache	apache
118	Guest	qwerty

5 Conclusion

Despite the availability of lots of IDPS tools and techniques, the attack over the network is still inexorable. The intruders and attackers are using state-of-the-art techniques and creating dangerous impact on the legitimate systems on the network. In this paper, we presented the framework to capture the attackers' activities. The observed attacks are recorded using virtual honeynet framework in real-time manner and analyzed to decode the strategy and intention of the attackers. The statistic details of attacks over the network presented here can be extended to specific kind of network attack in future with enhanced honeynet framework. This paper gives an insight about possible ways of eavesdropping the strategies and objectives of the attackers and helps the reader to design proper defensive mechanisms after secretly knowing the motive of the attackers.

References

1. Kreibich, C., Crowcroft, J.: Honeycomb: creating intrusion detection signatures using honeypots. ACM SIGCOMM Comput. Commun. Rev. **34**(1), 51–56 (2004)
2. Spitzner, L.: Honeypots: catching the insider threat. In: Proceedings of 19th Annual Computer Security Applications Conference. IEEE, Las Vegas, NV, USA (2003). https://doi.org/10.1109/csac.2003.1254322
3. Weiler, N.: Honeypots for distributed denial-of-service attacks. In: Proceedings. Eleventh IEEE International Workshops on Enabling Technologies: Infrastructure for Collaborative Enterprises. IEEE, Pittsburgh, PA, USA. https://doi.org/10.1109/enabl.2002.1029997
4. Alata, E., Nicomette, V., Kaaniche, M., Dacier, M., Herrb, M.: Lessons learned from the deployment of a high-interaction honeypot. Sixth European Dependable Comput. Conf. (2006). https://doi.org/10.1109/EDCC.2006.17,IEEE,Coimbra,Portugal
5. Liu, X., Peng, L., Li, C.: The dynamic honeypot design and implementation based on honeyd. In: Lin S., Huang X. (eds) Advances in Computer Science, Environment, Ecoinformatics, and Education. CSEE 2011. Communications in Computer and Information Science, vol. 214. Springer, Berlin, Heidelberg (2011)
6. Li, P.: Selecting and using virtualization solutions: our experiences with VMware and VirtualBox. J. Comput. Sci. Coll. **25**(3), 11–17 (2010)

7. Chowdhury, N.M.M.K., Bouta, R.: A survey of network virtualization. Comput. Netw. **54**(5), 862–876 (2010). Elsevier
8. Yan, L.K.: Virtual honeynets revisited. In: Proceedings from the Sixth Annual IEEE SMC Information Assurance Workshop. IEEE, West Point, NY, USA. https://doi.org/10.1109/iaw. 2005.1495957
9. Chamales, G.: The honeywall CD-ROM. IEEE Sec. Priv. IEEE. https://doi.org/10.1109/ msecp.2004.1281253
10. Provos, N.: Honeyd: A Virtual Honeypot Daemon", 10th DFN-CERT Workshop. Hamburg, Germany (2003)
11. www.vit.ac.in
12. Ding, J.-H., Chang, P.-C., Hsu, W.-C., Chung, Y.-C.: PQEMU: a parallel system emulator based on QEMU. In: IEEE 17th International Conference on Parallel and Distributed Systems. IEEE (2011). https://doi.org/10.1109/icpads.2011.102
13. Sochor, T., Zuzcak, M.: High-interaction linux honeypot architecture in recent perspective. In: Gaj P, Kwiecień A, Stera P. (eds) Computer Networks
14. Jiang, X., Wang, X.: Out-of-the-box monitoring of VM-based high-interaction honeypots. In: Kruegel C., Lippmann R., Clark A. (eds) Recent Advances in Intrusion Detection. RAID 2007. Lecture Notes in Computer Science, vol. 4637. Springer, Berlin, Heidelberg (2007)
15. Rogers, M.K., Goldman, J., Mislan, R., Wedge, T., Debrota, S.: Computer forensics field triage process model. J. Digit. Forensics Sec. Law **1**(2), Article 2 (2006). https://doi.org/10. 15394/jdfsl.2006.1004
16. Kenkre, P.S., Pai, A., Colaco, L.: Real time intrusion detection and prevention system. In: Satapathy S., Biswal B., Udgata S., Mandal J. (eds) Proceedings of the 3rd International Conference on Frontiers of Intelligent Computing: Theory and Applications (FICTA) 2014. Advances in Intelligent Systems and Computing, vol. 327. Springer (2015)
17. de Vivo, M., Carrasco, E., Isern, G., de Vivo, G.O.: A review of port scanning techniques. ACM SIGCOMM Comput. Commun. Rev. **29**(2), 41–48 (1999). ACM New York, NY, USA
18. CN.: Communications in Computer and Information Science, vol. 608. Springer (2016)
19. Wassermann, G., Su, Z.: Static detection of cross-site scripting vulnerabilities. In: Proceeding ICSE '08 Proceedings of the 30th international conference on Software engineering, pp. 171–180. ACM, New York (2008)
20. Hubczyk, M., Domanski, A., Domanska, J.: Local and remote file inclusion. In: Kapczyński, A., Tkacz, E., Rostanski, M. (eds.) Internet—Technical Developments and Applications, Advances in Intelligent and Soft Computing, vol. 118. Springer, Berlin (2012)
21. Xie, Y., Yu, F., Achan, K., Panigrahy, R., Hulten, G., Osipkov, I.: Spamming botnets: signatures and characteristics. ACM SIGCOMM Comput. Commun. Rev. **38**(4), 171–182 (2008)

Global and Local Signed Pressure Force Functions Active Contour Model Based on Entropy

Preeti Tiwari, Rajeev Kumar Gupta and Ramgopal Kashyap

Abstract Image segmentation is considered as a challenge in MRI images, synthetic and real images, because of intensity inhomogeneity which alters the final result of segmentation. Paper presents a novel method which deals with intensity inhomogeneous images. The proposed method introduces global and local fitting image energy function, where global fitting image is based on entropy and, moreover, presents global and local signed pressure function for stabling the gradient descent flow that is solving the energy function. Local signed pressure function has been established by multiplying entropy with local image difference and global signed pressure has been calculated by global image difference. Entropy used for estimating the bias field, global fitting term focused on homogeneous regions as well as provided the robustness to initialization of contour and local fitting term useful to detect objects with intensity inhomogeneity. Experimental result shows that in the presence of entropy, this method gives the superior performance concern to accuracy, time, and robust to initialization.

Keywords Segmentation · Active contour model · Intensity inhomogeneity · Local–global fitting image · Level set method

1 Introduction

Image segmentation is an important subject in image processing. An image is partitioned into multiple parts to detect interested object; this is called image segmentation [1]. There are different methods used for image segmentation like edge-based [2], clustering-based [3], region-based [4], etc. Image segmentation

P. Tiwari (✉) · R. K. Gupta
Department of Computer Science and Engineering, SISTec, Bhopal, India
e-mail: preetitiwari2107@gmail.com

R. Kashyap
Department of Computer Science and Engineering, Amity University,
Raipur, Chhattisgarh, India

© Springer Nature Singapore Pte Ltd. 2020
S. S. Dash et al. (eds.), *Artificial Intelligence and Evolutionary Computations
in Engineering Systems*, Advances in Intelligent Systems and Computing 1056,
https://doi.org/10.1007/978-981-15-0199-9_5

plays an essential part in medical science. Diagnosis of patients depends on the accurate object boundary detection of magnetic resonance images (MRI). Sometimes at the time of image acquisition, MR image, synthetic images contain intensity inhomogeneity [5]. Intensity inhomogeneity occurs because of imperfection of imaging instruments, external effects, etc. Image segmentation is quite difficult in the presence of intensity inhomogeneity. Many traditional intensity-based segmentation methods do not deal with intensity inhomogeneity because that assumes a uniform intensity, and so, the final segmentation result of image is not accurate.

The popular Mumford–Shah (MS) [6] model considers image intensity as a piecewise smooth; therefore, this model gives the good segmentation result, but the model is not good for those images which contain weak boundaries. The Chan–Vese (CV) [7] model was replacing the piecewise smooth function with piecewise constant function. Chan–Vese model uses intensity of the inner and outer regions of contour. It gives the efficient segmentation results in the presence of weak boundaries, but it gives poor results for intensity inhomogeneity because this method works on homogeneous regions of images. For solving the problem of intensity inhomogeneity, novel category of level set method has been introduced that is based on local term which uses the local region-based information into energy function. This level set method is called local region-based (LRB) Method. Li et al. [8] have introduced a local binary fitting (LBF) which uses kernel function for dealing with intensity inhomogeneity. This model reduces the limitation of CV model, but the major limitation of this method is sensitive with the location of initial curve. He et al. [9] initiated to improve the version of LBF model, and local entropy defines the weight function. This method is more robust with noise as compared to original LBF, but this is also sensitive to initialization.

The present paper proposes a novel global and local fitting image active contour method based on entropy. Using entropy for estimating the bias field, entropy performs gray-level distribution in better manner. Both global and local fitting-based energy functions are used in proposed work. Local fitting term used to detect objects with intensity inhomogeneity and global fitting term focused on homogeneous regions. Some key points are mentioned here:

- Global fitting term was formulated by using entropy. Here, entropy efficiently deals with the noise and provides more robustness to initialization of contour.
- For stabling the gradient descent flow that solves the energy function, global and local signed pressure force functions (SPF) have been introduced.
- Local SPF is the difference of local image which multiplies with entropy. Due to the uses of entropy here, getting result is much faster and accurate. Global SPF is global image difference.

2 Background

In LIF model [10], energy function of local image fitting model (LIF) is based on the difference from original image to fitted image. Energy function is represented as follows:

$$E = \frac{1}{2} \int |\text{Img}(x) - I_{\text{LFI}}(x)|^2$$
$$I_{\text{LFI}}(x) = f_1(x) M_1(\emptyset) + f_2(x) M_2(\emptyset)$$

where $f_1(x)$ and $f_2(x)$ are the local intensity means and $M_1(\emptyset) = H(\emptyset)$ & $M_2(\emptyset) = 1 - H(\emptyset)$. $H(\emptyset)$ is the Heaviside function.

A novel level set method, local statistical active contour model (LSACM) [11], where inhomogeneous objects separated using Gaussian distributions of different mean and variance. Pixel intensities are transformed from original image domain to another domain. It provides more robustness to noise. The mean intensity of Gaussian distribution is calculated by multiplying the original signal with bias field within Gaussian window.

Akram et al. [12] present improved LGFI active contour method, based on global and local fitted image (LGFI) energy function. Global and local signed pressure force (SPF) functions replaced global and local image differences, respectively; so, the gradient descent solutions of energy function is more stable and reduces level set curve convergence time. Gaussian kernel not only regularizes contour, but also avoids expensive computational re-initialization. The major drawback of this method is high time complexity because of local image fitted active contour and it does not give good results for different position of initial contour.

3 Proposed Method

3.1 Entropy

In information theory, the definition of entropy was introduced by Claude Shannon. The image information entropy Ep is represented as follows:

$$Ep(\text{Img}) = -\sum_{k=1}^{n} P_k \log P_k \tag{1}$$

where P_k is the probability of image(Img). Entropy of image is associated to complications hold in given neighborhood, which is usually described by structuring components. Entropy filter is as possible as to identify the fine variants in local gray-level distribution. Various parts of an image contain different information, and for accurate result of segmentation, it is must to compute the entropy of

(a)

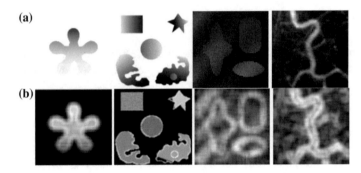

(b)

Fig. 1 Result of entropy on noisy and intensity inhomogeneous images. **a** Original images and **b** entropy images

each parts of an image. Ω_x is continuous domain of image, $\Omega_x \subset \Omega$. Entropy of image domain Ω_x is defined as follows [13]:

$$Ep(x, \Omega_x) = -\frac{1}{\log|\Omega_x|} \int_{\Omega_x} P(y, \Omega_x) \log P(y, \Omega_x) dy \tag{2}$$

where $P(y, \Omega_x)$ is gray-level distribution and expressed by

$$P(y, \Omega_x) = I(y) \int_{\Omega_x} I(z) dz \tag{3}$$

Proposed method first applies entropy filter that uses these two Eqs. (1) and (2). The result of entropy is shown in Fig. 1. It is effectively robust with intensity inhomogeneity and noise. Entropy image shows better visibility and effectively deals with intensity inhomogeneity.

3.2 New Active Contour Formulation

This paper proposes a novel active contour method based on entropy, where global and local fitting images are used to construct new energy function. Entropy defines degree of heterogeneity within segmented region of image and performs gray-level distribution in better manner. The new energy function (E_{NLGFI}) for segmenting intensity inhomogeneous images are represented as follows:

$$E_{NGLFI} = E_{GLFI}(\phi) + v A(\phi) + \mu L(\phi) \tag{4}$$

where $E_{GLFI}(\phi)$ is the energy function of global and local fitted images. $L(\phi)$ and $A(\phi)$ are length and area of contour, respectively. $v \geq 0$ and $\mu \geq 0$ are fixed parameters for area and length, respectively. Here, $E_{\text{GLFI}}(\phi)$, $A(\phi)$ and $L(\phi)$ are defined as follows:

$$E_{\text{GLFI}}(\phi) = \int_{\Omega}(\text{Img}(x) - \text{Img}_{\text{bLFI}}(x))(\text{Img}(x) - \text{Img}_{\text{GFI}}(x))dx$$

$$A(\phi) = \int_{\Omega} \delta(\phi)*|\nabla\phi|\, dx \quad L(\phi) = \int_{\Omega} g*H(\phi)\, dx \tag{5}$$

where edge indicator function g, has a range $[0, 1]$. K_{σ} is Gaussian kernel constant with standard deviation σ, and g is formulated as follows:

$$g = \frac{1}{1 + |I * \nabla K_{\sigma}|^2} \tag{6}$$

A. *Local fitted image*:

Ground truth image $J(x)$ is considered to be constituted by k-piecewise constants. Image $(\text{Img}(x))$ expressed as follows [8]:

$$\text{Img}(x) = b(x)\sum_{i=1}^{k} c_i M_i, \quad x \in \Omega_i$$

where c_i is a constant that represents intensity means for respected region Ω_i and M_i is the membership function of each region. $b(x)$ is bias field of the image. Local fitting term differentiates the small changes between foreground and background, and so, it is used to detect objects with intensity inhomogeneity. For two-phase model, the local fitted image equation is shown as follows:

$$\text{Img}_{\text{bLFI}}(x) = b(x) * (c_1 M_1 + c_2 M_2) \tag{7}$$

where

$$b(x) = \frac{\sum_{i=1}^{2} K_{\sigma} * [\text{Img}(x) * c_i * M_i]}{\sum_{i=1}^{2} K_{\sigma} * [c_i^2 * M_i]} \tag{8}$$

$$c_1 = \frac{\int K_{\sigma} * [\text{Img}(x) * b(x) * M_1]dx}{\int K_{\sigma} * \left[b(x)^2 * M_1\right]dx} \tag{9}$$

$$c_2 = \frac{\int K_{\sigma} * [\text{Img}(x) * b(x) * M_2]dx}{\int K_{\sigma} * \left[b(x)^2 * M_2\right]dx} \tag{10}$$

B. *Global fitted image*:

Global fitting term focuses on homogeneous regions. It is calculated with the help of entropy that provides the more robustness to initialization of contour & to noise and accurately detects interested segmented region in less time. Global fitted image equation is shown as follows:

$$\text{Img}_{\text{GFI}}(x) = Ep * (m_{c1}M_1 + m_{c2}M_2) \tag{11}$$

where piecewise constants m_{c1} and m_{c2} are inner and outer contour approximate intensities, respectively, and $M_1(\emptyset) = H(\emptyset)$ & $M_2(\emptyset) = 1-H(\emptyset)$. $H(\emptyset)$ is the Heaviside function. m_{c1} and m_{c2} values are evaluated by CV [10] model.

$$m_{c1} = \frac{\int [\text{Img}(x) * M_1] dx}{\int M_1 dx} \tag{12}$$

$$m_{c2} = \frac{\int [\text{Img}(x) * M_2] dx}{\int M_2 dx} \tag{13}$$

C. *Minimizing energy function*:

Minimizing the Eq. (5) with respect to $\emptyset$

$$\frac{\partial \emptyset}{\partial t} = \begin{bmatrix} \lambda_1 * (\text{Img}(x) - \text{Img}_{\text{bLFI}}(x)) * (m_{c1} - m_{c2}) \\ + \lambda_2 * b(x)(\text{Img}(x) - \text{Img}_{\text{GFI}}(x)) * (c_1 - c_2) \end{bmatrix}$$
$$* \delta(\emptyset) + v * g + \mu * \text{div}\frac{\nabla \emptyset}{|\nabla \emptyset|} * \delta(\emptyset) \tag{14}$$

In Eq. (14), larger values are generated by $(\text{Img}(x) - \text{Img}_{\text{bLFI}}(x)) * (m_1 - m_2)$ and $(\text{Img}(x) - \text{Img}_{\text{GFI}}(x)) * (c_1 - c_2)$ that cross the threshold value and the result is unstable contour. So, for normalizing the values to $[-1, 1]$, $(\text{Img}(x) - \text{Img}_{\text{bLFI}}(x))$ and $(\text{Img}(x) - \text{Img}_{\text{GFI}}(x))$ are replaced by F_{Lspf} and F_{Gspf} functions. For stabling the gradient descent flow, new global and local signed pressure force function (SPF) is introduced. Local SPF is the difference of local image multiply with entropy and it gives faster and accurate outcomes in the presence of entropy. Global SPF has been established by global image difference. The smooth gradient decent flow is given as follows:

$$\frac{\partial \emptyset}{\partial t} = [\lambda_1 * F_{\text{Lspf}} * (m_{c1} - m_{c2}) + \lambda_2 * b(x) * F_{\text{Gspf}}(c_1 - c_2)]$$
$$* \delta(\emptyset) + (v * g) + \mu * \text{div}\frac{\nabla \emptyset}{|\nabla \emptyset|} * \delta(\emptyset) \tag{15}$$

where proposed local and global SPF functions:

$$F_{\text{Lspf}} = \begin{cases} Ep * \left(\dfrac{\text{Img}(x) - \text{Img}_{\text{bLFI}}(x)}{\max |\text{Img}(x) - \text{Img}_{\text{bLFI}}(x)|} \right), & \text{Img}(x) \neq 0 \\ 0, & \text{Img}(x) = 0 \end{cases} \tag{16}$$

$$F_{\text{Gspf}} = \begin{cases} \left(\dfrac{\text{Img}(x) - \text{Img}_{\text{GFI}}(x)}{\max |\text{Img}(x) - \text{Img}_{\text{GFI}}(x)|} \right), & \text{Img}(x) \neq 0 \\ 0, & \text{Img}(x) = 0 \end{cases} \tag{17}$$

Local SPF $\left(F_{\text{Lspf}} \right)$ represents the positive values for outer border of contour, negative values for inner border of contour, and zero values in the remaining place. On the other hand, Global SPF $\left(F_{\text{Gspf}} \right)$ generates positive values for the outer region of the object, negative values for inner region of the object, and zero values on the object border.

4   Results and Discussion

The proposed method was run on 2.4 GHz Intel Core i3 with 4 GB RAM and implemented in MATLAB R2014b. Results are performed on both real MRI and synthetic images.

Segmentation Result on Synthetic images: Fig. 2 compares the experimental output of proposed method with other state-of-the-art methods. We use intensity variational images in this experiment. First image is based on homogeneous intensity object, but the segmentation is difficult when there is gradual changes in the distribution of intensity. Experiment clearly has shown that LIF [10] and LSACM [11] do not properly detect the object boundary of images, LGD [14] also does not detect the object boundary properly and contour stuck in background image. The result of LBF [8], improved LGFI [12] method, and proposed method gives good results, but in terms of time complexity, proposed method is better as compared to two other methods. Moreover, parameters set for both methods (improved LGFI and proposed method) are same, but proposed method takes less iteration and time. Table 1 shows the analysis of time complexity of different methods concern to iterations and CPU time. In this analysis, proposed method takes less iteration and CPU time for all images (Col 1–Col 5) shown in Fig. 2.

Segmentation Result on Medical images: Fig. 3 experiment shows the bias field, bias corrected image, and final contour of different methods like LBF [8], LSACM [11], improved LGFI [12], and proposed method. This experiment shows that LBF method that gives the poor result as compared to other methods. In this experiment, brain image is tested, and LBF method is unable to detect the upper and neck portion of brain image. LSACM method gives much better result compared to LBF, because this method detects the neck portion in better manner. Improved LGFI method is

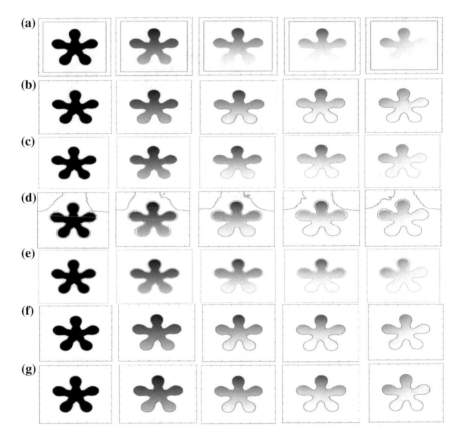

Fig. 2 Proposed method compared with different state-of-the-art-methods on intensity varied images. **a** Initial contour, **b–g** shows result of LBF, LIF, LGD, LSACM, improved LGFI, and proposed method, respectively

much better than LBF and LSACM, because it gives better segmentation result on brain image, and it covers upper portion of brain as well as neck portion. But the proposed method gives the best segmented result on this image, and it detects the better segmented portion of image that is required to diagnosis.

In this experiment, analysis of time complexity in terms of iteration and CPU time as follows: In terms of iteration, LBF [8] takes 500, LSACM [11] takes 300, improved LGFI [12] takes 110, and proposed method takes 37 iterations. In terms of CPU time, LBF takes 8.37 s, LSACM takes 88.93 s, improved LGFI takes 27.62 s, and proposed method takes 17.18 s. Here, proposed method gives the best result in only 37 iterations and 17.18 s.

Quantitative Analysis: Jaccard similarity (JS) [15] values are used for quantitative analysis of different methods on different intensity inhomogeneity images. JS value is dividing the intersection of these two images A and B by their union. It is nearby 1, where A is more similar to B. Figure 4 shown JS values of LBF [8], LIF

Table 1 Time complexity analysis, for example, is shown in Fig. 1.2

Images	LBF		LIF		LGD		LSACM		Improved LGFI		Proposed method	
	Iteration	CPU time	Iteration	CPU time	Iteration	CPU time	Iteration	CPU time	Iteration	CPU time	Iteration	CPU time
Col 1	300	5.68	250	6.87	500	50.34	50	110.95	44	6.57	20	3.96
Col 2	500	8.60	300	7.51	500	48.77	58	120.00	83	11.01	28	5.01
Col 3	750	20.97	400	9.45	500	49.10	60	130.90	131	15.73	37	5.85
Col 4	1000	32.91	1000	19.82	500	49.30	80	190.46	180	19.78	45	6.45
Col 5	1000	33.63	1000	20.17	500	50.69	130	210.45	190	20.50	45	6.23

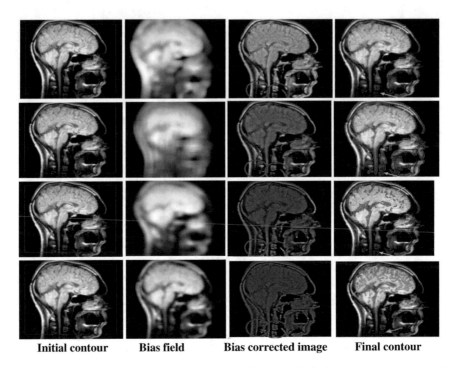

Initial contour   **Bias field**   **Bias corrected image**   **Final contour**

Fig. 3 Initial contour, bias field, bias corrected image, and final contour are represented columnwise, respectively. Bias correction and segmentation are performed by LBF (row 1), LSACM (row 2), improved LGFI (row 3), and proposed method (row 4)

[10], LGD [14], LSACM [11], improved LGFI [12], and proposed method which are calculated for Fig. 2. The JS value of proposed method is more nearby to 1 as compared to other methods.

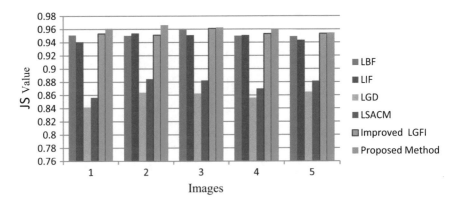

Fig. 4 JS values according to Fig. 2

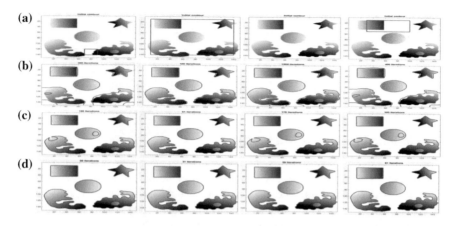

Fig. 5 Comparison results with different initialization on synthetic images. **a** Original image with different initialization. Segmentation performed by **b** LSACM, **c** improved LGFI, and **d** proposed method

Fig. 6 JS values according to Fig. 5

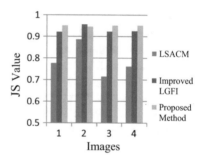

In Fig. 5, results are tested on four different positioning of initial contour. LSACM method does not give the good segmentation result on any of the initial contour which is tested, but in second initialization, the segmentation result is nearby actual result. The proposed method gives the correct results on all types of initial contour and it also takes very less iterations compared to others. It takes 59, 31, 59, and 61 iterations, respectively, on the given initializations.

JS values calculated to check the performance of different methods on different type of contours. Figure 6 shows the comparison chart of different methods with respect to Fig. 5 images. JS values of proposed method are more nearby to 1; Table 2 represents JS values of LSACM, improved LGFI, and proposed method.

Table 2 JS values as per Fig. 5 images

Images	LSACM	Improved LGFI	Proposed method
1	0.7768	0.9216	0.9514
2	0.8868	0.9567	0.9460
3	0.7150	0.9221	0.9513
4	0.7617	0.9246	0.9515

5 Conclusion and Future Scope

This paper presents new entropy-based active contour energy method for bias correction and image segmentation. Proposed method uses entropy for estimating bias field. Because of entropy, this method efficiently deals with the noise and intensity inhomogeneity. Proposed global and local SPF functions replaced the global and local image fitted differences for stabling gradient descent flow. To regularize the curve at each step and to avoid the computationally expensive re-initialization, Gaussian kernel is used.

Formulating the new energy function of our method from LIF and improved LGFI method and modification done by proposing new local and global SPF are based on entropy and local-global fitted image differences. Experimental result on medical and synthetic images shows proposed method takes less time and gives efficient results than others; so, this method is better than other state-of-the-art methods. But, proposed work does not work for color images.

In future scope, first enhance proposed method toward multi-phase image segmentation and plan to work on new method which takes much lesser time for segmentation and it gives the more accurate results.

References

1. Smistad, E., Falch, T.L., Bozorgi, M., Elster, A.C., Lindseth, F.: Medical image segmentation on GPUs—a comprehensive review. Medical Image Analysis (2015)
2. Wang, L., et al.: Active contours driven by edge entropy fitting energy for image segmentation. Signal Processing (2018)
3. Dhanachandra, N., Manglem, K., Chanu, Y.J.: image segmentation using K -means clustering algorithm and subtractive clustering algorithm. Procedia Comput. Sci. **54**, 764–771 (2015)
4. Kashyap, R., Gautam, P.: Modified region based segmentation of medical images. In: IEEE: International Conference on Communication Networks (ICCN), pp. 209–216 (2015)
5. Khadidos, A., Sanchez, V., Li, C.T.: Weighted level set evolution based on local edge features for medical image segmentation. IEEE Trans. Image Process. **26**(4), 1979–1991 (2017)
6. Mumford, D., Shah, J.: Optimal approximations by piecewise smooth functions and associated variational problems. Commun. Pure Appl. Math. **42**(5), 577–685 (1989)
7. Chan, T.F., Vese, L.A.: Active contours without edges. IEEE Trans. Image Process. **10**(2), 266–277 (2001)
8. Li, C., Kao, C.Y., Gore, J.C., Ding, Z.: Implicit active contours driven by local binary fitting energy. In: *Proceedings of the IEEE Computer Society Conference on Computer Vision and Pattern Recognition*, pp. 1–7 (2007)

9. He, C., Wang, Y., Chen, Q.: Active contours driven by weighted region-scalable fitting energy based on local entropy. Sig. Process. **92**, 587–600 (2012)
10. Zhang, K., Song, H., Zhang, L.: Active contours driven by local image fitting energy. Pattern Recognit. **43**(4), 1199–1206 (2010)
11. Zhang, K., Zhang, L., Lam, K.M., Zhang, D.: A level set approach to image segmentation with intensity inhomogeneity. IEEE Trans. Cybern. **46**(2), 546–557 (2016)
12. Akram, F., Garcia, M.A., Puig, D.: Active contours driven by local and global fitted image models for image segmentation robust to intensity inhomogeneity. PLoS One **12**(4) (2017)
13. Tang, J., Jiang, X.: A variational level set approach based on local entropy for image segmentation and bias field correction. Comput. Math. Methods Med. **2017** (2017)
14. Wang, L., He, L., Mishra, A., Li, C.: Active contours driven by local Gaussian distribution fitting energy. Sig. Process. **89**(12), 2435–2447 (2009)
15. Ji, Z., Xia, Y., Sun, Q., Cao, G., Chen, Q.: Active contours driven by local likelihood image fitting energy for image segmentation. Inf. Sci. (Ny) **301**, 285–304 (2015)

Optimization of Big Data Using Rough Set Theory and Data Mining for Textile Applications

I. Bhuvaneshwarri and A. Tamilarasi

Abstract Prediction of any property of the material has attracted the attention of many scientists all over the world. In order to produce better products, building construction, remote sensing, prediction of water availability, etc., information technology (IT) has played a dominant role. Textile industries are one of the sources to decide our economy of our country. Among various fabrics, single jersey finished cotton knitted fabrics are liked by many people because of its comfort. It has many comfort properties such as fabric width, fabric weight, fabric shrinkage, and fabric handle. Although attempts have been made to predict fabric width and fabric weight, the type of network that used was old and no new approaches have been made. In order to remedy this situation, new techniques such as rough set theory and data mining techniques have been applied with a view to predicting fabric width and this research paper addresses this issue in depth. We propose a new algorithm scalable rough priority prediction model (SrPPM) to predict fabric width of single jersey finished cotton knitted fabric with real-time textile big dataset over Hadoop MapReduce framework. Our results show that our proposed framework works well, in terms of time efficiency and scalability.

Keywords Big data · Data mining · Fabric width · Hadoop MapReduce framework · Rough set theory · Single jersey finished cotton knitted fabric

I. Bhuvaneshwarri (✉)
Department of Information Technology, Institute of Road and Transport Technology, Erode,
Tamil Nadu 638316, India
e-mail: pbw.irtt@gmail.com

A. Tamilarasi
Department of Computer Applications, Kongu Engineering College, Perundurai, Erode,
Tamil Nadu 638060, India
e-mail: angamuthu_tamilarasi@yahoo.co.in

© Springer Nature Singapore Pte Ltd. 2020
S. S. Dash et al. (eds.), *Artificial Intelligence and Evolutionary Computations
in Engineering Systems*, Advances in Intelligent Systems and Computing 1056,
https://doi.org/10.1007/978-981-15-0199-9_6

1 Introduction

In order to produce better and quality products, IT has played dominant role in terms of prediction. For example, the prediction of eclipses, rainfall, storm, thunder, availability of water, business stock markets, earthquake, weather forecasting, cyclone, typhoons (high-velocity winds) are the areas in which considerable amounts of work done to predict them. Fabric width is a very important parameter which affects single jersey finished knitted cotton fabric comfort properties and hence plays a vital role. Among the knitted garments, single jersey structure is prone to curling tendency. The fabric width is a complex parameter, which is affected by the multitude of factors such as diameter of machine, machine gauge, yarn diameter, yarn count, shrinkage, twist factor, and wales per centimeter.

The quality of data is important for prediction. In order to improve accuracy of prediction, data preprocessing is done to standardize or normalize the textile big dataset. Generally, the dataset comprises lot of attributes or features. But not all the features are used for data analysis or prediction, because some of the features are relevant and others are irrelevant for prediction. The irrelevant features affect the accuracy of the prediction. The feature selection methods are providing the solution to remove irrelevant features from dataset to achieve better prediction.

The big data analytics is now becoming an emerging field. The lot of attributes is involved in the prediction of the fabric width. So the massive textile dataset is generated. So it is necessary to manage or optimize big data effectively over Hadoop MapReduce framework.

This research paper is organized as the following sections. In Sect. 2, we discuss the related works that are already carried out in feature selection, predictive analysis, and managing or optimization of big data. In Sect. 3, we discuss our proposed framework and algorithm. In Sect. 4, we give the experimental results and discussion by applying the proposed algorithm on real-time textile big dataset and comparison of the proposed algorithm with other existing algorithms. The last section concludes the paper.

2 Related Work

The previous textile studies used various mathematical models [1, 2] to process textile dataset. The rough set theory-based feature selection methods were used on vague and uncertain dataset [3, 4]. The artificial neural network (ANN) was mainly used for prediction of many properties of knitted fabric such as bursting strength [5], performance [6], yarn price [7], air permeability [8], and shrinkage and fabric weight [9] in textile industries. But these earlier methods were unable to predict unknown features and complex relationships among features of dataset. And also, those methods are not scalable for massive dataset and take more time for computation.

In the existing studies, statistical models [9], neural networks [5–10] were used to predict various knitted fabric properties and data mining classification techniques [11–13]. There are tremendous works which have been carried out in the field of big data and managing of big data under Hadoop MapReduce framework such as proposed scalable classification system [11], scalable ensemble algorithm [12], and solution to big data classification problems [13].

The purpose of our research work to solve issues of knitted fabric manufacturers. In this research paper, we discuss the proposed scalable rough priority prediction model algorithm framework under Hadoop MapReduce framework to predict fabric width using new approaches such as rough set and data mining technique which have not been exploited by previous research workers.

3 Proposed Algorithm Framework

3.1 Proposed Scalable Rough Priority Prediction Model (SrPPM)

Input: T_D: Textile dataset, n: number of attributes
Output: SrPPM model

Step 1: Remove noisy and missing values using data preprocessing methods.
Step 2: Standardize the textile dataset using Z-score normalization.

$$NT_D = T_D - \mu_{T_D}/\sigma_{T_D} \tag{1}$$

Step 3: Select the optimal attribute for the prediction model from the large textile dataset.

    3.1 Calculate approximation values for all attributes using Eq. (2).

$$\sigma_{(P,Z)}(a) = \frac{(card(POS_p(Z)) - card(POS_{p-(a)}(Z)))}{card(U)} \tag{2}$$

    3.2 Compare the approximation values of each attribute a_j.
    3.3 If both attributes are highly associated, then remove the less approximation value attribute(s).

Step 4: Training Algorithm
Consider each row in input dataset as tuple $x_i = x_i^0, x_i^1, \ldots, x_i^{N-1}$ and each label y_i from $C_1, \ldots, C_{M-1} C_M$

Map_function
Associate with each tuple x_i its label y_i as key
Shuffle_function
Group tuples by label
Reducer_function
For every class C_j, j from 1 to M:

For each i from 0 to N–1:
Calculate sum of all x_i per group $\Rightarrow$ sum (x_{ij})
Calculate count of all x_i per group $\Rightarrow$ cnt (x_{ij})
Calculate mean of x_i per group $\Rightarrow (x_{ij}) = $ sum (x_{ij})/cnt (x_{ij})
Calculate sum of square difference from mean $\Rightarrow \sqrt{x_{ij}}$
Find variance of x_{ij} per group $\Rightarrow \sqrt{\sigma_{ij} x_{ij}}/Cnt(x_{ij})$

Step 5: Prediction Algorithm
To predict y for given $x = \{x_0, x_1, \ldots, x_{N-1}\}$

Map_ function

For every class C_j, j from 1 to M:
For every i from 0 to N–1:
Calculate conditional probability for each i, $P(x_i/C_j)$

$$P(x_i/C_j) = \frac{1}{\sqrt{(2\pi)\sigma_{ij}^2}} \exp^{\frac{(x_{ij}-\mu_{ij})^2}{2\sigma_{ij}^2}} \tag{3}$$

Shuffle_function

Group computed conditional probabilities by x

Reducer_function

Using naive assumption (conditional independence of variables):
Find full conditional probability

$$P(x/C_j) = \prod_{i=0}^{N-1} P(x_i/C_j) \tag{4}$$

Calculate $t_i = P(x/C_j) * P(C_j)$
Choose y $= C_j$ such that t_i is maximum

3.2 Description of Our Proposed SrPPM Algorithm

The proposed SrPPM algorithm framework is shown in Fig. 1. The textile big dataset is given to SrPPM algorithm under Hadoop MapReduce framework. It will give predicted fabric width as output. SrPPM algorithm has two parts: The first part of the proposed algorithm is to find optimal attributes among the various textile conditional attributes using rough set techniques. The attributes are to be classified based on the attribute significance value which we obtained through the computational program. From the output of the program, it should be clear that some of the conditional attributes such as stitch density, tightness factor, actual GSM are not considered because of their less importance during optimal attribute selection and also observed that the significant attribute score of the width shrinkage and length shrinkage is less when compared to the other conditional attributes. Based on the Rough set Core and Reduct concept, all the above-said attributes are removed from the input textile dataset. The second part of the proposed algorithm is developed using Gaussian Bayes MapReduce Algorithm. The algorithm consists of two phases such as training phase and prediction phase. Each phase has three functions named as Map_function, Shuffle_function, and Reducer_function.

In the training phase, Map_function assigns the label y_i as value for each tuple x_i. The Shuffle_function groups the tuples as per the labels. The Reducer_function computes all the statistical measure required to compute the Gaussian Bayes.

In the prediction phase, the Map_function computes the posterior probability of class with respect to data samples using Gaussian bayes approach. Then, the total probability of data sample for each class is computed using Shuffle_function. The Reducer_function computes the class of data sample based on the probability computed in the Map_function.

Fig. 1 SrPPM algorithm framework

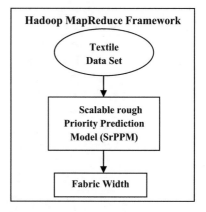

4 Experimental Results and Discussion

4.1 Dataset

The massive textile dataset was collected from the textile industry. The textile dataset has nearly 3 lakh fabric samples based on 5 different machine diameters ranging from 19 to 23. The summary of entire sample textile dataset for machine diameter 19 inches is shown in Table 1, which consists entirely of 16 textile attributes. The first 15 are conditional attributes, and fabric width (cm) is a decision attribute. The entire dataset was preprocessed to fill missing values and standardized using Z-score normalization to produce quality data.

4.2 Results and Discussion

The proposed SrPPM is implemented in Hadoop cluster. We have HPC cluster with six nodes. The efficiency of rough computational priority prediction model (RCPPM) is theoretically and empirically proved in our previous research work. In this research paper, we are concerned with the time efficiency of parallel version of RCPPM called SrPPM classification algorithm in big data environment. This research paper focuses on fabric width prediction using big data computational techniques.

Table 1 Summary of entire sample dataset for machine diameter 19 inches

S. No.	Features/attributes	High value	Low value
1.	Machine diameter (inches)	19	0
2.	Machine gauge (inches)	24	0
3.	Actual GSM	157.31	1.18
4.	Actual linear density (tex)	17	14.10
5.	Nominal linear density (tex)	14.76	14.76
6.	Length shrinkage (%)	5	0
7.	Width shrinkage (%)	5	0
8.	Lea weight (gms)	1.87	1.55
9.	Lea strength (lbs)	78.49	0
10.	Twist multiplier (tpcm* $tex^{0.5}$)	36.98	33.68
11.	Loop length (mm)	2.42	2.01
12.	Course per cm	27.98	21.67
13.	Wales per cm	19.88	16.48
14.	Stitch density (cm^2)	556.30	357.22
15.	Tightness factor ($tex^{0.5} * mm^{-1}$)	1.87	0.02
16.	Fabric width (cm)	88.91	80.28

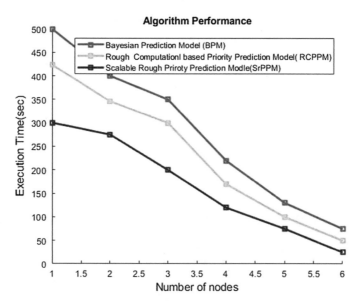

Fig. 2 Performance of SrPPM based on number of nodes

The performance of proposed scalable rough priority prediction model (SrPPM) is compared with the rough computation priority prediction model (RCPPM) and Bayesian prediction model (BPM) on single node. The scalability of the proposed SrPPM model is also tested in distributed parallel domain. The scalability evaluation includes two aspects: (1) performance with different numbers of nodes, and (2) performance with different sizes of training datasets.

Figure 2 illustrates the execution time of our proposed SrPPM model with different numbers of nodes when the number of records is 1, 2, and 3 lakhs, respectively; we have observed that the overall execution time decreases when the number of nodes increases. This indicates that the more nodes are involved in computing which increases the efficiency of the algorithm.

The speedup of a parallel algorithm is defined as the ratio of the parallel execution time and the sequential time. We tested the speedup of the SrPPM algorithm upon various numbers of data nodes with the increase of sample dataset, shown in Fig. 3. It can be seen that the more the data nodes or the larger the sample dataset, the larger the speedup, which is consistent with the general conclusion of parallel algorithms.

Our proposed learning algorithm is 10.5 times faster than the sequential algorithm. This means that the influence of sample dataset size on speedup is more than that of the data node number; that is, the larger the sample dataset, the more dramatic efficiency improvement will be achieved.

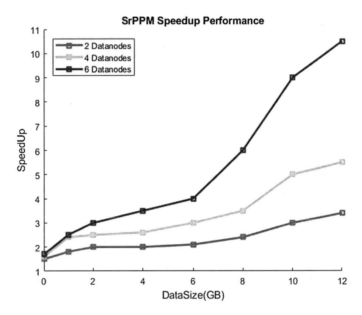

Fig. 3 Performance of SrPPM based on size of training dataset

5 Conclusion

Prediction of fabric width will aid the garment manufacturers and manufacturers of knitted fabrics in order to adjust the various parameters to achieve the optimum condition. It will also be useful for yarn manufacturers for supplying the most appropriate yarn for knitting purposes. The need of scalable fabric width prediction system is highly required to resolve the fabric-related problems for knitted fabric manufacturers in textile industry. Predicting and analyzing fabric width using scalable approach are complex task. In this research paper, scalable rough priority prediction model (SrPPM) is proposed to predict the fabric width using various textile parameters. Our existing proposed RCPPM model cannot fit to manage huge datasets. For example, as the size of training data grows, the process of building scalable classifier can be very time consuming. To solve the above challenges, SrPPM approach is proposed to improve the scalability of the model. We have compared the performance of the proposed approach with existing approach with respect to a number of nodes and size of dataset. The empirical results show that the proposed algorithm exhibits both time efficiency and scalability. In future works, the fabric width prediction will be studied using deep scalable computational approaches.

References

1. Yildirim, P., Birant, D., Alpyildiz T.: Data mining and machine learning in textile industry. Wiley Interdisc. Rev. Data Min. Knowl. Disc. **8**(1) pp. e1228 (2018)
2. Jaouachi, B., Khedher, F.: Evaluation of sewed thread consumption of jean trousers using neural network and regression methods. Fibres Text. Eastern Europes (2015); Thangavel, K., Pethalakshmi, A.: Dimensionality reduction based on rough set theory: a review. Appl Soft Comput. **9**(1), 1–2 (2009)
3. Thangavel, K., Pethalakshmi, A.: Dimensionality reduction based on rough set theory: a review. Appl. Soft Comput. **9**(1), 1–2 (2009)
4. Przybyła-Kasperek, M.: Feature selection based on the rough set theory and dispersed system with dynamically generated disjoint clusters. In: IEEE International Conference on INnovations in Intelligent SysTems and Applications (INISTA) pp. 223–228 (2017)
5. Ertugrul, S., Ucar, N.: Predicting bursting strength of cotton plain knitted fabrics using intelligent techniques. Text. Res. J. **70**(10), 845–851 (2000)
6. Gong, R.H., Chen, Y.: Predicting the performance of fabrics in garment manufacturing with artificial neural networks. Text. Res. J. **69**(7), 477–482 (1999)
7. Venkataraman, D., Vinay, N., Vardhan, T.V., Boppudi, S.P., Reddy, R.Y., Balasubramanian, P.: Yarn price prediction using advanced analytics model. In: IEEE International Conference on Computational Intelligence and Computing Research (ICCIC) pp. 1–8 (2016)
8. Matusiak, M.: Application of artificial neural networks to predict the air permeability of woven fabrics. Fibres Text. Eastern Eur (2015)
9. Anupreet, K., Roy, K.: Prediction of shrinkage and fabric weight (g/m^2) of cotton single jersey knitted fabric using artificial neural network and comparison with general linear model. Int. J. Inf. Res Rev. **03**(06), 2541–2544 (2016)
10. Furferi, R., Governi, L., Volpe, Y.: Modelling and simulation of an innovative fabric coating process using artificial neural networks. Text. Res. J. **82**(12), 1282–1294 (2012)
11. Sun, N., Sun, B., Lin, J.D., Wu, M.Y.: Lossless pruned naive bayes for big data classifications. Big Data Res (2018)
12. Baldán, F.J., Benítez, J.M.: Distributed FastShapelet Transform: a big data time series classification algorithm. Inf. Sci (2018)
13. Dagdia, Z.C.: A scalable and distributed dendritic cell algorithm for big data classification. Swarm Evol. Comput (2018)

Machine-Learning-Based Device for Visually Impaired Person

Tuhina Priya, Kotla Sai Sravya and S. Umamaheswari

Abstract The blind-assistive devices promote the detection of objects and alerting the user by a buzzer or an alarm. In this project, a camera is placed inside a cap which is used by the visually impaired person. The machine-learning algorithm is used for accurate detection of the object and provides an alarm. The ultrasonic sensor is used to measure the distance between the visually impaired person and the real-time object detected when the object is detected the alarm is actuated. Nowadays, the assistive devices include the involvement of both hardware and software section to assist the blind user. The proposed method for the visually impaired person aims to detect the object more accurately so that the visually impaired person can navigate to their full potential in real-time application.

Keywords Real-time object detection · Raspberry Pi 3 processor · Machine-learning · Yolo algorithm · Ultrasonic sensor · Visually impaired person

1 Introduction

The modern society comes across millions of visually impaired people in everyday life. The lives of these visually challenged people are tough and cannot be handled without any smart supportive techniques. Issues faced by these people are mainly in navigation as well as to be given an alert buzzer in order to be inattentive posture. The solution is to develop a device to perform the decision-making process and to inform the visually impaired person what object is toward the path. Only the

T. Priya · K. S. Sravya · S. Umamaheswari (✉)
Department of Electronics and Instrumentation Engineering, SRM Institute of Science and Technology, Kattankulathur, Chennai, India

T. Priya
e-mail: tuhina_ravi@srmuniv.edu.in

K. S. Sravya
e-mail: kotlasai_raj@srmuniv.edu.in

© Springer Nature Singapore Pte Ltd. 2020
S. S. Dash et al. (eds.), *Artificial Intelligence and Evolutionary Computations in Engineering Systems*, Advances in Intelligent Systems and Computing 1056, https://doi.org/10.1007/978-981-15-0199-9_7

hardware and software tools can help a blind person to do detection, such as object detection and image classification, this leads to the navigation system.

To assist a blind and visually impaired person, researchers have found several technologies. The proposed technique acquires the machine-learning concept to finalize and solve the ongoing problems faced by the blind and visually impaired person. When compared to existing ideas, machine learning is an innovative approach to obtain desired results.

Real-time object recognition approach for assisting blind person was proposed by Jamel S. Zraqou, Wissam M. Alkhadour, and Mohammed Z. Siam, This idea was based on expanding possibilities that people with vision loss can achieve their full potential in the real-time system. Object detection for blind people was proposed by Shivaji Sarokar, Sujit Gore, and Seema Udgirkau, this idea was focused on computer vision module of the smart vision system assisting the visually impaired person. Obstacle detection and warning system by acoustic feedback by Alberto Rodríguez, J. Javier Yebes, and Pablo F. Alcantarilla, this idea was based on an obstacle detection system for assisting a visually impaired person. A dense disparity map is computed from the image of a stereo camera supported by the user. Variable object-detection system for the blind by Alessandro Dionisi, Emilio Sardinia, and Mauro Serpelloni, this idea was based on an, i.e., device able to provide the blind some pieces of information about distances and simplify the search as soon as possible. The object recognition for the blind by Georgios Nikolakis, Dimitrios Tzovaras and Michael G. Strintzis, this idea was based on haptic application to allow a blind person to recognize the 3D object that exists in a virtual environment. This leads to permit a blind person to touch, grasp, and manipulate objects. A primary traveling assistant system of bus detection and recognition for visually impaired people was proposed by Hangrong Pan, Chucai Yi, Yingli Tian. This idea was based on a system which notifies the visually impaired person in form of speech warning about the coming bus in front of the person, also it notices the route number and similar information is notified to the visually impaired person in the form of text. Real-time vehicle classification method for multi-lanes roads was proposed by Mo Shaoqing, Liu Zhengguang, Zhang Jun and Wu Chen. This idea was based on vehicle categorization method for massive traffic motion multi-lanes roads, which differentiate vehicles into cars, bus, and truck, and three cameras are used in order to have a better view of the two lanes. Portable Camera-Based Assistive Text and Product Label Reading From Hand-Held Objects for Blind Persons was proposed by Chucai Yi, Yingli Tian, and Aries Arditi. This idea was on camera-based assistive text reading frame work, it helps blind persons in order to read text labels as well as product packaging via hand-held objects in real-time application. An oral tactile interface for blind navigation was proposed by Hui Tang, D. J. Beebe. This idea was designed, evaluated in order to produce an indication for direction an oral tactile interface for guidance in outdoor real-time application Navigation and space perception assistance for the visually impaired: The NAVIG project was proposed by S. Kammoun, G. Parseihian, and O. Gutierrez. This idea aims to contribute the mobility aids like a walking stick, guide dogs, etc., by adding distinctive features to limit some particular objects in the surrounding for easier navigation.

The machine learning is an innovative approach which adds novelty to the project. The real-time object detection assists the visually challenged person to get better vision and understanding of their Surroundings. In order to perform real-time object detection and alarm system, the machine learning via the Open CV Software is needed. The four parameters of the proposed methodology are center of bounding box, width, height, value C corresponding to a class of an object. When the algorithm is executed, classes and bounding box are predicted for the whole image. After real-time object detection, a warning is sent to the blind person that will assist the blind person to walk in different surroundings.

1.1 Related Work

1. Machine Learning:

Machine Learning is a category of algorithm which authorizes the software application to predict outputs without being specifically programmed. It works based on spontaneous characteristic and experience. This machine-learning initiation was beginning with observation. The automatic feature is the primary aim of this learning and is performed without any human interference. To train machine learning, there is a requirement of additional time and resources. The collaboration of machine learning with any other concept makes the result even more accurate and adds more volume to the information.

2. Ultrasonic Sensors:

Ultrasonic sensors are used in various applications since they have high sensitivity and high penetrating power which easily detects real-time objects. The ultrasonic sensor calculates the time interval between sending the signal and receiving the echo to determine the distance between the detected object and visually impaired person. To detect the condition of the obstacle detected by the ultrasonic sensor, comparison of echo sample with sound waves occurs.

2 Proposed Approach

In this proposed work, object detection is performed by various steps. The first step is to give the parameters of the desired object to the raspberry Pi 3 processor, by a machine-learning algorithm called YOLO Algorithm. The machine learning is performed in Open CV software. In this work, the features of the object present in the real-time surroundings can be detected. When the real-time object is detected,

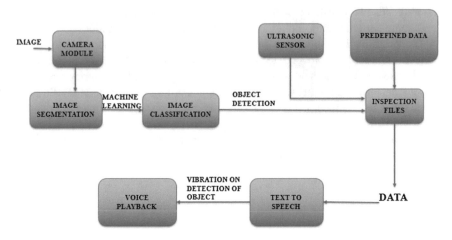

Fig. 1 Block diagram of proposed work

then the data transferring process occurs and an alert is given depending on the distance between the object and the visually impaired person.

The working model of this project is explained in Fig. 1 in which various processes are used to perform image processing when given a JPEG image as an input to the camera module. The further processing has been explained in detail below. In this Fig. 1 block diagram, we have illustrated the image classification and object detection.

Step 1: The camera module captures the JPEG format image which is given as an input.

Step 2: The image, i.e., obtained from the camera module is segmented. The image segmentation is performed to simplify or change the representation of the image so that it is easier to classify and analyze the data.

Step 3: The machine-learning algorithm, i.e., Yolo algorithm is implemented on the segmented image for further classification. In image classification, the entire image is classified with a single label. When performing image classification, given as an input image to the neural network, a single class label and an accuracy percentage with the class label has been obtained. This class label is meant to characterize the contents of the entire image in one label.

Step 4: The object-detection performance builds on image classification and seeks to localize exactly where in the image each object appears. Therefore, object detection allows us to present one image to the network and obtain multiple bounding boxes and class labels out.

Step 5: The ultrasonic sensor will calculate the time interval between sending the signal and receiving the echo to determine the distance between the detected object and visually impaired person.

Fig. 2 YOLO algorithm

Step 6: The data analysis of the predefined file existing in the library is compared and inspected via file inspection and the data sent; an alarm is given at the same time depending on the distance between the detected real-time object and visually impaired person.

In this methodology, the algorithm used is YOLO algorithm. The prediction of classes and bounding boxes for the whole image takes place when the algorithm is executed. The task is to predict a class of an object and the bounding box specifying object location (Fig. 2).

Each bounding box can be described using four descriptors:

- Center of a bounding box (b_x b_y)
- Width (b_w)
- Height (b_h)
- Value c is corresponding to a class of an object (i.e. car, traffic lights) (Fig. 3).

Fig. 3 Bounding box detected with the label class: car

Following equation has been used

$$(\mathbf{P_c}, \mathbf{W_x}, \mathbf{W_h}, \mathbf{W_w}, \mathbf{D}) \tag{1}$$

$$\mathbf{S} * \mathbf{S} * (\mathbf{B} * \mathbf{5} + \mathbf{C}) \tag{2}$$

where in Eq. (1),

1. P_c is the probability that c is corresponding to a class of an object.
2. W_x, b_y is center of a bounding box.
3. W_h is bounding boxes height.
4. W_w is bounding box width.
5. D is the value, i.e., corresponding to a class of an object.

where in Eq. (2),

1. $S * S$ is the grid size of the image.
2. B is the bounding boxes.
3. C is the class probabilities.

3 Experimental Result

3.1 Input Image

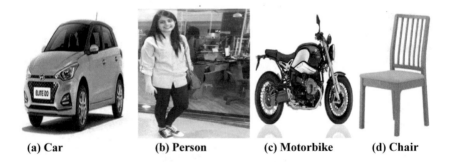

(a) Car   (b) Person   (c) Motorbike   (d) Chair

3.2 Output Image

car: 100.00%

(a) Car

(b) Person

motorbike: 99.95%

(c) Motorbike

(d) Chair

3.3 Object Detection

In Fig. 4, three objects are detected at once in the same image defining the different class of the object. The performance builds on image classification and seeks to localize exactly where in the image each object appears.

When performing object detection, we wish to obtain:

- (x, y)-coordinates for each object in an image
- The class label associated with each bounding box
- The accuracy percentage associated with each bounding box and class label.

Fig. 4 Label object class,
i.e., person and potted plant

Therefore, object detection allows us to:

- Present one image to the network
- Obtain multiple bounding boxes and class labels out.

   After the object is detected an alarm is actuated.
   Hence, real-time object detection is performed.

3.4 Observational Reading

See Table 1.

3.5 Graphical Representation

In Fig. 5, graph is plotted between accuracy and distance. The accuracy decreases
with the increase in the distance.

Table 1 Observational table

S. No.	Object detection			
	Image	Detection accuracy	Distance	Time period
1.	Label: Person	99.77%	20 cm	0.5 s
2.	Label: Dog	99.82%	15 cm	0.8 s
3.	Label: Person 1 (front) Person 2 (behind)	Person 1: 99.95% Person 2: 87.97%	Person 1: 30 cm Person 2: 1 m	1.5 s
4.	Label: Motorbike	99.95%	25 cm	1.2 s

where,
- Accuracy in (%), defines the confidence level of the real-time object detected
- Distance is in cm
- The Time period is in seconds (s)

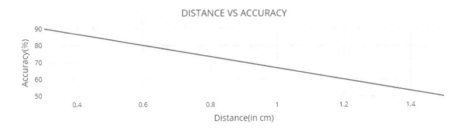

Fig. 5 Graph: accuracy versus distance

4 Conclusion

The real-time object detection is performed using machine learning. The raspberry Pi 3 model is used. The aim of the proposed work is to obtain real-time object detection and recognition with a proper accuracy level. The collection of data from the camera module is segmented and classified further then the object detection takes place. The file inspection is performed, depending on the data obtained, and the alarm is actuated at the same time. Thus, real-time object detection takes place.

Bibliography

1. Dakopoulos, D., Bourbakis, N.G.: Wearable obstacle avoidance electronic travel aids for blind: a survey. IEEE Trans. Syst. 25–35 (2010)
2. Dollár, P., Appel, R., Belongie, S., Perona, P.: Fast feature pyramids for object detection. IEEE Pattern Anal. Mach. Intell. (8), 1532–1545 (2014)
3. Li, B., Wu, T., Zhu, S.-C.: Integrating Context and Occlusion for Car Detection by Hierarchical and-or Model, pp. 652–667. Springer International Publishing, Switzerland (2014)
4. Lee, C., Kim, M., Park, J., Oh, J., Eom, K.: Design and implementation of the wireless RFID glove for life applications. Int. J. Grid Distrib. Comput. 3(3) (2010)
5. Sjostrom, C.: Designing haptic computer interfaces for blind people. In: Proceedings of ISSPA 2001, Kuala Lumpur, Malaysia, August (2001)
6. Pan, H., Yi, C., Tian, Y.: A primary traveling assistant system of bus detection and recognition for visually impaired people. In: IEEE International Conference on Multimedia and Expo Workshops a Jose, CA, USA, pp. 31–34 (2013)
7. Shaoqing, M., Zhengguang, L., Jun, Z., Chen, W.: Real-time vehicle classification method for multiple-lane road. In: IEEE International Conference on Industrial Electronics and Applications (ICIEA), pp. 960–964 (2009)
8. Yi, C., Tian, Y., Arditi, A.: Portable camera-based assistive text and product label reading from hand-held objects for blind persons. IEEE/ASME Trans. Mechatron. 19(3), 808–817 (2014)
9. Tang, H., Beebe, D.J.: An oral tactile interface for blind navigation. IEEE Trans. Neural Syst. Rehabil. Eng. 116–123 (2006)
10. Kammoun, S., Parseihian, G., Gutierrez, O.: Navigation: space perception assistance for the visually impaired: the NAVIG project. IRBM Numero special ANR TECSAN. 33(2), 182–189 (2012)

Forest Cover Classification Using Stacking of Ensemble Learning and Neural Networks

Pruthviraj R. Patil and M. Sivagami

Abstract Deforestation is one of the major issues, that is, being affecting the environment for the long time and there are few effective measures have been taken to withstand it and to maintain the pristine of the nature. One of them is preserving the wilder forests. The main motive of the proposed work is to classify the forest dataset so that it helps the authorities in maintaining the forests and protecting them by controlled deforestation and re-growing. The proposed classification technique introduces the stacking approach of Ensemble learning which uses random forests, extra trees with boosting and multilayered perceptron techniques for forest cover classification. The proposed model is evaluated using dataset from the UCI library. The proposed stacking approach shows the improvement in the quality of forest covers classification results and is shown using ROC curve analysis.

Keywords Data mining · Forest covers · Stacking · Random forest · Extra trees · Multilayered perceptron · Boosting · Principle component analysis

1 Introduction

Big data is the huge pool of data collections. The required information from this pool can be acquired from few welldefined processes. One of such processes is the process of knowledge discovery in databases (KDD). KDD is the process of discovering useful knowledge from a collection of data. KDD follows the sequence of steps, namely aggregation, preprocessing, data modelling and data analysis [1]. The processes of KDD is shown in Fig. 1 (*Source* Fayyad et al. [2]). For the data modelling, one requires the machine-learning algorithms from which the proper

P. R. Patil (✉) · M. Sivagami (✉)
School of Computing Science and Engineering, Vellore Institute of Technology,
Chennai, India
e-mail: pruthvirajr.patil2016@vitstudent.ac.in

M. Sivagami
e-mail: msivagami@vit.ac.in

© Springer Nature Singapore Pte Ltd. 2020
S. S. Dash et al. (eds.), *Artificial Intelligence and Evolutionary Computations in Engineering Systems*, Advances in Intelligent Systems and Computing 1056, https://doi.org/10.1007/978-981-15-0199-9_8

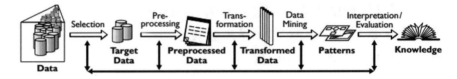

Fig. 1 KDD process (*Source* Fayyad et al. [2])

hypothesis can be framed. Machine learning is a subset of artificial intelligence (AI) which basically does not deal with the hardcoding on the machine. It makes the machine to learn by its own by training it with the mined datasets from the big data using different machine-learning algorithms. The results of the machine-learning algorithms are based on the extent to which they are being trained and the pre-ciseness shall be measured using certain measures of accuracy. In supervised learning while training the system, there is a clear demarked output. To train the model which has the vast dataset requires. By multiple rigorous approaches, one shall get reliable probabilities of the possible outcomes. But if such probabilities are further stacked, and trained using a meta classifier, it shall be expected that one gets a better output than the base rigorous models. Thus, stacking is the approach of increasing the accuracy by combining the probabilities of the outputs of the base models used where the training the outcomes of them in the form of probabilities using the meta classifier. The base models comprise of Ensemble learners or other rigorous approaches. Ensemble learning is rigorous method of machine learning. In the supervised learning that describes data in the apt representation like trees or graphs, there are vast numbers of approaches to form a hypothesis function that can meet the quality of the optimal hypothesis function. Based on the higher skewness and overlapping of the output classes, ensembles learning like random forests, extra trees as well as neural networks seem to produce promising results. One can also increase the accuracy of these ensemble models by boosting the ensemble learner by creating the classifier out of the weak intermediate learners or models during the learning phase. On the other hand, neural networks is another type of rigorous learning where the data has to surpass through the hidden layers defined under the constraint of different given activation functions.

2   Related Work

Most of the researchers have used machine-learning algorithms for classifying the forests and few of them have been mentioned as literature survey. Crain and Davis [3] worked over principle component analysis (PCA) on which they applied *k*-means clustering and over that multi classed support vector machines (SVM) to identify the forest cover types. Schmitt C. B., Belokurov A., et al. in the Global Ecological Forest Classification and Forest Protected Area Gap Analysis [4] clas-sified the forest types protected per given eco region. Gribel [5] has worked on two class decision forest, multiclass decision forest, multiclass neural networks to classify the forest covers. Jolliffe [6, 7] and Hwang et al. [8], worked on principle component analysis (PCA) that can be used to reduce the dimensionality of the dataset so that to remove the unwanted data of the data present on the different dataset. Pahlm and Sornmo [1], and Vehbi Olgac et al. [9] used feed forward artificial neural network model for predicting the forest cover type. Freund and Schapire [10] worked and analysed over the booster models over ensemble learning to increase the accuracy by using the factor of weak learners. Xie et al. [11], worked on another type of tree learner called extra trees that uses lesser time to compile. The main suggestion considered is that, the stacking using the probabilities of the corresponding outcomes of the models as the data points, predicting the final outcome using the meta classifier over was better than the outcome of the individual base models which work is presented by Magdalena Graczykl et al. over the dataset of Real Estate Appraisal [12].

3 Proposed Model

The proposed model uses the techniques of feature engineering, principle component analysis to preprocess the data and the technology of stacking involving the Ensemble approaches of random forests, extra trees with boosting as well as the neural network approaches. The results of the proposed stacking model for the forest cover classification are compared with its base models.

The main goal is to classify the forests into the seven predefined classes using the predefined dataset. The forest classification has been done as a process of knowledge discovery in databases. The blocks of KDD are shown in Fig. 1.

The preprocessing of data, data modelling for forest classification, and finally testing of the proposed model are the components of proposed knowledge discovery in databases approach for forest classification. The individual blocks of KDD of the proposed model have been explained in the below subdivisions. The substeps of data preprocessing or data aggregation and data modelling are shown in Fig. 2.

3.1 Data Aggregation

Given 15121 tuples with 54 features that initially as shown in Table 1 are present in the dataset taken.

Fig. 2 Data-flow graphical
representation

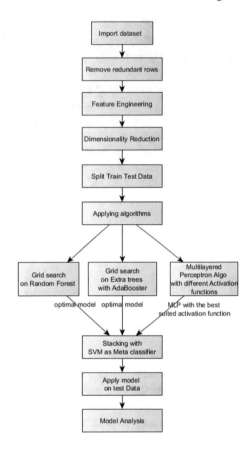

Table 1 Features in the
dataset

• Elevation
• Aspect
• Slope
• Horizontal distance to hydrology
• Vertical distance to hydrology
• Horizontal distance to roadways
• Hillshade 9 am
• Hillshade noon
• Hillshade 3 pm
• Horizontal distance to firepoint
• Wilderness level (four wilderness binary features)
• Soil types (40 soil types)

3.1.1 Pearson's Correlation Co-efficient

Pearson's correlation co-efficient is used to check how one feature depends on other and also if any zero-variant feature is present in the dataset. If yes, then that shall be eliminated as it contributes almost nothing significant to the given data. Pearson's co-efficient is the measure for correlation between two variables (x, y) is given by the Eq. 3.1.

$$r = \frac{1}{n-1} \sum \frac{(x_i - \bar{X})(y_i - \bar{Y})}{S_x S_y} \qquad (3.1)$$

The value of r ranges between $+1$ and -1; $r > 0$ refers a positive relationship of X and Y: as one attribute gets increased, the other too gets increased; $r < 0$ refers a negative relationship: as one attribute gets increased, the other attribute gets decreased; $r = 0$ refers that there is no correlation between the two attribute. Where in, S_x and S_y are the standard deviation of x and y, respectively. It was found that soil 7 and soil 15 are non-variant and their variance is zero. Hence, it was considered be better to remove those two features to make the hypothesis simpler.

3.1.2 Feature Engineering

To increase the predictive power of machine-learning algorithms by creating features from raw data that facilitates the machine-learning process. Here, 21 new features are being added to the dataset. The amenities are considered to be the hydrology, fire points and the roadways as they contribute more than other factors to the wilderness of the forests. Hence, the linear combination of such highly correlated variables seemed to be helpful for the prediction models. Table 2 gives the details about how new features have been obtained and their description.

3.1.3 Dimensionality Reduction Using Principle Component Analysis (PCA)

The dimensionality reduction shall be done using PCA. PCA solved the singularity problem from its process and could avoid an over-fitting. Given that there were 73 features after addition of new features using feature engineering. Then, PCA was applied on them and the number of eigen values that were greater than threshold 0 was 64 in number. Hence, only those top 64 data points are to be chosen.

Table 2 Details about new features created

New_Feature/s	Description
HR1, HR2	Sum and difference between values of features, namely "Distance between forest and nearest hydrology", "Distance between forest and nearest Roadways"
HF1, HF2	Sum and difference between values of features, namely "Distance between forest and hydrology", "Distance between Forest and nearest fire points"
FR1, FR2	Sum and difference between values of features, namely "Distance between forest and nearest fire points", "Distance between Forest and the nearest roadways"
Mean_HR, Neg_mean_HR	The feature HR1 divided by 2 and HR2 divided by 2, respectively, indicating the positive and negative means between features "Distance between forest and nearest hydrology", "Distance between forest and nearest Roadways"
Mean_HF, Neg_mean_HF	The feature HF1 divided by 2 and HF2 divided by 2, respectively, indicating the positive and negative means between features "Distance between forest and hydrology", "Distance between Forest and nearest fire points"
Mean_FR, Neg_mean_FR	The feature FR1 divided by 2 and FR2 divided by 2, respectively, indicating the positive and negative means between features "Distance between forest and nearest fire points", "Distance between Forest and the nearest roadways"
Ele_vert, Neg_ele_vert	Sum and difference between features, namely "Elevation" and "vertical distance from forest to hydrology", respectively
Mean_amenities	Mean of three features, namely "horizontal distance from forest to hydrology", "distance between forests and nearest roadways" and "Distance from forest to the nearest fire points"
Slope2	Square root of features [("horizontal distance to hydrology") to the power 2 + (vertical distance to hydrology) to the power 2]
Mean_hillshade	Mean of features namely "hill shade at 9 am", "hill shade at noon" and "hill shade at 3 pm"
Mean_fire_hyd	Mean of features namely "Vertical distance to hydrology" and "horizontal distance to fire point"
Elev_to_HD_Hyd	Difference between feature namely "Elevation" and 0.2*feature, namely "Horizontal distance from forest to nearest Hydrology"
Elev_to_HD_Road	Difference between feature namely "Elevation" and 0.05*feature, namely "Horizontal distance from forest to nearest roadways"
Elev_to_VD_Hyd	Difference between the feature namely "Elevation" and the feature, namely "Vertical distance from forest to the nearest hydrology"

3.2 Data Modelling

There are 64 features have been obtained as the output of the data preprocessing of the proposed forest cover classification approach. Now, the proposed model works with these 64 features.

3.2.1 Stacking

Stacking is the technology where the result of the set of different base learners at the level-0 is combined by a meta learner at the level-1. Here, support vector machines (SVM) is used as the meta learner. Radial basis function is used as the kernel trick (Eq. 3.2).

$$f(X_i, X_j) = e^{(-\gamma(X_i - X_j)^2)} \tag{3.2}$$

where γ (gamma) parameter is given by the following Eq. 3.3 and X_i is the target variable and σ is the free variable, q is the degree of polynomial function and theta is bias given in Eq. 3.4.

$$\gamma = -\frac{1}{2\sigma^2} \tag{3.3}$$

$$f(X_i, X_j) = (\gamma X_i, X_j + \theta)^q \tag{3.4}$$

3.2.2 Ensemble Approach

Ensemble approach is the rigorous approach in supervised learning. In case of ensemble learning, this allows flexible structures to be available in which the data is loaded into to train the model. Ensemble learning is picked because as the output classes seemed to be overlapping as well as the amount of data present is large in number.

Random Forest Approach

Random forest is the type of Ensemble learning for classification in which the multitude of decision trees are created and the average of their outcome is being taken. Grid search was used for the hyper parameter tuning (Table 3) purpose to use the best possible hyper parameters to be passed into the scikit inbuilt function GridSearchCV () with threefold cross validation as one of the parameters.

Table 3 Hyper parameters and values tested for random forest

Hyper parameters tuned	Possible values tested
Max_depth	3, 5, 7, None
min_samples_split	2, 3, 5, 7
Bootstrap	True, False
n_estimators	100, 500, 1000, 1500, 2000

The best model included the hyper parameters with the following values of bootstrap = False, max_depth = None, min_samples_split = 2, n_estimators = 500.

Extra Trees Approach

The extra trees algorithm includes another level of randomness to the mix by choosing cut points for each split in random while using the whole of the in-sample data to build the tree. The same values of Table 3 were used to tune the same hyper parameters.

Applying AdaBooster on Extra Trees

Ada Boosting combines its hypotheses with one final hypothesis so that it can achieve greater accuracy than the accuracy of weak learner's hypothesis. The main usage technique of the Ada Booster is that each and every example of the training set should act in a different role for discrimination at different training stages.

3.2.3 MultiLayered Perceptron Approach (MLP)

Multilayered perceptron is a class of artificial neural network (ANN). Here, the model is trained with the help of four non-linear activation functions. tanh, identity, relu and logistic. Tanh can be expanded as the ratio of the half-difference and half-sum of two exponential functions in the points x and $-x$ that is shown in the Eq. 3.5

$$\tan h(x) = \frac{\sin h(x)}{\cos h(x)} = \frac{e^x - e^{-x}}{e^x + e^{-x}} \tag{3.5}$$

In the case of relu, the rectifier is an activation function defined as the positive part of its argument given by the Eq. 3.6.

$$f(x) = x^+ = \max(0, x) \tag{3.6}$$

Similarly, A logistic function or logistic curve is a common "S" shape (sigmoid curve), with Eq. 3.7.

$$f(x) = \sigma(x) = \frac{1}{1 + e^{-x}} \tag{3.7}$$

Hence, the model is trained and was being tested with the MLP algorithm with these four activation functions using the hidden layers as (300, 150, 75 and 40) in number separately, the outputs are got as shown in the Fig. 4 where the values in x-axis are being indicated as 1, 2, 3 and 4 for the activation functions tanh, identity, logistic and relu, respectively, and y-axis for accuracy.

4   Experimental Results and Analysis

The dataset is being taken from the UCI library of datasets for the proposed work result analysis. For the performance analysis of the proposed model, the forest cover data set is divided into two parts. The training was done by using 80% of the dataset used and the testing was done on the remaining 20% of the dataset. The dataset consists of seven continuous features, three discrete features (0–255), and 44 binary features. The types of forest covers listed in Table 4. The wilderness_Areas include four binary columns, soil_type contains 40 binary columns indicating if that particular type of soil is present at that particular region or not. The dataset has seven types of forest covers and each represented as numbers from 1 to 7 respectively.

The given classes of cover types are carefully aggregated analysed and the model was trained using the different machine-learning algorithms. There are four wilderness areas included in the dataset. The distribution of seven output classes of forest covers among these wilderness areas is shown in Fig. 3.

To know the efficiency of the approach, the performance of the proposed model is analysed by using the methods given in Table 5 also in which accuracies of the respective models are listed in the column in the table based on the same dataset (Figs. 4 and 5).

Graphical representation of all the accuracy scores shall be seen in Fig. 5. In this figure, algorithms 1, 2, 3, 4 and 5 indicates the learners of random forests, extra trees (without booster), extra trees (with booster), MLP (with tanh as the activator), and finally stacking with SVM as the meta learner, respectively. Similarly, in Fig. 6, hamming loss is represented which is almost equals to (1-accuracy). The data flows according to the flow diagram in Fig. 2. Hence, this method gave the accuracy of 87.49%. To measure the completeness and the quality of the models, the recall measure of individual models are shown in Fig. 8.

Precision deals with how close the given hypothesis is with respect to the optimal hypothesis, bias is nothing but as the complexity of the features increases, the bias too increases hence it needs large number of data to get trained to increase its variance to get the hold over the data model in the precise manner. To check how good the model performs, Recall value deals with false positives and false negative.

Table 4 Seven classes of forest cover types	• Spruce/fir (1)
	• Lodgepole pine (2)
	• Ponderosa pine (3)
	• Cottonwood/Willow (4)
	• Aspen (5)
	• Douglas-fir (6)
	• Krummholz (7)

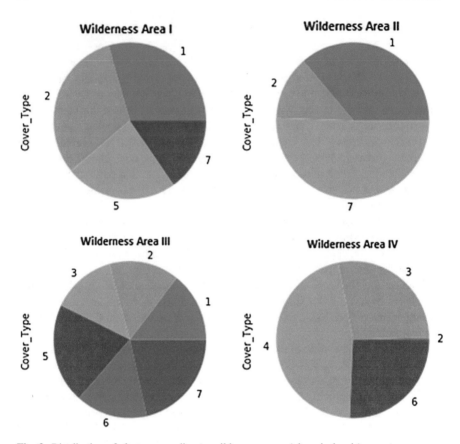

Fig. 3 Distribution of classes according to wilderness areas (given in levels)

Table 5 Models with their accuracies

S. No.	Model name	Accuracy (%)
1	Random forests over PCA	87.26
2	Extra trees classifier without Booster over PCA	86.73
3	Extra trees classifier with Booster over PCA	86.86
4	Neural networks	87.36
5	Stacking of all the above approaches (proposed model)	87.49

Hence, when compared with all the given models, the variation in the Precision, Recall and F-Score (weighted average of precision and recall) values of each and every output classes are given with respect to each models are shown in Figs. 7, 8 and 9, respectively. Wherein *X*-axis, there are the seven output classes and from them and *Y*-axis containing the precision or recall or f-scores in the respective graphs. It shall be inferred that cottonwood/willow type of tree cover is precisely

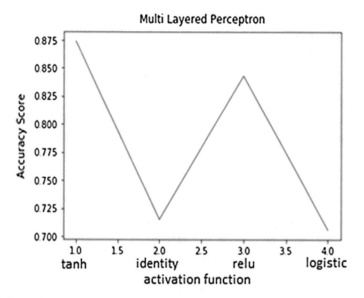

Fig. 4 MLP activation functions versus accuracy

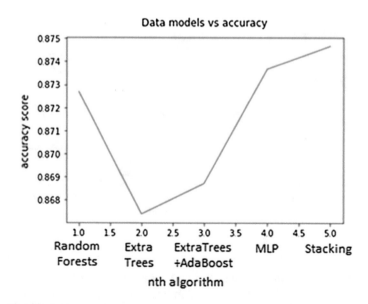

Fig. 5 Algorithms versus accuracy

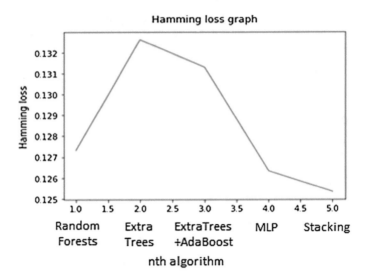

Fig. 6 Algorithms implemented versus hamming loss

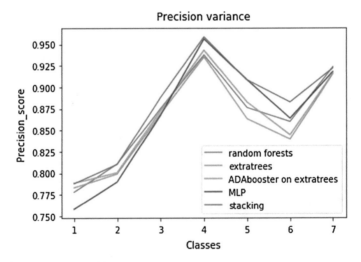

Fig. 7 Precision values for all output classes

predicted up to 95% and more ambiguity occurs in the prediction of spruce/fir type of tree cover type i.e. 75%. And it can be inferred from the graphs that the stacking approach seemed to be the viable model than that of its' base models.

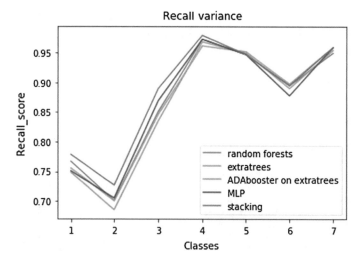

Fig. 8 Recall values for all output classes

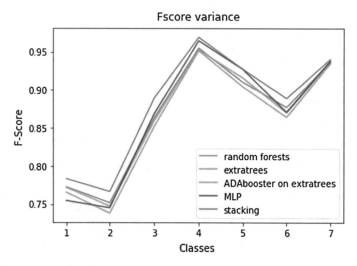

Fig. 9 F-score values for all output classes

5 Conclusion and Future Work

The proposed multiclassifier model introduces the stacking of the Ensemble learning and multilayer perceptron approach for the classification based on the forest cover dataset from the UCI library. The proposed model classifies seven different output classes of the forest cover dataset. Out of the models applied on the dataset, the best results were achieved by stacking all the data models that used

SVM as the meta classifier. This approach gave the accuracy of 87.49% when it is tested using the test data considered for the given work. Many other forms of artificial neural networks shall also be used to put the data model in the position to predict the classes in more precise manner.

References

1. Pahlm, O., Sornmo, L.: Software QRS detection in ambulatory monitoring—a review. Med. Biol. Eng. Comput. **22**, 289–297 (1984)
2. Fayyad, U.M., Piatetsky-Shapiro, G., Smyth, P.: The KDD process for extracting useful knowledge from volumes of data. Commun. ACM **39**(11), 27–34 (1996)
3. Crain, K., Davis, G.: Classifying Forest Cover Type Using Cartographic Features. Stanford University—CS 229: Machine Learning (2014)
4. Schmitt, C.B., Belokurov, A., Besançon, C., Boisrobert, L., Burgess, N.D., Campbell, A., Coad, L., Fish, L., Gliddon, D., Humphries, K., Kapos, V., Loucks, C., Lysenko, I., Miles, L., Mills, C., Minnemeyer, S., Pistorius, T., Ravilious, C., Steininger, M., Winkel, G.: Global Ecological Forest Classification and Forest Protected Area Gap Analysis. Analyses and Recommendations in View of the 10% Target for Forest Protection Under the Convention on Biological Diversity (CBD), 2nd revised edition, pp. 1–34. Freiburg University Press, Freiburg, Germany (1981, 2008, 2009)
5. Gribel, D.L.: Forest Cover Type Prediction. PUC-Rio (2015)
6. Jolliffe, I.T.: Principal Component Analysis. Springer, New York (1986)
7. Jolliffe, I.T., Cadima, J.: Principal component analysis: a review and recent developments. Philos. Trans. Ser. A Math. Phys. Eng. Sci. **374**(2065) (2016)
8. Hwang, W., Kim, T.K., Kee, S.C.: LDA with subgroup PCA method for facial image retrieval. In: The 5th International Workshop on Image Analysis for Multimedia Interactive Services (WIAMIS), Portugal, Lisbon, pp. 21–23 (2004)
9. Vehbi Olgac, A., Karlik, B.: Performance analysis of various activation functions in generalized MLP architectures of neural networks. Int. J. Artif. Intell. Expert Syst. **1**, 111–122 (2011)
10. Freund, Y., Schapire, R.E.: A decision-theoretic generalization of on-line learning and an application to boosting. J. Comput. Syst. Sci. **55**, 119–139 (1997)
11. Xie, X., Wu, S., Lam, K.-M., Yan, H.: PromoterExplorer: an effective promoter identification method based on the AdaBoost algorithm. Bioinformatics **22**(2), 2722–2728 (2006)
12. Graczyk, M., Lasota, T., Trawiński, B., Trawiński, K.: Comparison of bagging, boosting and stacking ensembles applied to real estate appraisal. In: Asian Conference on Intelligent Information and Database Systems. Wrocław University of Technology, Institute of Informatics, Wrocław, Poland, March 2010

Predicting the Trends of Price for Ethereum Using Deep Learning Techniques

Deepak Kumar and S. K. Rath

Abstract This study intends to predict the trends of price for a cryptocurrency, i.e. Ethereum based on deep learning techniques considering its trends on time series particularly. This study analyses how deep learning techniques such as multi-layer perceptron (MLP) and long short-term memory (LSTM) help in predicting the price trends of Ethereum. These techniques have been applied based on historical data that were computed per day, hour and minute wise. The dataset is sourced from the CoinDesk repository. The performance of the obtained models is critically assessed using statistical indicators like mean absolute error (MAE), mean squared error (MSE) and root mean squared error (RMSE).

Keywords Deep learning · Ethereum · MLP · LSTM · Cryptocurrency

1 Introduction

Cryptocurrency is new to the financial market but considered as an alternative to existing paper currency. Cryptocurrencies like Bitcoin, Litecoin, Ethereum, etc. are attracting significant investment at present. Ethereum is a cryptocurrency that is gaining prominence after Bitcoin. Unlike Bitcoin, Ethereum offers unique features like smart contract and decentralized application platform to its users [1]. The prediction of the cryptocurrency is a topic of interest, in particular for those who intend to invest in it. Prediction of price for cryptocurrency is a time series prediction problem which uses historical data to predict on time series scale [2].

In time series analysis, the purpose is to estimate the potential future value with the help of the past data. However, machine learning techniques are also considered

D. Kumar (✉) · S. K. Rath
Department of Computer Science and Engineering, National Institute of Technology, Rourkela, India
e-mail: deepak.acumen@gmail.com

S. K. Rath
e-mail: skrath@nitrkl.ac.in

© Springer Nature Singapore Pte Ltd. 2020
S. S. Dash et al. (eds.), *Artificial Intelligence and Evolutionary Computations in Engineering Systems*, Advances in Intelligent Systems and Computing 1056,
https://doi.org/10.1007/978-981-15-0199-9_9

as a better alternative applied to predict and classify on the basis of the accuracy of the time series problem. The available methodologies for time series forecasting are moving average, auto regression, autoregressive moving average, autoregressive integrated moving average artificial neural network (ANN), etc.

In this paper, two different deep learning models, multi-layer perceptron (MLP) and long short-term memory (LSTM), have been implemented and their performances are critically assessed for the Ethereum price prediction.

2 Literature Survey

Cryptocurrency like Ethereum is very new to the financial world with a highly volatile price. Analysis of price prediction of cryptocurrency using deep learning techniques especially for Ethereum can be considered as a potential one, because of its good number of advantages. Shah, D. and Zhang apply Bayesian regression to Bitcoin price prediction [3].

M. Matta et al. have made a study on the relationship between Bitcoin cost, the volume of tweets and views for Bitcoin. They have compared Bitcoin price trends with Google Trends and concluded that positive tweets may predict the price trends [4].

Greaves, Alex and Benjamin Au. have computed that the techniques like support vector machines (SVM) and ANN are successfully predicting the price of Bitcoins using Bitcoin blockchain data and achieved an accuracy of 55% for price trends. The authors have investigated that using only blockchain data is insufficient in price prediction [5].

There has been work on price prediction by sentiment analysis using support vector machines (SVM). I. Madan, S. Saluja and Aojia Zhao have implemented SVM, generalized linear model (GLM) and random forest on Bitcoin blockchain data and achieved the accuracy of 99% but without any validation of their models [6].

McNally Sean, Jason Roche and Simon Caton have compared recurrent neural network (RNN) and LSTM for predicting the price trends of Bitcoin. They have achieved the accuracy of 52% for the LSTM model and 50% for the RNN model. They have applied ARIMA with an accuracy of 50% and compared with RNN and LSTM models [7].

Earlier ANN has been used for stock market prediction (which is also a time prediction problem). A comparative study of different deep learning models of stock market prediction is presented in Ref. [8]. They have applied RNN and LSTM on stock price data of different companies and compared all three techniques.

Hiransha M., E. A. Gopalakrishnan, Vijay Krishna Menon and K. P. Soman have used MLP, RNN and LSTM models to predict price trends of the stock price of different companies' stock price. They compared the performance of the models on the basis of the mean absolute percentage error (MAPE) [9].

3 Methodology Adopted

3.1 Dataset Considered

Data were collected from the CoinDesk and CoinMarket repositories. Dataset collected in a comma-separated values (CSV) file. Models were compared on daily, hourly and minute basis data. The data on a daily basis from August 2015 to August 2018 which contain open price, closed price, high and low price and volume per day and hourly data from July 2017 to August 2018 have been considered. Daily data consist of 1000 data samples, hourly data consist of 1500 data samples and minute data consist of 400,000 rows of data.

3.2 Deep Learning Models

Neural networks can be implemented for price prediction problem due to a few distinctive characteristics. To begin with, neural networks have been a self-adjusting model based on training data, and they have the capability to look after the issue with only a little knowledge about its design and without any compulsion to the prediction model with the addition of any extra assumption. Secondly, neural networks have generalization capacity which implies that after training, they can identify the new patterns even if they have not any training dataset. Since in most of the pattern recognition issues, predicting future events is based on past data. Due to the high volatility in cryptocurrency, ANN is considered as a useful technique for this type of problem.

3.2.1 Multi-layer Perceptron

Multi-layer perceptron (MLP) is a feedforward artificial neural network (ANN) with more than one hidden layer. In feedforward network, information flows from the input layer to the output layer, and model is trained through the backpropagation technique. The backpropagation technique is used for weights and bias optimization from output node to its input node with respect to weight and bias. MLP can be used for pattern classification, prediction and approximation.

Figure 1 shows an MLP network inspired by biological neuron [10, 11]. The neuron has n inputs, and every neuron is connected through weight links. MLP has three sections of layers: the first input layer, second hidden layer and the last output layer [10]. Neurons in the networks are connected through neurons of the next subsequent layers.

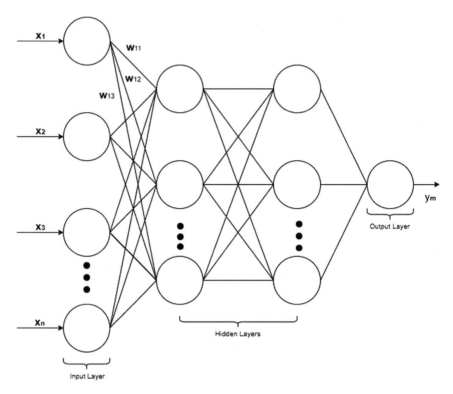

Fig. 1 Multi-layer perceptron model

3.2.2 Long Short-Term Memory

Long short-term memory [12] is a special kind of recurrent neural network [13]. RNN is a powerful and robust neural network as they have internal memory. Since they have internal memory, RNN can recall important concerning the input signal. The preciseness in prediction is the motivation behind why they are the favoured techniques for consecutive information like time series, speech, sound, video and climate. RNN works fine for short-term dependencies, but also suffers from exploding gradients [14] and vanishing gradients [15].

LSTM is specifically designed to prevent the long-term dependency issue. It has the ability to remember the information for longer-time period. The structure of LSTM is different to RNN, in RNN, there is a basic structure of neural network with the feedback loop between neurons of each hidden layers, whereas in LSTM it has memory cell that propagates the information through network and control as well. The key to LSTM structure is that it uses four gates to process the input information in the networks (Fig. 2).

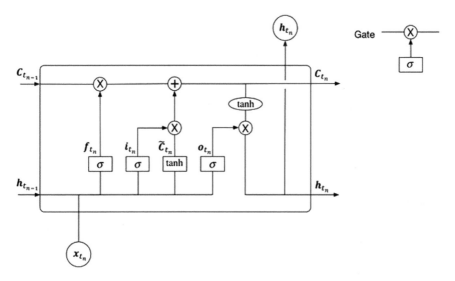

Fig. 2 LSTM architecture

Here,

- C_{tn-1} represents the old memory cell state.
- h_{tn-1} for the output of the previous cell.
- C_{tn} for present cell state which runs through the entire network and has the capacity to remove or add information with the assistance of gates.
- h_{tn} output of present cell.
- f_{tn} forget gate layer that decides portion of information to be permitted.
- i_{tn} is the input gate layer.
- $\tilde{C}_{tn}$ is the new values created by a tanh layer that generates a new vector, which will be added to the state.
- o_{tn} is an output of sigmoid gate layer and this layer creates numbers between zero and one, describing just how much of each element should be permitted through.

The cell state is updated based on the outputs from the gates that can be represented mathematically as follows.

$$f_{tn} = \sigma\left(W_f \cdot [h_{tm-1}, x_t] + b_f\right) \tag{1}$$

$$i_{tn} = \sigma\left(W_i \cdot [h_{tm-1}, x_t] + b_i\right) \tag{2}$$

$$C_{tn} = \tan h(W_c \cdot [h_{tm-1}, x_t] + b_c) \tag{3}$$

$$o_{tn} = \sigma\left(W_o[h_{tn-1}, x_t] + b_o\right) \qquad (4)$$

$$h_{tn} = o_{tn} * \tan h(C_{tn}) \qquad (5)$$

where

- x_t: input vector,
- h_{tn-1}: output vector,
- C_{tn}: cell state vector,
- f_{tn}: forget gate vector,
- i_{tn}: input gate vector,
- o_{tn}: output gate vector
- and W and b are the parameter matrix and vector.

4   Implementation

4.1   Data Preparation

In this paper, three datasets are used for model evaluation: first, daily Ethereum price; second, hourly price data; and third, minute wise price data. From this, the data have been extracted only day wise open price, hourly wise open price and minute wise open price, because the open price is preferred by the investor to decide whether to buy particular cryptocurrencies or not. The input data are usually normalized in neural networks applications in the range of (0, 1), and it depends on the activation function. So in this research, values of the open price have been normalized in the range of (0, 1) using the Eq. (6), and then models were used for evaluation using MLP and LSTM algorithms.

$$x(\text{norm}) = \frac{x - \min(x)}{\max(x) - \min(x)} \qquad (6)$$

where x(norm) is the normalized value, x(min) is minimum value and x(max) is the maximum value in the training dataset. These normalized data were given as the input to the network in a window size of 60 to predict days, hours and minutes based trend for prediction.

4.2   Evaluation Criteria

The price prediction problems are judged upon the values of performance measure such as mean absolute percentage error (MAPE), since it shows the error in the predicted model. In addition to the MAPE, three other measures are used to compare prediction methods which are mean squared error (MSE), root mean squared error (RMSE) and mean absolute error (MAE). The equation for calculating these measures are given below:

$$\text{MAPE} = \frac{1}{n} \sum_{i=1}^{n} \frac{|(\text{Actual}_i - \text{Predicted}_i)|}{\text{Actual}_i} \qquad (7)$$

$$\text{MSE} = \frac{1}{n} \sum_{i=1}^{n} (\text{Actual}_i - \text{Predicted}_i)^2 \qquad (8)$$

$$\text{RMSE} = \sqrt{\frac{1}{n} \sum_{i=1}^{n} (\text{Actual}_i - \text{Predicted}_i)^2} \qquad (9)$$

$$\text{MAE} = \frac{1}{n} \sum_{i=1}^{n} |\text{Actual}_i - \text{Predicted}_i| \qquad (10)$$

5 Simulation Results

The result of MLP and LSTM models are compared over daily, hourly and minute wise data.

5.1 Dataset: 1 (Per Day)

Dataset 1 consists of Ethereum price on the basis of daily wise. Approximately 1200 days of price data have been considered for prediction of Ethereum price by using of MLP and LSTM. The performance measurements are compared through various error functions. The results quantified are shown in Table 1.

Table 1 shows the various loss functions value for predicting the open price for 1000 days by using MLP and LSTM.

Table 1 Error function incurred during the prediction

Parameters	MLP	LSTM
MSE	0.021	0.018
MAE	0.114	0.013
MAPE	32.29	3.67
RMSE	21.3	20.53

5.2 Dataset: 2 (Per Hour)

This dataset consists of Ethereum price on an hourly basis (price per hour). It contains data from 01 July 2017 to 02 August 2018 that comprise of 1500 data sample. The results quantified are shown in Table 2.

Table 2 shows the result obtained from over 1500 h of data by using MLP and LSTM. Results show that both MLP and LSTM produce nearly the same result, but LSTM slightly performs better, since it has a better mechanism for long-term dependency.

5.3 Dataset: 3 (Per Minute)

The minute wise dataset contains 400,000 data sample with per minute price from 01 January 2018 to 30 October 2018. The results quantified are shown in Table 3.

Table 3 shows the result of one-minute data. Comparative analysis of the results is shown in Tables 1, 2 and 3. It is observed that the LSTM model shows better prediction over the MLP. Figures 3 and 4 show the trend analysis on daily price.

Figures 3 and 4 show the trend analysis of daily price by using the MLP and LSTM. As figures show, MLP is not that sufficient to capture the sudden and non-linearity in the price between 800 and 1000 days where LSTM has proven slightly better than MLP, but not completely satisfactory. Trends analysis obtained for hourly data is shown in Figs. 5 and 6 as mentioned below.

Figures 5 and 6 show hourly trends, and it is shown that LSTM works better than MLP. From Table 2, LSTM has better MSE than MLP. Trends analysis obtained for minute wise data are shown in Figs. 7 and 8 as mentioned below.

Table 2 Error function incurred during the prediction (per hour)

Parameters	MLP	LSTM
MSE	0.0120	0.0130
MAE	0.0048	0.0072
MAPE	8.856	1.38
RMSE	17.29	7.12

Table 3 Error function incurred during the prediction (per minute)

Parameters	MLP	LSTM
MSE	0.011	0.0024
MAE	0.0028	0.0014
MAPE	3.25	2.21
RMSE	90.06	18.16

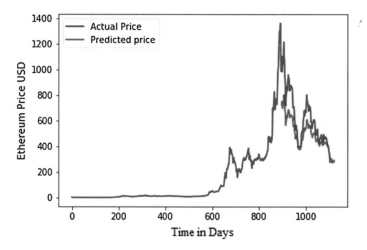

Fig. 3 Daily wise actual versus predicted price by using MLP

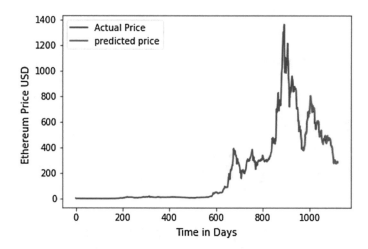

Fig. 4 Daily wise actual versus predicted price by using LSTM

Figures 7 and 8 show the price trends by the minute. Here, MLP and LSTM models perform better than previous data because the deep learning network performs better with a large amount of data. By evaluating both Figs. 7 and 8 and Table 3, it is observed that LSTM works far better than MLP with minimum losses.

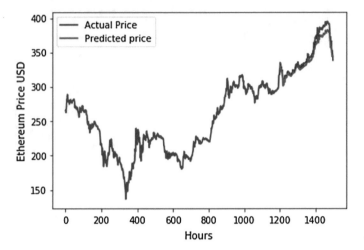

Fig. 5 Hourly wise actual versus predicted price by using MLP

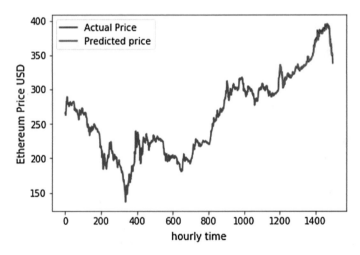

Fig. 6 Hourly wise actual versus predicted price by using LSTM

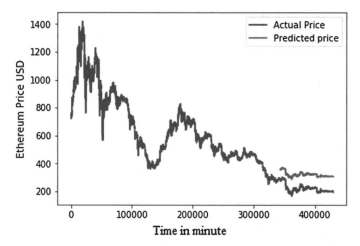

Fig. 7 Minute wise actual versus predicted price by using MLP

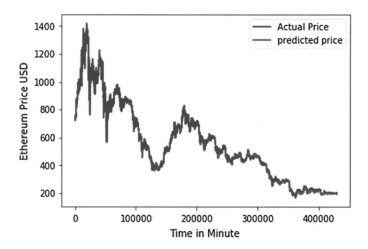

Fig. 8 Minute wise actual versus predicted price by using LSTM

6   Conclusion and Future Work

In this paper, it may be concluded that deep learning models are suitable techniques for capturing the price trends of cryptocurrencies for larger datasets. MLP and LSTM both can predict the price trends, but results show that the LSTM model is more robust and precise for long-term dependency as compared to MLP. Here, both MLP and LSTM models are applied on daily, hourly and minute wise data, and from the result, it is observed that neural network is able to predict more precisely if there is less deviation between two subsequent values in prices. From the results, it is clear that LSTM outperformed the MLP marginally but not very significantly. In future work, it is planned to study the performance of prediction using other prediction models under deep learning techniques. Also, impacts of parameters such as hash-rate and difficulty level may be considered to study the impact on performance.

References

1. Christidis, K., Devetsikiotis, M.: Blockchains and smart contracts for the internet of things. IEEE Access **4**, 2292–2303 (2016)
2. Georgoula, I., Pournarakis, D., Bilanakos, C., Sotiropoulos, D., Giaglis, G.M.: Using time-series and sentiment analysis to detect the determinants of bitcoin prices (2015)
3. Shah, D., Zhang, K.: Bayesian regression and bitcoin. In: 2014 52nd Annual Allerton Conference on Communication, Control, and Computing (Allerton), pp. 409–414. IEEE (2014)
4. Matta, M., Lunesu, I., Marchesi, M.: Bitcoin spread prediction using social and web search media. In: UMAP workshops (2015)
5. Greaves, A., Au, B.: Using the bitcoin transaction graph to predict the price of bitcoin (2015)
6. Madan, I., Saluja, S., Zhao, A.: Automated bitcoin trading via machine learning algorithms. http://cs229.stanford.edu/proj2014/Isaac%20Madan (2015)
7. McNally, S., Roche, J., Caton, S.: Predicting the price of bitcoin using machine learning. In: 2018 26th Euromicro International Conference on Parallel, Distributed and Network-based Processing (PDP), pp. 339–343. IEEE (2018)
8. Selvin, S., Vinayakumar, R., Gopalakrishnan, E.A., Menon, V.K., Soman, K.P.: Stock price prediction using LSTM, RNN and CNN-sliding window model. In: 2017 International Conference on Advances in Computing, Communications and Informatics (ICACCI), pp. 1643–1647. IEEE (2017)
9. Hiransha, M., Gopalakrishnan, E.A., Menon, V.K., Soman, K.P.: NSE stock market prediction using deep-learning models. Procedia Comput. Sci. **132**, 1351–1362 (2018)
10. Moghaddam, A.H., Moghaddam, M.H., Esfandyari, M.: Stock market index prediction using artificial neural network. J. Econ. Finance Adm. Sci. **21**(41), 89–93 (2016)
11. Menon, V.K., Vasireddy, N.C., Jami, S.A., Pedamallu, V.T.N., Sureshkumar, V., Soman, K. P.: Bulk price forecasting using spark over nse data set. In: International Conference on Data Mining and Big Data, pp. 137–146. Springer, Cham (2016)
12. Guresen, E., Kayakutlu, G., Daim, T.U.: Using artificial neural network models in stock market index prediction. Expert Syst. Appl. **38**(8), 10389–10397 (2011)
13. Jia, H.: Investigation into the effectiveness of long short term memory networks for stock price prediction. arXiv preprint arXiv:1603.07893 (2016)
14. Pascanu, R., Mikolov, T., Bengio, Y.: Understanding the exploding gradient problem. CoRR, abs/1211.5063 (2012)
15. Hochreiter, S.: The vanishing gradient problem during learning recurrent neural nets and problem solutions. Int. J. Uncertainty Fuzziness Knowl. Based Syst. **6**(02), 107–116 (1998)

Development of a Novel Embedded Board: CANduino

Ashish Daniel Cherian, S. Aman Nair, Aishwarya Janardhanan, Yashika Gupta and M. Sangeetha

Abstract This paper proposes the analysis of a self-designed embedded system based on the controller area network. Most automotive industries utilize CAN bus to reduce the intricacies of wiring in an automobile. To overcome the shortcoming and compatibility of different sensors in a particular communication network and its applicability in different fields for the current real-time monitoring system, we have developed a communication module CANduino, which creates a gateway linking different communication protocols in a single network. CANduino is an integration of a microcontroller and the CAN protocol leading to an ordered, secured, and a faster medium with higher capacity messages, thus increasing flexibility of the system. With this module, the user does not need to understand the aspects of CAN protocol, controller, and transceivers, thereby making it an optimum solution for innovators.

Keywords Controller area network · Embedded systems · Real-time systems · Monitoring system · Arduino · Serial peripheral interface

A. D. Cherian (✉) · S. Aman Nair · A. Janardhanan · Y. Gupta · M. Sangeetha
Department of Electronics and Communication Engineering,
SRM Institute of Science and Technology, Kattankulathur, Tamil Nadu, India
e-mail: ashishdaniel_che@srmuniv.edu.in

S. Aman Nair
e-mail: saman_msi@srmuniv.edu.in

A. Janardhanan
e-mail: aishwarajanardhanan_jan@srmuniv.edu.in

Y. Gupta
e-mail: yashikagupta_san@srmuniv.edu.in

M. Sangeetha
e-mail: sangeetha.m@ktr.srmuniv.ac.in

© Springer Nature Singapore Pte Ltd. 2020
S. S. Dash et al. (eds.), *Artificial Intelligence and Evolutionary Computations in Engineering Systems*, Advances in Intelligent Systems and Computing 1056,
https://doi.org/10.1007/978-981-15-0199-9_10

1 Introduction

This paper aims at designing an embedded system development board consisting of Atmel ATmega328p microcontroller featuring CAN protocol. Currently, there exist development boards like the Arduino and Raspberry Pi which are comparatively easy to use than other development boards in the market. But, separate shields must be employed to use the CAN protocol along with this development board. Till date, there is not a single product that integrates capabilities of a microcontroller and communication module. We have tried to bridge this gap between the microcontroller and communication protocol.

In today's fast pacing world, data is colossal, and data flow is cardinal. Static and rigid data flow is a major drawback which makes its application modular. CANduino is a programmable node which features the concept to change data flow from static to dynamic and rigid to flexible and robust. Thus, it ensures efficient communication which is expandable to an extensive range of applications according to the desire and need of the user. The main advantages of CANduino are high speed, high security, easy to program and bootload, error-free transmission, reduced wiring, low power consumption, and low cost.

This paper consists of six sections. Section 2 talks about the background of the paper. Section 3 reviews the existing systems. Section 4 explains overall system design followed by hardware design and software implementation. Section 5 verifies the feasibility of the proposed system with the help of experimental results. Section 6 draws the inference of the paper.

2 Background

BOSCH developed the controller area network (CAN) in 1980 to simplify communication specifically for an automobile. As a serial communications protocol (ISO 11898), it effectively supports distributed real-time control with high authenticity, minimal cost and access delay, and immense magnitude of security [1]. The two-wire bus connects all the nodes. The twisted-pair wires have a characteristic impedance of 120 Ω (nominal). These wires are known as CAN high and CAN low, and differential voltage technique is used for data transmission. A maximum of 2032 devices can be linked up to a single bus. But, hardware restrictions allow at most 110 nodes to be utilized to establish a single bus [2]. With the help of a CAN controller and its transceiver, one can implement and execute this protocol effectively. It performs and monitors data flow in the bus with the help of CAN libraries, which support the basic functioning and validation of the data flow throughout the bus.

3 Review of Existing Systems

To implement data transferring and controlling capabilities, one has to use a microcontroller along with a protocol shield which are two separate boards used to form a single node. For instance, a CAN bus shield used with Arduino gives power for interaction possibilities. There also exist chips that combine both functions of the controller and protocol but require the use of starter kits to enable application development and debugging. These boards usually demand separate USB to a micro b 2.0 converter to enable programming applications. Using these add-ons increases weight and board size by a higher margin.

There subsists a need to overcome these limitations by integrating the functionalities of microcontrollers, communication protocols, and programming facilities into a single development board. The proposed design not only reduces wiring requirements but also minimizes the board size considerably. Unlike previous models, our design uses the current version CAN 2.0. Also, semiconductor manufacturing companies offer economical prices for integration of a CAN microcontroller into one chip [3].

4 Proposed System Design

The entire system is integrated as a single module with dual-layer fabrication. Figures 1 and 2 show the design of front side and back side of the board, respectively. Tables 1 and 2 give its port description. It consists of 13 ports, 8 JST (1.25 pitch) connectors, 4 RJ12 connectors, and 1 USB micro b 2.0 connector. Figure 3 shows the overall system block diagram. Figure 4 shows the schematic.

Data acquisition and system control are the two major parts of the system. Data acquisition takes place through 8 analog inputs and 8 digital I/O (input/output) pins via sensors, actuators, and display modules connected to the microcontroller. System control is done by single-chip microcontroller ATmega328p. It controls the wired CAN module which consists of MCP2515 (CAN controller), MCP2561 (CAN transceiver), and CAN bus which is a pair of wires CAN high and CAN low. Power to the system is supplied via the mini USB jack which also contains the CH340 driver (A USB bus convert chip to convert USB to a serial interface) to burn the program into the module or through an external power source.

4.1 Hardware Design

ATmega328p: This CMOS microcontroller has AVR RISC architecture with the advantage of having both digital and analog pins. It is a low-power 8-bit microcontroller. It works at a speed of 1 Mbps and is equipped with 32×8 GPIO

Fig. 1 Front side of the board

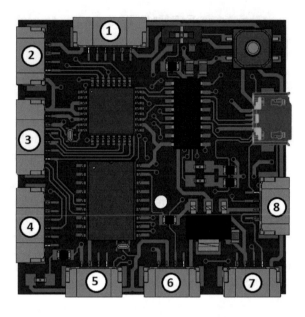

Fig. 2 Back side of the board

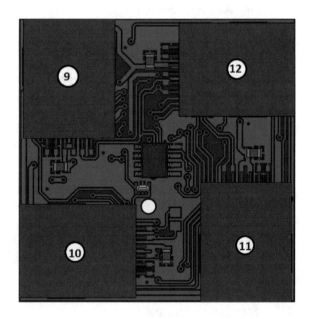

registers [4]. Its high endurance, non-volatile memory (EEPROM, SRAM, and FLASH), and unique feature of boot loading give it an edge over other micro-controllers. The microcontroller is burnt using Arduino IDE, thereby creating user-friendly software interface. A 32-lead TQFP package has been employed to

Table 1 Pinouts for CANduino (Front side)

Port	Type	Pin	Description	Port	Type	Pin	Description
1	JST	1	3.3 V/5 V	4	JST	4	D6
1	JST	2	GND	4	JST	5	RESET
1	JST	3	A5	4	JST	6	A3
1	JST	4	A4	5	JST	1	5 V
1	JST	5	D4	5	JST	2	GND
1	JST	6	D3	5	JST	3	A0
2	JST	1	3.3 V/5 V	5	JST	4	D9
2	JST	2	GND	6	JST	1	5 V
2	JST	3	D0/TX	6	JST	2	GND
2	JST	4	D1/RX	6	JST	3	A1
3	JST	1	5 V	6	JST	4	D8
3	JST	2	GND	7	JST	1	3.3 V/5 V
3	JST	3	D10/CS	7	JST	2	GND
3	JST	4	D11/MOSI	7	JST	3	D5
3	JST	5	D12/MISO	7	JST	4	A2
3	JST	6	D13/SCK	8	JST	1	3.3 V/5 V
4	JST	1	5 V	8	JST	2	GND
4	JST	2	GND	8	JST	3	D2/INT (Used for CAN)
4	JST	3	D7	8	JST	4	A6

Table 2 Pinouts for CANduino (Back side)

Port	Type	Pin	Description
9/10/11/12	RJ12	1	CAN HIGH
9/10/11/12	RJ12	2	CAN LOW
9/10/11/12	RJ12	3	5 V
9/10/11/12	RJ12	4	5 V
9/10/11/12	RJ12	5	GND
9/10/11/12	RJ12	6	GND

reduce board size. It does not have CAN bus protocol support, and hence, CAN controller and CAN transceiver are employed.

MCP2515: MCP2515 is a very easy CAN controller to work with, as it can be interfaced with the microcontroller through the serial peripheral interface (SPI). This controller implements CAN-bus protocol with the bit rate transfer of up to 1 Mbps [5]. It connects the microcontroller as a slave wherein it receives order from the master to either send the data to the bus or forward the message received from the bus. It checks and controls the transmission and reception of the data packets on the CAN bus.

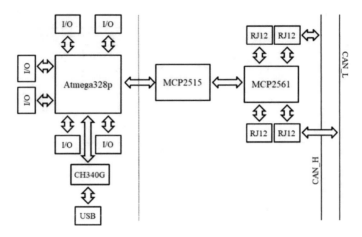

Fig. 3 Block diagram of the system

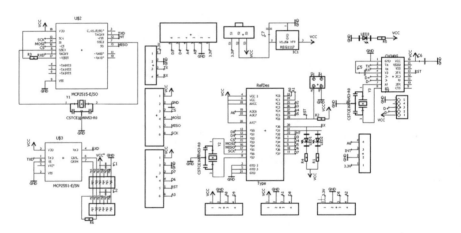

Fig. 4 Schematic of CANduino

MCP2561: It is a high-speed CAN transceiver with the purpose of serving as an interface between a CAN controller and the physical two-wire CAN bus. It is compatible with the ISO standards and provides differential transmit and receive functionality to MCP2515. Digital signals originating from the CAN controller are converted to appropriate differential signals for transmitting over the bus. It acts as a buffer for CAN controller and CAN bus, thereby protecting it from high-voltage spikes generated due to external sources.

4.2 Software Implementation

The microcontroller Atmel ATmega328p is programmed and burned using Arduino IDE (Integrated Development Environment). Arduino has its own development board which is programmable through the Arduino IDE and is very flexible. The programs can be sketched using any programming language as it is ultimately developed into binary machine code by the compiler for a selected processor. The IDE is a cross-platform which originates from processing and wiring language. The program is initially saved as a text file and then converted into a hexadecimal-encoded file with the help of a pre-existing program, avrdude. The avrdude program is then burned into the target board via a loader program in the target's firmware. The software programming utilizes modular programming ideas. The Arduino IDE converts the sketch to C++. This program is converted into machine code via the avr-gcc along with the Arduino standard functions such as digital/analog write() or print, resulting in a single hex file. This code is then burned to the target through the USB.

CAN bus connects the CAN transceiver (MCP2561) and is controlled by the CAN microcontroller (MCP2515). The CAN microcontroller is connected to the main microcontroller (ATmega328p) via SPI protocol. Thus, data traversal occurs to/from the ATmega328p through the bus. Real-time data is acquired by sensors attached to the analog/digital ports of ATmega328p. CAN message is prepared by converting data to digital form, and an ID is allocated to it according to message priority. The CAN bus software checks bus availability for an error-free transmission and reception, and this data is denoted by its ID, respective time-stamp, data length (DLC), and number of times the message has been transmitted or received. Figure 5 indicates data flow in the wired network.

5 Result

The experimental setup consists of a CANduino node connected to a microchip CAN bus analyzer. The analyzer detects the presence of CAN messages in a network. Figure 6 shows an experimental setup to test data transmission by CANduino. A series of random data generated is transmitted by CANduino, and this data when successfully received is accessed through microchip CAN bus analyzer software. The software shows messages according to the assigned ID, the number of bytes in each message along with each byte's data, and number of times the message from same ID has been received. Simulation result is shown in Fig. 7.

Fig. 5 Data flow in wired
network

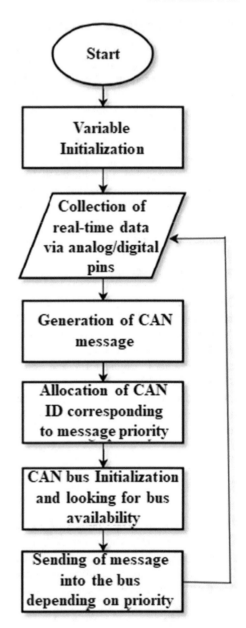

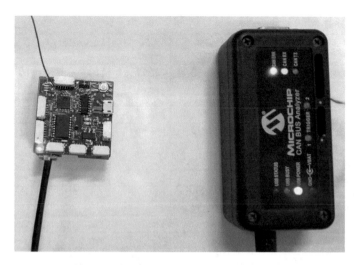

Fig. 6 Experimental setup for CANduino

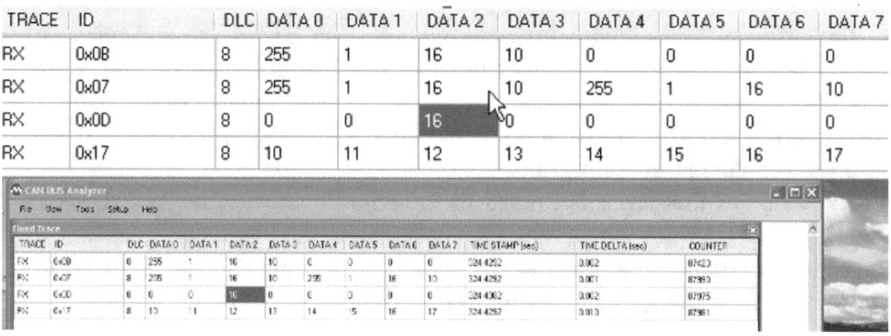

TRACE	ID	DLC	DATA 0	DATA 1	DATA 2	DATA 3	DATA 4	DATA 5	DATA 6	DATA 7
RX	0x0B	8	255	1	16	10	0	0	0	0
RX	0x07	8	255	1	16	10	255	1	16	10
RX	0x0D	8	0	0	16	0	0	0	0	0
RX	0x17	8	10	11	12	13	14	15	16	17

Fig. 7 Sample data successfully transmitted after simulation

6 Conclusion

There are wide varieties of development boards having different communication systems which require different add-ons to perform a single functionality. All these systems are integrated into a single module to achieve the same desired objectives and better performance measures. This paper describes the hardware and software design of a programmable communication module, CANduino. It consists of a microcontroller ATmega328p, an MCP2515 CAN controller, and MCP2561 CAN transceiver, the key devices essential for wired data transmission. The data from the sensors is processed and sent to the bus which transmits and receives real-time messages accurately and effectively.

Table 3 Comparative analysis between CANduino and other boards in the market

	Arduino UNO	Node MCU	MSP-EXP 430G2	PocketBeagle	CANduino
Clock speed (MHz)	16	80–160	0.0327	1000	16
Microcontroller used	ATmega328p	ESP8266	MSP430	OSD3358-SM	ATmega328p
Flash memory (kB)	32	16,000	256	521,000	32
Digital pins	14	10	2	44	14
Analog pins	6	1	8	8	8
Dimensions of the board (mm)	68 × 53.4	49 × 24.5	50 × 30	21 × 21	40 × 40
UART, I2C, SPI	Yes	Yes	Yes	Yes	Yes
CAN	No	No	No	Yes	Yes
Convenience	Easy	Moderate	Difficult	Difficult	Easy
Cost	Low	Low	High	Extremely high	Low

Table 3 describes a comparitive analysis between CANduino and other development boards. It suggests that even though PocketBeagle has better performance, it is costly and requires advanced knowledge to work with. On the other hand CANduino is a cost effective and power efficient development board that can be used for wide range of applications.

The scope of this paper is limited to the wired section of the board, which is enabled using CAN protocol. With further research, the functionality of this board can be extended for wireless connectivity, by incorporating RF module for real-time monitoring, thus having a greater impact on society.

References

1. Chen, F.L., Fan, X.Y., Ma, G.C.: Study on CAN-based lighting and identifying system for automobiles. In: Key Engineering Materials, pp. 779–782. Trans Tech Publications, Switzerland (2008)
2. Deodhe, Y., Jain, S., Gimonkar, R.: Implementation of sensor network using efficient CAN interface. IJCAT Int. J. Comput. Technol. **1**, 19–23 (2014)
3. Lee, K.C., Lee, H.-H.: Network-based fire-detection system via controller area network for smart home automation. IEEE Trans. Consum. Electron. **50**(4), 1093–1100 (2004)
4. Niveditha, A.T., Nivetha, M., Priyadharshini, K., Punithavathy, K.: IoT based distributed control system using CAN. In: Proceedings of the Second International Conference on Computing Methodologies and Communication, pp. 967–971 (2018)
5. Salunkhe, A.A., Kamble, P.P., Jadhav, R.: Design and implementation of CAN bus protocol for monitoring vehicle parameters. In: International Conference on Recent Trends in Electronics Information Communication Technology, India, pp. 301–304 (2016)

Sales Demand Forecasting Using LSTM Network

Balakrishnan Lakshmanan, Palaniappan Senthil Nayagam Vivek Raja and Viswanathan Kalathiappan

Abstract This article presents the model to sales forecast in marketplace and compares with different machine learning models to predict the demand in the future. With the recent advancement of deep learning architecture, it is possible to handle large voluminous market data. In this article, we have proposed long short-term memory (LSTM) network which takes the historical sales data of the products as input and forecasts the demand of each product for next three time series. The raw sales data are first preprocessed using various techniques, and then, the processed input is fed into the model. The sales data are fine-tuned periodically based on the actual demand. The real-time sales data are acquired from markets situated in and around southern state of India, and our work is validated with 60–40 split in which 60% data are used as training and 40% data are used as testing and gives best forecasting accuracy with lowest error value.

Keywords LSTM network · Time series · Sales demand · Recurrent neural network · Forecasting

1 Introduction

As market competition increases, many product's life cycles (from release to exiting the market) become shorter and shorter. The business world is growing rapidly with the advent of technologies, and the companies compete against each other to hold their position. It is very essential for a business organization to predict the sales, in order to meet the demands. Almost, all corporate giants use their data to forecast the

B. Lakshmanan (✉) · P. S. N. Vivek Raja · V. Kalathiappan
Mepco Schlenk Engineering College, Sivakasi, India
e-mail: lakshmanan@mepcoeng.ac.in

P. S. N. Vivek Raja
e-mail: vivekraja98_cs@mepcoeng.ac.in

V. Kalathiappan
e-mail: harishharish988_cs@mepcoeng.ac.in

© Springer Nature Singapore Pte Ltd. 2020
S. S. Dash et al. (eds.), *Artificial Intelligence and Evolutionary Computations in Engineering Systems*, Advances in Intelligent Systems and Computing 1056,
https://doi.org/10.1007/978-981-15-0199-9_11

demands of their products so that their inventory can be managed. The market value and the demand for the products fluctuate, and it is difficult to establish a pattern to meet the demands. On the other hand, the competitors pose a major threat to our product. Hence, price planning is as important as that of production. People with small- and medium-scale business cannot use the power of neural networks to forecast their sales. Therefore, it necessitates data analytic tool for prediction analysis. Sales forecast plays a crucial role in many areas like fashion industry, entertainment, retail stores, etc. Sometimes, inaccurate forecasts may result in overstock or understock results. So, even one or two percent increase in the performance of the sale prediction models may result in a huge change over in the performance of the system.

The organization of this article is as follows: Sect. 2 presents related works, Sect. 3 discusses the proposed system design, and Sect. 4 presents the results and discussions. Finally, Sect. 5 draws the conclusion and future work.

2 Related Works

Much of the focus on the literature has been on demand forecasting for consumer product companies. There has been less work done on demand forecasting for business products. There was not much work done on the sales demand forecast for small-scale and other companies. Sales forecast was available and currently implemented in many corporate companies, but there were no such forecast mechanisms available for small- and medium-scale business people. The small- and medium-scale businessmen are also in the need to forecast their needs and demands in this competitive world. Newer advanced techniques including deep learning approaches were used rarely. In addition, these factors like changing weather conditions, stock prices, trends, seasons, etc., make it practically impossible to forecast the sales accurately. Pai and Liu [1] proposed least square support vector regression (LSSVR) models that are used to deal with multivariate regression data and gives good accuracy. Ren et al. [2] introduced a novel panel data-based particle-filter (PDPF) model for sales forecasting. In this work, PDPF method results better forecasting in item-based sales when compared with color-based sales forecasting in fashion industry. In reference Tian et al. [3] worked on model-driven feature extraction approach. In this work, it is based on spatio-temporal analytics applied for feature selection for sales forecasting. In the work of Wu et al. [4] discussed ARIMA based statistical models or regression models on sales forecasting which results better forecasting accuracy with lowest residual sum of squares (RSS). Gurnani et al. [5] proposed hybrid models for forecasting of sales. In their work, the authors analyzes how composite model gives better results than an individual model. Most of these models [3, 6–12] are used to forecast sales and are not able to work on the dynamic changes in the real-world sales data. This article presents an approach for forecasting the sales with the deep learning model LSTM.

So, our objective is to power up the neural network model with the deep architecture model called LSTM network through which the model can reduce the error to a greater extent. The proposed model outperforms the existing models as the current multilayer perceptron models used for forecasting lacked in minimizing the errors and acquiring at the local minima finally which are overcome by our approach. The model used by a company to forecast their restaurant sales has achieved an accuracy of 92.36%, but the proposed system has outperformed it with an accuracy of about 96.77%.

3 Proposed System

As the forecast is based on a historical sales data and as it is a time series prediction, it would be much better if we had memory records of the past data. So, there is a need for something like memory-associated networks. So, LSTM is the most preferable option for a time series sales prediction [13].

To overcome the problem of fluctuation in sales and inability to meet the demand in business of all levels, a prediction tool that predicts the sales based on the historical sales data for every level of business is developed. Every businessman can feed their sales dataset and also log their sales. If the user wishes to predict his sales for his products, by the help of our proposed work, the sales forecast for the next three time series is predicted and displayed.

3.1 System Design

Figure 1 shows the system design of the proposed LSTM network. The data were manually acquired from the businessmen and converted into a suitable input data after preprocessing. The LSTM network takes the historical sales data and predicts the sales for the next three time series.

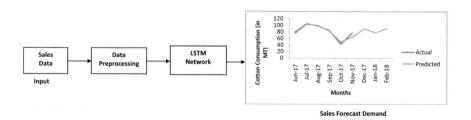

Fig. 1 Proposed LSTM network for sales forecast demand

The algorithm of the proposed approach to estimate sales demand is as follows:

Algorithm 1 LSTM Network for Sales Forecast Demand

Algorithm 1 LSTM Network for Sales forecast demand

Input: Time series of market data
Output: Built Model M, Accuracy A
Step 1: Split the data into test and train data
Step 2: Procedure fit_lstm(train,epoch,neurons)
    Fit the LSTM model to the training data
    $X < -$ train
    $Y < -$ train $- X$
    model $=$ sequential()
    model.add(LSTM(neurons),stateful=True)
    for each i range(epoch) do
        model.fit(X,Y)
        return model
Step 3: Fit the lstm model
    lstm _model $=$ fit_ lstm(train,epoch,neurons)
Step 4: Forecast the training dataset
    lstm _model.predict(train)
Step 5: Validate on test set
Step 6: Return Model M, Accuracy A

3.2 Dataset

Sales data acquisition: The sales data were collected from businessmen of all levels such as daily vegetable vendor, wholesale market dealer, retail slipper store, supermarket, and a cotton mill. The sales data of daily vegetable vendor in the southern district of Tamil Nadu consisted of around 15 products which were recorded manually in a ledger.

The wholesale market dealer manages a southern part region of vegetable market and records the daily incoming of vegetables from different cities as number of sacks of each vegetable which is manually recorded in a ledger.

The sales data from a retail slipper consisted of slippers of many sizes for various ages. The purchase and sales of the products were recorded monthly. The supermarket store in the southern district sells around 500 products of various brands. The supermarket is running for around 15 years in the area and has about 1000 customers per month. The sales data are recorded using computerized software used for billing.

The cotton mill is one of India's oldest cotton spinning mill situated in the southern region of Tamil Nadu State. The sales data are recorded using computer in which a computer operator enters the sales data every month for ten different cotton varieties.

Conversion of data into a dataset: Since most of the data were maintained manually, the data were properly converted into a suitable format before feeding the data into the LSTM network. A CSV file that consists of sales of different products with respect to each time series, for each businessman was created. This was done to ensure uniformity in processing data for further use.

Preprocessing of dataset: The historical sales data consist of many missing values due to improper maintenance of the ledger. The data preprocessing techniques were used to handle the missing data and noisy data.

3.3   LSTM Network for Sales Forecast Demand

The model used is LSTM network as shown in Fig. 2, and it is a kind of recurrent neural network. A recurrent neural network is a neural network that attempts to model time or sequence-dependent behavior [14]. This is performed by feeding back the output of a neural network layer at time t to the input of the same network layer at time "$t + 1$."

The first step for this combined input is for it to be squashed via a tanhlayer. The second step is that this input is passed through an LSTM network which has LSTM cell blocks in place of standard neural network layers. These cells have various components called the input gate, the forget gate, and the output gate. Notice first, on the left-hand side, we have our new word/sequence value x_t being concatenated to the previous output from the cell h_{t-1}. The first step for this combined input is for

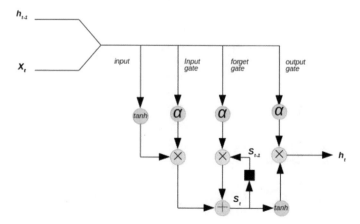

Fig. 2 LSTM cell network for sales prediction

it to be squashed via a tanhlayer. The second step is that this input is passed through an input gate. The next step in the flow of data through this cell is the internal state/forget gate loop.

LSTM cells have an internal state variable s_i. This variable lagged one time step; i.e., s_{t-1} is added to the input data to create an effective layer of recurrence. This addition operation, instead of a multiplication operation, helps to reduce the risk of vanishing gradients. However, this recurrence loop is controlled by a forget gate. This works the same as the input gate but instead helps the network learn which state variables should be "remembered" or "forgotten."

4   Implementation and Results and Discussions

The proposed LSTM network model is built using deep learning Keras libraries in Python. The MEAN stack is implemented using NodeJS which provides the database connectivity and PyShell collaboration. The database used in our work is MongoDB. We used different size of training datasets, whose sizes vary, and the model is built for testing. The accuracy of the proposed model was found to be 96.77% which has outperformed all the other machine learning models. The dataset was trained and tested on various conventional forecast prediction models like MLP regressor, linear regression, and multilayer backpropagation neural network, and their accuracy was recorded. The proposed model is tested with different ratios of train-test splits among which 60–40 split gave up the best accuracy, and its comparative analysis graph for various sales forecast machine learning models is shown in Fig. 3.

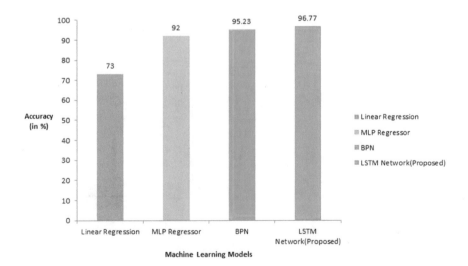

Fig. 3 Comparative performance analysis of various sales forecast models

Table 1 Comparative performances of various machine learning models

Models	Explained variance score	MSE	MSLE	R2 score
LSTM network (Proposed)	0.99	2.89	0.008	0.9777
Back propagation neural network	0.98	4.52	0.01	0.943
MLP regressor	0.96	6.72	0.014	0.9
Linear regression	0.91	12.1	0.019	0.873

The explained variance score, MAE, MSE, MSLE, R2 score were calculated and tabulated for each and every model and it is given in Table 1.

5 Conclusion and Future Work

This article proposes LSTM network to predict sales for a period of time in the future. The effectiveness of LSTM network is compared with various conventional machine learning models that were used for sales prediction analysis. The proposed model takes some historical data and current recorded sales data into account and presents the forecast results with good accuracy. A detailed study can be done on each and every product, and the features related to every product can be added to the model of that product for more accurate prediction. The model can be deployed in the cloud for access from anywhere. Our future work will be concentrated on the improvement in training, learning, and prediction of sales and also look into the appropriate optimization algorithms for selecting the optimal features for high prediction accuracy which helps to reduce the memory and CPU utilization.

References

1. Pai, P.-F., Liu, C.-H.: Predicting vehicle sales by sentiment analysis of twitter data and stock market values. IEEE Access **6**, 57655–57662 (2018)
2. Ren, S., Choi, T.-M., Liu, N.: Fashion sales forecasting with a panel data-based particle-filter model. IEEE Trans. Syst. Man Cybernatics **45**, 411–421 (2015)
3. Tian, C.H., Wang, Y., Mo, W.T., Huang, F.C., Dong, W.S., Huang, J.: Pre-release sales forecasting:a model-driven context feature extraction approach. IBM J. Res. Dev. **58** (2014)
4. Wu, L., Yan, J.Y., Fan, Y.J.: Data mining algorithms and statistical analysis for sales data forecast. In: Proceedings of Fifth International Joint Conference on Computational Sciences and Optimization, pp. 577–581 (2012)
5. Gurnani, M., Korke, Y., Shah, P., Udmale, S., Sambhe, V., Bhirud, S.: Forecasting of sales by using fusion of machine learning techniques. In: International Conference on Data Management, Analytics and Innovation (ICDMAI), pp. 93–101 (2017)

6. Raza, K.: Prediction of stock market performance by using machine learning techniques. In: International Conference on Innovations in Electrical Engineering and Computational Technologies (ICIEECT), pp. 1 (2017)
7. Fischera, T., Kraussb, C.: Deep learning with long short-term memory networks for financial market predictions. FAU Discussion Papers in Economics (2017)
8. Kaneko, Y., Yada, K.: A deep learning approach for the prediction of retail store sales. In: IEEE 16th International Conference on Data Mining Workshops (ICDMW), pp. 531–537 (2016)
9. Duncan, B.A., Elkan, C.P.: Probabilistic modeling of a sales funnel to prioritize leads. In: Proceedings of the 21th ACMSIGKDD Sydney, Australia (2015)
10. Xiong, R., Nichols, E.P., Shen, Y.: Deep learning stock volatility with domestic trends (2015)
11. Makridakis, S., Wheelwright, S., Hyndman, R.: Forecasting Methods and Applications, 3rd edn. Wiley, New York (1998)
12. Allera, S.V., McGowan, J.E.: Medium-term forecasts of half-hourly system demand: development of an interactive demand estimation coefficient model. IEE Proc. **133**, 393–396 (1986)
13. Colahs Blog: Understanding LSTM networks (2015)
14. Sak, H., Senior, A., Beaufays, F.: Long short-term memory based recurrent neural network architectures for large vocabulary speech recognition (2014)

Privacy-Enhanced Emotion Recognition Approach for Remote Health Advisory System

M. Thenmozhi and K. Narmadha

Abstract Speech recognition also known as voice recognition involves in identifying the words in the speech and translates it into machine-readable form. Speech recognition applications are currently being used for health care and special needs' assistance such as transcription of medical reports, conversational systems, command and control. The main benefit of this technology is that it can be accessed from anywhere at any time. Hence, it is now being popularly adopted for remote health advisory system especially for elderly patients. This system involves in sending a patient's voice from a remote location for whom medical advice is required. The caregiver or the therapists provide medical advice based on the recognized voice. In the literature, many techniques exist for automatic speech recognition. However, the existing works do not much concentrate on providing technique for speech analysis and classification useful for remote health advisory system. And moreover, the existing works have not focussed on the privacy issue that arises while sending patient's voice to a remote server. Thus, in this paper a privacy-enhanced voice recognition approach for health advisory system has been proposed to solve the limitations in the existing works.

Keywords Automatic speech recognition · Emotion recognition · Mel-frequency cepstral coefficients · Convolution neural network

M. Thenmozhi (✉) · K. Narmadha
Department of Computer Science and Engineering, Pondicherry Engineering College, Pondicherry, India
e-mail: thenmozhi@pec.edu

K. Narmadha
e-mail: knarmadha8@pec.edu

© Springer Nature Singapore Pte Ltd. 2020
S. S. Dash et al. (eds.), *Artificial Intelligence and Evolutionary Computations in Engineering Systems*, Advances in Intelligent Systems and Computing 1056, https://doi.org/10.1007/978-981-15-0199-9_12

133

1 Introduction

Automatic speech recognition is the process where a machine or program is able to identify words and phrases involved in a communication. Recognizing the emotional state of a speaker by analyzing his voice refers to the technique called as speech emotion recognition (SER). To improve the performance of a speech recognition system, SER could extract useful semantics from the speech signal [1]. In case of human–machine interactions where the system response depends upon the emotions of the user, SER plays an effective role [2]. There are different application areas where SER could be of vital use such as health care, e-learning, marketing, and entertainment. Based on the emotional state of the user, the machine could provide valuable suggestions in the relevant context [3].

One of the much required areas where the importance of SER is felt is remote health advisory system mainly focused for elderly people. Elderly people may be alone in home for most of the time and may also have health issues. Their isolation may lead to high depression, mortality, etc. Hence, an automatic speech and emotion recognition system can help such elderly patients to take advice from their caregiver or therapist even sitting at home. SER involves an in-depth analysis of speech signal which includes the representation of the signal and extraction of features which contains emotional information and applying appropriate pattern recognition techniques to indentify the emotion state [4]. Speech features for emotion recognition can be grouped into prosodic features such as pitch, energy, and zero-crossings [3, 5]; voice quality features such as harsh, breathy, and tense; spectral features like Mel-frequency cepstral coefficients (MFCCs), linear predictor coefficients (LPCs), and linear predictor cepstral coefficients (LPCCs) [6, 7]; and nonlinear features like Teager energy operator. Thus, the extraction of suitable features can efficiently characterize and predict different emotions. In order to generate an effective SER system, different techniques are available in the literature out of which deep learning methods have produced promising results [8–10]. Hence, in this paper convolution neural network (CNN) based on deep learning architecture has been adopted for performing SER which extracts voice features from spectrograms and classifies the emotion in the voice. The proposed work concentrates on providing privacy by using data perturbation technique for preserving the voice features.

2 Proposed System

The main objective of the proposed work is to provide privacy-preserving emotion recognition system for patient medical advice. Figure 1 represents the architecture of the proposed approach. There are two main phases in order to process the patient's voice, the analysis phase and the access phase. In the analysis phase, the patient's sample voices are collected using a smart device. The collected voice is

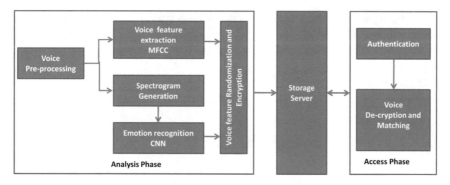

Fig. 1 Architecture of the proposed system

then sent to the preprocessing stage which identifies and separates the useful parts of the speech signal. In order to recognize the voice, it is necessary to extract the features. The extracted voice features are then encrypted and stored in the server. In parallel to the feature extraction process, the analysis phase involves in identifying the emotions in the patient's voice which helps the physician to provide better medical advice. In the access phase, the physician provides his credentials to access the storage server. The submitted patient's voice can then be decrypted. He can then submit his patient's prerecorded voice sample. The next step involves in finding a match between the voice sample and the voice received from the server. When a match is found, the physician is provided with the patient's voice along with the emotion information attached to it. Following are the phases of the proposed approach.

2.1 Voice Preprocessing

The first phase of an automatic speech recognition system is the voice preprocessing step. The recorded voice may contain non-speech or unvoiced parts. Hence, a voice activity detector is used in this step to divide the speech signal into voiced and unvoiced parts. Zero-crossing rate (ZCR) technique is adopted by the voice activity detector which is used to discern unvoiced speech. The voice signal changes value from positive to negative and vice versa. ZCR can be computed by finding total number of times change that is encountered in the signal divided by the length of the frame. Accordingly, Eqs. 1 and 2 define the ZCR:

$$Z(i) = \frac{1}{2N} \sum_{n=1}^{N} |\text{sgn}[x_i(n)] - \text{sgn}[x_i(n-1)]| \tag{1}$$

where *sgn* is the sign function, $x_i(n)$ is the sequence of voice samples of ith frame, and $n = 1$ to N samples.

$$\text{sgn}[x_i(n)] = \begin{cases} 1, & x_i(n) > 0, \\ -1, & x_i(n) < 0, \end{cases} \tag{2}$$

2.2 Feature Extraction Using MFCC

Once the voice part alone is separated from the recorded voice signal, it is sent as input to the feature extraction component. There are various feature extraction methods like LPCC, LPC, and MFCC for speech recognition. In this proposed work, MFCC has been adopted as it could perform well for extracting the feature vectors compared to other methods [6, 7]. Following are the steps used in the MFCC algorithm:

(i) *Pre-emphasis*: This step is used to increase the energy of the signal at a higher frequency. Here, the isolated sample is passed through a filter which emphasizes higher frequencies. This boosting helps in providing more information to the speech signal.

(ii) *Framing and blocking*: Here, the input signal is divided into smaller chunks. The signal that is segmented into small duration blocks or chunks is known as frames. The signal is blocked into small frames of N samples followed by the adjacent frames separated by M samples such that M is less than N. This results in the adjacent frames being overlapped by N-M. As the voice signal is continuous, the framing and blocking help to capture a portion of the signal which is stationary in order to perform an efficient spectral analysis.

(iii) *Windowing*: In order to smoothen the signal, windowing is applied. For the smoothening process, each of the frames obtained in the previous step is multiplied by a window function. This helps in reducing the discontinuity at the starting and at the end of the frame. Hamming window is a windowing function applied which is suitable for detecting formants. The window is defined as $W(n), 0 \leq n \leq N - 1$ where N is the number of samples in each frame. If $X(n)$ represents the input signal and $W(n)$ represents the Hamming window, then the output signal $Y(n)$ can be obtained using Eq. 3 as follows:

$$Y(n) = X(n)W(n) \tag{3}$$

Hamming window is represented using Eq. 4 as given below:

$$W(n) = 0.54 - 0.46 \cos(2\pi n/(N - 1)), 0 \leq n \leq N - 1 \tag{4}$$

(iv) *Fast Fourier transform (FFT)*: This step involves in converting the time domain into the frequency domain of the signal. The FFT algorithm is used to obtain the magnitude frequency response of each frame derived from the windowing process. The output of this step is a spectrum or periodogram "P" represented in Eq. 5.

$$P = \frac{|\text{FFT}(x_i)|^2}{N} \tag{5}$$

where, N is the samples and x_i is the ith frame of signal x.

(v) *Mel scale*: In order to find the approximation of energy at each spot of the spectrum, it is mapped onto the Mel scale by applying triangular filter banks. The Mel-frequency cepstrum can be calculated as represented in Eq. 6 from linear frequency using the following:

$$\text{mf} = 2595 \log 10 \left(\frac{f}{700} + 1 \right) \tag{6}$$

(vi) *Discrete cosine transform (DCT)*: In order to generate the cepstral coefficients, discrete cosine transform (DCT) is applied over the log of the spectrum. The output after applying DCT is known as MFCC given using Eq. 7.

$$C_n = \sum_{k-1}^{k} (\log D_k) \cos \left[m \left(k - \frac{1}{2} \right) \frac{\pi}{k} \right] \tag{7}$$

The output of the MFCC algorithm is the set of features required for classification. The extracted features along the identified emotion are next randomized and sent to the storage server for further processing.

2.3 Emotion Recognition

This step involves identifying the emotion in the patient's voice. When this information is available to the physician, it helps him to provide better medical advice for the patient. There are different types of emotion such as sad, angry, calm, and comfort. To identify the emotion, first the speech signal is converted to a spectrogram. A spectrogram is the visual representation of signal strength. The visual representation of the voice signal is then fed into convolution neural network (CNN) which could effectively classify the input to the appropriate emotion type. Following are the steps involved in emotion recognition:

1. *Generation of Spectrogram*

The voice signal is represented in a visual format called as spectrogram. It captures the strength of the signal at different frequencies over a period of time.

The spectrogram is basically a two-dimensional graph with time and frequency represented along the horizontal axis and vertical axis, respectively. The strength of the signal, that is, the amplitude of the frequency, is denoted using various colors by varied intensities. In order to represent the lower amplitudes, dark colors are used. Brighter colors are used to represent the stronger amplitudes. To generate spectrogram, first the input signal is broken into smaller blocks, where each block represented as columns of a 2D matrix. The fast Fourier transform (FFT) is applied over each column of the 2D matrix. The time and frequency ranges are computed. The graph can be plotted by using the positive frequencies.

2. *Emotion Identification Using Convolution Neural Networks*

There may be different types of emotion in the input signal such as sad, angry, and fear. In order to identify the emotion in the patient's voice, the proposed approach uses convolution neural network (CNN). The CNN consists of several layers such as convolution layer, pooling layer, and fully connected layer. In order to perform the classification and prediction, first the model needs to be trained. For this, in the initial convolution layer the low-level features from the spectrogram are extracted per frame and it is represented as a set of feature maps by convolving the input with different filters. This is followed by dimension reduction using a max pooling layer which helps in reducing the spatial resolution of the convolution feature maps. The purpose of this pooling layer is to accumulate only the maximum activation features from the feature maps. After multiple set of convolution and pooling layers, it is possible to extract features and reduce the number of parameters from the input represented as spectrograms. Since, it is required to produce the output in the form of different number of classes, an output layer called fully connected layer is used. The fully connected layer has neurons which have full connections to all activations in the previous layers. By a single forward and backward pass, a training cycle is completed. Once the model has been trained, it can used for obtaining prediction for each input voice signal represented as spectrograms. As a last step of the CNN architecture, a softmax function is applied in order to output the predictions in the form of probabilities for the different emotions. In order to update the belief values for all emotions, the predictions obtained from the model are used as evidence. By computing mean predictions from the gathered evidence, the probabilities for individual emotions can be determined. The emotion that is predicted for the given voice input is then tagged along with the voice feature extracted using MFCC.

2.4 Voice Feature Randomization and Encryption

In order to preserve the patient's voice submitted to the storage server, randomization technique is used here. Random rotation perturbation is a randomization technique that is used for perturbation of the feature matrix F. First, the input matrix F needs to be rotated in a particular angle in order to generate the rotation matrix R. The perturbed matrix P is generated by multiplying the feature matrix with

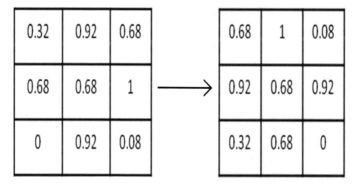

Fig. 2 Rotation matrix

Fig. 3 Perturbed matrix

1.2624	1.408	0.872
1.408	1.8224	0.68
0.872	0.68	0.864

rotation matrix, i.e., $P = F \times R$. Figure 2 represents the steps of rotation perturbation. Here, the feature matrix is rotated to 90°. Figure 3 represents the perturbed matrix which is constructed by multiplying the feature matrix and rotated matrix. The obtained randomized voice features are encrypted and stored in the storage server for further processing.

2.5 Access Phase

The second phase is the access phase, where the caregiver or the therapist of the corresponding patient can retrieve the voice submitted for providing appropriate advice. The caregiver or the therapist has list of all voice samples for their patients. On receiving an alert related to the submitted voice from the storage server, caregiver or the therapist can submit his account credentials and decrypt the patient's voice. The caregiver then submits received patient's voice and the sample voice to the storage server for matching process performed using Euclidean measure.

The server returns the similarity value to the caregiver or the therapist device. Based on the similarity score, the caregiver or the therapist ensures his patient's voice and can provide medical advice according to the emotion information attached to the voice.

3   Related Works

Mohammad H et al. proposed an approach for healthcare applications [11]. Here, they have applied MFCC algorithm to extract voice features and used privacy preservation technique to preserver voice features. But they have not concentrated on emotion identification required for health care. Ravanelli M et al. proposed an architecture for distant speech recognition using deep neural networks (DNNs). They considered speech enhancement and speech recognition DNNs for their proposed architecture [12]. Harshita Gupta and Divya Gupta performed a detailed study on feature extraction techniques in the analysis of speech signals. They compared the performance of linear predictive coding (LPC) and linear predictive cepstral coefficients (LPCCs) [13]. Sonawane A et al. proposed emotion recognition from speech which helps in improving the effectiveness of human–machine interaction. In this work, they applied MFCC for feature extraction and used support vector machine (SVM) classifier for emotion recognition [14]. Mirsamadi S, Barsoum E, Zhang C in their work presented a modified recurrent neural network for identifying emotions in speech signal. Their proposed architecture helped in learning both frame-level characterization and temporal aggregation [15].

4   Results and Discussion

The performance of the proposed approach can be measured in terms of the quality metrics such as accuracy. The experiment was conducted using EMODB dataset which is used for speech emotion recognition. Figure 4 represents the performance accuracy of the CNN models that were trained and tested. Here, epoch denotes the number of training iterations. The network learned the training data with accuracy 100%. Increasing the number of training improved the performance of the model on the test data. Table 1 represents the numerical confusion matrix of the CNN. The percent of each emotion that was correctly identified is represented in the diagonal numbers, and the remaining values represent the emotions that were incorrectly predicted.

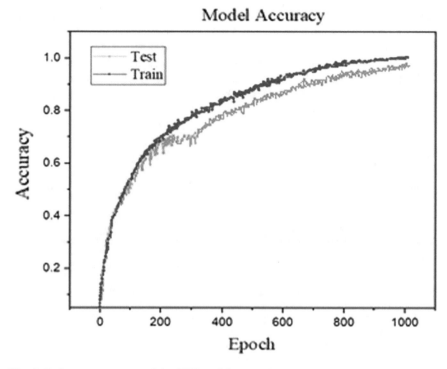

Fig. 4 Performance accuracy of the CNN model

Table 1 Confusion matrix of CNN

		Predicted labels				
		Sad	Angry	Scared	Bored	Neutral
Actual labels	Sad	96.80	0	0	0	3.13
	Angry	0	100	2.15	0	0
	Scared	0	2.7	97.50	0	0
	Bored	0	0	0	92.60	3.50
	Neutral	0	0	0	1.26	97.50

5   Conclusion and Future Work

Remote health advisory system is one of the important areas where the application of automatic speech recognition plays an important role. Identifying emotions is very vital for providing proper medical advice to the elderly patients in the healthcare domain. Deep learning techniques have been providing promising results in the area of speech emotion recognition. In this paper, we attempt to solve the problem of speech emotion recognition based on convolution neural networks.

Another important issue to be addressed in remote health advisory system is the patient's privacy. As the patient's voice data may be available in a storage server, there is possibility of misuse. Hence, the proposed work also addresses the privacy and security issues related to access of sensitive information. Thus, the proposed work provides a comprehensive approach for remote health advisory system by applying speech recognition, deep learning, and privacy preservation techniques.

References

1. Pierre-Yves, O.: The production and recognition of emotions in speech: features and algorithms. Int. J Human Comput. Stud. **59**, 157–183 (2003)
2. Hansen, J.H.L., Hasan, T.: Speaker recognition by machines and humans: a tutorial review. IEEE Sig. Process. Mag. **32.6**, 74–99 (2015)
3. Zhang, Z. et al.: Cooperative learning and its application to emotion recognition from speech. IEEE/ACM Trans. Audio Speech Lang. Process. (TASLP) **23.1**, 115–126 (2015)
4. Saini, Preeti, Kaur, Parneet: Automatic speech recognition: a review. Int. J. Eng. Trends Technol. **4**(2), 1–5 (2013)
5. Chen, L. et al.: Speech emotion recognition: features and classification models. Digit. Sig. Process. **22.6**, 1154–1160 (2012)
6. Dave, N.: Feature extraction methods LPC, PLP and MFCC in speech recognition. Int. J. Adv. Res. Eng. Technol. **1**(6), 1–4 (2013)
7. Gupta, K., Gupta, D.: An analysis on LPC, RASTA and MFCC techniques in automatic speech recognition system. In: 2016 6th International Conference-Cloud System and Big Data Engineering (Confluence). IEEE (2016)
8. Shaw, A., Vardhan, R.K., Saxena, S.: Emotion recognition and classification in speech using artificial neural networks. Int. J. Comput. Appl. **145.8**, 5–9 (2016)
9. Chandna, P. et al.: Monoaural audio source separation using deep convolutional neural networks. In: International Conference on Latent Variable Analysis and Signal Separation, pp. 258–266. Springer, Cham (2017)
10. Zhang, Y. et al.: Towards end-to-end speech recognition with deep convolutional neural networks. arXiv preprint arXiv:1701.02720 (2017)
11. Hadian, M. et al.: Privacy-preserving voice-based search over mhealth data. Smart Health (2018)
12. Ravanelli, M. et al.: A network of deep neural networks for distant speech recognition. In: 2017 IEEE International Conference on Acoustics, Speech and Signal Processing (ICASSP), pp. 4880–4884. IEEE (2017)
13. Gupta, H., Gupta, D.: LPC and LPCC method of feature extraction in speech recognition system. In: 2016 6th International Conference- Cloud System and Big Data Engineering (Confluence), pp. 498–502. IEEE (2016)
14. Sonawane, A., Inamdar, M.U., Bhangale, K.B.: Sound based human emotion recognition using MFCC & multiple SVM. In: 2017 International Conference on Information, Communication, Instrumentation and Control (ICICIC), pp. 1–4. IEEE (2017)
15. Mirsamadi, S., Barsoum, E., Zhang, C.: Automatic speech emotion recognition using recurrent neural networks with local attention. In: 2017 IEEE International Conference on Acoustics, Speech and Signal Processing (ICASSP), pp. 2227–2231. IEEE (2017)

An Effective Approach for Plant Monitoring, Classification and Prediction Using IoT and Machine Learning

Kumari Shibani, K. S. Sendhil Kumar and G. Siva Shanmugam

Abstract Agriculture has been the most important sector of our country. As the world is growing and moving in a fast pace, and depending on the automation of most of the things there is also a need to maintain plants by doing some automation. In general, every plant has some specific needs that have to be addressed for its survival. Therefore, a system must be developed where plants can communicate with the user through IoT. By monitoring these parameters, we can ensure that the plants are healthy. There is also the need to analyse, collect and make the best use of the parameters for classifications about their state. In this project, the focus is mainly on two parts. The first part aims to model a system where we keep track of the requirements of the plants using sensors and IoT. Data collection is done through sensors and sent to the Blynk app. Automatic water controller activates only when the moisture content falls below a certain threshold. The second part aims to analyse the Blynk collected data to classify the plant based on its conditions, i.e. healthy or unhealthy. Finally, we compare the performance results of support vector machine, random forest and logistic regression for classification.

Keywords Sensors · IoT · Blynk · Machine learning · Comparative analysis

1 Introduction

With the emergence of different innovative and latest technologies, life has become simpler and easier in almost all aspects. Nowadays, automatic systems are being preferred over manual operating system. The web is a part of every user life over

K. Shibani · K. S. Sendhil Kumar (✉) · G. Siva Shanmugam
School of Computer Science and Engineering, Vellore Institute of Technology, Vellore, India
e-mail: sendhilkumar.ks@vit.ac.in

K. Shibani
e-mail: shibani15.shibani@gmail.com

G. Siva Shanmugam
e-mail: sivashanmugam.g@vit.ac.in

© Springer Nature Singapore Pte Ltd. 2020
S. S. Dash et al. (eds.), *Artificial Intelligence and Evolutionary Computations in Engineering Systems*, Advances in Intelligent Systems and Computing 1056,
https://doi.org/10.1007/978-981-15-0199-9_13

decades thereby rapidly increasing Internet users across the world. IoT is the latest Internet technology developing in today's world. IoT being emerging network providing data sharing among objects and complete tasks while on is preoccupied with work activities [1]. IoT has connected devices and sensors transferring data sensed from the environments. The type of application decides over protocols for communication, network requirements, connection requirements, etc. It has the ability of exchanging and providing different data efficiently. This is achieved through electronic sensors used in devices. A proper sensing and monitoring system is needed to achieve efficient IoT usage for a given application. The different sensors like pressure sensor, temperature sensor, light intensity sensor, humidity sensor, gas sensor, etc. are needed for sensing the data and help in monitoring various electrical parameters. IoT will enable the world to become smart [2].

Plant growth mainly depends on certain environmental factors. Hence, keeping track of these factors provides user with all the necessary factors about the plants growth to maximize production. Important point to note is to determine and interpret data through this knowledge. As for the health of plant, soil moisture is the foremost important factor as it supplies plant with sufficient nutrients. This paper aims to monitor the plants and its health condition. The complete visualization is done using Blynk app helping user monitor their plants from remote location. The sensors have been placed in the plant for collecting data, and thus, data is interpreted in the Blynk platform.

Another aspect of the study, which has been a recent trend, now is to embed IoT with machine learning [3]. Sensor data helps to monitor plants and maintain it but to extract information that can further be used to individuals as well as farmers it is best to apply machine learning to it. Machine learning mainly deals with the classification and the prediction of different events based on the train and test models. This paper is divided into two parts monitoring using IoT and classification and prediction using Machine learning algorithms.

2 Literature Survey

Ezhilazhahi and Bhuvaneswari [4] developed a remote monitoring system for continuously monitoring the soil moisture of the plant. Their main objective was combing IoT with WSN. Exponential weighted moving average was adopted for increasing lifetime in the research. Their aim was to develop a plant health monitoring system maintained by users. Data collection is done via Zigbee technology transferring it to server. Event detection algorithm adopted to increase the lifetime of sensor network.

Siddagangaiah and Srinidhi [5] discussed implementing a monitoring system for plant health that controls certain environment monitoring parameters that affect plants like plant temperature, plant humidity, and plant light intensity, etc. and recover the soil moisture as well. IoT cloud platform used for data transfer is

Ubidots. If there are any differences in the value of the sensor-stored notification to users phone will be sent.

Thamaraimanalan et al. [6] had an opinion of adopting automatic systems rather than a manual one. Creation of mobile application to monitor environmental parameters to automate the process of watering. They used Firebase to transfer the data collected by the sensors. Maintaining garden was assured by automating the system.

Rajesh et al. [7] developed an automated system to detect the state of the plant. The system would detect the disease automatically using values of the sensors like colour, humidity and temperature. Arduino along with cloud platform is used to analyse the data collected via sensors. Comparison of collected data and predetermined data determines the plant conditions.

Badage [8] proposed a system providing solution for monitoring the fields and area under cultivation made use of the images of remote sensing for disease detection automatically. Main objective was to detect the disease early without affecting further production. With the first step dealing with the training of datasets, the second step deals with the maintenance and monitoring of the crop using machine learning algorithm.

Patil and Thorat [9] in their work towards early detecting the grape disease developed a monitoring system using HMM with the provision of sending alerts to experts and farmers. The overall system is comprised of temperature sensor, moisture sensor, relative humidity sensor and sensor for leaf wetness and also for transmission of data, ZigBee.

3 Proposed Plant Monitoring System

As a monitoring system, in case of hardware implementation the sensors—temperature and humidity sensor, soil moisture sensor, light intensity sensor, touch sensor and buzzer—were connected with the Node MCU board and are placed in the plant. Each sensor has its specific task and purpose. The sensed values from the sensors will be displayed onto the mobile screen through the Blynk app, which shows LIVE data, and accordingly, the control actions needed is done. As for water controlling system, water pump is connected to the plant as well as relay, which is connected to the Arduino board. Whenever the soil moisture decreases beyond a certain threshold value, the plants are watered through the water pump.

The second part of this paper deals with the classification in which plant samples are taken under consideration. We find the optimal conditions for temperature and humidity for the plants. We then export the data from the Blynk application that we use as a part of IoT server and prepare a dataset. According to the conditions, plants are classified as healthy and unhealthy. We then take into considerations three machine learning classification algorithm and find accuracy to determine which algorithm best classifies a plant.

3.1 Proposed Architecture

Figure 1 shows the proposed architecture of Plant monitoring system. In this, the left section named IoT deals with the connection of sensors and the visualization of sensor data through Blynk with the help of Node MCU. The Arduino used serves the purpose of maintaining the plants as the touch sensor deployed in the model sounds a buzzer when one tries to destroy a plant. In addition, the relay connected to the board as well as water pump is used to automatically water the plant when the soil content falls beyond threshold. The right section in the figure focuses on comparison of performance of machine learning algorithms support vector machine, random forest and Logistic over a dataset that is in CSV format exported from Blynk app and imported as an input dataset for the analysis.

3.2 Workflow of Plant Monitoring System

Figure 2 shows the workflow of the Plant monitoring system . Top level includes the plants, which we need to monitor through the sensors using a well-defined technology Internet of things. Blynk app is used for data collection which is then exported in CSV format. After removing the not required fields, the dataset is imported in *R* to be used for comparison of classification algorithm. Visualization of results is done via graphs and plots in *R*. Accuracy is calculated at the end. From the Blynk data, if the threshold of moisture is reached, plants are watered automatically.

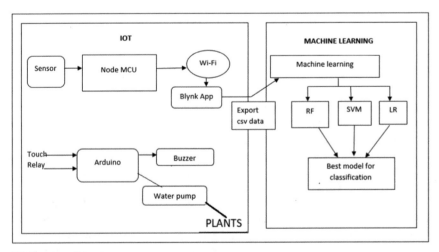

RF- Random Forest
SVM- Support Vector Machine
LR- Logistic Regression

Fig. 1 Proposed architecture of Plant monitoring system

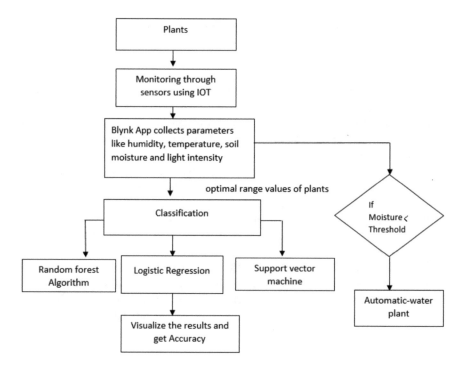

Fig. 2 Workflow of the plant monitoring system

3.3 IoT Monitoring

The sensors are connected to NodeMCU. Here, we use two boards for convenience NodeMCU and Arduino. The code for implementation is written in Arduino IDE with necessary Blynk libraries for connecting the application that is used. The Blynk application that we use gives LIVE readings of the values of the sensors. The temperature sensor and humidity sensor DHT11, soil moisture sensor and light intensity sensor are interfaced with NodeMCU and the plant. The soil moisture sensor's reading is taken as analog reading, whereas temperature and light intensity reading is taken as digital value in Blynk [10]. The touch sensor and buzzer connected to Arduino works as an alarming system when one attempt to touch or destroy the plant. The water pump also connected to Arduino works as automatic watering system [11] when the soil moisture level falls below a threshold. The observations are made with different samples of same plant to gather data at different intervals of time. Finally, after collecting data over a certain period we have a dataset for classification as to determine health condition of the plants, i.e. healthy or unhealthy.

3.4 Machine Learning Classification

- **SVM**: The algorithm is applied to the dataset with classifier added as last column classified according to the temperature and humidity values. Dataset is divided into train and test data. The algorithm runs to find in space with N number of features a hyperplane, which is decision boundary to classify the data to different classes. Confusion matrix is computed and accuracy is determined accordingly.
- **Logistic Regression**: The algorithm is applied to the dataset with classifier added as last column classified according to the temperature and humidity values. Classifier is built with sample of training sets, and the model is tested on the data under test. Next prediction is done to determine the accuracy of the model. For binary classifier performance, ROC curve can be used and AUC can be calculated.
- **Random Forest**: The algorithm is applied to the dataset with classifier added as last column classified according to the temperature and humidity values. Creation of forest consists of a certain number of decision trees. First step towards this algorithm is forest creation followed by prediction.

4 Results and Simulation

Figure 3 shows the readings and visualization of sensor data seen as LIVE graph in Blynk application. The variation in readings is captured effectively.

Figure 4 shows the plot as visualized using the support vector machine algorithm. The accuracy of the algorithm for the classification prediction was found to be 0.936. The area under the blue colour is the area of healthy plants and that of pink is the area of the unhealthy plants. We visualize the separating hyperplane in with two features: temperature and humidity.

Figures 5, 6 and 7 show the different plots of random forest. The first plot depicts the error for different classes (coloured) and out-of-bag samples (black) over the amount of trees. Classes are in the same order as the results from printing the model. The second plot shows a histogram depiction of number of nodes in the tree model and size of it. This is a histogram for time-series data. The bars represent the range of values, and their height indicates the frequency. Since it is a time series with a gradual seasonality and trend, most of the values are towards the lower end of the spectrum. That is why the histogram shows a decreasing trend as the values increases. The third and fourth plots show train and test data for multidimensional

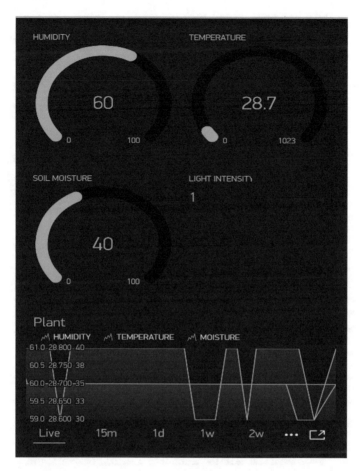

Fig. 3 Visualization and reading

scaling. It shows proximity matrices derived from the analysis of random forest, with x-axis and y-axis as multidimensional scaling coordinates. The accuracy of the algorithm for the classification prediction was found to be 0.98.

Figures 8 and 9 shows the plot for logistic regression classification. The plots are the visualization of the two features under consideration plotted against class response variable. The accuracy of the algorithm for the classification prediction was found to be 0.90.

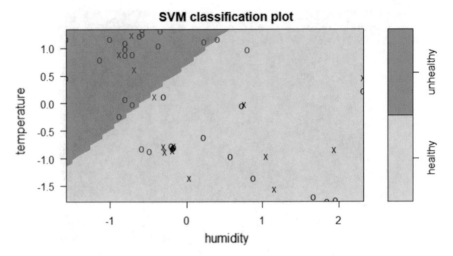

Fig. 4 SVM plot for classification results

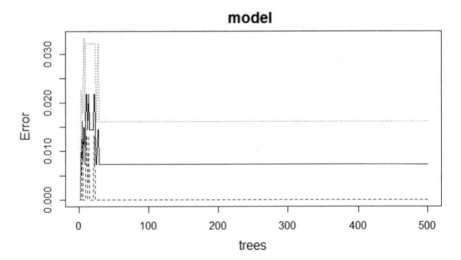

Fig. 5 Errors

Fig. 6 Train data for n-D scaling

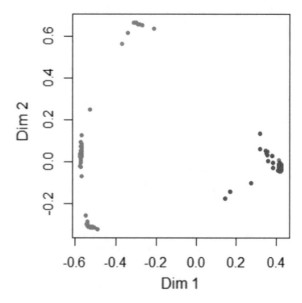

Fig. 7 Test data for n-D scaling

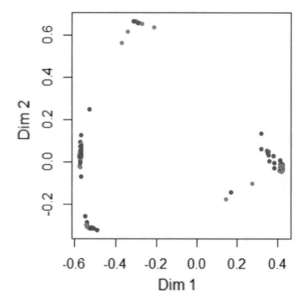

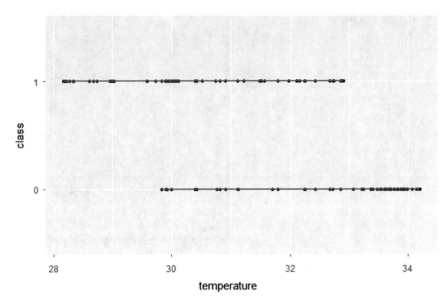

Fig. 8 Class versus temperature

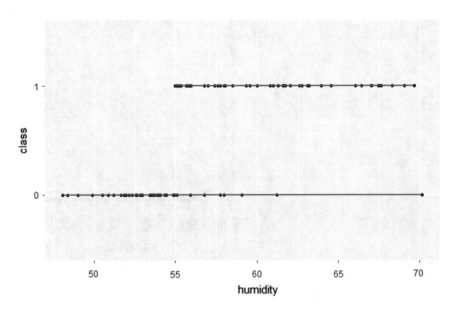

Fig. 9 Class versus humidity

5 Discussion

The sensors deployed in the plant sense the four environment parameters, i.e. plant temperature, plant humidity, light intensity and soil content. The readings are viewed as well as visualized in the Blynk application that we used. The data also gets stored in the Blynk server to be restored later. The application provides a feasible solution for the user to know the status of the plant from anywhere anytime. The status of plant soil moisture also lets you know when the plants are being watered.

For classifying and predicting the plants with respect to its conditions, the sensor data values from different plants of same species are collected to act as a dataset input for the machine learning algorithms. The accuracy measured by the three algorithms is compared, and we come to conclusion that random forest is the best algorithm to classify the health of a plant based on given parameters of temperature and humidity. It gives the best accuracy with the given dataset and should be preferred for classification and prediction purpose

By analysing these three algorithms, it can be inferred that the accuracy of random forest was the highest for the classification of the dataset into healthy and unhealthy plants, whereas support vector machine and logistic regression had a close margin among them. So the performance results of random forest algorithm being the highest among the three, it is the best algorithm that can be used as classification and prediction for this type of dataset.

6 Conclusion and Future Work

The objective of monitoring and classification is achieved using advanced features of Internet of things and machine learning. Knowing the status from anywhere helps people know about their plants effectively. The dataset is trained, tested and compared with the use of three different models to find the accuracy. This helps to compare the performance results of the algorithms used. Farmers can also get to know the status of the crops they are cultivating with this set-up. The future scope lies in the detection of soil nutrient content. It can be used to determine the types of acids, alkalis or salts present in the soil. Machine learning algorithms can be optimized to get a better accuracy than before.

References

1. Lavanya, M., Muthukannan, P., Bhargav, Y.S.S., Suresh, V.: Iot based automated temperature and humidity monitoring and control. J. Chem. Pharm. Sci. **10**(5), 86–88 (2016)
2. Al-Omary, A., AlSabbagh, H.M., Al-Rizzo, H.: Cloud based IoT for smart garden watering system using Arduino Uno. In: Smart Cities Symposium 2018, pp. 1–6, IET, April 2018

3. Moraru, A., Pesko, M., Porcius, M., Fortuna, C., Mladenic, D.: Using machine learning on sensor data. In: Proceedings of the ITI 2010, 32nd International Conference on Information Technology Interfaces, pp. 573–578. IEEE (2010)
4. Ezhilazhahi, A.M., Bhuvaneswari, P.T.V.: IoT enabled plant soil moisture monitoring using wireless sensor networks. In: 2017 Third International Conference on Sensing, Signal Processing and Security (ICSSS), pp. 345–349, IEEE, May 2017
5. Siddagangaiah, S.: A novel approach to IoT based plant health monitoring system. Int. Res. J. Eng. Technol. 3(11), 880–886 (2016)
6. Thamaraimanalan, T., Vivekk, S.P., Satheeshkumar, G., Saravanan, P.: Smart garden monitoring system using IoT. Asian J. Appl. Sci. Technol. (AJAST) 2(2), 186–192 (2018)
7. Yakkundimath, R., Saunshi, G., Kamatar, V.: Plant disease detection using IoT. Global J. Comput. Sci. Technol. 8(9), 18902–18906 (2018)
8. Badage, A.: Crop disease detection using machine learning: Indian agriculture. Int. Res. J. Eng. Technol. (IRJET) 5(9), 866–869 (2018)
9. Patil, S.S., Thorat, S.A.: Early detection of grapes diseases using machine learning and IoT. In: 2016 Second International Conference on Cognitive Computing and Information Processing (CCIP), pp. 1–5. IEEE (2016)
10. https://www.blynk.cc
11. Kishore, K.K., Kumar, M.S., Murthy, M.B.S.: Automatic plant monitoring system. In: 2017 International Conference on Trends in Electronics and Informatics (ICEI), pp. 744–748. IEEE (2017)

Encrypted Transfer of Confidential Information Using Steganography and Identity Verification Using Face Data

Parth Agarwal, Dhruve Moudgil and S. Priya

Abstract In last few years, information security has become a significant part of data communication. In order to address this crucial point in data communication to secure the confidential data, steganography and cryptography can be combined to give better results. Steganography is a practice to hide any confidential or sensitive information into another information that plays as the cover for the information, adding a layer of security on top of science of secret writing, known as cryptography. In this paper, we discuss an improved LSB steganography technique to optimize the algorithm hiding the bits in different channels selected dynamically using key. We further propose a platform leveraging cryptographic algorithms and verification using face data in pipeline to help keep data more secure.

Keywords Cryptography · Steganography · Facial recognition · Face data · Encryption · Security · AES encoding

1 Introduction

Advent use of Internet has resulted in a data explosion in all the fields. In last few years, there has been an exponential increase in amount of information stored on electronic medium. This data is mostly confidential and knowledge is required for information retrieval. No one wants some second person's hands on their confidential data, and this gives rise to information security. Thus, a safe way to communicate and transmit the message is the necessity of the hour. To address this crucial point, we propose a new system combining the power of cryptography with

P. Agarwal (✉) · D. Moudgil · S. Priya
CSE, SRM IST, Kattankulathur, India
e-mail: ag.parth11@gmail.com

D. Moudgil
e-mail: dhruve.moudgildm@gmail.com

S. Priya
e-mail: priyas3@srmist.edu.in

© Springer Nature Singapore Pte Ltd. 2020
S. S. Dash et al. (eds.), *Artificial Intelligence and Evolutionary Computations in Engineering Systems*, Advances in Intelligent Systems and Computing 1056, https://doi.org/10.1007/978-981-15-0199-9_14

our improvised algorithm based on LSB steganography and on top of that added facial verification to ensure data is being retrieved by its intended receiver. Steganography is the art of hiding any confidential or sensitive data into another information that acts as the cover for the original data. Steganography already has been into existence for long and has a lot of implementations as discussed in [1, 2]. Inspired by modifications discussed in [3], we further discuss optimized techniques to make it more secure and have a significant increase in capacity. As an added security measure, cryptography, the art of secret writing can be incorporated with the existing defined architecture. Cryptography has been there from a long time and has developed over and over to yield sophisticated algorithms that require very high computation power and time to breakthrough. We aim for a system where we use steganography to secure the soft point of cryptography which is the key to decrypt it. Further in paper, we propose an end-to-end pipeline, which uses cryptography algorithm along with steganography to secure the key and wrap it up with facial verification to ensure its delivery to intended receiver. Facial verification using face data has a lot of algorithms in existence, each with their pros and cons as discussed in [4]. As our primary concern is security of all over system, as a trade-off, accuracy is given preference over speed. The system's pipeline has evolved from the already existing system by targeting the flaws or drawbacks.

2 Literature Survey

Steganography is used to hide data into some other data or image. The most commonly used basic technique for steganography with image as discussed in [5] is hiding the information in the least significant bits of the image's pixels. Altering the hex color code by unit value does not make a significant change enough to be noticed by naked eye, thus enabling us to secretly hide the information. Information can be stored in more than one bit of pixel as done in [6], where a nibble is used. Increasing the number of bits to store information significantly increases the payload but on the contrary its trade-offs the detectability factor and makes it a little more obvious that image has been tampered with to naked human eye. Various other methods are used which include discrete cosine transformations, pixel value prediction, etc., as discussed and analyzed in [1, 2]. It is discussed in [7] that the human eye is less sensitive to variations in pixels with sharp edges as compared to light ones so it is suggested to detect the edges and then make alterations. In all the earlier discussed techniques, there is embedding of information linearly and sequentially which makes it easier for the hacker to break it. Therefore, to add a level of dynamic nature, a hash function technique is discussed in [8] where information is embedded in the least two bits and the first six bits of the pixel along with hash function are used to pick out the next location where data is supposed to be hidden. To make the process more dynamic, [3] uses a secret key to get the next location where the data is hidden which makes it difficult for hackers to crack. Steganography can be combined over with cryptography to make it more difficult for hackers to crack as it adds up a layer of security as done in [6, 9]. The basic model used for communication over public channel is discussed in [9]. For the sake of adding an additional security layer, we use facial recognition to authenticate the receiver and decrypt the image to extract the hidden information. Various techniques for facial recognition are discussed in [4]. From [4], we can conclude that convolutional neural networks yield highest accuracy for this purpose. The related work and the various paper descriptions are shown in Table 1.

Table 1 Related work concept, algorithm, and their drawbacks

S No.	Year of publication	Authors	Concept	Algorithm	Drawbacks
1	2018	Mehdi Hussain, Ainuddin Wahid et al.	Review of all the image steganography techniques in spatial domain emerged in last 5 years	Uses various algorithms to perform steganography and then perform analysis over them	No. technique proved to be ideal for steganography. Each had their own pros and cons
2	2016	Sahar A. El Rahman	Comparative analysis of images with data hidden into them using DCT	Uses discrete cosine components to hide data into LSB Uses low and middle frequency for analyzing performance with PSNR and MSE	DCT based on low frequency has high distortions
3	2018	Kumar Gaurav and Umesh Ghanekar	Embedding of secret text in sharp edges of the image	Uses algorithms based on local reference of edge detection	Sensitive to image processing techniques
4	2017	Dr. Naveen Kumar Gondhi and Er. Navleen Kour	Does a comparative analysis of various face recognition techniques	Uses various algorithms to perform recognition and then compares analysis	Each technique had their own pros and cons and none proved to be ideal. Application dependent
5	2016	Soumen Bhowmik and Arup Kumar Bhaumik	Improved payload and security using a new technique for embedding data	Stores data into last two LSBs and uses first 6 and hash function to look out for next point to store data	Since is included alteration of last two bits, it sometimes look tampered with naked human eye

(continued)

Table 1 (continued)

S No.	Year of publication	Authors	Concept	Algorithm	Drawbacks
6	2016	Utsav Sheth and Shiva Saxena	Stores data into image so as to maximize the capacity of the data stored in image Increases security using AES encryption of data	Stores data into least significant nibble giving four times the capacity and increases security to some extent by using AES encryption	As more bits are tampered here, it turns really easy to detect image edits with bare eyes and hence low security
7	2014	Deshmukh Priyanka Pravin and SmitaKasar	Stores secret image into cover image trying to improve security using a blend of cryptography and steganography	Hides image data into another compressing the original file using AES to increase the security factor	Highly sensitive to image processing techniques
8	2011	S. M. Masud Karim, Md. Saifur Rahman and Md. Ismail Hossain	Improves security and complexity of algorithm used to store data making it more secure	It hides data into a channel dynamically selected based on XOR value of secret key with LSB of red channel	Sensitive to image processing techniques and data is always stored in a channel one of green or blue
9	2006	Deshpande Neeta, Kamlapur Snehal and Daisy Jacobs	Runs an evaluation on different number of bits for LSB technique	Uses LSB technique on 2, 4, and 6 bits for storing secret data and runs and evaluation on cover image	More the number of bits, less secure the image gets as it shows high variance to original image

3   Existing System

The existing system checks the least significant bit (LSB) of the red (R) in the RGB layer. A single bit of the secret key (K) is taken and the XOR (R, K) is performed. The value of the XOR determines if the next info is to be hidden in green or blue of the RGB. If the XOR gives 1, for the next step, the LSB of green is replaced by 1 bit of the information to be hidden. Otherwise, if the value is 0, the LSB of blue is replaced by 1 bit of the data to be hidden. For every successful replacement of the data to be hidden in the green or blue of RGB, the counter variable (initially 0), is increased by 1. The value of counter variable is constantly checked with the hidden information bitstream length. The above process is continued till the value of counter is less than the value of the hidden information bitstream length.

The above-mentioned technique is proved to be better than the ones discussed in [5–8] in complexity and performance while maintaining the quality of image.

(A) Encoded-Image Creation Process

The encoded-image is the image which contains the data to hidden embedded in the cover image. It can be then sent through the public channel to the receiver. The data to be hidden is first compressed. Then that compressed data is encrypted using a key to generate the entity to be embedded.

Cover image: A cover image is an image which acts as the base for the data to be hidden. The cover image bits are manipulated to embed the secret data. The secret entity generated after the encryption process is then embedded inside the cover image to generate the public channel.

Transmission channel: The generated image can be sent through any channel to the receiver. Since the data embedded is already encrypted using a private key, the encoded-image can be sent through insecure channels as well.

(B) Recovering Original Data from the Encoded-Image

The Encoded-image received by the receiver is then used for extracting the secret entity embedded by following the reverse process. The secret message is then decrypted using the decryption key to retrieve the compressed hidden message. This compressed message is then decompressed to obtain the original message which was hidden in the cover image. Thus, the process ends with the secure retrieval of the original data.

(C) Drawbacks of the Current System

The current system is used to hide image data into another image. The process used to perform steganography in [9] is relatively simpler to decode if used with text data. Moreover, the technique used is sensitive to image processing techniques. A key is used as a way to secure the data and to make sure only people who are expected to, access the data but as the key is the most crucial part and if the key is hacked or taken over, it will be relatively simpler to extract the data so an extra added layer of security might be helpful.

4 Proposed Method

The proposed method discussed below shows enhanced method of algorithm displayed in [1, 7]. The algorithm shows a new version of steganography that uses a key to hide the bits of information in channels selected dynamically. The flowchart in Fig. 1 shows the steganography process. Algorithm expects an input image which is converted into RGB channels. Each character in secret information is represented in 8-bit binary format. The image is represented into 3D matrix and bits are encoded into one of the channels of every cell of the gird. The least significant bit of one of the channel is changed to bit of the information while the decision for

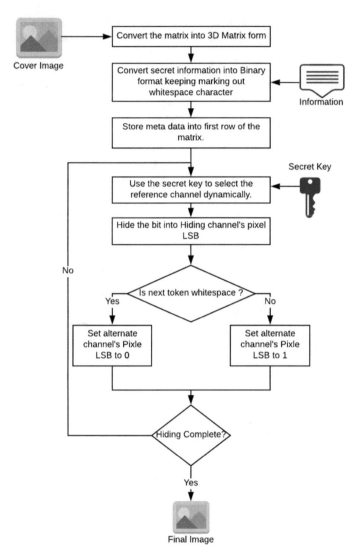

Fig. 1 Proposed method flowchart (Steganography)

the channel is made dynamically using the key. The metadata is then stored in the first N pixels of each channel. The metadata includes all the specifications of the text which is hidden into the image and helps in decryption process. It holds information like text length, sensitivity, etc., to have a proper retrieval of hidden text at receiver's end. The secret key is used as a mask to select the channel for embedding information bit dynamically. The secret key is also converted into bits. The bits then are converted into group of three. Two of the bits are used to decide the channel which is used as a mask to decide the final hiding channel while the one left is used as a bit for masking the channel to decide in which channel the bit is finally to be embedded. The LSB of the channel is then changed to that of the data bit and we move on to the next pixel on the grid for embedding the data bit. The process is then repeated until all the data bits are hidden. This adds to complexity of the hiding mechanism. The complexity can be calculated as given in Eq. 1.

Complexity Equation for Algorithm

$$complexity = \#pixels * (\#channels\ for\ masking * \#probable\ hiding\ channels) * \gamma \tag{1}$$

#pixels is the number of pixels available in the image. #channels refer to the count of channels which are being used for masking in calculation of hiding channel. #probable hiding channels refer to the potential channels which can be used for hiding the bits. γ here refers to the complexity of hiding one bit in LSB of pixel's selected channel. Equations 2 and 3 show the calculation of complexity for the existing and new algorithm. It can be verified that the updates in algorithm make it fairly difficult to break out increasing the complexity by three-folds.

Complexity for the Old Algorithm

$$complexity_{old} = \#pixels * (1 * 2) * \gamma \tag{2}$$

Complexity for Updated Algorithm

$$complexity_{new} = \#pixels * (3 * 2) * \gamma \tag{3}$$

Further enhancement is done by increasing the capacity of the network. A whitespace character is represented by 32, which is then converted into 8-bit representation. Therefore, we leverage the different channels to overcome the same and use a mechanism for hiding whitespace characters and helping us to save 8 bits for a whitespace character. While parsing the data bits, we check if the next token in whitespace character. If so, we change the bit for one left out channel of three (one used for masking and other for hiding) to 1, otherwise 0 and hence saving 8 bits for a whitespace character which increases the capacity of the image to save the data in itself as shown in Eq. 4.

Update in Capacity of the Network

$$capacity_{new} = capacity_{old} + (\#whitespaces * 8)bits \tag{4}$$

5 Proposed System Architecture

(A) Pipeline

Pipeline is shown in Fig. 2. The workflow for the pipeline is as follows:

1. Cryptography is performed on the confidential information file with a randomly generated 256-bit key.
2. The key is then hidden into the image using a user-defined key with the help of enhanced steganography technique defined in the previous section.
3. Then the user is returned with encrypted file with an image with key hidden into it.
4. Both the files are then transferred over to the intended receiver.
5. When receiver tries to decrypt the file, verification is done using the face data.
6. After successful verification, the user is prompted to provide with user-defined key.
7. User-defined key is then used for decoding the image and extraction of cryptography key is done.
8. Cryptography key is then used to decrypt the data in encrypted file which results in confidential information.

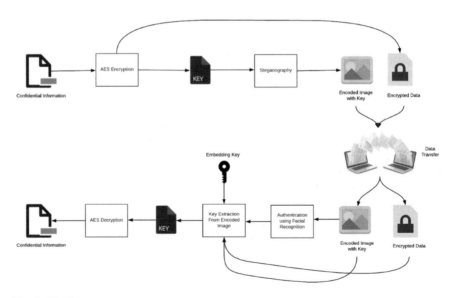

Fig. 2 Pipeline

(B) Modules

Encryption:

The encryption process is shown below in Fig. 3. The confidential information is encrypted using AES algorithm with a randomly generated key. This key is then hidden into a cover image using steganography with a user-defined key. The outputs from this module are: an encrypted data file along with an image with cryptography key hidden into it. Both the files cannot be transferred over to the intended user. Receiver's ID is also stored in database which is used later on for verification purpose.

Decryption:

The receiver here is required to provide with encrypted file along with image and steganography key. The receiver's identity is verified first using face data and upon successful authentication of the same, the key provided by the user is used for decoding the image and extracting the cryptography key from it. After cryptography key is extracted from image, and it is used for decryption of the confidential information as shown in Fig. 4.

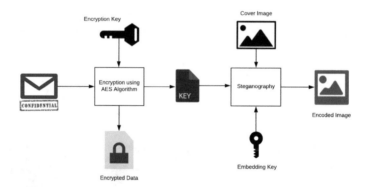

Fig. 3 Encryption module

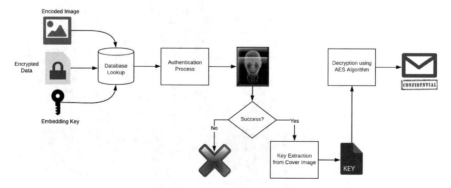

Fig. 4 Decryption module

6 Results and Discussions

The steganography algorithm was implemented and run-on different files with varying data length and 'Lena' image was used as a cover image. Table 2, shown below displays the peak signal-to-noise ratio in dB. It can be observed that as the length of the data is increased, PSNR values start to decrease as more the data, higher is the noise introduced. PSNR for data with 1 Kb is calculated to be 66.61828 dB which is typically optimal.

Figure 5 displays 'lena.png'. Steganography algorithm is applied on data file with 1024 bytes and the resultant image is shown in Fig. 7. It can be noticed that almost no difference is observed in the image with naked eyes. Histograms are obtained for both the images and shown in Figs. 6 and 8, respectively. On observing the histograms, a little variance at certain points can be observed which is due to the change in values of the bits.

As we can see, the images are still indistinguishable with naked eye and the complexity of algorithm is increased fairly to make it difficult to break. Also, the

Table 2 Data length and PSNR values

Data length (bytes)	PSNR (dB)
128	75.46329
256	72.58264
512	69.59433
1024	66.61828

Fig. 5 Original Lena image

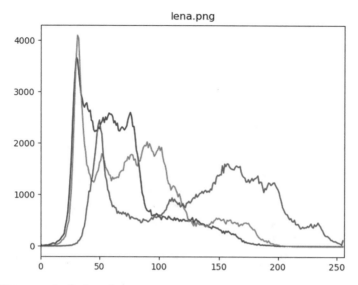

Fig. 6 Histogram plot for Lena image

Fig. 7 Lena image with 1024 bytes of data steganography

capacity of the image has been increased significantly. However, the technique makes use of least significant bits of the image pixels is fairly dependent on its matrix values, and hence the technique is sensitive to any kind of image processing techniques like compression, cropping, resizing, etc.

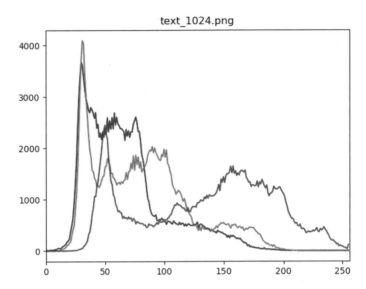

Fig. 8 Histogram plot for Steganography Lena image

References

1. Hussain, M., Wahab, A.W.A., Idris, Y.I.B., Ho, A.T.S., Jung, K.-H.: Image steganography in spatial domain: a survey. Sig. Process. Image Communi. (2018)
2. El_Rahman, S.A.: A comparative analysis of image steganography based on DCT algorithm and steganography tool to hide nuclear reactors confidential information. Comput. Electr. Eng. (2016)
3. Masud Karim, S.M., Rahman, M.S., Hossain, M.I.: A new approach for LSB image steganography using secret key. In: Proceedings of 14th International Conference on Computer and Information Technology (ICCIT 201 I), 22–24 Dec 2011, Dhaka, Bangladesh
4. Gondhi, N.K., Kour, E.N.: A comparative analysis on various face recognition techniques. In: International Conference on Intelligent Computing and Control Systems, Madurai, India (2017)
5. Neeta, D., Snehal, K., Jacobs, D.: Implementation of the LSB steganography and its evaluation for various bits. In: International Conference on Intelligent Computing and Control Systems (ICICCS), Bangalore, India, 6 Dec 2006
6. Sheth, U, Saxena, S.: Image steganography using AES encryption and least significant Nibble. In: International Conference on Communication and Signal Processing, Melmaruvathur, India, 6–8 Apr 2016
7. Gaurav, K., Ghanekar, U.: Image steganography based on Canny edge detection, dilation operator and hybrid coding. J. Inf. Secur. Appl. (2018)
8. Bhowmik, S., Bhaumik, A.K.: A new approach in color image steganography with high level of perceptibility and security. In: International Conference on Intelligent Control Power and Instrumentation (ICICPI), Kolkata, India, 21–23 Oct 2016
9. Pravin, D.P., Kasar, S.: Security improvisation in steganography using AES 128/192/256. Int. J. Eng. Res. Technol. **3**(11) (2014)

Intelligent Wireless Sensor Networks for Precision Agriculture

D. Anitha, Vaibhav D. Shelke, C. G. Anupama and Pooja Rajan

Abstract Wireless sensor networks (WSNs) today generate a large amount of data which is then analyzed on a centralized server to get actionable insights from the data. As the entire IoT infrastructure is expensive for the user, especially in the case of a farmer, precision agriculture improves the overall cost of farm products. Precision agriculture helps farmers by making farming efficient, and due to this, less wastage occurs during the process. This paper proposes an approach to get these actionable insights from the data within the WSNs. The analysis generated in the network will be shown on specified intelligent nodes which will handle all the processing. Due to this, the cost of maintaining a centralized server is reduced and the overall infrastructure cost for the analysis of crop monitoring system is affordable by a farmer. The analysis done in this paper is mainly on soil moisture data which helps in predicting the growth of the plant. By knowing the growth rate of the plant, a farmer can tweak this rate with the amount of water he pours and a prediction is provided for the growth of the plant. This way there is no excess amount of water used on the plant which may harm the plant. This analysis will help the farmer to monitor growth rate of the farm in a controlled manner.

Keywords Crop monitoring · Plant growth · Precision agriculture ·
Soil moisture · Transpiration and wireless sensor networks

D. Anitha (✉) · V. D. Shelke · C. G. Anupama · P. Rajan
Department of Software Engineering, SRM Institute of Science and Technology,
Chennai, India
e-mail: anitha.d@ktr.srmuniv.ac.in

V. D. Shelke
e-mail: vaibhavshelke017@gmail.com

C. G. Anupama
e-mail: anupama.g@ktr.srmuniv.ac.in

P. Rajan
e-mail: prajan199728@gmail.com

© Springer Nature Singapore Pte Ltd. 2020
S. S. Dash et al. (eds.), *Artificial Intelligence and Evolutionary Computations
in Engineering Systems*, Advances in Intelligent Systems and Computing 1056,
https://doi.org/10.1007/978-981-15-0199-9_15

1 Introduction

Wireless sensor networks (WSNs) are widely used in agriculture to provide farmers with a large amount of information. A large amount of data is generated during the monitoring of the crops. As the sensors are intentionally and densely deployed, the data gathering events are possible to concurrently trigger the responding actions from a portion of sensors.

Since, in the normal case direct data transmission from source nodes to the sink node leads to high data redundancy and communication load. This project presents a proposed framework to employ a WSN as a hardware computation platform. Sending all the raw data to a base station for centralized processing is very expensive in terms of energy consumption and often not practical for large networks because of the scalability problem of wireless networks transport capacity [1, 2]. Embedding intelligent nodes in the network will allow to compute the data in a distributed manner and hence will reduce the cost of a centralized system [3].

This project aims to recognize patterns and give a prediction approach to the data that is generated during the crop monitoring process. The large amount of data generated in the network will be analyzed within the network, and prediction will be done on the INs that will be embedded into the WSN. The network is made intelligent by embedding INs which are capable of aggregating data from all the source nodes and deriving a pattern from the data [2, 3].

Polynomial regression models will be implemented into the INs to gain insights from the data generated in the network for pattern recognition and prediction tasks. This will allow the farmer to reduce the overall central system cost and will be able to give insights of the data directly on the target nodes [4].

The main aim of the paper is to reduce the overall cost of the setup needed to analyze the farm data. The system explained in the paper will be scalable, easy to install, and cheap. By keeping in mind the cost of the system, the algorithms used for analysis will be efficient, fast, and light. The algorithm used will be applied to real-time data for providing real-time analysis which will be less computationally intensive [4].

The paper is structured into 11 sections as follows. We look at the preliminary details required in Sect. 2. We describe our proposed methodology in Sect. 3—data aggregation, Sect. 4—data transmission, Sect. 5—data cleansing, Sect. 6—data analysis, and Sect. 7—data prediction. Results and analysis are shown in Sect. 8. In Sect. 9, we look at related works, and finally, we present our conclusions and future works in Sect. 10 and ending it with references used for this paper.

2 Preliminaries

In this section, we explain the various hardware components required for the project. We also explain the symbols used and their description.

2.1 Notations Used

Table 1 lists all the symbols used in the paper and their meanings.

2.2 Hardware Components

The components used for this project are shown in Fig. 1 which include a DHT11 sensor which gathers temperature and humidity data, a light intensity sensor which helps determine whether it is day or night by giving a digital output, and a soil moisture sensor which is directly placed in the soil near the plant roots to collect moisture data of the respective plant. All these sensors are connected to a node MCU which collects and sends the data using its Wi-Fi module. A Raspberry Pi is used to create an IN which does all the analysis and prediction process for the project and generates the desired output. The sensors along with the node MCU are placed on a breadboard to keep the connections intact. The node MCU is outsourced by a battery for continuous transmission of data, for real-time data analysis.

Table 1 Notations

Symbol	Description
V_i	Input value
V_o	Output value
p_{sm}	Rolling mean value
p_m	Raw sample value
n	Number of samples
J	Transpiration (%)
SM	Soil moisture (%)
AH	Absolute humidity (%)
$C_D = 300$	Diffusion constant
d_{sensor}	Sensor distance (cm)
AO	Analog output of soil moisture sensor
RH	Relative humidity (%)
t	Temperature (°C)
x	Time stamp
y	Soil moisture (%)
m	Slope
c	Constant
INs	Intelligent (sink) nodes
IoT	Internet of things
PA	Precision agriculture
SNs	Sensor nodes
UDP	User datagram protocol
WSNs	Wireless sensor networks

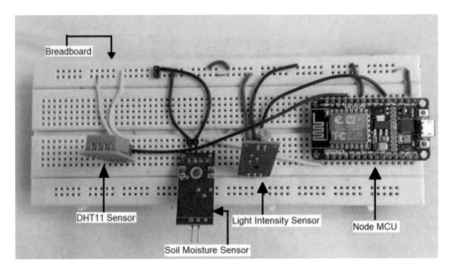

Fig. 1 An initial setup of components

3 Data Aggregation

The data is gathered from a farm consisting of many plants. Each plant has 3 primary sensors which will sense light intensity, soil moisture, temperature, and humidity from the setup as shown in Fig. 2 [3]. The sensors are chosen in such a way to record the basic features needed for predicting the relative growth of plants [1]. The data samples are taken at every 3.5-s interval to make the data smooth and real time.

The sensors communicate with the microcontroller by giving digital and analog values. The data generated by these sensors are compressed by converting the 10-bit values to 5-bit values (1). This decreases the error resolution and helps in sending data fast and with lesser bandwidth.

Algorithm for Compression

$$V_o[1] = (((V_i \gg 5)\%32) > 9)?(V_i \gg 5)\%32 + 55 : (V_i \gg 5)\%32 + 48;$$
$$V_o[2] = ((V_i\%32) > 9)?(V_i\%32) + 55 : (V_i\%32) + 48;$$

$$(1)$$

The above algorithm successfully compresses the 10-bit values to 5-bit characters which reduce the overall size of the packet. The total power consumed by the data aggregator is approximately 5–6 V and 1.0 A.

Fig. 2 The overall setup for a plant with battery source

4 Data Transmission

A large amount of data is generated by the SNs. This data is collected at INs where the data is further processed (refer Fig. 3).

The protocol used in data transmission is user datagram protocol (UDP). This protocol helps to keep the data packets light in size. The protocol is also very helpful in sending the real-time data with almost no lag in transmission. Broadcasting data from SNs is also very easily achieved using UDP.

The UDP makes the system scalable as any number of SNs can send data packets to the INs in real time [1, 2].

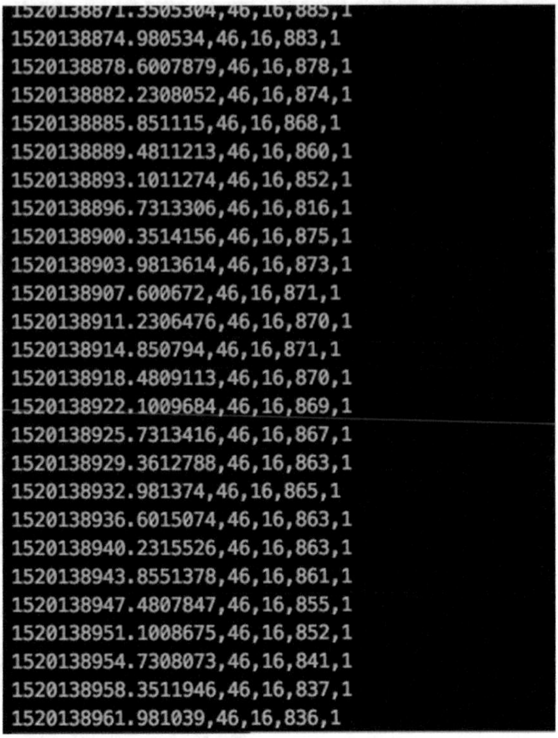

Fig. 3 Data transmission
using UDP communication

```
1520138871.3505304,46,16,885,1
1520138874.980534,46,16,883,1
1520138878.6007879,46,16,878,1
1520138882.2308052,46,16,874,1
1520138885.851115,46,16,868,1
1520138889.4811213,46,16,860,1
1520138893.1011274,46,16,852,1
1520138896.7313306,46,16,816,1
1520138900.3514156,46,16,875,1
1520138903.9813614,46,16,873,1
1520138907.600672,46,16,871,1
1520138911.2306476,46,16,870,1
1520138914.850794,46,16,871,1
1520138918.4809113,46,16,870,1
1520138922.1009684,46,16,869,1
1520138925.7313416,46,16,867,1
1520138929.3612788,46,16,863,1
1520138932.981374,46,16,865,1
1520138936.6015074,46,16,863,1
1520138940.2315526,46,16,863,1
1520138943.8551378,46,16,861,1
1520138947.4807847,46,16,855,1
1520138951.1008675,46,16,852,1
1520138954.7308073,46,16,841,1
1520138958.3511946,46,16,837,1
1520138961.981039,46,16,836,1
```

5 Data Cleansing

The data generated by the sensors contains a lot of noise which needs to be removed before analyzing the data (refer Fig. 4). As the cleansing process will run on a low-computational power node, i.e., our IN, so, we need to keep algorithms light and efficient [1].

To overcome this problem, we use rolling mean algorithm to remove the noise in the data (2). The algorithm also smoothens the curve which further simplifies our data analysis part (refer Figs. 5, 6, and 7).

Rolling Mean Algorithm

$$\underline{p_{SM}} = \frac{p_M + p_{M-1} + \cdots + p_{M-(n-1)}}{n} = \frac{1}{n}\sum_{i=0}^{n-1} p_{M-i} \tag{2}$$

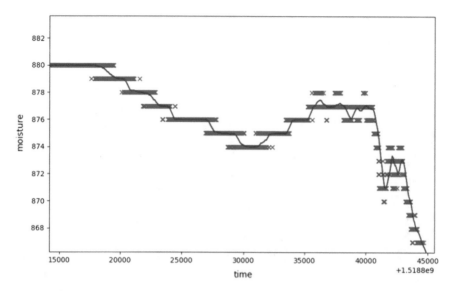

Fig. 4 Zoomed-in graph of soil moisture versus time which shows noise in the data is marked with a red cross. The smoothened data is depicted using the blue continuous line, which uses rolling mean algorithm

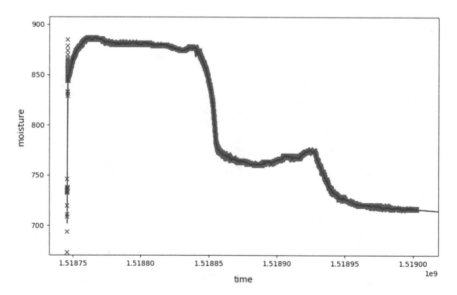

Fig. 5 This graph shows the soil moisture variation for two continuous days and its cleansed data

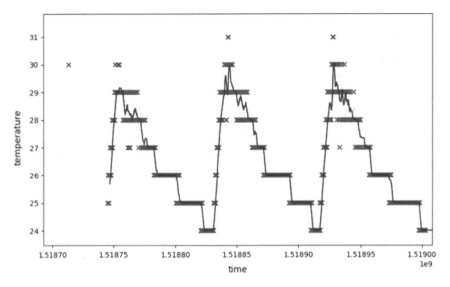

Fig. 6 This graph shows the temperature variation for two continuous days and its cleansed data

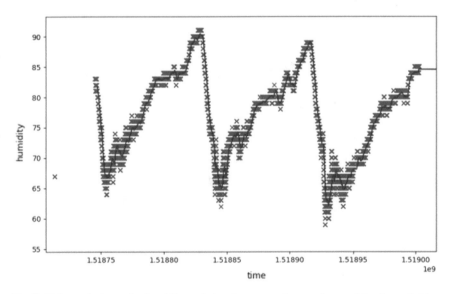

Fig. 7 This graph shows the humidity variation for two continuous days and its cleansed data

6 Data Analysis

The main objective of the project is to predict the growth of the plant by analyzing the soil moisture data. This section of the project involves generating insights from the data that is collected [1, 4]. This will provide the farmer with a rate of growth and the approximate time to water the plants.

Water is used by plants for transpiration, translocation, maintaining rigid structure, and photosynthesis. But, only a small amount of it is used for growth and metabolism processes. The rest 97–99.5% is lost into the atmosphere by transpiration.

Lack of soil moisture content leads to slow transpiration and hence ends up showing signs of premature aging like browning or curling of leaves. Transpiration occurs more during the daytime than at night. As the relative humidity rises, transpiration rates fall as it is easier to transpire in dry air than saturated air. Warmer temperatures cause higher rates of transpiration as they can hold more moisture. Due to transpiration, the amount of water used will be increased and thus, the plant grows. Hence, we can say that the rate of transpiration is directly proportional to plant growth. But, excess transpiration leads to stunted growth of plants. The rate of transpiration shown in Fig. 8 is determined using the given formula (3).

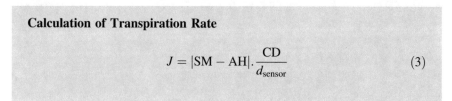

Calculation of Transpiration Rate

$$J = |SM - AH| \cdot \frac{CD}{d_{sensor}} \tag{3}$$

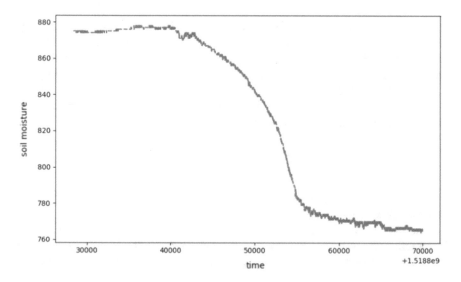

Fig. 8 Transpiration rate in a day

where

$$SM = AO/1023.00$$

$$AH = 216.7 \cdot \frac{RH}{100} \cdot 6.112 \cdot \frac{e^{\frac{17.62 \cdot t}{243.12 + t}}}{273.15 + t}.$$

The analysis consists of recording the rate at which the soil moisture decreases during the day. This rate or the value of the slope is recorded for the given day. The process is repeated for months. By doing this, a gradual decrease in the rate of soil moisture used by the plant is seen. This resembles the relative growth of the plant occurred during the recorded months.

Polynomial regression technique is used to fit the data of soil moisture versus time for given day (4) [1, 4]. The following equation is used to fit the line:

Polynomial Regression Technique

$$y = mx + c \tag{4}$$

where

m $\frac{\underline{x \cdot y} - \overline{xy}}{\underline{x^2} - \underline{x}^2}$

c $\underline{y} - m\underline{x}.$

We can see that degree 2 polynomial (refer Fig. 10) fits the data more accurately than degree 1 polynomial (refer Fig. 9). The polynomial with degree 1 is used in the project, so as to keep the calculation of slopes simple. Also, most of the data generated in the day by soil moisture sensor is linear, so it is best to use degree 1 polynomial for the calculations.

The relative growth of the plant is measured by taking the ratio of previous and current day slopes (5). The equation is given as:-

$$Relative\,Growth\% = \frac{slope\,of\,n - 1th\,day}{slope\,of\,nth\,day} \tag{5}$$

The above equation can be used to get the relative growth of a plant over a month's time by taking subsequent ratios of the slopes.

The temperature factor also needs to be taken care of which is responsible for evaporating the water from the soil. So, we can subtract the $\alpha \cdot$temperature value from the predicted value to make the prediction more accurate where alpha is temperature constant.

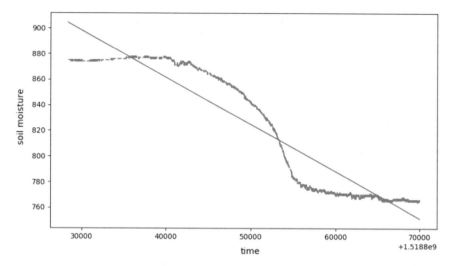

Fig. 9 Fitting degree 1 equation on transpiration curve

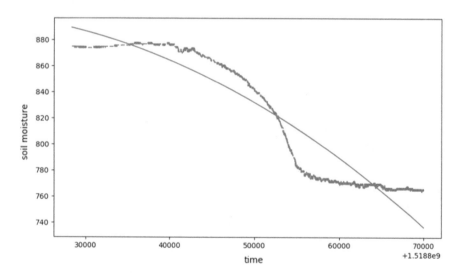

Fig. 10 Fitting degree 2 equation on transpiration curve

7   Data Prediction

The data analyzed from the previous section is used to predict the growth of the plant by analyzing the patterns. The dataset of slopes generated in the previous section is used for predicting the growth of the plant (refer Fig. 11).

The prediction line explains that if the same amount of water is continually used to grow the plant, then the plant will have a reduced growth rate [1, 4]. To increase the growth rate of the plant in a calculated manner, a greater amount of water is to be put which will reduce the slope of the curve and increase the growth rate eventually (refer Fig. 12).

8   Results and Analysis

The most important feature of the system is monitoring soil moisture sensor in PA. The water is intact within the soil pores. Soil water is the major component of the soil responsible for plant growth. If the moisture content of the soil is optimum for plant growth, then the plants can readily absorb soil water. This rate at which the soil moisture gets used up by the plant is directly proportional to the growth of the plant. It is shown in Fig. 12 that the rate of transpiration decreases as the plant

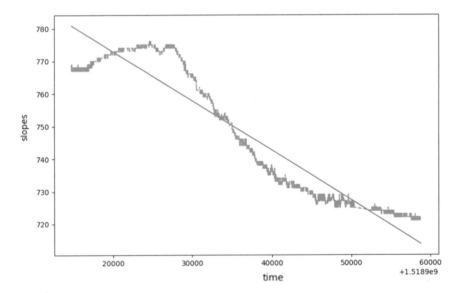

Fig. 11 This dataset shows us that there is a consistent decrease in the overall slope of the data. It tells us that the rate at which the plant is using up the soil moisture is increasing. Hence, being indicative of the relative growth of the plant that has happened in the past month

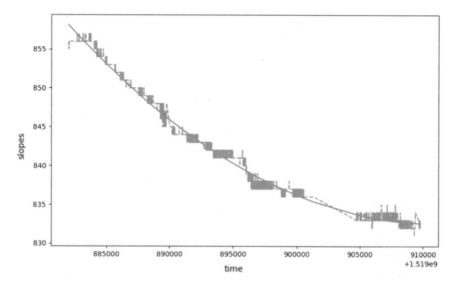

Fig. 12 The above graph shows that by increasing the amount of water in a controlled way, we can alter the growth rate of the plant. By increasing the moisture content, the growth rate of the plant gradually increases

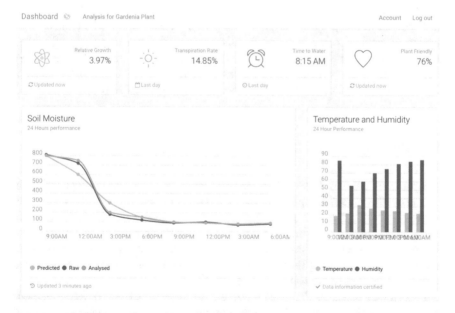

Fig. 13 An overview of the dashboard

grows. To control the growth of the plant, a perfect time to water is provided to the farmer in a dashboard (refer Fig. 13).

An analysis done on the project justifies that the project is highly efficient as it helps the farmer to gain more insights of the farm, collectively and in real time, by using prediction techniques. Also, it is manageable and reusable, as the farmer does not have to bear the infrastructure cost and can be set up at any farm with one IN doing the processing and showing the target values. Hence, it is easy to deploy any number of SNs to any farm with no server issues, making it scalable for large agricultural farms also.

9   Conclusions and Future Works

From this project, we have seen that how polynomial regression techniques can be applied in gaining insights from the data that was generated by the plant. An interesting observation that was seen is that the growth of a single plant is directly proportional to the rate of soil moisture getting used up. By this process, we can easily predict the growth of the plant just by measuring the soil moisture content of the soil.

The future work on this can be applying new machine learning models like artificial neural networks to gain a deeper understanding of the soil moisture data. Also, temperature and humidity are responsible for the growth of the plant which are recorded in this project but are not used in the calculation of the relative plant growth. Therefore, other affecting factors can also be used to measure the plant growth more accurately. The method provided in this paper for measuring plant growth is time-efficient, less expensive, and more reliable than the conventional techniques of monitoring plants with various types of cameras.

10   Related Works

In a paper of 2014 by Kassim et al. [3], the main focus was on the topic WSN in PA. They theoretically applied WSN in farming for improving quality and production of crops. The data generated in WSN was analyzed, and decisions were made accordingly in the form of actuating the SNs by normalizing the parameters. The decisions taken were computed centrally and analyzed for further normalization of parameters. Using these findings, a prediction system can be implemented on the data generated from the network.

In a similar paper by Stamenkovic et al. [2] in 2016, advanced WSN methods were reviewed for agricultural purposes. In this, a survey was given to implement advanced machine learning approaches in WSN, which, in turn, helped in bringing new machine learning techniques to current WSN model.

This was done by implementing artificial neural networks in WSN [1]. The unsupervised learning methods for clustering of sensory inputs helped in implementing ART and FuzzyART algorithms to lower the communication costs and save on energy, by processing the data on a computational intensive node within the network.

To make this possible in multiple independent sets, Hopfield neural network was used to embed in WSNs using parallel and distributed computation [4]. With the help of distributed computation on WSN as a hardware computation platform, a graph can be generated for maximum independent sets, hence making it highly optimized for prediction of sensor data in WSN.

References

1. Kulakov, A., Davcev, D., Trajkovski, G.: Implementing artificial neural-networks in wireless sensor networks (2005)
2. Stamenkovic, Z., Randjic, S., Santamaria, I., Pesovic, U., Panic, G., Tanaskovic, S.: Advanced wireless sensor nodes and networks for agricultural applications (2016)
3. Kassim, M.R.M., Mat, I., Harun, A.N.: Wireless sensor network in precision agriculture application (2014)
4. Serpen, G., Li, J.: Parallel and distributed computations of maximum independent set by a Hopfield neural net embedded into a wireless sensor network (2011)

Mining of Removable Closed Patterns in Goods Dataset

V. S. Amala Kaviya, B. Valarmathi and T. Chellatamilan

Abstract Production factories have many items produced with various parts to bring out the complete product at the end. Every product that is produced in the factory obtains the company some amount of income. The manufacturing of these products also costs a huge sum for purchase and production. When a situation arises for the company to shut down its manufacturing or when there is some crisis, the industry will not have money to buy the parts to finish the product. The problem of removable pattern mining is to identify the patterns which can be removed to reduce the loss to the factory's profit under certain conditions. In such situations, the production managers need an alternate to reduce the cost of production without decreasing the profit rate. This paper proposes removable closed patterns to represent and compress the mined removable patterns without much profit reduction. Erasable closed pattern mining algorithm is implemented to mine all the erasable closed patterns by reducing the resources needed for production without affecting the information loss. The removable itemset with one pattern along with their element occurrence pair is taken into consideration. Further the erasable patterns are mined and used for useful extraction of members. Then the erasable closed patterns are mined using the module that checks the closeness of the patterns in the product datasets.

Keywords Information systems applications · Data mining · Collaborative filtering · Pattern mining · Closed patterns · Product management

V. S. Amala Kaviya · B. Valarmathi (✉) · T. Chellatamilan
Department of Software Engineering, School of Information Technology and Engineering,
Vellore Institute of Technology, Vellore, India
e-mail: valargovindan@gmail.com

V. S. Amala Kaviya
e-mail: amalakaviyavs@gmail.com

T. Chellatamilan
e-mail: chellatamilan@gmail.com

© Springer Nature Singapore Pte Ltd. 2020
S. S. Dash et al. (eds.), *Artificial Intelligence and Evolutionary Computations in Engineering Systems*, Advances in Intelligent Systems and Computing 1056,
https://doi.org/10.1007/978-981-15-0199-9_16

1　Introduction

Numerous technologies and inventions have arisen for making sure the time and money of customers are efficiently saved in today's world. It is vital to mine knowledge from existing datasets in order to help industries know more about the repeated patterns and the unwanted patterns. Data mining is very handy now in order to bring out helpful patterns that can enable attaining knowledge efficiently. The knowledge obtained can help in many fields, like decision and recommendation systems. Numerous issues in data mining have created new research ideas.

Frequent pattern mining extracts all the patterns that occur repeatedly in the database. When coming to removing all the erasable patterns, extra efforts are to be given in order to bring out patterns that help in removing the patterns that occur least among the components. Consider a dataset DS of n transactions. Every transaction has n items present in DS. The support calculated in each itemset is calculated by finding all the transactions that have a pattern or itemset. The final patterns are obtained by checking if the support is more than the threshold that is expected to be present minimum. Taking these patterns from the dataset does absorb a lot of time as the count is more than expected and not precise. Now they are removed to reduce the count of patterns. Similarly, to extract the removable patterns, the threshold value is taken and confirmed if the loss caused by removing a part doesn't increase. To even simplify the number, closed concept is used to prevent supersets being repeated in subsets. Table 1 shows the contrasting facts between frequent pattern mining and erasable itemset mining.

2　Related Works

Two new algorithms, Apriori and AprioriTID, were proposed for discovering all the efficient association rules among large databases of transactions [1]. Experiments prove that these algorithms have excelled than those than existed earlier. The rules discovered had a single consequent item and many antecedent items [2]. NC_sets for containing reduced, important facts of itemsets in datasets were proposed. This algorithm makes it simpler to calculate the profit of an itemset via multiple strategies on NC_sets and removes unnecessary data by itself. This was efficient

Table 1 Difference between erasable itemset mining and frequent pattern mining

Erasable itemset mining	Frequent pattern mining
Computes using the value of itemsets	Computes using count
Value must be less than or equal to the threshold	The count should be more than or equal to the threshold
Used in product's production planning in the manufacturing industry	Used in the collection of items in retail trade

enough and better than Meta [3]. Extracting top-rank-m repeating patterns is an eminent task, finding the patterns inside the dataset belonging to the m first ranks with respect to support. BTK algorithm has been proposed to clear all the existing cost issues. TB-tree was used to hold important facts about frequently occurring patterns [4].

PID_list tracks the ids of components making up an itemset. Regarding this, an algorithm to extract top-k-ranked erasable patterns has been proposed. It takes a lot of time to extract or generate more candidates after many scans. Preventing this would increase performance [5]. Mining erasable itemsets, a new problem is put forth that is obtained from planning in the manufacturing industry. These algorithms indicate an increase in efficiency for synthetic database [6]. Utility mining on-shelf having negative values was derived for using negative values along mining algorithms, making it easier to complete the mining tasks. The synthetic product datasets showed that the algorithm is effective and gave better results leading to a successful extraction [7].

Top-rank-k patterns are frequently mined effectively with an N-list structure. Effectiveness of the algorithm has increased tremendously [8]. An approach for detecting sequential patterns with weights from sequence datasets has been made [9]. An algorithm to mine erasable itemsets (MEI), using a strategy called divide-and-conquer and the changes in the Pidset has been done to get the erasable itemsets fully [10]. Comparison between the methods has been made in terms of all possible performance parameters. So far that was the best for this purpose. dMERIT + should be used while less memory is needed [11]. In this paper, the MERIT+ algorithm, which is an upgraded version of MERIT, was the main base to bring deMERIT+. dNC-Sets are also used for this purpose. Tests prove that deMERIT+ is better than MERIT+ in running time [12].

The subsume concept is used to identify the data of a huge erasable itemset without the normal calculation charges. This concept was used in various algorithms [13]. This paper proposes an algorithm for extracting patterns considering the unique support of every item [14]. Frequent closed patterns can be mined using this algorithm called NAFCP. Structures based on N-list give a very simple and compact visualization that is much better than the previous ways of representation also increasing the performance measures [15].

After analyzing the above works, implementation of the main algorithms for frequent patterns along with the erasable closed pattern mining algorithms is to be made to show its importance in today's industrial world and for its profit. The ECPat is the best strategy for sparse datasets. The dNC-ECPM algorithm beats ECPat algorithm and an adjusted mining erasable itemsets calculation in terms of the mining time and memory use for every outstanding dataset.

3 Experimental and Computational Details

We consider this product database, which consists of product numbers, the components which constitute that product, and its profit value. This product database is shown in Table 2.

In order to find all the effective removable itemsets or patterns from this database set, we need to understand few definitions.

Definition I Considering a database DS and assuming a threshold value represented as ξ, a pattern called as Z can be termed removable if:

$$f(Z) \leq P \times \varepsilon$$
$$f(Z) = \sum Pm.\text{Val}$$
$$\{Pm|Z \cap Pm.\text{Item!} = \emptyset\}$$

$$T = \sum Pm.\text{Val}$$
$$\{Pm \in \text{DB}\} \tag{1}$$

Where

- $f(Z)$ is pattern Z's profit;
- P is the total profit of product dataset.

A structure for extracting removable patterns was proposed by Bay and Tuong [12]. This structure goes by the name dPidset and can be used for removing the itemset increasing the loss.

Definition II Let the pattern taken into consideration be Z. Pidset $p(Z)$ in a database DS of a particular pattern Z is shown in the Eq. (2).

$$p(Z) = \cup p(A)|A \in Z \tag{2}$$

Table 2 Goods dataset for mining removable patterns

Products	Components	Value
1	p, q	990
2	p, q, t	190
3	r, t	140
4	q, s, t, u	40
5	r, s, t	90
6	s, t, u, w	190
7	s, w	140
8	s, u, w	90

where:

- A is one item present in the pattern Z;
- $p(Z)$ is the product set that has A in it.

Definition III Consider ZP and ZQ to be the patterns derived from merging P and Q to Z pattern. The dPidset of ZPQ is defined in Eq. (3).

$$dP(ZPQ) = p(ZQ)\backslash(ZP) \tag{3}$$

The dPidset is the product variable names of patterns that are in ZQ and not in ZP.

Theorem I If ZP and ZQ be two patterns and $dP(ZP)$ and $dP(ZQ)$ are the dPidsets of ZP and ZQ, respectively, then the dPidset of ZPQ is given by Eq. (4).

$$dP(ZPQ) = p(ZQ)\backslash(ZP) \tag{4}$$

Apart from this, they have also proposed a good method for subtracting two dPidsets ($d1$ and $d2$) which can be used alone with dNC-Sets to find the erasable itemsets too but not as significant as the following algorithm.

From the selected items, each component is singled out and the support or Pidset of that particular component in all other components is found. Now different candidates are generated, and each of their profits is cross-checked with the threshold value for losing profit. This makes the experiment succeed in realizing which components are of less benefit to the factory and the rest can be used properly.

Theorem II The following properties are very important in finding the different combinations of components.

1. When $dP(ZP) = dP(ZQ), g(ZP) = g(ZQ) = g(ZPQ)$.
2. When $dP(ZP) \subset dP(ZQ), g(ZP) \neq g(ZQ)$ but $g(ZQ) = g(ZPQ)$.
3. When $dP(ZP) \supset dP(ZQ), g(ZP) \neq g(ZQ)$ but $g(ZP) = g(ZPQ)$.
4. When $dP(ZP) \not\subset dP(ZQ)$ and $dP(ZQ) \not\subset dP(ZP), g(ZP) \neq g(ZQ) \neq g(ZPQ)$.
   While finding the combinations, if there exists the same combination with different profits, then one of the following takes place.

1. When $dP(ZP) \subset dP(ZQ)$, ZQ is removed and ZPQ is placed instead of ZP.
2. When $dP(ZQ) \subset dP(ZP)$, ZP is replaced with ZPQ.
3. Otherwise, neither ZP nor ZQ are changed.

```
Input: product dataset DB and threshold ξ
Output: E_result, which is the set of all ECPs
 1  Scan DB to determine its total profit (T), and the erasable 1-
    patterns (E_1) with their dPidsets
 2  Let Hashtable = ∅ be a hashtable for storing the indexes of ECPs
 3  Sort E_1 according to the length of dPidsets in decreasing order
 4  If E_1 has more than one element, call Expand_E(E_1).

 1  Procedure Expand_E(E_v)
 2  Sort E_v according to the length of dPidsets in decreasing order
 3  For i ← 0 to |E_v| do
 4  Begin for
 5     E_next ← ∅
 6     For j ← i+1 to |E_v| do
 7     Begin For
 8        dP(ECP) = dP(E_v[j]) \ dP(E_v[i])
 9        If g(ECP) ≤ ξ × T then
10           If dP(E_v[i]) = dP(E_v[j]) then
11              E_v[i] = E_v[i] ∪ E_v[j]
12              Update E_next
13              Remove E_v[j]
14              j--
15           Else if dP(E_v[i]) ⊂ dP(E_v[j]) then
16              E_v[i] = E_v[i] ∪ E_v[j]
17              Update E_next
18           Else
19              ECP = E_v[i] ∪ E_v[j]
20              Add ECP to E_next
21     End for;
22     If Check_Closed_Property(E_v[i]) then
23        E_result ← E_v[i];
24        Add E_v[i] to Hashtable with g(E_v[i]) as the key
25     If |E_next| ≥ 1 then
26        Expand_E(E_next)
27  End for

 1  Function Check_Closed_Property(EP)
 2  ECPs ← Hashtable[g(EP)]
 3  If ECPs is not null then
 4     For each ECP in ECPs do
 5        If EP ⊂ ECP then
 6           Return false
 7  Return true
```

The algorithm that finds all the closed removable itemsets is given above. The datasets have been inputted into the code created by using Java Eclipse IDE, Intel Core i5 2.3-GHz CPU, and 12 GB RAM run on Windows 8.1.

4 Results and Discussion

The itemsets which have profit values less than the threshold are given below in Table 3. These components can be removed, whereas the others are vital and needed for the profit of the product.

Table 3 All ECPs for DBE with ξD 40%

ECPs	Values
r	230
u	320
ur	550
w	420
wr	650
wu	460
swu	550
swur	690
tr	650

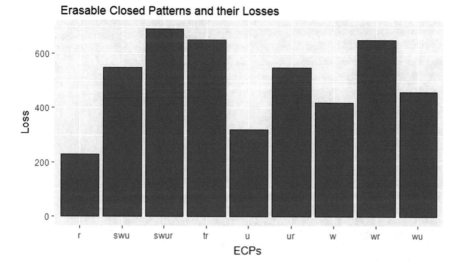

Fig. 1 Removable patterns obtained from the goods dataset

For the dataset that we considered, assume $\xi = 0.4$ for finding the removable patterns. After complete extraction, you can find that t and tsuw are two removable patterns because $g(t) = g(tr) = 650 < T \times 0.4 = 748$. All the itemsets existing and generated are compared with this value, and if lesser than that, they are considered to be removable. Using this idea, all the datasets have been analyzed same as this example. Figure 1 shows the itemsets to be removed with their profits. Notice that, t cannot be an ECP because tr happens to be its superset with the same profit.

4.1 Memory Usage

Accidents dataset (real-time dataset) has been used for analyzing the algorithm and the results have been compared with the existing MEI algorithm. It is found to be extremely memory efficient to extract the erasable closed patterns. Figure 3 shows the memory usage plot opposing the threshold value.

4.2 Mining Time

Figure 4 shows the mining time plotted in the experiment. The ECPat algorithm was found to have better results and speed in the accidents.dat dataset. The algorithm has been showing extreme performance speed using this algorithm.

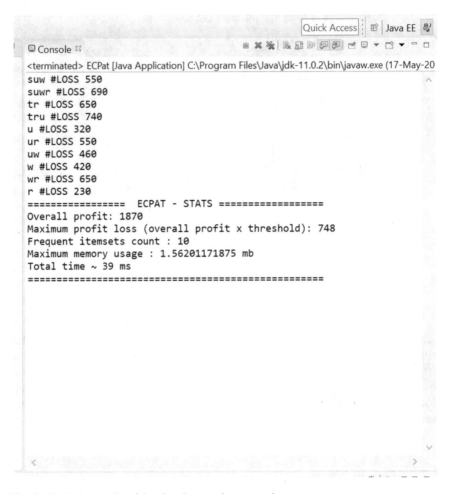

Fig. 2 Simulation results of the closed sets to be removed

5 Conclusion

Among the algorithms for faster determination of the erasable closed patterns, ECPat is considered to be the best approach to mine the close patterns. Tests have been done to analyze the performance parameter in ECPat and the existing MEI algorithm (all erasable patterns are mined by MEI) and then closed patterns are filtered from the erasable patterns using the algorithms mentioned. This in turn shows efficient ways for pattern mining. In the future, we can study more erasable patterns to mine from huge datasets and for mining maximal erasable patterns.

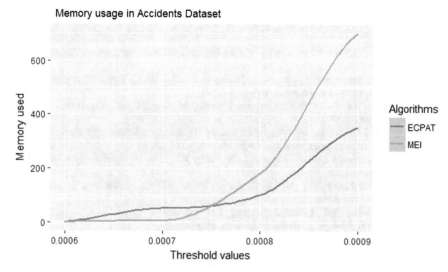

Fig. 3 Memory usage plotted for the accidents dataset

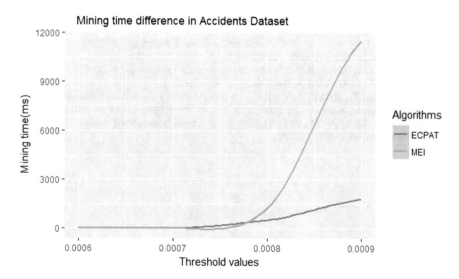

Fig. 4 Mining time plotted in accidents dataset

References

1. Agrawal, R., Srikant, R.: Fast algorithms for mining association rules. Proc. VLDB **94**, 487–499 (1994)
2. Agrawal, R., Imielinski, T., Swami, A.: Mining association rules between set of items in large databases. Proc. SIGMOD **93**(2), 207–216 (1993)
3. Deng, Z.-H., Xu, X.-R.: Fast mining erasable itemsets using NC_sets. Expert Syst. Appl. **39** (4), 4453–4463 (2012)
4. Dam, T.-L., Li, K., Fournier-Viger, P., Duong, Q.-H.: An efficient algorithm for mining top-rank-k frequent patterns. Appl. Intell. **45**(1), 96–111 (2016)
5. Deng, Z.H.: Mining top-rank-k erasable itemsets by PID_lists. Int. J. Intell. Syst. **28**(4), 366–379 (2013)
6. Han, J., Pei, J., Yin, Y.: Mining frequent patterns without candidate generation. Proc. SIGMOD, 1–12 (2000)
7. Lan, G.C., Hong, T.P., Huang, J.P., Tseng, V.S.: On-shelf utility mining with negative item values. Expert Syst. Appl. **41**(7), 3450–3459 (2014)
8. Huynh, T.L.Q., Le, T., Vo, B., Le, B.: An efficient and effective algorithm for mining top-rank-k frequent patterns. Expert Syst. Appl. **42**(1), 156–164 (2015)
9. Lan, G.C., Hong, T.P., Lee, H.Y.: An efficient approach for finding weighted sequential patterns from sequence databases. Appl. Intell. **41**(2), 439–452 (2014)
10. Le, T., Vo, B.: MEI: an efficient algorithm for mining erasable itemsets. Eng. Appl. Artif. Intell. **27**, 155–166 (2014)
11. Le, T., Vo, B., Nguyen, G.: A survey of erasable itemset mining algorithms. WIREs Data Min. Knowl. Discov. **4**(5), 356–379 (2014)
12. Le, T., Vo, B., Coenen, F.: An efficient algorithm for mining erasable itemsets using the difference of NC-sets. In: 2013 IEEE International Conference on Systems, Man, and Cybernetics, vol. 13, pp. 2270–2274, June 2013
13. Nguyen, G., Le, T., Vo, B., Le, B.: EIFDDAn efficient approach for erasable itemset mining of very dense datasets. Appl. Intell. **43**(1), 85–94 (2015)
14. Yun, U., Shin, H., Ryu, K.H., Yoon, E.: An efficient mining algorithm for maximal weighted frequent patterns in transactional databases. Know. Based Syst. **33**, 53–64 (2012)
15. Le, T., Vo, B.: An N-list-based algorithm for mining frequent closed patterns. Expert Syst. Appl. **42**(19), 6648–6657 (2015)

Comparative Analysis and Performance of DSTATCOM Device Using PI and Second-Order Sliding Mode Control

K. Swetha and V. Sivachidambaranathan

Abstract In distribution systems, more number of power quality problems are there; out of them, harmonics plays major role; these are caused by the nonlinear devices used in the distribution network. In distribution systems, it is around 75% problems which are caused due to especially harmonics. Harmonics exists due to nonlinear devices used in distribution system. These harmonics are compensated by a distribution static compensator (DSTATCOM). By using this injection or absorption of reactive power, voltage and current fluctuations, harmonics elimination can achieve. It is shunt connected device by using this, shunt current compensation can be performed. In this paper detailed explanation about DSTATCOM and the amount of injected current at source side, total harmonic distortion in injected currents are explain with total harmonic distortion values by using prosperous control technique of DSTATCOM is compared with conventional PI controller. Simulation results are obtained by using MATLAB/SIMULINK software.

Keywords Second-order sliding mode control (SOSMC) · Power quality (PQ) · Proportional and integral (PI) control · Total harmonic distortion (THD) · Distribution static compensator (DSTATCOM)

1 Introduction

Power quality is the foremost criteria in power systems. Power quality can be improved by using various FACTS devices out of them; distributed static synchronous compensator will place a major role. Because of switch mode power supply, adjustable speed drives and other nonlinear loads cause more amount of

---

K. Swetha (✉) · V. Sivachidambaranathan
Department of EEE, Sathyabama Institute of Science and Technology, Chennai, India
e-mail: swethasathyabama@gmail.com

V. Sivachidambaranathan
e-mail: sivachidambaram_eee@yahoo.com

© Springer Nature Singapore Pte Ltd. 2020
S. S. Dash et al. (eds.), *Artificial Intelligence and Evolutionary Computations in Engineering Systems*, Advances in Intelligent Systems and Computing 1056,
https://doi.org/10.1007/978-981-15-0199-9_17

non-sinusoidal components are injected into the distribution system. Power curative is required to improve the system performance and PQ. This incorporates the elimination of harmonics and compensation of reactive power by balancing the source currents. DSTATCOM is used to deteriorate current-related power quality issues.

DSTATCOM is having the major advantages like:

- Compensates negative sequence currents.
- Compensates reactive power for balanced/unbalanced and linear/nonlinear loads.
- Effectively eliminates harmonics due to nonlinear loads.
- Load balancing harmonic current compensation.
- Improves current harmonic distortion.
- Improves power quality.

DSTATCOM is a combination of voltage source inverter along with power electronic devices fed by small DC capacitance. This device will operate based on the controlling pulses generated from the controllers. Total harmonic distortion in injected currents given in (Shahgholian and Azimi in Electronics [1]). More number of control techniques are available to generate controlling signals to device. Sometimes, these will generate by the consideration of reference signal along with controller output and other soft computing techniques, etc. [2]. In this paper, SOSMC controller is developed to enhance the compensation of reactive power and along with the elimination of harmonics in three-phase distribution system is explained.

FEATURES of the Controller:

- The convergence speed of the controller is very high.
- It is having much robustness with respect to output load variation and internal parameter variations.
- For voltage fluctuations, this controller is less sensitive.

2 Overview About Paper

This paper mainly focuses on control techniques used for DSTATCOM. It is having voltage source inverter along with power electronic devices like MOSFETs, IGBTs; these devices will turn on based on triggering pulses, and to run the circuit, most important part is the generation of firing pulses. More number of controllers is present to generate these firing pulses out of all controllers. This paper deals comparison between with conventional or proportional integral controller; with robust or second-order sliding mode controller, it is having more advantages over PI controller. Here by using this controller, injected currents and THD values at source side are mentioned, along with source voltages and source currents. And also how voltages and currents' effects with these controllers are explained in detailed. With these controllers, how much is the reactive powers and active powers are described. Differential evolution based tuning of proportional integral controller for modular multilevel converter STATCOM [3]. The simulation Of STATCOM for distribution systems using a mathematical model approach given in [4] grid with abnormal condition is explained in [5] distribution network with distributed PV generation given in [6]. Boost converter with MPPT algorithm is proposed [7] DSTATCOM for loss minimization in radial distribution system explained [8] Intelligent control technique for DC micro grid given in [9] reduced harmonic with multilevel inverter technique introduced [10].

3 Device Explanation

Distribution Static Synchronous Compensator (DSTATCOM) is connected in parallel with the distribution system at point of common coupling. This can perform load compensation, source current compensation, harmonic filtering, load balancing etc. It is connected at point of common coupling. In the distribution system, the bus voltage constant can be maintained by using DSTATCOM against any unbalance or distortion loads. It is able to infuse an unbalanced currents and voltage harmonics caused by nonlinear parameters which can be eliminated by compensating source currents. Out of all custom power devices, this is one and only one device connected in parallel with power system having voltage or current source inverters. The basic parameters of DSTATCOM are as shown in Fig. 1; it is a combination of voltage source inverter which is in series with filter that is used to filters the harmonics caused by inverter circuit. DC link capacitor connected to inverter circuit will give supply at starting time of inverter operation. Inverter is connected in series with the coupling transformer in order to transfer or observe responses from device to system.

Equations related to DSTATCOM model are

$$L(di_r/dt) + Ri_r = u_{sr} - u_{tr}$$
$$L(di_y/dt) + Ri_y = u_{sy} - u_{ty}$$
$$L(di_b/dt) + Ri_b = u_{sb} - u_{tb}$$

After converting the above equation into $\chi\gamma$ coordinate method, then the equations arewhere

$$L(di_\chi/dt) + Ri_\chi = u_{s\chi} - u_{t\chi}$$
$$L(di_\gamma/dt) + Ri_\gamma = u_{s\gamma} - u_{t\gamma}$$

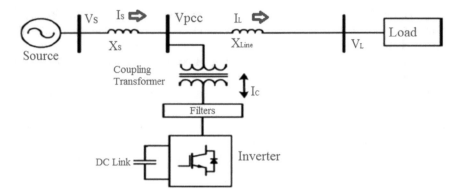

- If $V_{device} = Vsys(V_s,)$ no transfer of reactive power between device and gird.
- $V_{device} > V_{sys}$, DSTATCOM generate a capacitive reactive power.
- $V_{device} < V_{sys}$, DSTATCOM absorbed an inductive reactive power.

Fig. 1 Basic Parameters of DSTATCOM

where

$$[i_\chi i_\gamma]^r = M_{ryb/\chi\gamma}[i_r \quad i_y \quad i_b]^T$$

$$[u_{s\chi}u_{s\gamma}]^r = M_{ryb/\chi\gamma}[u_{sr} \quad u_{sy} \quad u_{sb}]^T$$

$$[u_{t\chi}u_{t\gamma}]^r = M_{ryb/\chi\gamma}[u_{tr} \quad u_{ty} \quad u_{tb}]^T$$

$$T_{ryb/\chi\gamma} = \sqrt{\frac{2}{3}} \begin{bmatrix} 1 & \frac{-1}{2} & \frac{-1}{2} \\ 0 & \frac{\sqrt{3}}{2} & -\frac{\sqrt{3}}{2} \end{bmatrix}$$

4 Proportional Integral controller

In order to maintain constant voltage, magnitude at load side at any type of disturbance condition PI controller is used. Inverters will work based on firing pulses; these firing pulses will generate by using pulse width modulation techniques, and one of the PWM techniques is SPWM; by using this, sine wave is compared with reference triangular signal; when the magnitude of reference signal is greater, then the sine wave that duration pulse will generate with these pulses inverter will operate. With the help of high switching frequency converter, efficiency is going to be improved that can be possible by using PI controller. By using this, comparisons between error signal from reference current and terminal current will be done and

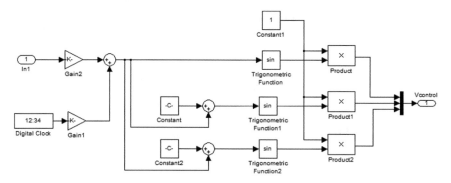

Fig. 2 PI controller for DSTATCOM Generation of reference control signals for PWM controller

resultant angle will generate to trigger the inverter circuit, but while generating firing angle, this controller will try to compensate error signal to zero, so by using this load current is equal to reference current value (Fig. 2).

5 Second-order Sliding Mode Controller

Second-order sliding mode control (SOSMC) is one of the control techniques used to overcome drawbacks in first-order sliding mode controller, and it is nonlinear control technique. It is having features like better accuracy, robustness, tuning is easy and implementation is also don't have any complications [11]. Bi-Directional Three Phase Parallel Resonant High Frequency AC Link Converter is given in [12], Enhancement of Active Power Filter is given [13], Closed loop Control with LLC Series-Resonant convert discussed in [14], A single switch parallel Quasi resonant converter topology explained [15], Cascaded Multi-Level Inverter-Based STATCOM for High Power Applications [16].

SoSMC working principle is explained below, where τ^*_{EQ} denotes the equivalent part of the SoSMC.

$$\tau^* = \tau^*_{EQ} + A\tau^*_{dis}$$

where ζ_{EQ} represents the SoSMC part of equivalent. $g_s\mathrm{sign}(\dot{\theta})$ represents the system's solid friction torque and sign indication of the signum function.

$$\zeta + \lambda A\phi^{\bullet\bullet}_{ES} + (g_v - \lambda A)\phi^{\bullet} + g_s\mathrm{sign}(\phi^{\bullet}) - \zeta_g\cos(\phi)$$

Here, ζ_{dis} is the indication of the controller discontinuous part.

The super-twisting approach is given below

$$SU =$$

where

$$
\begin{cases}
SU = \lambda e + \dot{e} + e_1;\ i = 1, 2, 3 \text{ with:} \begin{cases} \theta_d = \theta_{ES} \\ e = \theta_{ES} - \theta \\ \dot{e} = \dot{\theta}_{ES} - \dot{\theta} \end{cases} \\
\sigma_2 > \dfrac{\varphi_0}{I_m} \ge 0, \\
\sigma_1^2 \ge 4 \dfrac{\varphi_0}{I_m^2} \dfrac{\sigma_2 I_m}{\sigma_2 m_U} + \dfrac{\varphi_0}{\varphi_0} > 0 \text{ (required conditions)}
\end{cases}
\tag{13}
$$

The formulation of τ_{EQ} is diagrammatically illustrated in Fig. 1. Here, the load current is passed to transformation, and the output is then given to filter, and the filtered current is then passed to compensator, from which τ_{EQ} is accomplished. The estimation of minimized reactive power is diagrammatically illustrated. Sliding surface is nothing, but SOSMC systems are designed to drive the system states onto a particular surface in the state space. Once the sliding surface is reached, sliding mode control keeps the states on the close neighbourhood of the sliding surface. Sliding mode control is having a two-part controller design.

- First one is the design of a sliding surface so that the sliding motion satisfies design specifications.
- Second one is the selection of a control law that will make the switching surface attractive to the system state.

There are two main advantages of sliding mode controllers:

- System dynamic behaviour of may be tailored by the particular choice of the sliding function.
- The closed loop response becomes totally insensitive to some particular uncertainties.

This principle helps to model parameter uncertainties, disturbance, and nonlinearity that are bounded. From a practical point of view, SOSMC allows for controlling nonlinear processes subject to external disturbances and heavy model uncertainties.

Section V

See Fig. 3.

Section VI

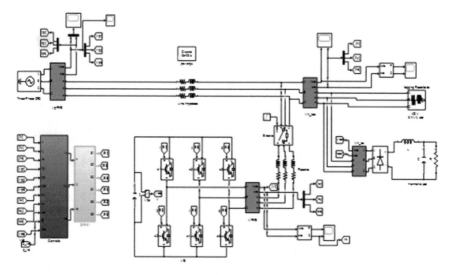

Fig. 3 Simulation circuit of distribution system with controller

6   Simulation Results with PI Controller

See Figs. 4, 5 and 6.

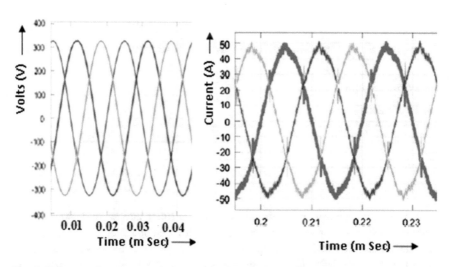

Fig. 4 Voltage and current waveforms by using PI controller current is having more harmonics at 100% load

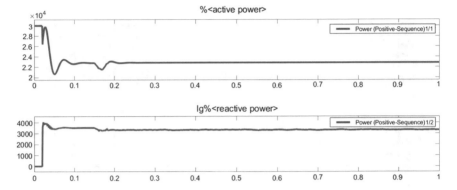

Fig. 5 Active and reactive powers by using PI controllers at 100% load

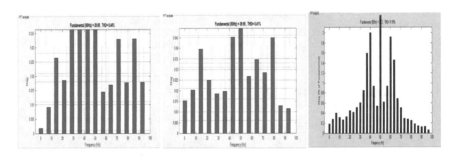

Fig. 6 Current harmonics of three-phase supply THD having 6.4 by using PI controller at 100% load

7 Simulation Results with SOSMC Controller

See Figs. 7, 8 and 9.

Load Parameters

S. No.	100% of load	150% of load	200% of load
R	20	30	40
L (mH)	0.1	0.15	0.2
C (μF)	500	750	1000

Comparison between proportional integral controller and second-order sliding mode controllers is explained above out of both PI having more THDs in source currents 6.4 and SOSMC having 0.84. In source, current fluctuations are more in PI controller by using SOSMC those are reduced. Distortions in active and reactive powers are reduced by using SOSMC.

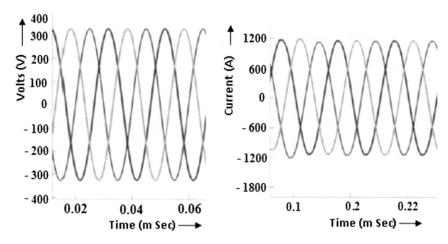

Fig. 7 Voltage and current waveforms by using SOSMC controller current is having less harmonics at 100% load

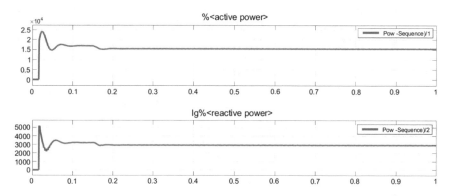

Fig. 8 Active and reactive powers by using SOSMC controllers 100% load

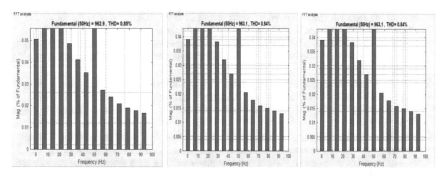

Fig. 9 Current harmonics of three-phase supply THD having 0.84 by using SOSMC controller 100% load

Experiment conducted for different loads changes the output compensated source currents for three phases which are tabulated below.

100% of load		150% of load		200% of load	
PI	SOSMC	PI	SOSMC	PI	SOSMC
6.47	0.85	7.69	0.58	9.18	0.47
6.43	0.84	7.77	0.57	9.34	0.45
6.45	0.85	7.61	0.58	9.25	0.47

Section VII

8 Conclusion

PI controller and SOSMC-based DSTATCOM are presented for mitigation of harmonic current in an electrical distribution system. Control of instantaneous active and reactive power, PWM technique, is used to generate the reference currents need by the model predictive controller to compensate harmonic currents in the distribution system. By using SOSMC, total harmonic distortion is decreased to 0.84 and active and reactive powers effectively generated. Current harmonics are decreased. Simulation results using MATLAB/Simulink presented in this paper demonstrate that the proposed DSTATCOM was able to minimize the source current harmonics and improve power quality in distribution system.

References

1. Shahgholian, G., Azimi, Z.: Analysis and design of a DSTATCOM based on sliding mode control strategy for improvement of voltage sag in distribution systems. Electronics **5**, 41 (2016). https://doi.org/10.3390/electronics5030041; www.mdpi.com/journal/electronics
2. Patel, A.A.: Application of DSTATCOM for harmonics elimination and power quality improvement. INSPEC Accession Number: 17874363. https://doi.org/10.1109/icecds.2017.8390008,publisher:ieee
3. Kumar, L. S., Kumar, G. N., Prasanna, P. S.: Differential evolution based tuning of proportional integral controller for modular multilevel converter STATCOM. In: Computational Intelligence in Data Mining—Volume 1, pp. 439–446 (2016). Springer, New Delhi
4. Xie, W., Duan, J.M.: The simulation Of STATCOM for distribution systems using a mathematical model approach. In: Huang, B., Yao, Y. (eds.) Proceedings of the 5th International Conference on Electrical Engineering and Automatic Control, vol. 367. Springer, Berlin, Heidelberg
5. Beniwal, N.: DSTATCOM with i-PNLMS based control algorithm under abnormal grid conditions. In: IEEE Transactions. IEEE (2018). https://doi.org/10.1109/tia.2018.2846739

6. Rohoumal, W., Balog, R.S.: D-STATCOM for a distribution network with distributed PV generation. 978-1-5386-7538-0/18/$31.00 ©2018 IEEE
7. Babu, A.R., Raghavendiran, T.A.: Performance enhancement of high voltageGain two phase interleaved boost converter Using MPPT algorithm. J. Theor. Appl. Inf. Technol. **68**(2), 360–368 (2014). ISSN 1992–8645
8. Shanmugasundram, P., Babu, A.R.: Application of DSTATCOM for loss minimization in radial distribution system. In: Proceedings of the International Conference on Soft Computing Systems, Advances in Intelligent Systems and Computing, vol.397 pp.189–198 (2015)
9. Babu, A.R., Raghavendiran, T. A.: High voltage gain multiphase interleaved DC-DC converter for DC micro grid application using intelligent control. Comput. Electr. Eng. **74**, 451–465 (2019). ISSN:0045–7906
10. Selvamuthukumar, K., Satheeswaran, M., Babu, A.R.: Single phase thirteen level inverter with reduced number of switches using different modulation techniques. ARPN J. Eng. Appl. Sci. **10**(22) 10455–10462 (2015). ISSN 1819–6608
11. Raymond, A.: A quick introduction to sliding mode control and its applications, in the year of 2008 in semantic scholar
12. Mary, P.P., Sivachidambaranathan, V.: Design of new bi-directional three phase parallel resonant high frequency AC link converter. Int. J. Appl. Eng. Res. **10**(4), 8453–8468. ISSN 0973-4562
13. Mary, M.P.P., Sivachidambaranathan, V.: Enhancement of active power filter operational performance using SRF theory for renewable source. Indian J. Sci. Technol. **8**(21), 1–7 (2015). ISSN 0974-6846
14. Indira, D., Sivachidambaranathan, V., Dash, S.S.: Closed loop control of hybrid switching scheme for LLC series-resonant half-bridge DC-DC converter. In: Proceedings of the Second International Conference on Sustainable Energy and Intelligent System (SEISCON 2011), pp. 295–298, IET Chennai and Dr. MGR University, July 20–22 (2011)
15. Geetha, V., Sivachidambaranathan, V.: A single switch parallel quasi resonant converter topology for induction heating application. Int. J. Power Electron. Drive Syst. (IJPEDS) **9**(4), 1718–1724 (2018). ISSN 2088-8694
16. Jeyanthi, M., Sivachidambaranathan, V.: Cascaded multi-level inverter-based STATCOM for high power applications. Int. J. Appl. Eng. Res. **10**(17 Special issues), 13652–13757 (2015). ISSN 0973-4562

Stock Market Prediction Using Long Short-Term Memory

M. Ferni Ukrit, A. Saranya and Rallabandi Anurag

Abstract Stock markets have been an integral part of our socioeconomic society. People invest a lot of monetary funds into them so as to earn gains. But that is not the case every time due to the ever wavering nature of the markets. To minimize the risk of loss due to drastically changing market, people have come up with many predictive models to simulate the future of stock markets. This paper presents a model that can predict the stock market. The use of stacked Long Short-Term Memory gives the model an advantage over the conventional machine learning models to provide better MSE.

Keywords Recurrent neural networks (RNN) · Long Short-term memory (LSTM) · Stock market · Prediction · Google · Mean squared error (MSE)

1 Introduction

Stock market is a major factor that defines how successful a company is. Google, Apple, and others are major players in the market and how these companies fare on the market can impact other companies. So, how these companies do on the market in the future should be of great importance. This neural network attempts to do the same for Google's stock price from beginning of 2012 to the end of 2016 and predict upon data it has not been trained upon.

A specific kind of neural network called the recurrent neural network has been used for the purpose. The RNN is a specially designed neural network that is used

M. Ferni Ukrit (✉) · A. Saranya · R. Anurag
Department of Software Engineering, SRM Institute of Science and Technology,
Kattankulathur, India
e-mail: fernijegan@gmail.com

A. Saranya
e-mail: aksaranya@gmail.com

R. Anurag
e-mail: anuragrallabandi@gmail.com

© Springer Nature Singapore Pte Ltd. 2020
S. S. Dash et al. (eds.), *Artificial Intelligence and Evolutionary Computations in Engineering Systems*, Advances in Intelligent Systems and Computing 1056,
https://doi.org/10.1007/978-981-15-0199-9_18

for time-series analysis, viz. each layer predicts its output based on the output of the previous layer [1].

When RNNs were first developed, it faced a problem. With each iteration as the gradient is passed back for updating the weights, the process is slower for far off layers and the input that is transmitted for the next iteration is through untrained neurons which affect the output. This is known as the Vanishing Gradient Problem [2].

This was rectified by Long Short-Term Memories that do not use the long-term dependencies for computing the outputs [3–5]. This is the network used for this particular model. The formula for a RNN is specified in Eq. (1).

$$h(t) = f_H(W_{1H}x(t) + W_{HH}h(t - 1)) \qquad (1)$$

It basically says the current hidden state **h(t)** is a function **f** of the previous hidden state **h(t − 1)** and the current input **x(t)**. The **theta** are the parameters of the function f.

A study of the stock market is integral for building a prediction model [6]. It helps in understanding the nuances of the said market, figuring out the features that are required for the model and edging out the features that is not required [7].

2   Methodology Used

This model consists of the following steps

Step 1: **Data Preparation**

The model will predict based on date and opening value. The opening value is the value the market opens at for that particular trade day.

The data collected was normalized for values between 0 and 1 using Eq. (2).

$$x_{new} = x - x_{min}/x_{max} - x_{min} \qquad (2)$$

The scaled data is then recreated to a data structure with 40 and 60 time steps, and 1 output will be used as an input for the LSTM in which the output is decided by the previous 80 inputs and so on. Then the data is reshaped based on three dimensions: number of stock prices, the number of time steps and the number of indicators.

Step 2: **Building the model**

In the neural network model, a LSTM with 50 input neurons and add a Dropout layer with 20 neurons (forty percent) will be ignored during training for regularization.

For the second and the third layer, the model mimics the input layer with the same parameters thereby creating a stacked LSTM (Fig. 1).

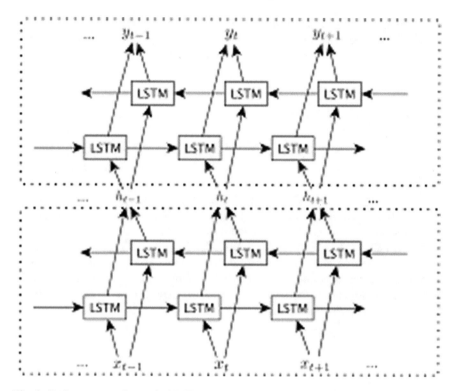

Fig. 1 Basic structure of a stacked LSTM

The next layer is the output layer with one neuron. The output is computed based on an optimizer and loss function. The optimizer used is the ADAM or the Adaptive Moment Estimation Optimizer provided in Eqs. (3) and (4).

$$Mt = \beta 1mt - 1 + (1 - \beta 1)gt \tag{3}$$

$$vt = \beta 2vt - 1 + (1 - \beta 2)g2t \tag{4}$$

where Mt is the first moment and vt is the second moment of the gradient and the loss function (error metric) used is mean squared error (MSE).

Step 3: Fitting the model

In this step, data is fed to the neural network and trained for prediction assigning random biases and weights.

Step 4: Output

The output generated is compared with target values. The error obtained during each epoch is reduced through back propagation which adjusts the weights and biases of each neuron (node) present in the subsequent layers.

Lemmas, Propositions, and Theorems. The numbers accorded to lemmas, propositions, theorems, etc., should appear in consecutive order, starting with Lemma 1, and not, for example, with Lemma 11.

3 Results and Discussion

The hardware used is Intel i5200U CPU clocked at 2.20 GHz as a training platform. The model is implemented in Keras for Python as front end and tensor flow backend as coding environment.

The training data set used is Google's stock price from beginning of 2012 to the end of 2016 collected for each financial day for each month of the financial year ranging from January 3, 2012 to December 16, 2016.

The test set used is Google's stock price for January 2017.

The loss function mean square error (MSE) or mean square deviation (MSD) of an estimator (of a procedure for estimating an unobserved quantity) measures the average of the squares of the errors or deviations. Large errors are easily identified and reduced using this metric using Eq. (5).

$$\text{MSE} = 1/n \sum_{i=1}^{n} [y_i - y_i^{\sim}]^2 \tag{5}$$

For training the optimizer used is ADAM, and the data was normalized. Date and open values are the parameters for various numbers of epochs and a batch size of 32 to calculate MSE. Tables 1 and 2 show the MSE achieved for different epochs for 40 and 60 time steps.

Figures 2 and 5 show more MSE because it was only trained for 100 epochs, and moreover, it is the first training.

In Figs. 3 and 6, the model shows better MSE due to increase in the number of epochs.

In Figs. 4 and 7, the model predicts closest to the real value because the number of epochs is increased.

From Tables 2 and 3, it is inferred that the more the number of time steps, the better the MSE achieved.

Table 1 MSE achieved for different epochs for 40 time steps	S. No.	Number of epochs	MSE achieved
	1	100	0.0030
	2	200	0.0019
	3	300	0.0017

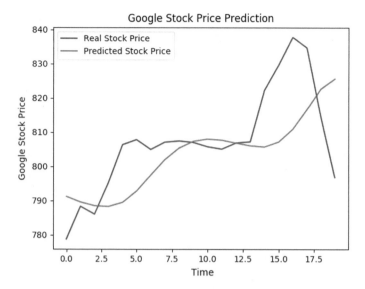

Fig. 2 Real versus predicted stock price for 100 epochs for 40 time steps

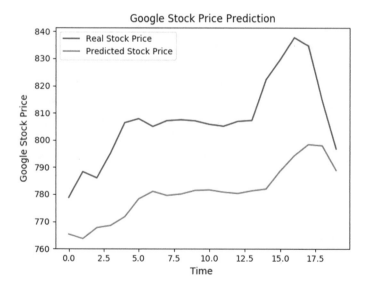

Fig. 3 Real versus predicted stock price for 200 epochs for 40 time steps

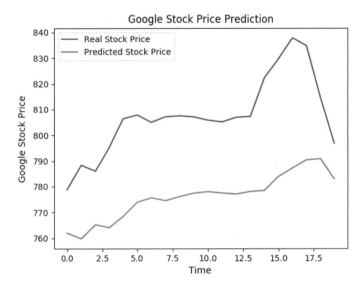

Fig. 4 Real versus predicted stock price for 300 epochs for 40 time steps

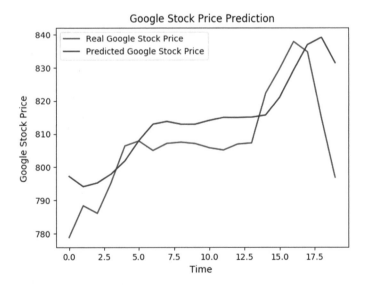

Fig. 5 Real versus predicted stock price for 100 epochs for 60 time steps

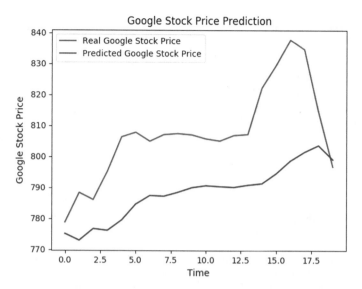

Fig. 6 Real versus predicted stock price for 200 epochs for 60 time steps

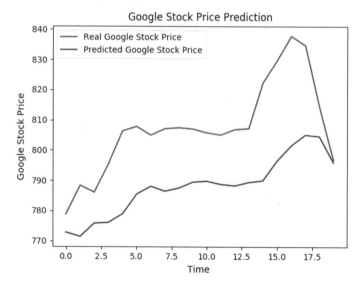

Fig. 7 Real versus predicted stock price for 300 epochs for 60 time steps

Table 2 MSE achieved for different epochs for 60 time steps

S. No.	Number of epochs	MSE achieved
1	100	0.0027
2	200	0.0017
3	300	0.0016

4 Conclusion

The increasing importance of stock markets is pushing researchers into finding newer methods for prediction which not only helps researchers but traders alike. This model focuses on prediction using LSTM. As the number of epochs increases, the model shows better mean square error up to a certain number until convergence is achieved.

References

1. Sutskever, I.: Training recurrent neural networks. Ph.D. thesis, University of Toronto (2012)
2. Pascanu, et al.: On the difficulty of training recurrent neural networks. ICML **3**(28), 1310–1318 (2013)
3. Sherstinsky, A.: Fundamentals of recurrent neural network (RNN) and long short-term memory (LSTM) network. CoRR abs/1808.03314 (2018)
4. Greff, K., Srivastava, R.K., Koutník, J., Steunebrink, B.R., Schmidhuber, J.: LSTM—a search space odyssey. IEEE Trans. Neural Netw. Learn. Syst. **28**(10), 2222–2232 (2015)
5. Li, Y., Ma, W.: Applications of artificial neural networks in financial economics: a survey. In: 2010 International Symposium on Computational Intelligence and Design (ISCID 2010), vol. 10, pp. 211–214 (2010)
6. Chen, K., Zhou, Y., Dai, Y.: A LSTM-based method for stock returns prediction: a case study of China stock market. In: IEEE International Conference on Big Data. ISBN: 978-1-4799-9926-2
7. Chavan, P.S., Patil, S.T.: Parameters for stock market prediction. Int. J. Compu. Technol. Appl. **4**(2), 337–340

Dragonfly Algorithm for Optimal Allocation of D-STATCOM in Distribution Systems

V. Tejaswini and D. Susitra

Abstract In distribution systems, parameters like outage of components, sudden or continuous load growth may cause power quality problems like excessive power losses, voltage instabilities, etc. To compensate power quality problems, usage of custom power devices is most effective approach. To get the best performance, custom power devices must place in optimal locations. Hence, this paper proposes a new swarm intelligence-based dragonfly algorithm (DA) to determine optimal placement of D-STATCOM in the radial distribution systems. DA algorithm is inspired by unique and rare swarming behaviors of dragonflies in nature. Power losses minimization and voltage profile improvement are considered as main objectives for solving optimization problems. The proposed DA model is compared with conventional methods like genetic (GA) and PSO in terms of convergence and cost analysis.

Keywords Power quality-STATCOM · Dragonfly algorithm · GA · PSO

1 Introduction

In recent days, power quality improvement has been considered for both power distribution companies and different types of voltage customers [1]. There are many reasons for more attention to power quality of power distribution companies such as outage of components, sudden or continuous load growth, increasing harmonics, excessive power losses, voltage instabilities, etc. To compensate power quality issues and to provide protection to the sensitive equipments, placing of proper custom power devices in distribution systems is necessary [2]. Custom power

V. Tejaswini · D. Susitra (✉)
Electrical Enigineering, Sathyabama Institute of Science and Technology
(Deemed to Be University), Chennai 600119, Tamil Nadu, India
e-mail: susitra.eee@sathyabama.ac.in

V. Tejaswini
e-mail: tejusateeshkumar@gmail.com

© Springer Nature Singapore Pte Ltd. 2020
S. S. Dash et al. (eds.), *Artificial Intelligence and Evolutionary Computations in Engineering Systems*, Advances in Intelligent Systems and Computing 1056,
https://doi.org/10.1007/978-981-15-0199-9_19

devices [3–5] are a kind of static feature that is utilized for distribution systems to rise constancy, flexibility with respect to controllability, faults and to improve power transfer ability in PQ networks. By optimizing the allocation of custom power devices [6, 7] in PQ networks, losses and economic cost in the distribution system for custom devices [8, 9] would be minimized, and also total power transmission between buses would be considerably affected. D-STATCOM is a member of custom power devices, which is connected in parallel to the system through coupling transformer at PCC. D-STATCOM used frequently for compensation of the power quality problems like load variations, power losses and voltage variations [10, 11]. Economical visibility, required quality, reliability and availability define the optimal location and sizing of D-STATCOM [12]. Only less contribution is there up to now in terms of optimal placement of D-STATCOM.

Some optimization models have been developed to provide solutions to the designing issues of custom power devices. In [13] Dec 2018, Salman and Ali used genetic algorithm to determine the optimal location and sizing of custom power devices. The main objective functions are power losses reduction, voltage profile enhancement, treatment of power flow in overloaded transmission lines and minimized total cost of power system. In [14] 2018, M. Laxmidevi Ramanaiah, M. Damodar Reddy presents Moth flame optimization for identifying the optimal location of UPQC in distribution system. In large distribution systems, steady-state compensation capability of UPQCs provides a solution to the reactive power compensation. Main objectives are voltage profile improvement, real power loss reduction. In [15] 2018, Atma and Ashwani presented the effect of diverse loading model on optimal distribution of D-STATCOM in radial distribution system for power loss minimization along with overall energy saving. Based on reduction in power losses, the energy saving has been portrayed. In 2017, Kavitha and Neela [10] have established a scheme to accomplish the optimal sizing and location of the custom power devices that play a major role in selecting the level to which the objective of increasing the system performance could be attained in a cost-effective approach. Moreover, the implemented scheme BBO, WIPSO and PSO has been investigated in terms of improving security, underneath enhanced system loading conditions. In [16] Sep 2016, Elazim and Ali used Cuckoo search algorithm to find optimal location of STATCOM in multi mission power system. It is inspired from life of bards. By using STATCOM, the overall voltage profile and system load ability can be improved. To determine the STATCOM location, PV curves are used, and optimization problems are solved by using CS algorithm. Compare to GA, CS algorithm damps out the system oscillations for different loading conditions.

This paper provides a way to solve the optimal location and sizing issues existing in the D-STATCOM devices in distribution system. A novel swarm intelligence optimization technique called dragonfly (DA) is developed for identifying the optimal placement of D-STATCOM in radial distribution systems. Implemented DA algorithm is compared with other conventional methods like GA, PSO in terms of convergence analysis and cost analysis.

The paper is sectionalized as follows. Section 2 explains problem identification from previous methods. Section 3 describes the arithmetical model of load growth, load model, cost of energy loss and cost of D-STATCOM. Section 4 defines the indices for optimal location of D-STATCOM. Section 5 explains D-STATCOM modeling. Section 6 explains DA algorithm and objective function, Sect. 7 results, and Sect. 8 concludes the paper.

2 Problem Identification

GA was implemented in [13] that offers minimizing the total power losses and power system stability improvement. But it requires more run time, requires long revisiting of optimal solutions. In addition, PSO was implemented in [10] for voltage profile improvement and power loss reduction but it having slow convergence and get trapped in local areas. CS algorithm is developed in [16], and it enhances the voltage profiles and damps out the system oscillations. However, it is difficult to solve multi-objective problems. Thus, it needs to implement intelligent algorithms to improve the performance of optimal placement of custom power devices.

3 Arithmetical Model of Load Growth, Load Model, Cost of Energy Loss

A. **Load Growth**:

The load growth is calculated by Eq. (1)

$$\text{LOAD} = \text{LOAD} \times (1 + a)^n \tag{1}$$

a indicates annual growth rate, n indicates the period of plan to which the feeder which obtains the load. Here, $a = 10\%$ and n = 5 [17].

B. **Load Model**:

Load varies based on time. A system has different types of loads like residential, industrial and commercial. At each and every bus, percentages of residential, industrial and commercial load can be indicated as α, β and γ, respectively [17].

$$P = P_o \left[\alpha \left(\frac{V}{V_o} \right)^{m_{pr}} + \beta \left(\frac{V}{V_o} \right)^{m_{pi}} + \gamma \left(\frac{V}{V_o} \right)^{m_{pc}} \right] \tag{2}$$

$$Q = Q_o \left[\alpha \left(\frac{V}{V_o} \right)^{m_{qr}} + \beta \left(\frac{V}{V_o} \right)^{m_{qi}} + \gamma \left(\frac{V}{V_o} \right)^{m_{qc}} \right] \tag{3}$$

Here, P_o indicates the real power, and Q_o defines the reactive power at V_o.

C. **Cost of Energy Loss (CEL)**:

The CEL is defined by Eq. (4) in which EN_c indicates the energy rate, TI points out the time in (hours) and accordingly $t = 8760$ h and $E_e = 0.06$ $/kW h [17].

$$\text{CEL} = (\text{Total Real power loss}) * (E_e * t)\$ \tag{4}$$

D. **Cost of D-STATCOM**:

Cost of the reactive power caused by D-STATCOM represented by Eq. (5). The cost is offered to the devices which are dependent on cost savings obtained owing to loss reduction with D-STATCOM [17].

$$\text{COST}^{D-\text{STATCOM}} = \text{Investment cost}^{D-\text{STSTCOM}}$$
$$\times \frac{[1 + \text{Rx}]^{\text{mDST}} * \text{Rx}}{[1 + \text{Rx}]^{\text{mDST}} - 1} \tag{5}$$

mDST = life of D-STATCOM (years), Rx = rate of return.
  Accordingly, mDST = 30 years, Rx= 0.1, Investment cost = 50$/kVAl.

E. **Voltage Stability Margin (VSM)**:

VSM [17] is defined for the all buses via Eq. (6). Enhancement of VSM is described with D-STATCOM, and at each bus, results are compared, for voltage stability margin improvement. Each bus VSM is usually a count that lies in between zero and one.

$$\text{VSM}(\text{RE}(i)) = V(\text{SE}(i))^4 - 4(P(i)x(i) - Q(I)l(i))^2$$
$$- 4\left(V(\text{SE}(i))^2(P(i)l(i) + Q(i)x(i))\right) \tag{6}$$

4 Sensitivity Indices for Optimal Placement of D-STATCOM

To find best D-STATCOM locations in the system, following indices are utilized.

a. Fast Voltage Stability Index (FVSI)
b. Combined Power Loss Sensitivity (CPLS)
c. Voltage Stability Index (VSI)
d. Voltage Sensitivity Index (VSEI)
e. Proposed Stability Index (PSI).

A. *Fast Voltage Stability Index (FVSI)*:

FVSI is calculated in between sending and receiving node as given in Eq. (7)

$$FVSI = \frac{4Z_{ij}^2 Q_j}{V_i^2 L_{ij}} \tag{7}$$

Line impedance magnitude is indicated by Z, line reactance is indicated by L, receiving end reactive power indicated by Q_j, sending end voltage indicated by V_i

The bus which has high FVSI value is considered as more instable bus. Therefore, for D-STATCOM location, high FVSI value bus will be selected [17].

B. *Combined power loss sensitivity*:

Both real and reactive power losses exist in system due to the placement of D-STATCOM. For finding CPLS, both losses should be calculated by using below eqs.

$$\frac{\partial Ploss}{\partial Q_2} = \frac{2 * Q_{2*\widehat{R}|j|}}{V_2^2} \tag{8}$$

$$\frac{\partial Qloss}{\partial Q_2} = \frac{2 * Q_{2*L|j|}}{V_2^2} \tag{9}$$

$$CPLS \text{ in terms of real power} = \frac{\partial \widehat{S}loss}{\partial P_2} = \frac{\partial Ploss}{\partial P_2} + j\frac{\partial Qloss}{\partial P_2} \tag{10}$$

$$CPLS \text{ in terms of reactive power} = \frac{\partial \widehat{S}loss}{\partial Q_2} = \frac{\partial Ploss}{\partial Q_2} + j\frac{\partial Qloss}{\partial Q_2} \tag{11}$$

$$Loss \text{ sensitivity matrix} = \begin{vmatrix} \frac{\partial Ploss}{\partial P_2} & \frac{\partial Qloss}{\partial P_2} \\ \frac{\partial Ploss}{\partial Q_2} & \frac{\partial Qloss}{\partial Q_2} \end{vmatrix} \tag{12}$$

Power flow analysis is used for obtaining the loss sensitivity matrix. The bus which has peak CPLS value is considered as best bus for the location of D-STATCOM.

C. *Voltage Stability Index (VSI)*:

VSI calculated by using Eq. (13)

$$VSI = 4\left[Q_2 L + \frac{Q_2 \widehat{R}}{L}\right] \frac{[1 - \cos 2\phi]}{2V_1^2 \sin^2(\delta_1 - \delta_2 - \phi)} \tag{13}$$

For normal loading conditions, VSI should be less than one. Therefore, the bus which has highest VSI value will be chosen as best bus for the location of D-STATCOM.

D. *Voltage Sensitivity Index (VSEI)*:

VSEI [17] is calculated by using Eq. (14) under normal load conditions, and VSEI values are less than unity. The bus which has more VSEI value becomes more instable. Hence, for the location of D-STATCOM peak, VSEI value bus is considering as best bus.

$$\text{VSEI} = \frac{4L}{V_1^2} \left(\frac{P_2^2}{Q_2} + Q_2 \right) \le 1 \tag{14}$$

E. *Proposed Stability Index (PSI)*:

The mathematical design of this stability index [17] is shown as follows, and the branch current is calculated as per Eq. (15).

$$I_{ij} = \left[\frac{P_j + jQ_j}{V_j \angle \delta} \right]^* \tag{15}$$

The bus voltage of receiving end is formulated as follows:

$$V_j \angle \delta = V_i \angle 0 - \left(\widehat{R} + jL \right) I_{ij} \tag{16}$$

Replace Eq. (15) in Eq. (16).

$$V_j \angle \delta = V_i \angle 0 - \left(\widehat{R} + jL \right) \left[\frac{P_j + jQ_j}{V_j \angle \delta} \right]^*, \tag{17}$$

$$V_j \angle \delta = V_i \angle 0 - \left(\widehat{R} + jL \right) \left[\frac{P_j - jQ_j}{V_j \angle -\delta} \right] \tag{18}$$

$$V_j^2 = V_i V_j \angle -\delta - \left(\widehat{R} + jL \right) \left(P_j - jQ_j \right) \tag{19}$$

$$V_j^2 = V_i V_j \cos \delta - j V_i V_j \sin \delta - \left(\widehat{R} + jL \right) \left(P_j - jQ_j \right) \tag{20}$$

$$V_j^2 + \left[P_j \widehat{R} + Q_j L + j \left(P_j L - Q_j \widehat{R} \right) \right] = V_i V_j \cos \delta - j V_i V_j \sin \delta \tag{21}$$

Divide the real and imaginary part as shown in Eq. (21).

$$V_j^2 + P_j\widehat{R} + Q_jL = V_iV_j\cos\delta, \tag{22}$$

$$P_jL - Q_j\widehat{R} = -V_iV_j\sin\delta \tag{23}$$

Consider $\delta \approx 0$

$$V_j^2 + P_j\widehat{R} + Q_jL = V_iV_j, \tag{24}$$

$$P_jL - Q_j\widehat{R} = 0 \tag{25}$$

$$L = \frac{Q_j\widehat{R}}{P_j} \tag{26}$$

Replace Eq. (26) in Eq. (22)

$$V_j^2 + P_j\widehat{R} + Q_j\frac{Q_j\widehat{R}}{P_j} = V_iV_j, \tag{27}$$

$$V_j^2 - V_jV_i + \left(\frac{Q_j^2}{P_j} + P_j\right)\widehat{R} = 0 \tag{28}$$

The novel stability index is obtained for the stable bus voltage that is given in Eqs. (29), (30) and (31).

$$V_i^2 - 4\left(\frac{Q_j^2}{P_j} + P_j\right)\widehat{R} \geq 0, \tag{29}$$

$$1 \geq \frac{4\widehat{R}}{V_i^2}\left(\frac{Q_j^2}{P_j} + P_j\right), \tag{30}$$

$$\text{PSI} = \frac{4\widehat{R}}{V_i^2}\left(\frac{Q_j^2}{V_i^2} + P_j\right) \leq 1 \tag{31}$$

For normal load condition, value of PSI has to be less than one. For stable system, PSI value is near to zero. The value of PSI is high, the system will become sensitive to instability. Thus, the greatest PSI value bus is selected as best bus for D-STATCOM location.

5 Modeling of D-STATCOM

For the distribution system load flow analysis, D-STATCOM static model is utilized. D-STATCOM provides reactive power support to the bus at which it is connected. Therefore, bus voltage is enhanced. This D-STATCOM also affects the neighboring buses voltages, and due to this, power loss will be minimized. V'_n indicates voltage of the candidate bus, and $V'_{\hat{m}}$ indicates voltage of the previous bus. $I'_{\hat{m}}$ indicates the current which is equal to sum of $I_{\hat{m}}$ and I_{DS}. current injected by the D-STATCOM indicated by I_{DS}.

Static model of the D-STATCOM has been utilized for the distribution system load flow analysis and the reactive power support is provided by D-STATCOM at the connected bus. The voltage profile of the bus is enhanced, where the device is connected, and voltage of other buses will also be improved due to the reactive power support and the losses reduction. Thus, the D-STATCOM will also affect the voltages of the neighboring buses. The V'_n indicates new voltages at the candidate bus, and $V'_{\hat{m}}$ indicates the new voltage at previous bus. The current $I'_{\hat{m}}$ is sum of $I_{\hat{m}}$ and I_{DS}. Here, IDS is the current injected by D-STATCOM and is in quadrature with the voltage.

New voltage after installing D-STATCOM can be calculated by using Eq. (32)

$$V'_n \angle \theta'_n = V'_{\hat{m}} \angle \theta'_{\hat{m}} - \left(\widehat{R}_{\hat{m}} + jL_{\hat{m}}\right)(I_{\hat{m}} \angle \delta) - \left(\widehat{R}_{\hat{m}} + jL_{\hat{m}}\right)\left(I_{DS} \angle \left(\frac{\pi}{2} + \theta'_n\right)\right) \quad (32)$$

The accomplishment of Eq. (33) is by splitting both the real and imaginary part of Eq. (32)

$$T = \frac{-B \pm \sqrt{D}}{2A} \quad (33)$$

$$\text{Where: } T = \sin \theta'_n \quad (34)$$

$$A = (k_1 k_3 - k_2 k_4)^2 + (k_1 k_4 + k_2 k_3)^2, \quad (35)$$

$$B = 2(k_1 k_3 - k_2 k_4) \cdot \left(V'_m\right)(k_4) \quad (36)$$

$$C = \left(V'_m \cdot \widehat{R}_{\hat{m}}\right)^2 - (k_1 k_4 + k_2 k_3)^2 \quad (37)$$

$$D = J^2 - 4HK \quad (38)$$

$$k_1 = \text{real}\left(V'_m \angle \theta'_{\hat{m}}\right) - \text{real}(Z_{\hat{m}} \cdot I_{\hat{m}} \angle \delta), \quad (39)$$

$$k_2 = \text{imag}\left(V'_m \angle \theta'_{\hat{m}}\right) - \text{imag}(Z_{\hat{m}} \cdot I_{\hat{m}} \angle \delta) \quad (40)$$

$$k_3 = -X_{\hat{m}}, \tag{41}$$

$$k_4 = -\widehat{R}_{\hat{m}} \tag{42}$$

Two roots exist for T. Eq. (43) explains the boundary factors that are selected for illustrating the appropriate value of root.

$$V'_n = V_n \Rightarrow I_{DS} = 0 \,\&\, \theta'_n = \theta_n \tag{43}$$

The results reveal the desired root of Eq. (33), $T = \frac{-J + \sqrt{E}}{2H}$. Therefore, the phase angle and magnitude of D-STATCOM current and reactive power support provided by D-STATCOM are shown in Eqs. (44) and (45), in which $*$ indicates the complex conjugate.

$$\angle I_{DS} = \frac{\pi}{2} + \theta'_n = \frac{\pi}{2} + \sin^{-1} T, \tag{44}$$

$$|I_{DS}| = \frac{V'_n \cos \theta'_n - k_1}{-k_4 \sin \theta'_n - k_3 \cos \theta'_n} \tag{45}$$

$$jQ_{DS} = \left(V'_n \angle \theta'_n\right) \cdot \left(I_{DS} \angle \left(\frac{\pi}{2} + \theta'_n\right)\right)^* \tag{46}$$

6 Proposed Algorithm: Dragonfly Optimization

This paper introduced a novel swarm intelligence-based Dragonfly algorithm. Dragonfly algorithm is inspired from the static or feeding and dynamic or migratory swarming behaviors of dragonflies in nature. Optimization has two phases those are exploration and exploitation. These phases are designed based on the social interaction of dragonflies in nature for different purposes like navigating, searching for foods and avoiding enemies. Dragonflies can swarm dynamically or statistically. Dragonflies divided in small groups and fly in different areas in a static swarm, which is the main consideration of the exploration phase [18]. In the dynamic swarm, dragonflies form bigger swarms and fly in only one direction, which is favorable in the exploitation phase.

Three principles are followed by the swarm's behavior. (1) Separation (2) Alignment (3) Cohesion.

The separation criteria are calculated by using Eq. (47), Eq. (48) describes the alignment, and cohesion is evaluated by Eq. (49)

$$S_i = -\sum_{j=1}^{N} M - M_j, \tag{47}$$

$$AL_i = \frac{\sum_{j=1}^{N} W_j}{N}, \tag{48}$$

$$CH_i = \frac{\sum_{j=1}^{N} M_j}{N} - M \tag{49}$$

where M_j indicates the jth position of neighboring individual, M denotes the present position of the individual, N indicates the count of neighboring individuals, and W_j denotes velocity of jth neighboring individual.

Attraction to food resource is computed by Eq. (50), in which M^+ indicates the food source position, and M denotes current individual position.

$$F_i = M^+ - M \tag{50}$$

Distraction away from an enemy is evaluated by Eq. (51), where M^- implies the position of the enemy, and M implies the position of the current individual.

$$E_i = M^- + M \tag{51}$$

Position vector (M) and step vector (ΔM) are measured by using Eqs. (52), (53) for updating the position of the dragonflies in exploration space and direction of dragonflies movements.

$$M_{t+1} = M_t + \Delta M_{t+1} \tag{52}$$

$$\Delta M_{t+1} = \left(pS_j + bAL_j + cCH_j + fF_j + eE_j\right) + w\Delta M_t \tag{53}$$

S_j indicates the separation of jth individual, p indicates the separation weight, b deplicates the alignment weight, AL implies the alignment of jth individual, CH signifies the jth individual cohesion, cohesion weight is indicated by c, F_j defines food resource of jth individual, food factor is indicated by f, e denotes enemy factor, w implies the inertia weight, E_j describes the position of enemy of jth individual, and t indicates the iteration counter.

For improving the stochastic behavior, exploration and randomness of the artificial dragonflies, they should fly around the exploration space by deploying an arbitrary walk, while there is a nonexistence of neighboring solutions. Under such situations, the dragonfly position is updated by Eq. (40), in which dimension of the position vectors is indicated by z, and the current iteration is indicated by "t."

$$M_{t+1} = M_t + \text{Levy}(z) \times M_t \tag{54}$$

The Levy flight is calculated by using Eq. (55), in which, β defines a constant, r_1 and r_2 indicates two arbitrary numbers which exist in between [0,1], and δ can be calculated by using Eq. (42), in which $\Gamma(x) = (x - 1)$.

$$\text{Levy}(x) = 0.01 \times \frac{r_1 \times \delta}{|r_2|^{\frac{1}{\beta}}}, \tag{55}$$

$$\delta = \left(\frac{\Gamma(1+\beta) \times \sin\left(\frac{\pi\beta}{2}\right)}{\Gamma\frac{(1+\beta)}{2} \times \beta \times 2^{\left(\frac{\beta-1}{2}\right)}} \right)^{\frac{1}{\beta}} \tag{56}$$

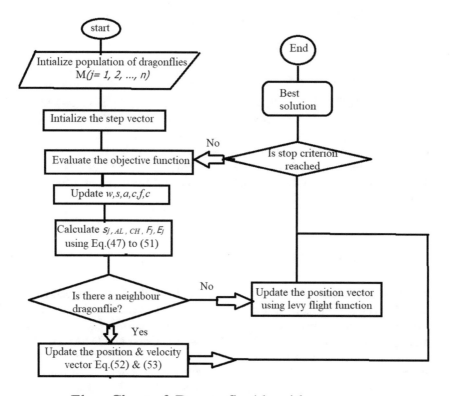

Flow Chart of Dragonfly Algorithm

Objective function:

Proposed algorithm objective function is defined by Eq. (57). Here, certain constraints are given like Voltage Stability Index, Demand meet, Cost of

Fig. 1 Solution encoding

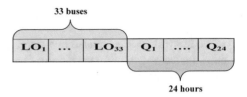

D-STATCOM, Cost of Energy loss (CEL) and Voltage Stability Margin (VSM). The corresponding demand should meet from the developed load model, and if not meet, then add penalty. Then, Voltage Stability Index should be in the variation 0.96–1.1. If not meet, then add penalty.

$$OB = min\left(\begin{array}{l} \text{Demand meet, VSI Meet} \\ \text{CEL, Cost of DSTATCOM, VSM} \end{array}\right) \qquad (57)$$

Solution representation:

To place the D-STATCOM in used bus system, corresponding location LO and size Q_0 should be known, and respective solution given as input for the proposed algorithm to get the optimal solution as shown in Fig. 1. Four experiments are done, in that, th first experiment requires single location and single size. Likewise the second experiment requires two locations and two sizes. The third experiment requires three locations and three sizes. The fourth experiment requires four locations and sizes.

7 Results and Discussions

A. **Simulation process**: The proposed DA model for optimal location and sizing of D-STATCOM in radial distribution system is implementing in MATLAB 2015a. The experiments as done by in IEEE 33 bus system. Four experiments were performed: (i) Using one D-STATCOM, (ii) Using two D-STATCOMs, (iii) Using three D-STATCOMs and (iv) Using four D-STATCOM. The developed design was compared with the other conventional algorithms like GA [13], PSO [14] in terms of convergence analysis and cost analysis.
B. **Optimal location and sizing of the proposed and conventional methods**:
Best solutions of proposed DA method and conventional GA and PSO methods are given in table in terms of location and sizing.

Hours	GA [13]	PSO [14]	DA
	18	33	33
1	7.8524	20	20
2	8.5263	17.625	20
3	9.5501	9.5971	20
4	13.222	19.846	20
5	3.2857	18.162	20
6	7.5585	20	20
7	15.097	19.097	20
8	16.025	19.955	20
9	16.318	0.86407	20
10	20	19.692	20
11	15.238	19.943	20
12	14.341	18.959	20
13	5.207	19.164	20
14	13.011	19.864	20
15	5.8079	3.474	20
16	14.466	18.794	20
17	17.356	19.813	20
18	19.544	19.738	20
19	0.46432	17.809	20
20	12.983	20	20
21	5.0578	4.8667	20
22	9.8658	20	20
23	13.834	0.65095	20
24	3.7727	19.163	20

C. Convergence analysis:

The convergence analysis of the proposed algorithm over other conventional methods is given in Fig. 2. The convergence model of experiment 1 was obtained from Fig. 2a, where at 20th iteration; the presented design cost function is 0.39% better than GA, 0.21% better than PSO algorithms. In addition, from Fig. 2b, at 100th iteration, the proposed model achieves its minimum cost function, which is 0.8% superior to GA, 0.53% superior to PSO algorithms. Further, from Figs. 2c, d, the implemented method for experiment 3 and experiment 4 achieves minimum cost function when distinguished with other conventional approaches. Thus, reduced cost function could be achieved by the proposed algorithm over other conventional methods as proved by the analysis.

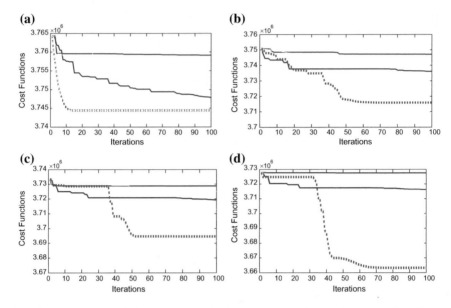

Fig. 2 Convergence analysis of the proposed scheme and conventional schemes for **a** Experiment 1, **b** Experiment 2, **c** Experiment 3, **d** Experiment 4. In Fig, Red Line indicates DA, Brown Line PSO, Blue Line indicates GA

D. **Cost analysis**:

The proposed method in terms of cost analysis when compared with other conventional methods for the entire four experiments is given in Fig. 3. The simulation on cost was done for all 33 bus systems for every five hours namely, 5, 10, 15, 20 and 24 h. From Fig. 3a, it is confirmed that for the entire stated hours, the presented scheme has accomplished minimum cost over other traditional schemes, and the cost of implemented technique is in the interval of 9300 for the entire corresponding hours. The investigation on cost for experiment 2 reveals that the suggested model obtains minimum cost; however, in certain hours (between 5 and 10), it has accomplished slight variation in cost when compared with other techniques. Similarly, the cost analysis of the remaining experiments is revealed in Fig. 3b, c and d.

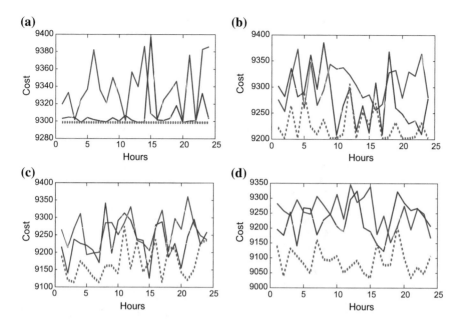

Fig. 3 Cost analysis of proposed model over other conventional methods **a** Experiment 1, **b** Experiment 2, **c** Experiment 3, **d** Experiment 4. In Fig, Red Line indicates DA, Brown Line PSO, Blue Line indicates GA

8 Conclusion

In this paper, Dragonfly algorithm is implemented to solve the location and sizing problems of the D-STATCOM in radial distribution system. To solve both the localizing and sizing crisis, solutions were encoded with two bound factors. Thus, by exploiting the DA algorithm, placement of D-STATCOM could be predicted in a precise manner. Performance of the system is tested by using simulation in terms of convergence and cost analysis. Results are compared with conventional GA, PSO algorithms. From the comparison, proposed DA algorithm minimizing the cost function effectively than the conventional methods. Thus, proposed DA algorithm minimizes the power losses, and also the system performance has been improved.

References

1. Bagherinasab, A., Zadehbagheri, M.: Optimal placement of D-STATCOM using hybrid genetic and ant colony algorithm to losses reduction. Int. J. Appl. Power Eng. **2**(2) (2013). ISSN: 2252-8792
2. Farthoodnea, M., Mohamed, A.: Optimum D-STATCOM placement using firefly algorithm for power quality enhancement. PEOCON-2013. 978-1-4673-5074-7/13$

3. Agrawal, R., Bharadwaj, S.K., Kothari, D.P.: Population based evolutionary optimization techniques for optimal allocation and sizing of thyristor controlled series capacitor. J. Electr. Syst. Inf. Technol. (2018)
4. Girón, C., Rodríguez, F.J., de Urtasum, L.G., Borroy, S.: Assessing the contribution of automation to the electric distribution network reliability. Int. J. Electr. Power Energy Syst. **97**, 120–126 (2018)
5. Zhai, H.F., Yang, M., Chen, B., Kang, N.: Dynamic reconfiguration of three-phase unbalanced distribution networks. Int. J. Electr. Power Energy Syst. **99**, 1–10 (2018)
6. He, Jia, Yang, Hai, Tang, Tie-Qiao, Huang, Hai-Jun: An optimal charging station location model with the consideration of electric vehicle's driving range. Transp. Res. Part C: Emerg. Technol. **86**, 641–654 (2018)
7. Hooshmand, R.-A., Morshed, M.J., Parastegari, M.: Congestion management by determining optimal location of series FACTS devices using hybrid bacterial foraging and Nelder–Mead algorithm. Appl. Soft Comput. **28**, 57–68 (2015)
8. Elmitwally, A., Eladl, A.: Planning of multi-type FACTS devices in restructured power systems with wind generation. Int. J. Electr. Power Energy Syst. **77**, 33–42 (2016)
9. Bhattacharyya, B., Kumar, S.: Loadability enhancement with FACTS devices using gravitational search algorithm. Int. J. Electr. Power Energy Syst. **78**, 470–479 (2016)
10. Kavitha, K., Neela, R.: Optimal allocation of multi-type FACTS devices and its effect in enhancing system security using BBO, WIPSO & PSO. J. Electr. Syst. Inf. Technol. (2017)
11. Ravi, K., Rajaram, M.: Optimal location of FACTS devices using improved particle swarm optimization. Int. J. Electr. Power Energy Syst. **49**, 333–338 (2013)
12. Sirjani, R., Jordehi, A.R.: Optimal placement and sizing of distribution static compensator in electric distribution networks: a review. Renew. Sustain. Energy Rev. **77**, 688–694 (2017)
13. Khunkitti, S., Siritaratiwat, A.: A hybrid DA-PSO optimization algorithm for multiobjective optimal power flow problems. Energies **11**, 2270 (2018). https://doi.org/10.3390/en11092270, www.mdpi.com/journal/energies
14. Lakshmi Devi, M., Reddy, D.: Optimal unified power quality conditioner allocation in distribution systems for loss minimization using Gray wolf optimization. Int. J. Eng. Res. Appl. **7**(11, Part-3), 48–53. ISSN: 2248-9622
15. Gupta, A.R., Kumar, A.: Impact of various load models on D-STATCOM allocation in DNO operated distribution network. Procedia Comput. Sci. **125**, 862–870 (2018)
16. Inkollu, S.R., Kota, V.R.: Optimal setting of FACTS devices for voltage stability improvement using PSO adaptive GSA hybrid algorithm. Eng. Sci. Technol. Int. J. **19**(3), 1166–1176 (2016)
17. Gupta, A.R., Kumar, A.: Optimal placement of D-STATCOM using sensitivity approaches in mesh distribution system with time variant load models under load growth. Ain Shams Eng. J. (2016)
18. Mirjalili, S.: Dragonfly algorithm: a new meta-heuristic optimization technique for solving single-objective, discrete, and multi-objective problems. Nat. Comput. Appl. (2015)

SRF Control Algorithm for Five-Level Cascaded H-Bridge D-STATCOM in Single-Phase Distribution System

Adepu Sateesh Kumar and K. Prakash

Abstract This paper presents a novel method to control a single-phase five-level cascaded H-bridges-based D-STATCOM in distribution system using SRF theory (synchronous frame theory) and inverse park transformation to enhance power quality. Reference current extraction and D-STATCOM gating pulses generation can be done by using SRF theory. To generate quadrature component of current, inverse park transformation is used in proposed method. In the proposed method, reactive power compensation, harmonic elimination, power factor improvement and total harmonic distortion (%THD) are considered as the main objective functions. The proposed technique is investigated through MATLAB. The obtained results are compared with conventional SPWM (Sinusoidal pulse width modulation) control technique.

Keywords D-STATCOM · Power quality · SRF theory · THD · SPWM · Cascaded H-bridge · Distribution system

1 Introduction

In electric power systems, electric power is generated at generation station, and that generated power is transmitted to the distribution systems through the high-voltage transmission lines. From the distribution system, finally power is supplied to customers [1]. In recent days, power demand is increased more due to increasing of customer usages. To meet the power demand, power system size and complexity are grown. Power system must provide uninterrupted and reliable power supply to the customers, especially distribution systems. Generally distribution system has more number of nonlinear loads, which affect the quality of power [2]. For

A. S. Kumar · K. Prakash (✉)
Department of EEE, Vaagdevi College of Engineering, Warangal, Telangana, India
e-mail: prakashkam@ieee.org

A. S. Kumar
e-mail: adepu.sateesh@gmail.com

© Springer Nature Singapore Pte Ltd. 2020
S. S. Dash et al. (eds.), *Artificial Intelligence and Evolutionary Computations in Engineering Systems*, Advances in Intelligent Systems and Computing 1056, https://doi.org/10.1007/978-981-15-0199-9_20

customer satisfaction and electrical equipments, safety power quality becomes more important constraint.

Different power quality problems are existed in distribution systems like unbalanced voltages, voltage sag and swell, voltage fluctuations, harmonics, dc offsets, noise, notches, etc. To compensate power quality problems, different conventional devices are used previously like, series voltage regulators and shunt capacitors are used for voltage profile maintenance in the power system [3]. Similarly, passive filters are used for reactive power compensation and harmonic elimination. But these conventional devices having some disadvantages, series generator cannot generate reactive power, it operates in step by step manner, and due to this reason, it has slow response. Shunt capacitors cannot provide continuous reactive power supply to the system, and it is having oscillatory behavior with inductive loads. Passive filters having high cost, resonance and fixed compensation. To compensate these problems, FACTS/custom power devices play main role in distribution system.

Custom power devices are more effective devices for voltage profile improvement, power loss minimization, harmonic compensation, reactive power compensation, etc. Different types of custom power devices are DVR, D-STATCOM and UPQC. Among all these devices, D-STATCOM is the best device for reactive power and harmonic compensation.

2 Literature Review

In [4], adaptive Neuro-fuzzy inference system (ANFIS) is developed to control the D-STATCOM, and the main objective functions are reduction of %THD, reactive power and −ve sequence component of currents induced by utilization of nonlinear loads and unbalanced loads.

In [5], predictive-fuzzy logic hybrid controller design is developed for D-STATCOM in distribution system. To generate the gate pulses of the voltage source inverter (VSC), indirect current control system with synchronous reference frame theory is used. Compensation of reactive power and harmonic elimination are the objective functions.

In [6], fuzzy controller is used to reduce the disturbances and %THD in the system. In [7], SPWM controller implemented for multilevel cascade H-bridge-based D-STATCOM, and main objective functions are compensation of reactive power and harmonics in the system.

A novel adaptive passivity-based control is designed for reference current tracking. This nonlinear passivity-based control is implemented for cascaded multilevel converter-based D-STATCOM. Medium-voltage reactive power can be compensated by using this method [8]. In [9], active power filters are used to minimize the harmonics in the system.

Over the conventional two-level voltage source inverter, multilevel inverter has the following advantages:

1. It generates the output voltages with very low distortion.
2. It reduces the dv/dt stresses.
3. Reduce the harmonic distortion to greater extent.
4. It draws input current with low distortion.
5. Multilevel inverter can operate at both fundamental and high switching frequency PWM.
6. It has less switching losses and higher efficiency.

Because of all the above advantages, multilevel inverter is used in this proposed work.

In this article, Sect. 3 explains the design of cascaded H-bridge-based D-STATCOM, Sect. 4 describes proposed configuration and control algorithm, and Sect. 6 implies the simulation results and discussion. Sect. 7 concludes the paper.

3 Design of CHB-Based D-STATCOM

The distribution static compensator (D-STATCOM) is a voltage source inverter-based device. It is used for mitigation of voltage sags, reactive power compensation and power loss reduction. It is connected in shunt with the distribution network through a coupling transformer. The D-STATCOM produces variable inductive or capacitive shunt compensation up to its maximum MVA rating. The D-STATCOM, with respect to reference ac signal, continuously checks the line waveform; this device can give lagging or leading reactive current compensation to decrease the voltage changes. It consists of an inverter module (Cascaded), dc capacitor, a transformer to match the line voltage and inverter output and a PWM control strategy [10].

In high-power medium-voltage (MV) applications, cascaded H-bridge (CHB) multilevel inverter is used most popularly. Generally, cascaded H-bridge consists of units of single-phase H-bridge cells. To obtain medium-voltage and low harmonic distortion operation, H-bridge cells are normally connected in cascade manner. The inverter requires a number of isolated dc supply to feed each H-bridge power cell. The number of voltage levels in a CHB inverter can be calculated by using $n = (2H + 1)$.

Where n indicates voltage level, and it is always an odd number for the CHB inverter, but in other multilevel topologies such as diode-clamped inverters, voltage level value can be either an even or odd number. H indicates the number of H-bridge cells per phase leg.

The voltage source converter (VSC) changes the dc voltage across the capacitor into a set of phase output voltages. These voltages are in phase and coupled with the ac system through coupling transformer. Better adjustment of the magnitude and phase of the D-STATCOM output voltages accepts good control over active and reactive power exchanges between the ac system and D-STATCOM.

The voltage source converter connected in shunt to the ac system for the following usages:

1. Voltage regulation and compensation of reactive power;
2. Elimination of current harmonics;
3. Correction of power factor.

The controller operates the inverter in such a way that phase angle between the line voltage (VL) and the inverter voltage (V_i) is dynamically adjusted so that the D-STATCOM absorbs or generates the required VAR at the point of common coupling.

If V_i = VL, D-STATCOM does not generate or absorb reactive power.
If VL > V_i, D-STATCOM absorbs inductive reactive power.
When V_i > VL, D-STATCOM generates capacitive reactive power.

4 Proposed Configuration and Control Algorithm

In cascaded H-bridge multilevel inverter, multiple units of H-bridge cells are connected in series to produce high ac voltages. A typical five-level CHB inverter configuration is shown in Fig. 2, where two H-bridge cells powered by two isolated dc supplies of equal voltage E are placed in each phase. Multipulse diode rectifiers normally provide the dc supplies. Different level of phase voltages can be determined, by switching the switches in proper conduction. Some of voltage levels can be determined by more than one switching state [11]. The switching action repetition is a general phenomenon in multilevel converters. It gives a good flexibility for switching pattern design.

The CHB-based D-STATCOM is shown in Fig. 1. Two H-bridge voltage source inverters are used to generate five-level output across the inverter as shown in Fig. 2. Every single H-bridge is formed with two-leg voltage source converter (VSC), and VSC consists of four IGBT switches. D-STATCOM is connected in shunt with system through an interfacing inductor L at the point of common

Fig. 1 Line diagram of the proposed system

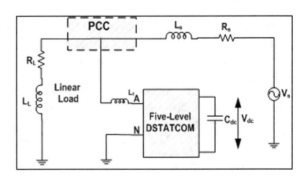

Fig. 2 Cascaded H-bridge
multilevel inverter

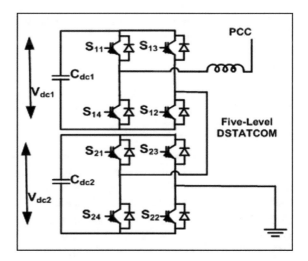

Fig. 3 Inverse park
transformation

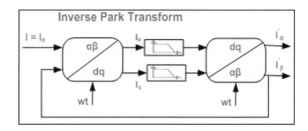

coupling. Proposed controller of D-STATCOM eliminates the harmonics in source
current for maintaining the total harmonic distortion (THD%). Implemented controller of D-STATCOM can also do power factor correction, reactive power
compensation. The performance of the controller is tested under different nonlinear
conditions. Required active and reactive power can be injected for D-STATCOM
operation by changing the magnitude and phase of the system.

A. *Inverse Park Transformation*

The model of inverse park transformation is shown in Fig. 3. Two loops are formed
in this method, and both the loops are nonlinear and interdependent. To eliminate
the algebraic loops in the system, first-order low-pass filters are used for each d and
q signal. Park transformation is done (i.e., $\alpha\beta/dq$'s). For the operation of inverse
park transformation, outputs of the park transformation are given as inputs as
shown in Fig. 3.

$$\begin{bmatrix} V_\alpha \\ V_\beta \end{bmatrix} = \begin{bmatrix} \cos\hat{\theta} & \sin\hat{\theta} \\ -\sin\hat{\theta} & \cos\hat{\theta} \end{bmatrix} \begin{bmatrix} V_d \\ V_q \end{bmatrix} \tag{1}$$

After the transformation, low-pass filters are used to filter out any noises or harmonics exist in V_d, V_q.

$$\text{LPF}(s) = \frac{w_c}{s + w_c} \tag{2}$$

where LPF is low-pass filter, w_c is cut-off frequency.

B. *Reference Current Generation:*

In Fig. 4, maximum amplitude of active component of current is calculated. To generate quadrature signals ($I_{L\alpha}$ and $I_{L\beta}$), the load current is observed and supplied to inverse park transformation, and it is transformed back to I_d and given to a low-pass filter. The reference active component of current ($I_{LP} + I_{CD}$) is produced by sum of the output of the low-pass filter and the output generated by the DC voltage control loop. Two DC capacitors voltages (V_{dc}) which are measured at each DC capacitor are added and compared with the DC bus reference voltage (V_{dc}^*). The error of the signal is calculated by using Eq. (3)

$$V_d(m) = V_{dc}^*(m) - V_{dc}(m) \tag{3}$$

where $V_d(m)$ indicates error of the voltage. To regulate the DC bus voltage of D-STATCOM, voltage error is given to proportional-integral controller. The output of the PI controller at mth sampling instant is as follows:

$$I_{cd}(m) = I_{cd}(m-1) + k_p\{V_{dce}(m) - V_{dcer}(m-1)\} + k_i V_{dce}(m) \tag{4}$$

where K_p denotes proportional gain, and K_i indicates integral gains. $V_{dce}(m)$ and $V_{dce}(m-1)$ indicate the DC bus voltage errors in mth and ($m-1$)th instant, and the amplitudes of active component of currents are indicated by $I_{cd}(m)$ and $I_{cd}(m-1)$ at the fundamental reference current.

Output of the PI controller (I_{cd}) and average magnitude of current (I_{LP}) is added and that added value is changed in the form of reference source current (I_α) from

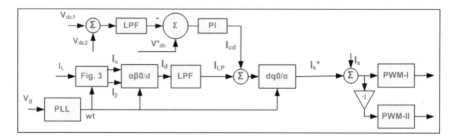

Fig. 4 Block diagram of reference source current generation

dq component. Reference source current is compared with the actual source current, and the resultant becomes error signal. This error signal is fed to the PWM controller-I for the first CHB. Negation is provided to the second CHB inverter to generate the gating pulses for both the inverters.

5   Simulink Model

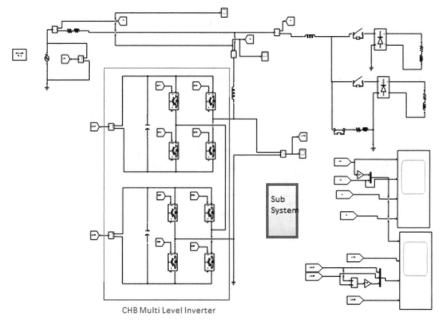

CHB Multi Level Inverter

Simulink Model of Proposed System

6   Simulation Results

Proposed model is designed in MATLAB/Simulink. Experiment is done on single-phase distribution system under nonlinear conditions. Figure 5 shows the performance of the system, Fig. 6 shows the output voltage of the device. Figures 7 and 8 show the current waveform and %THD. THD (%) of proposed method is compared with the conventional SPWM (sinusoidal pulse width modulation) method (Figs. 9 and 10).

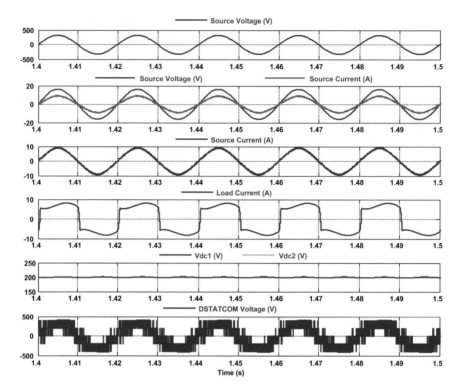

Fig. 5 Presentation of CHB when nonlinear load

Fig. 6 Device output voltage

7 Conclusion

This paper has implemented a new method, which uses SRF theory and inverse park transformation to control the single-phase, five-level cascaded H-bridge-based D-STATCOM in distribution system. The proposed system has operated under nonlinear load conditions. Reactive current, harmonic compensation and power factor correction are effectively done by using this new method. The proposed controller has compared its performance with conventional SPWM controller in terms of %THDs. From the comparison, proposed method has better performance than the conventional method.

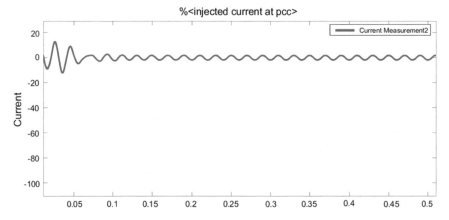

Fig. 7 Current waveform of proposed controller at PCC

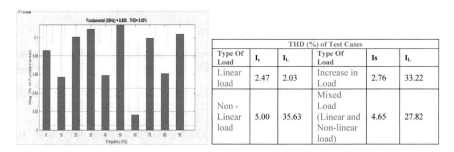

Fig. 8 Current harmonics at PCC with 5.00% THD

THD (%) of Test Cases					
Type Of Load	I_s	I_L	Type Of Load	Is	I_L
Linear load	2.47	2.03	Increase in Load	2.76	33.22
Non - Linear load	5.00	35.63	Mixed Load (Linear and Non-linear load)	4.65	27.82

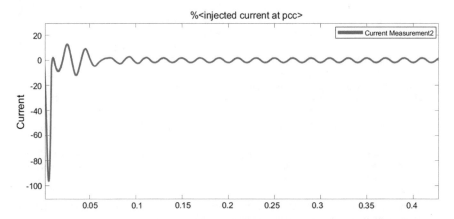

Fig. 9 Current waveform of SPWM controller

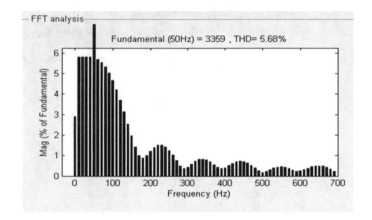

Fig. 10 Current harmonics with 5.68% THD for SPWM controller

References

1. Suryajitt, B., Sudhakar, G.: Power quality improvement using cascaded H-bridge multilevel inverter based DSTATCOM. J. Eng. Res. Appl. **4**(11, Version 3), 31–37 (2014). www.ijera. com, ISSN: 2248-9622
2. Swetha, R., Hassainar, S.: Voltage compensation for a 11 kV distribution system using multilevel inverter. Int. J. Eng. Res. Technol. (IJERT) **4**(04) (2015). ISSN: 2278-0181
3. Ram, A., Kumar, A.: Optimal placement of D-STATCOM using sensitivity approaches in mesh distribution system with time variant load models under load growth. Ain Shams Eng. J. (2016)
4. Kumar, A.S., Prakash, K.: Harmonic mitigation & neutral line currents reduction of distribution systems using adaptive neuro-fuzzy inference system based D-STATCOM. J. Eng. Appl. Sci., 5517–5522 (2018). ISSN: 1816-949X
5. Yanmaz K., Altas I.H., Mengi, O.O.: A novel model predictive-fuzzy logic hybrid controller design for D-STATCOM in wind energy distribution systems. Global NEST J. **20**, Copyright© 2018 Global NEST Printed in Greece
6. Dhanaya Sri, C., Prabakara Rao, K.: Cascaded fuzzy controller based multilevel STATCOM for high-power applications. Int. Res. J. Eng. Technol. (IRJET). e-ISSN: 2395-0056
7. Pawar, R., Borse, D.G.: Simulation of cascaded H-bridge multilevel inverter based D-STATCOM for power quality improvement with switching modulation index. In: 7th IRF International Conference, 27 April 2014, Pune, India. ISBN: 978-93-84209-09-4
8. Chen, Y., Wen, M.: Passivity-based control of cascaded multilevel converter based D-STATCOM integrated with distribution transformer. Electr. Power Syst. Res. **154**, 1–12 (2018)
9. Kumar, A.S., Prakash, K.: Analysis of split capacitor based DSTATCOM with LCL filter using instantaneous symmetrical components theory. In: 2nd IEEE International Conference (AEEICB-2016). With an ISBN Number: 978-1-4673-9745-2 ©2016 IEEE explore (Ref. no. 37745)
10. Kayasth, J., Hasabe, R.P.: Cascaded H-bridge multilevel inverter based DSTATCOM for power quality improvement. IJMTER (2017). ISSN: 2249-9745
11. Kumar, A.S., Prakash, K., Sharma, S.: A five-level cascaded DSTATCOM for single-phase distribution system. In: National Conference & Journal on Challenges and Issues in Operation of Competitive Electricity Markets (CIOCEM-2016) CPRI, Bangalore 2016

A Robust Multi-unit Feature-Level Fusion Framework for Iris Biometrics

A. Alice Nithya, M. Ferni Ukrit and J. S. Femilda Josephin

Abstract In this paper, an iris biometric framework based on feature-level multi-unit fusion, using both left and right iris images, is proposed. The fusion helps by taking advantage of the two iris units of an individual. For both the units, segmentation is performed using an improved iris segmentation methodology (ISM) precisely localizes iris region of interest (ROI) and processes noise factors like occlusions, specular reflections, off-axis and blurring in reduced search space with low computational time, from the input eye image. Feature extraction is performed using first-order and second-order statistical measures that accurately characterize the unique textural patterns from the localized iris ROI without converting from the polar to the cartesian space. The statistical features obtained are then fused using sum method and mean method of fusion. Back-propagation neural networks are used for classification. Experimental results tested on iris images acquired datasets like CASIA V3-Interval, CASIA V3-Lamp, MMU V1 and MMU V2 iris datasets show significant performance improvement in mean rule-based multi-unit feature-level fusion system when compared to single modal systems and sum rule-based multi-unit feature-level fusion system.

Keywords Iris recognition · Feature-Level fusion · Segmentation · First-order statistical measures · Second-order statistical measures · Back-Propagation neural network

1 Introduction

Biometric system relying on the evidence of a single source of information for recognition is called a single modal system [1]. These systems have to compete with several problems like noisy sensor data, mislocalization, inter-class similarities and intra-class variations. Hence, a single modal biometric system is not anticipated

A. Alice Nithya (✉) · M. Ferni Ukrit · J. S. Femilda Josephin
School of Computing, SRMIST, Chennai, India
e-mail: a.alicenithya@gmail.com

© Springer Nature Singapore Pte Ltd. 2020
S. S. Dash et al. (eds.), *Artificial Intelligence and Evolutionary Computations in Engineering Systems*, Advances in Intelligent Systems and Computing 1056, https://doi.org/10.1007/978-981-15-0199-9_21

239

to meet necessities like availability; cost; accuracy; etc. [2]. A single modal biometric system is not adequate to meet the diversified necessities imposed by quite a lot of large-scale recognition systems [3].

Some general challenges which affect the performance of a single modal biometric system are as follows [4]:

1. Circumvention is very easy in single biometric system. Iris images can be spoofed by showing fake synthetic iris structure.
2. Absence of biometric data for a set of population. For example, if there is a pathological problem in eye, it will be difficult to acquire iris images.
3. Acquired biometric sample may be affected by noise or vary over a period of time.

A potential solution to the aforementioned problems is found by experts in the fusion of the biometric samples generally referred to multimodal biometric systems [5]. Multimodal biometric systems utilize more than a single source of evidence for recognition. A fusion algorithm is classified into the following types based on the nature of evidence available [2]:

a. multi-sensor,
b. multi-unit,
c. multi-instances,
d. multimodal, and
e. multi-algorithm.

A multi-sensor model uses two or more sensors to acquire biometric modality from an individual. In a multi-unit model, multiple units (left unit and right unit) of the same modality are used for performing recognition task. Several images or instances of the same biometric modality are acquired using a single sensor and fused in a multi-instances model.

A multimodal model is a combination of more than one biometric modality acquired from an individual. For a single biometric modality, if more than one feature extraction algorithms or classifiers are combined, then it is called multi-algorithm model. In all the above-mentioned models, a multi-unit fusion model could improve the recognition performance exclusive of any additional software or hardware cost.

Depending upon the level of fusion, these approaches are further grouped into following types [5]:

a. Sensor-Level Fusion (SLF),
b. Feature-Level Fusion (FLF),
c. Match Score-Level Fusion (MSLF) and
d. Decision-Level Fusion (DLF).

Raw biometric modality acquired from several sensors (multi-sensor) or several instances of the same biometric modality acquired using a single sensor (multi-instance) is fused in a SLF model.

In MSLF, multiple biometric classifier algorithms are used to find match scores, and these match scores are then combined to generate a consolidated score. Similarly, in DLF, final decisions are fused to build up a multimodal model.

Features extracted for a single unit using several feature extraction algorithms (multi-algorithm) or features extracted from left and right units of same individual (multi-unit) are fused in a FLF model. Thus, FLF is characterized by the information richness present in it with ease of implementation.

1.1 Fusion in Iris Biometrics

Iris recognition is an important application in the field of biometric science. Iris recognition framework using single unit or instance may have challenges like noisy sensor data, occlusion caused by eyelids and eyelashes, mislocalization, spoof attacks and effect of cataract (eye disease).

Figure 1 shows examples where the use of only single iris unit may cause incorrect classification results, whereas the use of both the iris units may provide correct and improved classification results.

In the first row, left unit's upper portion of iris region of interest (ROI) is missing due to the presence of eyelashes occlusion. Similarly, in the second row, upper

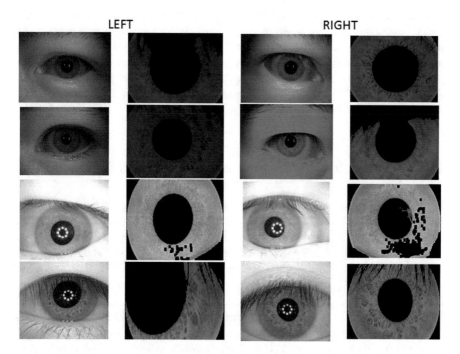

Fig. 1 Sample instances where single unit fails, but both the units can solve the problem

portion of right unit's iris ROI is missing due to the presence of eyelid occlusion. Due to the presence of noisy sensor data like specular reflections and non-uniform illumination, mislocalization occurs, and lower portion of iris ROI is also missing in the third row's right unit. In the fourth row, the left unit has pupil dilation effect and occlusions resulting in mislocalization of iris ROI.

The above-mentioned limitations are taken care by fusing left and right units of iris, which is of use when a single unit has noisy or unavailable data. By considering the advantages of the multi-unit fusion at biometric trait level and feature-level fusion at recognition framework level, this work contributions have been made to develop a feature-level multi-unit fusion framework for iris recognition.

The proposed paper is organized as follows. Initially, experimental methodology is discussed in Sect. 2. Section 3 focuses on describing the results obtained using the method described in Sect. 2 followed by a detailed discussion. Section 4 concludes the study.

2 Experimental Methodology

In this paper, an iris biometric system based on feature-level fusion (FLF) approaches is developed due to its suitability for improving accuracy and reducing computational cost. The fusion helps by taking advantage of both the units and compensates for the noisy or absence of data in a single unit system. An integration of the data at the feature level with multi-unit model is chosen in this work. A multi-unit model is chosen over other models, as the mathematical computations and number of sensors required to acquire the units are reduced. Similarly, feature-level fusion is characterized by the richness of information present in it with an ease of implementation.

2.1 Segmentation and Feature Extraction of Iris Units

In this proposed feature-level multi-unit iris biometric system, initially iris region of interest (ROI) is localized using the improved segmentation methodology proposed in [6] for both the left and right units of same individual. This methodology concentrated on taking advantage of both circular and spatial properties of the acquired images to precisely localize the iris region of interest (ROI). The iris-sclera boundary is identified using the shape property of iris using circular Hough transform [7] followed by identifying noises and pupil region inside the iris-sclera boundary region using their spatial properties [8]. This methodology handled noises like occlusions, specular reflections, half-closed eye images, off-axis images, non-uniform illuminations and pupil dilation effects efficiently. The search space involved in processing noises was also reduced by considering the spatial property of the noises. This methodology improved segmentation accuracy and minimized the computational cost also.

As a second step of the proposed iris biometric system, texture based feature extraction proposed in [9] was used in this work. As segmented iris ROI is characterized by the recurring occurrence of distinct textural patterns, texture-based feature extraction methods were used to extract the feature descriptors from both the iris units. Two prevailing sets of statistical measures-based feature descriptors are combined to perform iris recognition task in this paper. They are first-order statistical measures namely mean, standard deviation, skewness, kurtosis, entropy and coarseness index and second-order statistical measures namely autocorrelation, contrast, correlation, cluster prominence, cluster shade, dissimilarity, energy, entropy, homogeneity, max probability, sum of squares: variance, sum average, sum variance, sum entropy, difference variance, difference entropy, information measure of correlation 1, information measure of correlation 2, inverse difference moment normalized and chi-square value.

Feature descriptors calculated using first-order statistical measures and second-order statistical measures reveal textural information, smoothness and coarseness with lessened feature vector (FV) size illustrating better discrimination capability when given as input to a classifier [10, 11]. In this method, relative distance and relative orientation were altered in this method while building GLCM from segmented iris images to achieve scale and rotation invariance. Feature vectors were extracted when the relative distance is 2 for all possible orientations, and the (FV) are stored as a file.

Let $X_L = \{X_L^1, X_L^2, \ldots, X_L^{26}\}$ and $X_R = \{X_R^1, X_R^2, \ldots, X_R^{26}\}$ denote the FV extracted from the left iris unit and the right iris unit of an individual.

2.2 Multi-unit Feature-Level Fusion Framework

In this proposed feature-level multi-unit iris biometric system, initially iris region of interest feature vectors (FV) X_L and X_R computed using the above-mentioned method were combined using the mean method of fusion. The arithmetic mean or mean or average value is used to indicate the central tendency or a sum value of a number set divided by the total number of elements in the set. This is explained using Eq. 1.

$$X_M = \frac{1}{2} \sum_{i=1}^{n} X_i \tag{1}$$

In this case, for the the left unit and right unit feature descriptors obtained when $n = 2$, mean is calculated as follows $(X_L + X_R)/2$, for all the twenty six feature descriptors and the new FV is stored in a file.

Back-propagation neural network (BPNN) explained in [12] with 15 hidden neurons in the hidden layer was used for recognition task. When a test or probe

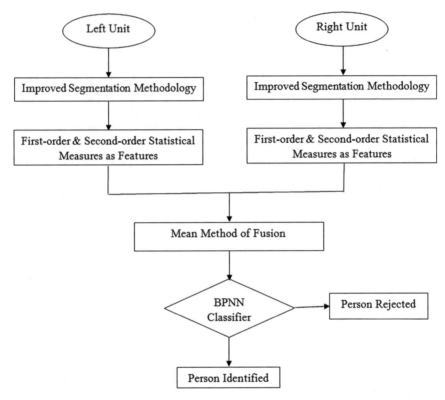

Fig. 2 Block diagram of the proposed multi-unit feature-level fusion-based iris recognition framework

image is given, this feature-level multi-unit fusion model will identify or reject the person accordingly.

Figure 2 shows the block diagram of the proposed feature-level multi-unit fusion model of iris recognition framework.

3 Experimental Results and Discussion

In this paper, a mean rule-based canonical form of multi-unit feature-level fusion technique was proposed, and the results were evaluated on publicly available iris datasets having both left and right iris sample of same individual like CASIA V3-Interval [13], CASIA V3-Lamp [13], MMU V1 [14] and MMU V2 [14]. The mean of the feature-level outcomes, i.e., the reconstructed feature descriptors are identified to improve recognition accuracy over other rule-based methods like sum rule.

Table 1 Comparison of genuine acceptance rate (GAR) for the left unit, right unit, sum-rule fusion unit and mean-rule fusion unit for iris datasets

Dataset used	Genuine acceptance rate (GAR)			
	Left unit (%)	Right unit (%)	Sum-rule fusion (%)	Mean-rule fusion (%)
CASIA V3-Interval [13]	97.28	95.2	98.8	99.125
CASIA V3-Lamp [13]	96.47	96.75	97.45	98.623
MMU V1 [14]	96.09	96.35	97.12	97.953
MMU V2 [14]	94.1	93.21	95.67	96.012

Genuine acceptance rate (GAR) was calculated for the left and the right iris units using the proposed framework. Table 1 gives a comparison of recognition accuracy for left unit, right unit and sum rule-based and mean fusion models for CASIA V3-Interval [13], CASIA V3-Lamp [13], MMU V1 [14] and MMU V2 [14] iris datasets.

From the results, the following observations were made:

i. A feature-level multi-unit fusion model helps in improving the recognition accuracy when compared to a single unit model.
ii. The proposed mean rule-based fusion technique outperforms sum-rule fusion technique with improvement in recognition accuracy as mean-rule model indicates the central tendency of the data.

4 Conclusion

In this paper, an iris biometric framework based on feature-level multi-unit fusion, using both left and right iris images, was proposed. The fusion helps by taking advantage of the two units. For both the units, segmentation is performed using an improved iris segmentation methodology (ISM) followed by feature extraction using texture-based feature extraction method to improve recognition performance. The statistical features obtained are then fused using sum method and mean method of fusion. Back-propagation neural networks are used for classification. Experimental results tested on iris images acquired datasets like for CASIA V3-Interval, CASIA V3-Lamp, MMU V1 and MMU V2 datasets show significant performance improvement in mean rule-based multi-unit feature-level fusion system when compared to uni-modal systems and sum rule-based multi-unit feature-level fusion system.

References

1. Obaidat, M.S., Rana, S.P., Maitra, T., Giri, D., Dutta, S.: Biometric security and internet of things (IoT). Biometric-based physical and cybersecurity systems, pp. 477–509. Springer, Cham (2019)
2. Singh, M., Singh, R., Ross, A.: A comprehensive overview of biometric fusion. Inf Fusion (2019)
3. Elhoseny, M., et al.: Multimodal biometric personal identification and verification. Advances in soft computing and machine learning in image processing, pp. 249–276. Springer, Cham (2018)
4. Dharavath, K., Talukdar, F.A., Laskar, R.H.: Study on biometric authentication systems, challenges and future trends: a review. In: 2013 IEEE international conference on computational intelligence and computing research. IEEE 2013
5. Mehrotra, H.: On the performance improvement of Iris biometric system. In: Ph.D. dissertation (2014)
6. Nithya, A., Lakshmi, C.: Towards enhancing non-cooperative iris recognition using improved segmentation methodology for noisy images. J. Artif. Intell. **10**, 76–84 (2017)
7. Liu, M., Zhou, Z., Shang, P., Xu, D.: Fuzzified image enhancement for deep learning in Iris recognition. IEEE Trans. Fuzzy Syst. (2019)
8. Han, Y.L., Min, T.H., Park, R.H.: Efficient iris localisation using a guided filter. IET Image Process. **9**(5), 405–412 (2014)
9. Nithya, A.A., Lakshmi, C.: On the performance improvement of non-cooperative iris biometrics using segmentation and feature selection techniques. Int. J. Biometrics **11**(1), 1–21 (2019)
10. Rasti, P., Morteza, D., Gholamreza, A.: Statistical approach based iris recognition using local binary pattern. DYNA-Ingeniería e Ind. **92**(1) (2017)
11. Hu, Y., Sirlantzis, K., Howells, G.: Optimal generation of iris codes for iris recognition. IEEE Trans. Inf. Forensics Secur. **12**(1), 157–171 (2017)
12. Nithya, A.A., Lakshmi, C.: Enhancing iris recognition framework using feature selection and BPNN. Cluster Comput, 1–10 (2018)
13. CASIA Iris Image Database. Available http://biometrics.idealteast.org/
14. Multimedia University (MMU) Iris Image Database. Available http://www.cs.princeton.edu/ ~andyz/irisrecognition

Providing Data Security in Deep Learning by Using Genomic Procedure

R. Vasanth and Dinesh Jackson Samuel

Abstract Cryptanalysis is the technique for putting a difficult arithmetic and rationality to give a heavy-duty policy procedure to pelt information named by way of enciphering and then also reclaim novel information again named as deciphering. The main determination of cryptanalysis is to communicate information among one hub to another and also avoid the listener problems, and mainly we want a strong procedures and strategy and also need good encipherment techniques. For this, introduce a new concept of DNA deep learning cryptography that ensures hiding an information by use of DNA terms and deep learning. In this method, every ABCs are transformed to various mixtures of the four bases. This is, respectively, *A*, *C*, *G* and *T*. These base pairs are making a humanoid deoxyribonucleic acid (DNA). In this paper to begin with, routine besides of execution intended for key creation grounded on DNA algorithm with Needleman–Wunsch (NW) algorithm. And furthermore, the genetic procedures transcription, translation, DNA sequencing and deep learning are used to encrypt and decrypt the information.

Keywords Deep learning · DNA cryptography · Transcription · Translation · Encrypt and decrypt

1 Introduction

A growth in the field of info safety has ever evolved. There are mainly three elementary consequences are used to provide a great approaches to the info safety (i.e., info availability, reliability and privacy). Cryptography provides great enci-

R. Vasanth (✉)
Research Scholar, Department of CSE, SRM Institute of Science and Technology, Kattankulathur, Kancheepuram 603203, India
e-mail: Vasanth14795@gmail.com

D. J. Samuel
Vellore Institute of Technology, Chennai, India
e-mail: jacksoncse@gmail.com

© Springer Nature Singapore Pte Ltd. 2020
S. S. Dash et al. (eds.), *Artificial Intelligence and Evolutionary Computations in Engineering Systems*, Advances in Intelligent Systems and Computing 1056,
https://doi.org/10.1007/978-981-15-0199-9_22

pherment techniques for hiding a data and as well as retrieve a data or info. Additionally, deep learning method is combined to the crypto process. A cipher and decipher methods are obtained from deep learning style by the use of various layers [1]. Crypto analysis is the study of science happening nearby 100 years in the past. And then, deep learning is the recent emerging technology. For enormous memory abilities and infinite parallelism based on info packing and shield by the use of genetic cryptanalysis with deep learning, gene computing provides ability to work in similar computing problem with deep learning methods aimed to improver presentation. The DNA cryptography uses the various DNA computing logics like DNA annealing and synthesis [2]. The transcription and translation process have two genetic structures for repeating the gene and also translating gene into protein to encryption. Meanwhile, there is not any molecular procedure to changing protein to DNA. And then, deciphering process is sensibly ended by inverse translation and transcription correspondingly. In this concept, encipherment of genetic order is done by deep learning [3]. In this paper, DNA cryptography and deep learning methods are used, respectively, aim and execution of digital phase, improver presentation of procedures, toward enciphering the info by the terms of gene sequence. The nucleotide bases have a four char in DNA to denote the info $£ = (A, C, T$ and $G)$ that indicates for assumed bundle of hundred words and then which generate 222 mixture by the way associated with $£ = (0, 1)$ in out-fashioned systems making just 213 mixtures. From now, this idea gathers a robust encryption algorithm, and the results are permanent and incomprehensible. In philosophy, 12×109 nucleotide bases are presented in solo aspect of gene. Numerically, four nucleotide bases are presented in one byte of numeric info. And one nucleotide base has only two bits of info. A single unit of DNA encloses 3000×106 bytes or 3000 MB of info. The work wished for, the gene procedures produce completely unplanned and permanent key. This algorithm joined through Needleman–Wunsch (NW) algorithm that is used to discover variation in different gene standards. The fundamental key and DNA crypto algorithm are recycled besides on deep learning, which provides totally novel crypto to entire design with genetic acts and humanoid performance and possible to apply (Fig. 1).

Deep learning gets lots of attention lately and for good reason. It makes impacts in areas such as computer vision and natural language processing [4]. Deep learning is a machine learning practices that study topographies besides duty in a straight line from facts. The facts can be images, texts and sounds. Deep learning offered efforts to end- to-end learning. Let us look an example, there is a set of images, and we want to recognize which category of objects which images belong to cars, trucks or boards. We start with a labeled set of images or training data. Labels correspond to the desired output of the data. The deep learning algorithm needs these labels as the algorithm said that specifics features and data present in the image. The deep learning algorithm then learns how to classify input images into the desired categories. We use the end-to-end learning because the task is learning directly from data. Another example is a robot learning how to control the movement of its hand to pick up a specific object (Fig. 2).

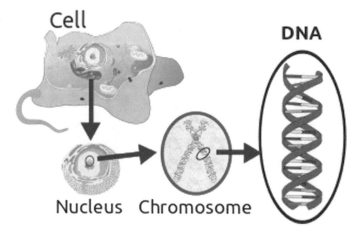

Fig. 1 Biological structure of DNA

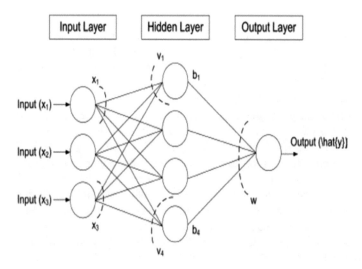

Fig. 2 Deep learning architecture

Another example is a robot learning how to control the movement of its hand to pick up a specific object. In this case, the task is learning how to pick up an object from given input image. Many of these techniques used in deep learning today [5]. For example, deep learning is used directly in hand-written codes in the service since the 1990s. The use of deep learning has searched in last five years. Primarily, it has three factors first, and deep learning methods are not more accurate than people classifying images. Second, GPUs are enabled to known networks and less

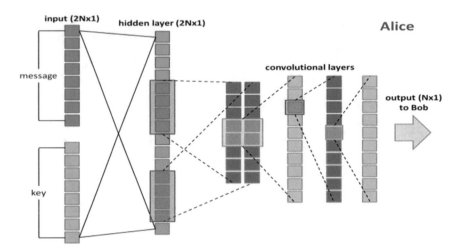

Fig. 3 Deep learning encryption process

time. And finally, large amount of label data requires to for deep learning to become excisable over the last few years. Most deep learning methods used in neural network architectures. The term deep considers a number of hidden layers in the network, and then, traditional n/w contains two or three hidden layers there as some reason it has 10s to 100s hidden layers (Fig. 3).

Nowadays, ecosphere partakes with large number of webs; statistics injury is located as mutual incidence. Deep learning possibly will support en route for recovering the data that lost cutting-edge of the web communiqué. Here stand dissimilar styles on the road to deep learning alike controlled, unconfirmed and bottomless strengthening education. The minute to join genetic material crypto and deep learning, control and productivity of together crypto and deep learning is robotically rise safekeeping [6]. To guard the treasured info and up to date all god creatures, the microelectronic safety skills are used in cryptosystem. Wild expanding of present era, particular initial encipher approaches encompassed substitution and the Vigenere's Cipher, original text are replaced by specific supplementary character set. This method is originated as pathetic and straightforwardly fracturable, first and foremost as contained characteristics of original text semantic. The fashionable cryptosystem resides the upcoming ideas:

- Plain data wants to be defines as original text.
- Secret Text: Data that is purely being situated agreed by anticipated individual or else scheme known as Secret text.
- Enciphering: changing original info obsessed by secret text by means of key.
- Deciphering: changing secret text keen on original text via key.
- Key: Recipe of numbers, characters, chief numeric text and unusual icons.

2 By Means of Genomic Algorithm

The genomic procedure is taking place in Charles Robert Darwin's philosophy of progression [1]. This is one of the adaptive investigative survey algorithms centered on mechanism concept normal choice and normal heredities. Evolutionary procedure is grounded on the household of genomic algorithms.

Stepladders Convoluted in Genomic Algorithm

Footstep I: A casual creator remains hand-me-down toward making the preliminary populace consuming P genes. Existence and replica of specific gene cutting-edge to populace be located encouraged through removal of impractical structures and worthwhile valuable manners.

Footstep II: Replica of populace has been attained thru iterative presentation bundle of stochastic workers that typically contains alteration, crossover and miscellany.

Footstep III: Miscellany stands typically centered to capability role that computes suitability of all separable genetic material in populace, at the positions of actual digit. Crossover border inherited operative hand-me-down in the direction of make dissimilarity as of unique peer group en route for alternative. Border degree specifies correspondence of the genes elect awake meant for borders.

Crossover: Crossover is a hereditary machinist hand-me-down to encourage difference since single age bracket to an additional. Crossover percentage points the similarity of the genetic material chosen for boundaries. For sample, if the boundary level is 0.7%, which worth in inhabitants of 100 genetic material, probabilities of boundary is 0.7 [7]. These requirements are to be situated completely in the meantime occasionally, and crossover can disrupt the regularity and can realize unwanted results. Henceforth, it consumes to remain realistic in equability.

Mutation: Transformation is a chromosomal operative, functional to a group of inhabitants to uphold its genomic mixture. This is typically completed by changing one or two genetic factors in a separate DNA. Mutation amount is the accidental that a DNA segment inside a genetic material will be rehabilitated or tossed [8].

Selection: Assortment is a method of choosing DNAs which spirit helper and recombine to generate off mainspring for the following age group. In command to hand-picked a DNA of consuming an extraordinary unplanned pair off and circulating its story to the following age group, a selection approach containing of suitability job is applied. Hence, an appropriate DNA in excess of the others in the populace is designated, developing healthier personages over period [9].

3 Key Generation by Means of Genomic Algorithm with DNA Computing

In this paper, genomic procedure is the base of key generation which will make an exclusive key, secondhand far along with the prearranged original text for enciphering. An unplanned digit producer is used to produce the early populace of DNAs [5]. Meanwhile in genomic process, simply the rightest DNA endures; there is a necessity to describe a suitability utility; it calculates the appropriateness of the inhabitants produced by putting on the stochastic processes. In this investigation, casualness formulates the base of individuality for the key, and henceforth, the suitability process to be used must be clever to compute the unpredictability of a separate DNA.

Early Inhabitants of random digit producer is secondhand to create an early populace of arbitrary dual strings, identified by way of DNAs. Dual strings of sixty-four-bit keys are there situated, created fifty-six bits and additionally generated the eight parity bits [6]. In the future result, an early populace of 100 genetic materials is formed casually. The inhabitants are successfully converted into a new-fangled age group of the inhabitants by putting on obviously going on genomic processes as shown in Fig. 4.

Fig. 4 Key generation flowchart

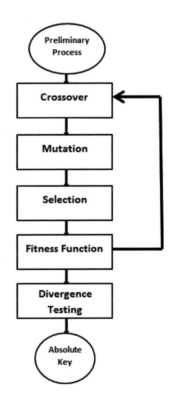

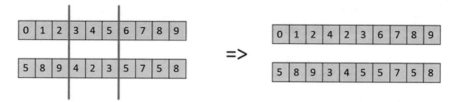

Fig. 5 Crossover operations

In the future result, an early populace of 100 genetic materials is formed casually. The inhabitants are successfully converted into a new-fangled age group of the inhabitants by putting on obviously going on genomic processes as shown in Fig. 4.

Crossover Two descendants DNAs are formed by merging the parental DNAs at the boundary limits. There are numerous methods of achieving the crossover.

The one hand-me-down at this time is K-Point boundary. K-Point boundary practices extra than single boundary limit toward crop dual descendants DNAs [10]. Dual blood relation DNAs are designated, and a numeral by the side of arbitrary stands produced as the statistics of boundary points as presented in Fig. 5.

The amount of crossovers with an exact crossover percentage is strong-minded by the equation:

$$NC = CR * NB * NK \cdot \ddot{} \ 100$$

Mutation Transformation might be situated distinct by way of a minor accidental twist happening in the DNA, in the direction of becoming an innovative result (Fig. 6).

It is rummage-sale to uphold and familiarize variety in the hereditary populace besides is typically used through a minimum-slung prospect. Whether the prospects are in height, the aforementioned is successful en route for exit the process near an unsystematic hunt. A transformation limits are carefully chosen on casual, and the bit is upturned that incomes 0 turn into 1 and 1 turn into 0 leave the additional bits presented in Fig. 4.

Fig. 6 Mutation operations

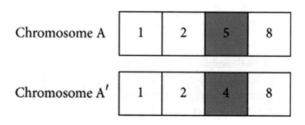

The amount of transformations through an explicit transformation percentage is strong-minded through the equation:

$$NM = MR * NB * NK \cdot\cdot\ 100$$

Fitness Function Fitness function computes the processes of in what way right the consequential DNA is. The fitness function growth the technique proceeds, demonstrating that the consequences are receiving healthier & enhanced. Run test consumes in the direction of computing the casualness in the produced DNA; meanwhile, these DNAs help to resolve the cryptographic key. Subsequently, this amount of scores aimed at every DNA is computed, that is a pointer of casualness.

Run Test Haphazardness is tough to classify, as it is precise problematic. To basically appearance at statistics & regulate whether or not it is casual. The run test is an examination of consequence or supposition test. The practice for this examination is founded upon the amount of turns, which stand structures of statistics that ought to a specific feature. At what time the interpretations are extra than twenty, then the dissemination of the experiential no. of turns would just about keep an eye on usual circulation. Henceforth, the fitness is intended as of the formulation in given equation.

$$f = a - \mu a/\sigma a$$

Finishing Selection Needleman–Wunsch procedure arrangement position remains the method of ordering 2 or else additional orders of charms just before classify areas of correspondence. Needleman–Wunsch is a procedure secondhand in the turf of environmental science and bioinformatics to relate organic orders for resemblance. It habits a counting scheme to compute the grade of resemblance or difference in the two gene orders. Preferably, the superior the groove, inordinate is the resemblance. These techniques are hand-me-down to compute the difference in the both key. The key with the uppermost aptness task is coordinated in contradiction of both key in the source. Keys through the least mark are preferred and XOR-ed. The concluding key is stowed in a concluding source, standing by. On the road to be occupied by way of the key for encipher. Thus, procedures are repeated ended hundred periods.

4 DNA Work Out

1. **Genetic material Ordering**: Gene ordering is the procedure of defining the order of nucleotide bases (A, T, C and G) in a part of chromosome.
2. **Data Encoding**: Every single English letters and arithmetical number cast-off in the basic text statistics is prearranged as a genetic material order. A casual order producer is used to allocate a random arrangement of four gene nucleotides to

Table 1 Encoding text using DNA computing assigning random sets of nucleotides to alphabets (*a*–*z*) and numbers (0–9)

Value of *a* is	Value of *m* is	Value of *y* is
Value of *b* is	Value of *n* is	Value of *z* is
Value of *c* is	Value of *o* is	Value of 1 is
Value of *d* is	Value of *p* is	Value of 2 is
Value of *e* is	Value of *q* is	Value of 3 is
Value of *f* is	Value of *r* is	Value of 4 is
Value of *g* is	Value of *s* is	Value of 5 is
Value of *h* is	Value of *t* is	Value of 6 is
Value of *i* is	Value of *u* is	Value of 7 is
Value of *j* is	Value of *v* is	Value of 8 is
Value of *k* is	Value of *w* is	Value of 9 is
Value of *l* is	Value of *x* is	Value of 0 is

the letters and arithmetical numbers. A casual order producer is used to crop an accidental string of nucleotides. By the side of an assumed point in period, an unsystematic order of chromosome nucleotides for English letters and numeric numbers is shown in Table 1.

3. **Transcription**: The procedure of transcript initiates once an enzyme named ribonucleic acid polymerase (RNA pol) accords the aforementioned to the pattern DNA element and activates to catalyze the invention of the balancing ribonucleic acid, termed the mRNA. A reproduction of a solitary gene strand is shaped, prepared in place of the procedure of translation.

4. **Translation**: Translation is a procedure of changing the hereditary cryptogram as of its DNA form containing a loop of four iterating knowledge to a concluding protein creation is made up of amino acids. Protein difficult particles called ribosomes ascribe themselves to adapted mRNA aspect, fashioned beginning transcription and transform the element mad about a restraint of protein molecule. That is consummate through transmission ribonucleic acid molecules (tRNA) that are the balancing element en route for the mRNA, transmits exact amino acids on the road to the ribosomes wherever nucleotides are declaimed and coordinated through exact amino acids.

Steps-By-Step enciphering routine:

Stage I: Complimentary element signifies the mRNA that consumes the distributed chunks of 64 bits, respectively. The situation additionally catches distributed chunks of eight bits each.

Stage II: Complementary element tRNA brings the amino acids. Amino acid signifies the cryptographic key and produced before, toward the mRNA (T).

Stage III: The chunks of 64 bits mRNA, tRNA and amino acid come to be alienated mad about chunks of eight bits for every single.

Stage IV: All the mRNA and amino acid developed and transformed keen on number method.

Stage V: A Conditions are made via deducting the fraction transformation of mRNA and tRNA (either huge) and permuting it.

Stage VI: The tRNA gained in Stage IV is at that time XOR operation in the direction of the amino acid (K). The outcome is deposited.

Stage VII: M is formerly rehabilitated to the aforementioned hexadecimal correspondent, secret message text.

5   Results

A in-depth explanation and possibility of the procedure have been there recognized and carried out by means of PHP intended for the key generation as well as DNA computing. A random initial populace of 100 DNA is generated, and different controllers are functioned for the key generation. Then, populace is turned into when the boundary point 2.5 is applied to the crossover operator. Finally, population became 426 when the mutation operator implemented with the mutation rate 0.5. By using the run test, the fitness value of all the genetic materials is stored. This is used to find the casualness. Each and every gene materials, mean and variance, is calculated, and hereafter, fitness values are gained. The Needleman–Wunsch procedure is used to check the dissimilarity of all the chromosome fitness value. The DNA with last score has greatest dissimilar with the DNA has the uppermost value. The final DNA was gained by XOR operation with the obtained two DNA patterns, and hereafter, it was added to the resource. It was in use as the key fits are aimed for enciphering and deciphering the procedure.

The each and every procedure is repeated hundred times to find the resource of the uppermost casual and nonrepeated cryptography keys. By means of DNA computing, A, T, C, G are intended for A–Z and 0–9. A long order of gene material nucleotide bases enciphers from the plaintext. This generated sequence is additionally converted by means of binary using nucleotide to binary transformation. The chunks of 64 bits are distributed from the long binary string. The each and every chuck is padded through VALUELESS. Transcript and interpretation processes by means of amino acid numerically occupied such as key aimed at cryptosystem remained functional besides the original manuscript is enciphered. Deciphering method shadowed the precise inverse of the enciphering method and the overall process implementation. The investigational examination completed by the number of symbols/letters occupied on behalf of decipherment progress and the time period taken by the procedure in the completion.

6 Conclusion

Genetic material and neural network (deep learning) approaches are offered cutting-edge of this method. The encipher techniques are abstractly established taking place the procedure of transcript and interpretation, that present two biotic actions for copying the chromosome, and at that time, translating genetic material hooked on protein correspondingly. Subsequently, here is nope molecular progression of translating protein into DNA, and deciphering is understandably finished through inverse translation and inverse transcription correspondingly by means of encouraging consequences (Fig. 7).

The yet to come of DNA deep learning cryptography is attractive greatly substantial. The future exertion has unlocked novel opportunities to examination additional on merging dissimilar arenas of AI similar to deep learning through genetic material cryptosystem toward doing well ended the conservative cryptographic procedures. Furthermore, to enhance additional coating of safety, the future procedure possibly will be present hand-me-down in combination through current cryptosystem procedures. By way of contemporary biosphere of attackers, treatment huge statistics of enormous inhabitants in addition to safeguarding that huge statistics is a contest. From now, developing methods similar to genetic cryptography and deep learning could play an essential part in that one. This is stagnant a portion desired to remain completed with esteems to the price and period efficient in this method. The inquiries education is done abstractly. That one necessitates detailed expertize, price and period to really implement this practically in the lab.

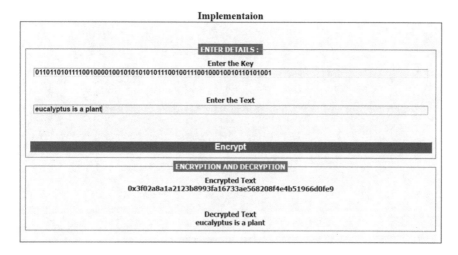

Fig. 7 DNA decipherment process implementation

References

1. Mishra, S., Bali, S.: Public key cryptography using genetic algorithm. Int. J. Recent Technol. Eng. **2**(2), 150–154 (2013)
2. Kumar, P., Sharma, A.: Data security using genetic algorithm in wireless body area network. Int. J. Adv. Stud. Sci. Res. **3**(9) (2018)
3. Zhou, Z., Chen, Z.: Dynamic programming for protein sequence alignment. Int. J. Bio-Sci. Bio-Technol. **5**(2) (2013)
4. Choudhary, R., Abrol, P.: Genetic algorithm based image cryptography to enhance security. Int. J. Adv. Res. Comput. Eng. Technol. (IJARCET) **6**(6) (2017). ISSN 2278-1323
5. Umbarkar, A.J., Sheth, P.D.: Crossover operators in genetic algorithms: a review. ICTACT J. Soft Comput. (2015)
6. Chen, C., Xiang, H., Qiu, T., Wang, C., Zhou, Y., Chang, V.: A rear-end collision prediction scheme based on deep learning in the internet of vehicles. J. Parallel Distrib. Comput. (2017)
7. Xiao, G., Lu, M., Qin, L., Lai, X.: New field of cryptography: DNA cryptography. Chinese Sci. Bull. **51**(12), 1413—1420. https://doi.org/10.1007/s11434-006-2012-5
8. Maind, S.B., Wankar, P.: Research paper on basic of artificial neural network. Int. J. Recent Innov. Trends Comput. Commun. **2**(1), 96–100 (2014). ISSN 2321-8169
9. Nazeer, M.I., Mallah, G.A., Shaikh, N.A., Bhatra, R., Memon, R.A., Mangrio, M.I.: Implication of genetic algorithm in cryptography to enhance security. (IJACSA) Int. J. Adv. Comput. Sci. Appl. **9**(6) (2018)
10. Vargas, R., Mosavi, A., Ruiz, L.: Deep learning: a review. In: Advances in intelligent System and Computing (2017)

Fault Identification in Deadline Sensitive Applications in Cloud Computing

Sodyam Bebarta, Tanajit Sutradhar and Bibhudatta Sahoo

Abstract Cloud Computing is a term that has risen exponentially over the last decade. These days when everyone has a handheld device, one wants to lap up the entertainment streaming on to their mobile directly from the cloud. The Service Level Agreement ensures that the quality of the content that is being provided to the consumers is error free. Thus making no compromise on the part of error that could disrupt the ongoing processes of the cloud computing. In case a failure occurs, the main objective is that the system should not stop functioning. The fault-tolerant system is mostly described as a strategy to build a secondary file along with the primary one for any mishap that might occur during the transaction, the secondary could take place of the primary one, and the consumer does not miss out anything. The challenge for the cloud service providers in this era is to provide a seamless and uninterrupted service. In this paper, we have discussed the issues that are generally quite common in terms of fault tolerance and proposed a model that tracks the fault once the task is loaded on to the virtual machine.

Keywords Cloud computing · Service level agreement · Fault tolerance · Virtual machines

S. Bebarta (✉) · T. Sutradhar · B. Sahoo
Department of Computer Science and Engineering, National Institute of Technology,
Rourkela, India
e-mail: sodyam@gmail.com

T. Sutradhar
e-mail: tanajit.kgp21@gmail.com

B. Sahoo
e-mail: bibhudatta.sahoo@gmail.com

© Springer Nature Singapore Pte Ltd. 2020
S. S. Dash et al. (eds.), *Artificial Intelligence and Evolutionary Computations
in Engineering Systems*, Advances in Intelligent Systems and Computing 1056,
https://doi.org/10.1007/978-981-15-0199-9_23

1 Introduction

A considerable lot of us ponder with the topic of cloud computing. In an innovatively propelled world, the term cloud computing ought to be well-known to the greater part of us, yet the inverse is valid. Cloud computing is a utility that shares assets, programming and data that are in the end gotten to by end clients. Ordinarily, facilitation of these assets happens on virtual servers accessible to the client through the web. Fundamentally, it is an act of using these servers for capacity, the board and handling of information, instead of utilizing the nearby computer to store information. In seeing distributed computing, one needs to comprehend the advancement of individualized computing.

As the web became an indispensible part of our lives, a system design outlines the web as a cloud somewhat to shroud the complexities of this framework and to clarify its nature. Basically, cloud is really an illustration depicting the web as a space that permits registering in a preinstalled situation that exists as an available support of end clients. A progressively specialized comprehension of this engineering is that distributed computing is a pioneer innovation utilized in joining other figuring arrangements, for example, parallel processing, lattice registering, dispersed processing and an entire host of other virtualization advances that work with utility figuring to achieve end-client administrations [1]. With regards to cloud computing, the comprehension of stages is required. In this paper, we have discussed the issues that are prevalent in the fault tolerance, elaborated on the two algorithms that we have used to identify the faults.

2 Literature Survey

Anju Bala and Inderveer Chana expounded on the current adaptation to internal failure methods in distributed computing dependent on their strategies, apparatuses utilized and investigation difficulties. They also proposed a cloud virtualized system architecture. In the proposed framework, autonomic adaptation to internal failure was actualized. The outcomes showed that the proposed framework can manage different programming issues for server applications in a cloud virtualized condition [2].

J. Huang, C. Wu and J. Chen proposed a security driven various leveled planning calculation for a cloud-based video gushing framework to accelerate the video transcoding process for ongoing administration applications. It masterminds the structure undertakings and intensely changes the amount of spaces so the cloud packs can execute the methodology even more profitably. Results demonstrated that the proposed planning curbs techniques kept up great framework load adjusting [3].

M. K. Gokhroo, M. C. Govil and E. S. Pilli emphasized on the need to build a robust system for cloud computing to convey the prescribed dimension of unwavering quality. They proposed a novel fault location and alleviation approach. The methodology lied in the technique for identifying the fault dependent on running status of the activity. The identification figuring once in a while checked the progression of work on virtual machines and declared ceased tasks due to inefficient VM to blame inactivity tolerant administrator. This decreased the assets wastage as well as guaranteed convenient conveyance of services [4].

J. Lee, H. Han and M. Song set forth a plan that catered to seamless viewers experience against transcoding vitality. They proposed two algorithms that figured out which adaptations ought to be transcoded with the point of augmenting the general QoE. Exploratory outcomes demonstrated that the proposed plan can successfully constrain the measure of transcoding vitality utilization [5].

P. Guo and Z. Xue's essential reinforcement show is broadly used to acknowledge adaptation to non-critical failure by copying an undertaking into two duplicates—an essential one and a reinforcement one. The excess presented by reinforcement duplicates acquired additional overhead for cloud frameworks. They proposed a continuous robust scheduling algorithm with improvement in cloud frameworks [6].

H. Han, W. Bao, X. Zhu, X. Feng and W. Zhou put forth a robust scheduling calculation named ARCHER for hybrid undertakings in cloud which coordinated the customary reinforcement model and checkpoint innovation and which could adaptably decide the execution time of the reinforcement duplicates of errands, so it incredibly upgrades the asset use and creates additional vacancies to execute assignments however many as could be allowed [7, 8].

3 Problem Statement

- The primary issue that cloud service providers encounter is to establish an effective, inexpensive and moreover a trust-worthy answer for an uninterrupted reach to the consumers.
- Choosing the type of VMs in the VM positioner to analyze the transformation so that the transcoding is done in an efficient manner and the cost incurred is also less.

4 Proposed Methodology

Cloud computing regularly comprises a few segments, for example, datacenter, servers, storage, network, middleware, software and applications and so on. It is a testing errand to deal with the far-reaching rundown of flaws for every one of the segments. System to deal with every one of the shortcomings in cloud is hard to assemble. Consequently, we have considered a specific instance of taking care of

VMs faults. VM is the fundamental part in cloud. Real disappointments happen because of powerlessness to deal with VMs faults [9]. The real parts incorporate monitoring module (to check the framework parameters) and problem screening section [10, 11]. Every part has a particular reason and is quite dependent to its past stage for information yield/input necessities [12]. The main sections of the methodology are explained as follows.

4.1 Host Machine

Cloud server dispenses the undertakings to various hosts. The host is in charge of dealing with solicitations coming to it and after that designating those solicitations to machines under its control.

4.2 Datacenter

The data center comprises cloud servers, computers, network devices, resources which are required to set up a cloud.

4.3 Cloud Server

Cloud server has indistinguishable work from that of a regular server, yet the functionalities may vary. Individuals utilizing cloud lease the cloud server instead of buying the cloud server. Every single demand of customers is made through these cloud servers. It is not in charge of acquiring every one of the solicitations yet additionally reacting to them.

4.4 Virtual Machine

These host machines are additionally separated into virtual machines to complete diverse errands all the while. Virtual machines are fundamentally the division of one single machine equipment astute chiefly. In the proposed methodology, each virtual machine has certain memory saved for putting away the work done in last checkpoint length. At the point when the edge of a virtual machine is achieved which means that machine is going to over-burden very soon, live movement process begins without stopping the machine, and henceforth, downtime is decreased [13]. In the event that a fault happens amid relocation then the information is duplicated from this held memory. The VM provisioner decides which

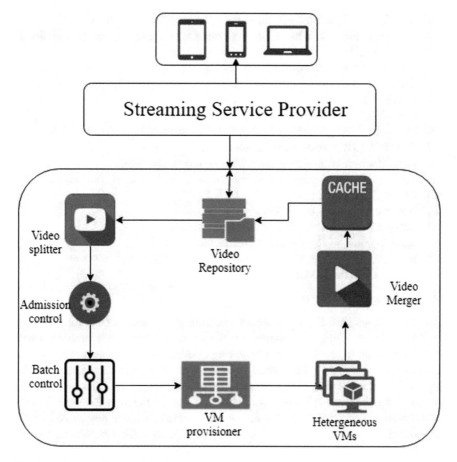

Fig. 1 Virtual machine provisioning

task to be allocated to what virtual machine [14]. It is at this point that the fault detection that can benefit from wastage of resources (Fig. 1).

i: **Fixed time Checkpoint**

This method checks for the faults with a fixed interval that has been pre-decided before allocating any jobs to the VM. The broker marks the start time and the scheduled finish time. If any problem occurs during the transaction, it can be found out by the fixed interval checkpoint, and the time of the detection is noted by the Broker.

Algorithm: Allocating Jobs on the Virtual Machines
Input: VMs = {VM$_1$, VM$_2$, VM$_3$ .... VM$_n$}
Output: Robust Virtual Machine VMs = {VM$_1$, VM$_2$, VM$_3$ .... VM$_n$}

1 Initialize: A fixed time is assumed for the checkpoint of the VMs
2 The total time taken by a job is divided into segments for the checkpoint calculation
3 Considering all Virtual Machines Vi that are a part of the execution process does
4 Check if the status of the VMi. = SUCCESS then
5 FailedVMList = Boolean value zero
6 else
7 FailedVMList = Boolean value one
8 Host Machine = on which the VM was allocated initially
9 Taking all Virtual Machines V_i that are allocated on Host Machine does
10 Check if the status of the VMi = FAILED then
11 HostMachineFailedList = Boolean value one
12 Publish Failed VM V_i and failed Host Machine
13 Save the time of recognition of the failure
14 Enlist the failed VM V_i to the index of failedVMList and failed host h to the index of failedHostMachineList

ii: **Increasing time monitoring**

The problem with the fixed time check-pointing is that some of the resources are kept busy for checking the progress all the time, and hence, we propose a new algorithm called increasing time monitoring (ITM). The ITM does not go for the check every now, and then instead, it checks in increasing time intervals. Say suppose the total time is 300 s, in this, the fixed checkpoint of (let us say) 20 s would go for the check 15 times, while the ITM checks the time doubling the time of the previous check. Let us assume the initial time is 5 s, the next would be 10, 20, 40…and so on. The total times it would check can roughly be around 7 times which are less than half of the fixed checkpoint algorithm.

Algorithm: Allocating Jobs on the Virtual Machines
Input: VMs = {VM$_1$, VM$_2$, VM$_3$ …. VM$_n$}
Output: Robust Virtual Machine VMs = {VM$_1$, VM$_2$, VM$_3$ ……. VM$_n$}

1 Initialize: The time is fixed by doubling the previously taken time for checking N
2 The total time taken by a job is divided into segments for the checkpoint calculation
3 for each VM V_i in the execution process where time t = N * 2 do
4 Check if the status of the VMi. = SUCCESS then
5 FailedVMList = Boolean value zero
6 else
7 FailedVMList = Boolean value one
8 Host Machine = on which the VM was allocated initially
9 Taking all Virtual Machine V_i that are allocated on Host Machine does

10 Check if the status of the VMi = FAILED then
11 HostMachineFailedList = Boolean value one
12 Publish Failed VM V_i and failed Host Machine
13 Save the time of recognition of the failure
14 Enlist the failed VM V_i to the index of failedVMList and failed host h to the index of failedHostMachineList

5 Implementation and Results

Cloudsim is chosen as a simulation tool for the above-mentioned algorithms. Cloudsim is a reenactment stage that can be reached out by clients to test their strategies, approaches and instruments for overseeing cloud frameworks.

The VM list and the jobs list are initially put up in the broker. The dedicated broker then allocates the tasks to the required VMs dependent on occupations request, cost and VMs accessibility parameters. Each one of the tasks is planned and completed effectively on the six virtual machines, and the simulation results are stored in the Table 1 as displayed. It indicates attributes of all tasks which have run splendidly fine with no faults and getting totally executed with conclusive success status (Figs. 2, 3; Table 2).

The simulation results after providing the algorithm with faults which can be detected by it.

Table 1 Tabulation of start time and finish time before identification of faults

Cloudlet id	Status	Data center id	Virtual machine id	Total time	Start time	Finish time
0	SUCCESS	01	0	323.3	0.1	323.4
5	SUCCESS	01	0	323.7	0.1	323.8
10	SUCCESS	01	1	324.1	0.1	324.1
1	SUCCESS	01	1	324.5	0.1	324.6
6	SUCCESS	01	2	324.9	0.1	325
2	SUCCESS	01	2	325.3	0.1	325.4
11	SUCCESS	01	4	324.7	0.1	324.8
7	SUCCESS	01	4	325.1	0.1	325.2
4	SUCCESS	01	5	325.5	0.1	325.6
9	SUCCESS	01	5	325.9	0.1	326
3	SUCCESS	01	3	326.3	0.1	326.4
8	SUCCESS	01	3	326.7	0.1	326.8

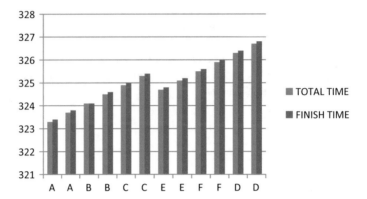

Fig. 2 Comparison between the total time and finish time

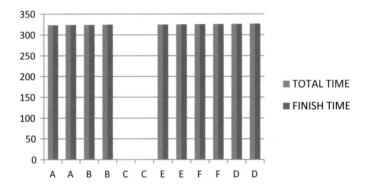

Fig. 3 Comparison between the total time and finish time with faults with the cloudlet IDs in the X-axis and the time of completion in the Y-axis

6 Conclusion and Future Work

Early location and relief of flaws in the distributed computing condition is an indispensable issue. On the off chance that the deficiencies go undetected, at that point they may transform into lethal disappointments. It might prompt loss of business, and thusly, legitimate activities may transform into weighty fiscal punishment and loss of validity for cloud specialist organizations. Continuously evaluating the frameworks, the accuracy of results depends on the logicality of the outcomes as well as on the time at which the outcomes are created. We have identified the faults that were detected the algorithms that have been discussed in this paper. The identified faults can be further mitigated to the nearest neighboring virtual machines.

Table 2 Tabulation of start time and finish time after identification of faults

Cloudlet id	Status	Data center id	Virtual machine id	Total time	Start time	Finish time
0	SUCCESS	01	0	323.3	0.1	323.4
5	SUCCESS	01	0	323.7	0.1	323.8
10	SUCCESS	01	1	324.1	0.1	324.1
1	SUCCESS	01	1	324.5	0.1	324.6
6	FAILED	01	2	0.1	0.1	0.2
2	FAILED	01	2	0.1	0.1	0.2
11	SUCCESS	01	4	324.7	0.1	324.8
7	SUCCESS	01	4	325.1	0.1	325.2
4	SUCCESS	01	5	325.5	0.1	325.6
9	SUCCESS	01	5	325.9	0.1	326
3	SUCCESS	01	3	326.3	0.1	326.4
8	SUCCESS	01	3	326.7	0.1	326.8

References

1. Egwutuoha, I.P., Chen, S., Levy, D., Selic, B.: A fault tolerance framework for high performance computing in cloud. In: 2012 12th IEEE/ACM international symposium on cluster, cloud and grid computing (ccgrid 2012), pp. 709–710, Ottawa, ON (2012)
2. Bala, A., Chana, I.: Fault tolerance-challenges, techniques and implementation in cloud computing. In: 2015 IEEE 23rd international symposium on quality of service (IWQoS), June 2015
3. Huang, J., Wu, C., Chen, J.: On high efficient cloud video transcoding. In: 2015 international symposium on intelligent signal processing and communication systems (ISPACS), Nov 2015
4. Gokhroo, M.K., Govil, M.C., Pilli, E.S.: Detecting and mitigating faults in cloud computing environment. In: 2017 3rd international conference on computational intelligence communication technology (CICT), May 2107
5. Lee, J., Han, H., Song, M.: Balancing transcoding against quality-of-experience to limit energy consumption in video-on-demand systems. In: 2017 IEEE international symposium on multimedia (ISM), Dec 2017
6. Guo, P., Xue, Z.: Real-time fault-tolerant scheduling algorithm with rearrangement in cloud systems. In: 2017 IEEE 2nd information technology, networking, electronic and automation control conference (ITNEC), May 2017
7. Han, H., Bao, W., Zhu, X., Feng, X., Zhou, W.: Fault-tolerant scheduling for hybrid real-time tasks based on CPB model in cloud. In: IEEE Access (2018)
8. Pak, D., Kim, S., Lim, K., Lee, S.: Low-delay stream switch method for real-time transfer protocol. In: The 18th IEEE international symposium on consumer electronics (ISCE 2014), June 2014
9. Egwutuoha, I.P., Cheny, S., Levy, D., Selic, B., Calvo, R.: Energy efficient fault tolerance for high performance computing (HPC) in the cloud. In: 2013 IEEE sixth international conference on cloud computing, pp. 762–769, Santa Clara, CA (2013)
10. Song, B., Ren, C., Li, X., Ding, L.: An efficient intermediate data fault-tolerance approach in the cloud. In: 2014 11th web information system and application conference, pp. 203–206, Tianjin (2014)

11. Antony, S., Antony, S., Beegom A.S.A., Rajasree, M.S.: Task scheduling algorithm with fault tolerance for cloud. In: 2012 International conference on computing sciences, pp. 180–182, Phagwara (2012)
12. Soniya, J., Sujana, J.A.J., Revathi, T.: Dynamic fault tolerant scheduling mechanism for real time tasks in cloud computing. In: 2016 International conference on electrical, electronics, and optimization techniques (ICEEOT), pp. 124–129, Chennai (2016)
13. Mohammed, B., Kiran, M., Awan, I., Maiyama, K.M.: Optimising fault tolerance in real-time cloud computing IaaS environment. In: 2016 IEEE 4th international conference on future internet of things and cloud (FiCloud), pp. 363–370, Vienna (2016)
14. Fan, J., Li, R., Zhang, X.: Research on fault tolerance strategy based on two level checkpoint server in autonomous vehicular cloud, In: 2017 7th IEEE international conference on electronics information and emergency communication (ICEIEC), pp. 381–384, Macau (2017)

Web Service Selection Mechanism in Service-Oriented Architecture Based on Publish–Subscribe Pattern in Fog Environment

Amit Ranjan and Bibhudatta Sahoo

Abstract Fog computing reduces latency and increases the throughput by processing data near the body sensor network. This paper considers a service-oriented architecture based on publish–subscribe pattern for web service selection in fog environment. With the advancement in fog environment, a large number of web services in fog environment provide same functionality but differ in their quality of service (QoS) parameter. This paper focuses on an approach based on genetic algorithm for the selection of a workflow which comprises different fine-grained web services. This paper only considers the non-functional aspects of web services mentioned in service-level agreement (SLA). In this paper, we propose a fitness function which uses quality of services and user preference in consideration while computing the fitness value. Moreover, genetic algorithm is used as an optimization algorithm to get an optimum workflow considering service-level agreement and user preference assigned to each QoS parameters. The conducted experiment shows that better result is obtained from the new fitness function.

Keywords Service-oriented architecture · Genetic algorithm · Service selection · Publish–subscribe pattern

1 Introduction

Fog computing is a revolutionary technology which is responsible for changing the whole IT environment. Its influence can easily been seen on various business applications. Applications are delivered in the form of web services over the

A. Ranjan (✉) · B. Sahoo
Department of Computer Science and Engineering, National Institute of Technology, Rourkela, India
e-mail: amitranjan717@gmail.com

B. Sahoo
e-mail: bdsahu@nitrkl.ac.in

© Springer Nature Singapore Pte Ltd. 2020   269
S. S. Dash et al. (eds.), *Artificial Intelligence and Evolutionary Computations in Engineering Systems*, Advances in Intelligent Systems and Computing 1056,
https://doi.org/10.1007/978-981-15-0199-9_24

Internet in fog ecosystem. In current scenario, many users are using Internet to acquire various resources in the form of web services.

A web service comes into the picture, when we talk about implementing service-oriented architecture (SOA). SOA is basically a new paradigm in designing systems. SOA is a concept to centralize the design in a specific suit of web services. Web services are outlined as a collection of various standards like Simple Object Access Protocol (SOAP), Universal Description Discovery Integration (UDDI), Web Services Description Language (WSDL) that enable an adaptable way for different applications to interact with each other. SOAP is a protocol used by different applications to intercommunicate with each other. UDDI is basically a registry containing list of various services which are published by different providers. WSDL is used to mark out the capabilities of web services and also the interface to refer it. WSDL is a self-descriptive document which ensures that customer can look at the performance of service at runtime and automatically invoke the service by generating corresponding code. All these different set of standards are XML-based (Extensible Markup Language) that allows different applications to connect with each other over network. Different services communicate by passing XML form of messages from one direction to opposite direction. The first implementation of SOA uses the Request–Response Message Exchange Pattern (MEP) called XML web services.

In this paper, we used publish–subscribe pattern for SOA, which is a bidirectional messaging approach in which the service provider publishes their data on a topic and service consumer subscribes to the topic provided by the provider. In this model, the service provider and consumer do not directly communicate with each other. The communication is done through an intermediary broker known as publish–subscribe broker. The appraisal of using a Publish-Subscribe pattern as a communication protocol is that the Publish-Subscribe pattern uses a lightweight communication protocol that is designed for devices which have limited power and computational resources.

When a message is published by publisher to the broker, the publisher marks the message with a topic identifying the message. The recipient sends a tag asking the broker to send a message related to the message. This marking on the message is called a message topic. The recipient must know the message topic that the sender uses to mark their messages. When the broker receives a message from the publisher, the broker sends the message to all customers who requested messages marked with the topic related to the sender. A simple example of Publish-Subscribe pattern is shown in Fig. 1.

In a fog environment, consumers can submit their request for availing some web services. Most of the time, it is observed that consumer requests for a complex service which is not provided in the form of a single service. In such a scenario, the consumer's request must be accomplished by a set of abstract services, which include pre-designed personal services [1]. Because many IT companies tend to provide web services, each application workflow of each abstract service has the same function, but many candidates with different quality of service (QoS) functions such as availability, reliability, cost and delay [2]. Therefore, the service

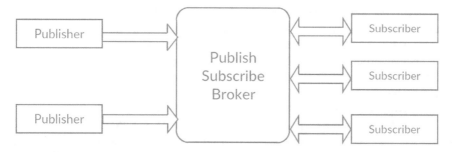

Fig. 1 Example of publish–subscribe connection

selection system needs to select only those services which are among the highest quality abstract services. Moreover, service quality of the picked web services must not surpass the quality constraints mentioned by consumer which are specified on the service-level agreement [3, 4].

One of the most challenging tasks related to web service is growing more and more number of service providers. Also, there must be a smart way to run service selection mechanism in the absence of human involvement. The mechanism must be able to cope and take decisions similar as human do. In that search space, traditional search algorithms are non-practice. It is almost impossible to find each and every combination of different web services to get the best solution. In fact, the optimization problem of finding out the combination of different web services is a non-deterministic polynomial time hard (NP-hard) problem [5].

The proposed algorithm in this paper is designed to make better the configuration process by identifying optimum sequence of services and decreasing the overhead of the customer to search for the best web services in the fog environment.

2 Related Work

Web service selection based on QoS parameters has acquired its popularity in recent years. Many researches were made on the web service selection in fog environment.

In [6, 7], they deal with the problem of choosing web services and offering computational approaches of scalable QoS based on a heuristic algorithm, which cuts the problem of optimizing small sub-problems that can be more effectively resolved than the original problem.

In [8], tensor factorization method is introduced to meet the scalability and response time for Internet services. In this work, only response time and throughput of services are considered as quality of service parameters, and also availability and reliability are not considered while choosing a web services.

In [9], the authors divide the QoS calculation model into four basic models, and each model is solved using a backtracking algorithm to optimize the selection. In combination with integrated quality of service and service technology, the authors adopted algorithms to find suboptimal services at prices based on quality of service constraints. Although the backtracking algorithm gives quicker result compared to genetic algorithm, optimality of GA is more.

In [10], the authors analysed the concept of adaptive trust management to support reliable service applications and SOA-based IoT systems. Each device captures user satisfaction experience with the device with which it has interacted and collects reliable feedback from other devices sharing social benefits. They just took only the community of friendship, social relationships and social benefits into account to choose trust feedback. The authors have not considered any other quality of service attribute into account while making their decision.

In [11], the author's approach is to return the majority of service composition problems to restricted satisfaction problems. They used integer programming method for constraint analysis. The researchers used a linear programming technique that best suits for smaller problems. But the complexity of this method got increased exponentially when it comes to higher level of problems.

In [12], a novel two-phase approach for service composition (TPASC) based on QoS records is presented to address the problem of service composition. The work done in this paper comprises two phases, i.e. heuristic QoS decomposition algorithm and Mixed Integer Programming (MIP)-based service selection.

In [13], the author is more focused about user that they have a single optimization goal; an efficient algorithm, consisting of particle swarm optimization (PSO) and Niche technologies, is presented to solve the problem of selecting services. In addition, the Niche particle swarm optimization (NPSO) and the highly intelligent PSO were used to generate a set of optimal Pareto composite services by optimizing several objective functions simultaneously.

In this paper, to solve the problem of service selection, we propose a variation of genetic algorithm with a new fitness function which includes the different QoS parameters along with their preference weight given by user based on their choice.

3 Overview of the System

As an implementation of SOA, we have used an architecture based on publish–subscribe model for web service selection. In this model, there is a bidirectional messaging approach that supports two-way messaging in which different data sources publish services based on a topic and the potential data user who subscribed to that topic will avail those services. In this model, data publisher and data consumer do not communicate directly with each other. The interaction between publisher and subscriber is performed via an intermediary system called message broker. In this architecture, the publishers that are registered to the system are regularly kept under observation. The parameters of every service are gathered on a regular basis that includes availability, reliability, cost and delay of the service. Then again, the users that are connected to the system can make their requests. Service requests are inclusive of a contract among the user and the service provider, mentioning the quality of service parameters of a workflow which is composed of different services. This contract is called as service-level agreement (SLA). SLA should not be invaded throughout the process of selection of the services. When a sender sends a message to the broker, the sender labels the message with a subject that pertains to the message. The publisher sends a label to the broker asking him to send a message attached to that label. The label is called the topic of the message. The receiver should know on what topic the sender uses to tag his messages. After the broker receives a message from the sender, the broker sends a message to all customers requesting messages tagged in the topic related to the sender. An example of Publish-Subscribe connection is graphically represented in Fig. 1.

The following are the important terms that are used in this paper are as follows:

3.1 Web Service Composition

Web service composition refers to the combination of appropriate abstract service provided by different service provider clustered together in order to serve the user request. For example, let us consider a real-life example of smart home. When a person approaches towards his house, he must be aware of the traffic and when he reached his home the doors got opened, turned on the lights, then after coffee maker starts preparing coffee and the air conditioner switched on. For execution of these requests, user requires different services. In order to fulfil user's request, usually a sequence of different services is required. This sequence of specific services for a task is called workflow. A basic example of a workflow selection is shown in Fig. 2.

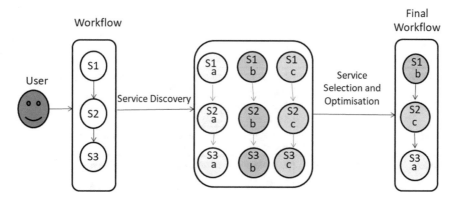

Fig. 2 Example of web service composition

As mentioned in Fig. 2, for execution of a workflow user requires to execute each of the abstract services. The very first step of serving the user request is to discover out the services which are sufficient enough to execute the abstract services. Service discovery [14, 15] phase include to find out the list of different service provider providing the appropriate service for execution of the task. A service assigned to a task during the composition phase is selected by the functional details of the services. Afterwards, service selection is followed by optimization. An optimized workflow is generated in this step from the set of services which are found in step one.

3.2 Web Service's QoS (Quality of Service)

Quality of service can be defined in terms of non-functional characteristics of the web services given by the service providers. These QoS are calculated based on execution and user performance, or retrieved from user feedback in terms of QoS quality criteria. In this paper, we have used four QoS parameters to evaluate atomic services, i.e. reliability, availability, cost and delay, as mentioned in Table 1.

3.3 Service-Level Agreement

When we talk about service-oriented architecture (SOA), a service-level agreement (SLA) can be defined as a legal contract between the service provider and the service consumer. This is defined when the workflow increases end-to-end quality of service requirements such as reliability, availability, cost and delay. In this paper, the SLA is now determined by the delay, cost, availability and reliability of the interconnected services.

Table 1 Types of QoS parameters used in this paper

QoS attribute	Unit	Description
Availability	Percentage	The likelihood of accessing a service at anytime and anywhere
Reliability	Percentage	The likelihood of accessing a service at anytime and anywhere
Cost	Rupee	The likelihood of accessing a service at anytime and anywhere
Delay	Milliseconds	The likelihood of accessing a service at anytime and anywhere

3.4 Problem Statement

The main problem in this scenario is that there are many service providers providing services for a particular task. The main goal of this work is to find a unique workflow with different services with different quality of service parameter, i.e. availability, reliability, cost and delay, which the overall performance is optimized and users get satisfied.

4 Approach Description

Our approach is based on genetic algorithm for finding an appropriate workflow for a service request made by customer. As mentioned in introduction, this work aims to design an approach based on GA to determine a composite services composing of several abstract services. The resulted set of composite service must have:

- To meet the QoS parameter mentioned in SLA.
- An optimized function based on a user preference weight assigned to each QoS parameter.

In Fig. 3, a flowchart of genetic algorithm is shown. First, the algorithm creates an initial population with a group of individuals. After it calculates the fitness value of each individual, it tries to find the best among them by evaluating the individuals with the help of fitness function. And then termination condition is checked, i.e. the optimum composition or a certain number of iteration is reached or not, if it is, then the algorithm terminates and gives the desired result otherwise two best individuals are then selected having higher fitness value. These individuals are used to reproduce to create one or more offspring by crossover procedure and mutation procedure. This continues until the required solution has been found.

In this section, various operations of genetic algorithm are explained in greater detail.

4.1 Fitness Function

Problem solving by genetic algorithm and fitness value of individual should be a step towards selecting an individual as a bellwether for any type of population. The fitness value is evaluated based on SLA. There are basically two types of quality of service criteria, i.e. positive criteria and negative criteria. Raising values of positive criteria, i.e. availability and reliability, is profitable to users, and decreasing the value for negative criteria, i.e. delay and cost, is also beneficial to users. There must be a fitness function that promotes the increase in the value of positive criteria and decrease in the value of negative criteria. Also the fitness function must consider the

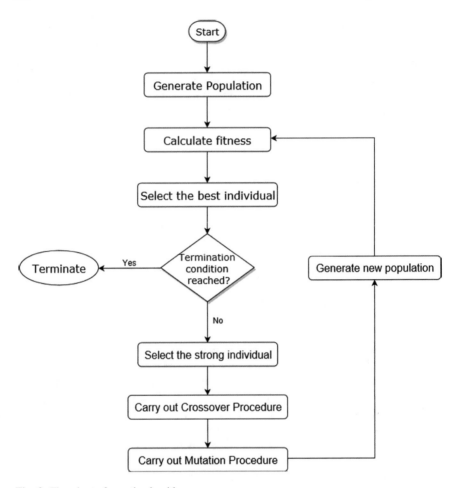

Fig. 3 Flowchart of genetic algorithm

preference for different QoS criteria. According to our approach, QoS criteria are weighted to represent user preferences and the weight is defined by the user.

To calculate the fitness value, first normalization of QoS criteria is required for uniform measurement as the scales of different QoS attributes are different. It is independent of the unit, and all the attributes are normalized into a range from 0 to 1. QoS parameters are normalized using Eqs. (1)–(4).

$$Q_{NA} = \frac{Q(A) - SLA(A)}{SLA(A)} \tag{1}$$

$$Q_{NC} = \frac{SLA(C) - Q(C)}{SLA(C)} \tag{2}$$

$$Q_{\text{NR}} = \frac{Q(R) - \text{SLA}(R)}{\text{SLA}(R)} \tag{3}$$

$$Q_{\text{ND}} = \frac{\text{SLA}(D) - Q(D)}{\text{SLA}(D)} \tag{4}$$

where Q_{NA}, Q_{NC}, Q_{NR} and Q_{ND} are the normalized values of availability, cost, reliability and delay. And SLA (A), SLA (C), SLA (R) and SLA (D) are the values of availability, cost, reliability and delay mentioned in service-level agreement. The proposed fitness function used to calculate the fitness value is mentioned in Eq. (5).

$$F = \frac{\alpha1 \cdot Q_{\text{NA}} + \alpha2 \cdot Q_{\text{NR}}}{\alpha3 \cdot Q_{\text{NC}} + \alpha4 \cdot Q_{\text{ND}}} \tag{5}$$

where $\alpha1$, $\alpha2$, $\alpha3$ and $\alpha4$ are the weights based on the user preferences, and also sum of these weights must be equal to one, i.e. $\alpha1 + \alpha2 + \alpha3 + \alpha4 = 1$.

4.2 Selection Operator

In our algorithm, we had used roulette-wheel selection algorithm for the selection of individual to produce the population of next generation via mutation and crossover operation. In this algorithm, a cumulative probability is assigned to an individual along with its fitness value. This algorithm ensures that best individual have more possibility to be chosen.

4.3 Crossover Operator

Crossover operator in genetic algorithm is used to generate new individual using the property of existing individual. In order to perform crossover operator for generations, algorithm went through three steps. The first step is to choose two parents from the existing population to generate the offspring. The next step is to select a random cut-off point. Then the tail parts of both the parents are interchanged for producing a new pair of offspring. A simple example of crossover operator is shown in Fig. 4.

Fig. 4 Example of crossover operator

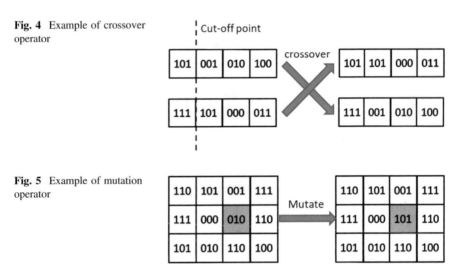

Fig. 5 Example of mutation operator

4.4 Mutation Operator

Mutation operator in genetic algorithm is responsible for generation of providing an opportunity for participation to the potential individual who could not participate in the early generation. Mutation operator first selects one of the potential individual from the population and selects a point of mutation. Then the value of that point is mutated. A simple example of mutation operator is shown in Fig. 5, where the binary bits get inverted after mutation.

5 Simulation and Evaluation

All experimental results were made on a Windows 10 PC with Intel Core i5 2.53 GHz CPU and 8 gigabyte of RAM, and coding for the algorithm is implemented in MATLAB. For simulation, above-mentioned four QoS were considered and user preference, i.e. the $\propto$ values provided by user for availability, reliability, cost and delay are 0.3, 0.3, 0.2 and 0.2 respectively. The range of availability, reliability, cost and delay were [2, 7] [5, 9] [0.3, 0.8] and [0.2, 0.9] in order. The population size was taken as 50, the mutation probability was 0.2, the crossover probability was taken as 0.8, and maximum iterations allowed were 500.

In our simulation, we have compared fitness value of the proposed fitness function (Eq. 5) and the previously used linear fitness function in which all the four QoS parameters are linearly added.

In the above, Fig. 6a gives best result as 3.09 and Fig. 6b gives its best result as 1.43, which clearly shows that our proposed fitness function results in higher fitness value with the same parameter as an input.

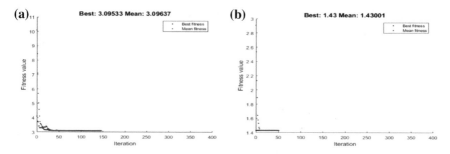

Fig. 6 **a** and **b** Comparison between fitness value obtained from two different fitness functions

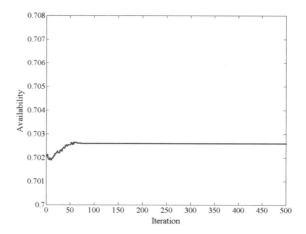

Fig. 7 Availability per iteration

Figures 7 and 8 show the graph for positive criteria, i.e. the reliability and availability gained from our algorithm per iteration. Since the maximum number of iterations is limited to 500, the graph shows its results only up to 500 iterations. Units of different QoS parameters were mentioned in Table 1.

Figures 9 and 10 show results for negative criteria, i.e. delay and price. As the number of iteration increases, the value of negative criteria decreases, and after a specific point, the graph shows the constant value, which means that we have completed the position of our stop and the graph will not fall further.

6 Conclusion and Future Work

In this paper, it can be concluded that genetic algorithm can be used for optimizing QoS parameters in the web service composition to produce better optimized results. In this paper, we have only considered the sequential composition of web services, but it can be implemented in parallel and branched composition of web services.

Fig. 8 Reliability per iteration

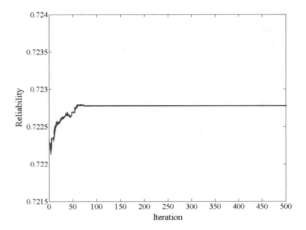

Fig. 9 Delay per iteration

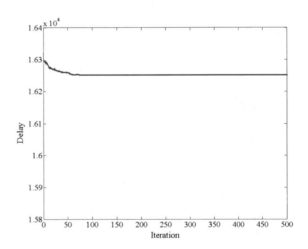

Fig. 10 Price per iteration

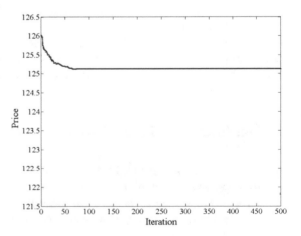

The new fitness function used in this paper gives high fitness value, and it can be further supported with other optimization algorithms to achieve the desired results. The selection of QoS parameters can be further researched, because in this paper we have only adopted SLA values and used it in our algorithm, which may result in incorrect output as compared to real-world scenario as in practical implementation, there are many server fluctuations which create hindrance in getting the desired result.

References

1. Chiu, D.A.:. A dynamic approach toward QoS-aware service workflow composition. In: Web Services, ICWS 2009, IEEE, pp. 655–662 (2009)
2. Rosenberg, F.A.: An end-to-end approach for QoS-aware service composition. In: Enterprise distributed object computing conference, 2009. EDOC'09. IEEE international, IEEE, pp. 151–160 (2009)
3. Teixeira, M.A.: A quality-driven approach for resources planning in service-oriented architectures. Expert Syst. Appl. **42**, 5366–5379
4. Baset, S.A.: Cloud SLAs: present and future. ACM SIGOPS Oper. Syst. Rev. **46**, 57–66 (2012)
5. Cremene, M.A.: Comparative analysis of multi-objective evolutionary algorithms for QoS-aware web service composition. Appl Soft Comput. **39**, 124–139 (2016)
6. Alrifai, M.A.: A scalable approach for Qos-based web service selection. In: International conference on service-oriented computing, pp. 190–199. Springer (2008)
7. Qi, L.A.: Combining local optimization and enumeration for QoS-aware web service composition. In: Web services (ICWS), IEEE, pp. 34–41 (2010)
8. Zhang, W.A.: An incremental tensor factorization approach for web service recommendation. In: Data mining workshop (ICDMW), IEEE, pp. 346–351 (2014)
9. Ming, Z.A.: QoS-aware computational method for IoT composite service. J. China Univ. Posts Telecommun. 35–39 (2013)
10. Chen, R.A.: Trust management for service composition in SOA-based IoT systems. In: Wireless communications and networking conference (WCNC), IEEE, pp. 3444–3449 (2014)
11. Aggarwal, R.A.: Constraint driven web service composition in METEOR-S. In: Services Computing, IEEE, pp. 23–30 (2004)
12. Yong, Z.A.: A novel two-phase approach for QoS-aware service composition based on history records. In: Service-oriented computing and applications (SOCA), IEEE, pp. 1–8 (2012)
13. Fan, X.-Q.: A decision-making method for personalized composite service. Expert Syst. Appl. **40**, 5804–5810 (2013)
14. Guinard, D.A. Interacting with the soa-based internet of things: Discovery, query, selection, and on-demand provisioning of web services. IEEE Trans. Serv. Comput. 223–235 (2010)
15. Gomes, P.R.: A QoC-aware discovery service for the internet of things. In Ubiquitous Computing and Ambient Intelligence, pp. 344–355. Springer (2016)

Heterogeneous Defect Prediction Using Ensemble Learning Technique

Arsalan Ahmed Ansari, Amaan Iqbal and Bibhudatta Sahoo

Abstract One of the quite frequently used approaches that programmers adhere to during the testing phase is the software defect prediction of the life cycle of the software development, this testing becomes utmost important as it identifies potential error before the product is delivered to the clients or released in the market. Our primary concern is to forecast the errors by using an advanced heterogeneous defect prediction model based on ensemble learning technique which incorporates precisely eleven classifiers. Our approach focuses on the inculcation of supervised machine learning algorithms which paves the way in predicting the defect proneness of the software modules. This approach has been applied on historical metrics dataset of various projects of NASA, AEEEM and ReLink. The dataset has been taken from the PROMISE repository. The assessment of the models is done by using the area under the curve, recall, precision and F-measure. The results obtained are then compared to the methods that exist for predicting the faults.

Keywords Software defect prediction (SDP) · Within-project defect prediction (WPDP) · Cross-project defect prediction (CPDP) · Heterogeneous defect prediction (HDP) · Ensemble learning

A. A. Ansari (✉) · A. Iqbal · B. Sahoo
Department of Computer Science and Engineering, National Institute of Technology, Rourkela, India
e-mail: arsalanansari17@gmail.com

A. Iqbal
e-mail: amaaniqbal2786@gmail.com

B. Sahoo
e-mail: bibhudatta.sahoo@gmail.com

© Springer Nature Singapore Pte Ltd. 2020
S. S. Dash et al. (eds.), *Artificial Intelligence and Evolutionary Computations in Engineering Systems*, Advances in Intelligent Systems and Computing 1056, https://doi.org/10.1007/978-981-15-0199-9_25

1   Introduction

Software defect prediction is an approach to find out the defected modules of a software much earlier than the testing phase of SDLC, so that the testing resources can be utilized efficiently. There are three major categories of SDP: Within-project defect prediction (WPDP), cross-project defect prediction (CPDP), and heterogeneous defect prediction (HDP).

In WPDP, the model is trained with the labelled instances of a project and predicts new instances of the same project. As the industry is evolving with a rapid pace, new types of projects are building; in those cases, the labelled instances of the same projects are not available for the training of a model. So a developer seeks some expertise from his or someone else's past experiences and tries to predict the defect proneness of modules of new project. This concept is generally called as transfer learning (TL). Based on TL, researchers have proposed CPDP, in which model predicts defect proneness for new projects lacking in historical data by taking the advantage of expertise available from other projects. The only limitation of CPDP is that it only works with same metrics set between projects. It is a challenging task to collect same metrics data due to the frequent occurrence of paradigm shifts in the software industry.

HDP [14] is an approach to predict defects across projects even with heterogeneous metric sets (no common metrics/features between source and target projects). HDP is one of the most recent research areas in the field of software engineering.

Figure 1 shows that there are two different projects having heterogeneous metrics set. Each block consists of many blocks which contain the metrics/features values. The last column represents module label as defective or non-defective. Non-defective is represented using grey colour box while the defective is represented using black. The question mark represents the unknown labels. As shown in the figure the module is trained using the labelled instances of a project and predicts defects on unlabelled instances of other projects having heterogeneous metrics set.

In this paper, the ensemble learning classifier (consists of 11 different classifiers) have been implemented and their performances are assessed with the previous existing approach.

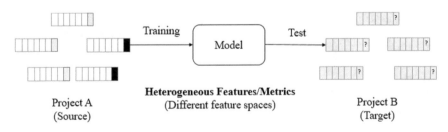

Fig. 1 Illustration of HDP

2 Literature Survey

Software defect prediction using machine learning has been in the foray since 1985 when Shen et al. [1] proposed the first prediction model using regression, although at that time the defect prediction was much easier because the cross project and heterogeneous defect predictions were not present earlier. Since then the researchers have been continuously researching in this area. Munson and Khoshgoftaar [2] claimed that regression techniques were inefficient so they suggested a new classification model which divided the components into two different sections: high risk and low risk and achieved an accuracy of 92% on their subjective system.

Lessmann et al. [3] conducted an experiment to benchmark the 22 classification models for software defect prediction. They concluded that not a single technique outshine the other.

Machine learning heavily depends on the historical data, what if the previous data of the task undertaken is not available. To resolve this issue, researchers proposed a new technique called cross-project defect prediction which exploits the historical data of one or multiple already existing projects and predicts defect of new project which is lacking in historical data [4–6].

Sheppard et al. [7] conducted an analysis of six hundred experimental outcomes taken from 42 initial findings and concluded that there is no uniformity amongst the researchers in reporting the performance of classification. One of the interesting thing they found that only 1% variability in the performance of prediction system would be credited to the algorithms the rest 99% is for the researcher.

Challagulla et al. [8] did an experimental analysis of machine learning-based software defect prediction techniques but they did not consider cross project and heterogeneous defect prediction. They applied various machine learning techniques such as linear regression, pace regression, support vector regression, neural network for continuous goal field, support vector logistic regression, neural network, logistic regression, naive Bayes, instance-based learning, K-nearest neighbours, J48 trees and 1-rule. Amongst all the techniques, instance-based learning and 1-rule outshines the others.

Challagulla et al. [8] also noticed that classifying the modules as defective or non-defective is a step ahead than forecasting the actual error in dataset. They also noticed that there was an irregularity in the rank of learning techniques in terms of forecast accuracy across different data sets which means it all depends on the datasets. Some techniques performed for some dataset(s) but not for all. But they tried to give the best possible techniques and from their analysis they concluded that naive Bayes, instance-based learning (IBL) and neural networks are the improved prediction models as compared to the other models. Almost all the machine learning techniques have been applied by the researchers for software defect prediction. Some of them performed good for some dataset some of them are unstable. Gan and Zhang [9] noticed that ability to forecast based on SVM is precarious so they proposed a new technique where they are using grey relational analysis before support vector machine (GRA-SVM) which improves the overall accuracy of the software defect prediction.

Cross-project defect prediction works only when the metrics/features are identical between two projects. The evolution in the industry is leading to very brand new software systems, which sometimes leads to heterogeneous metrics dataset for defect prediction algorithm.

3 Methodology Adopted

3.1 Dataset Considered

In this paper, four groups of datasets are used: NASA, AEEEM, ReLink and SOFTLAB. These datasets are standard dataset that has been used by many researchers for the defect prediction studies over time.

Table 1 shows the details about the dataset. It shows that there are four groups which contain three datasets each, having combination of both defective and non-defective instances. For example, the dataset JM1 is provided by NASA which contains the total of 9593 instances out of which 1759 are labelled as defective instances and it has 21 metrics/features.

Table 2 shows the common metrics between a group of dataset. Half of the pairs of datasets do not have a single common metrics, for example, NASA dataset has three common metrics with ReLink dataset. Those common metrics is removed from both the groups of dataset so that the dataset becomes heterogeneous in nature.

Table 1 Four groups of dataset with instance and metrics detail

Organization	Dataset	Number of instances		Number of metrics
		All	Defective	
NASA [10]	JM1	9593	1759	21
	MC2	127	44	39
	KC3	200	36	39
AEEEM [11]	EQ	324	129	61
	JDT	997	206	
	ML	1862	245	
ReLink [12]	Apache	194	98	26
	Safe	56	22	
	ZXing	399	118	
SOFTLAB [13]	ar4	107	20	29
	ar5	36	8	
	ar6	101	15	

Table 2 Common metrics between of group of dataset

Dataset pair		Number of common metrics
NASA	SOFTLAB	28
NASA	ReLink	3
NASA	AEEEM	0
SOFTLAB	ReLink	3
SOFTLAB	AEEEM	0
ReLink	AEEEM	0

3.2 Feature Selection

There are various feature selection techniques like chi-square, recursive feature elimination, recursive feature elimination with cross-validation, gain ratio, f-classif, significance attribute evaluation and mutual info classif. After conducting an experiment on each technique for selecting the most significant metrics, chi-square outshines the other. The top 30% metrics are selected from both source and target dataset. Chi-square test has been applied to find the dependency between features and its labels on both source and target dataset and then selected top 30% of metrics. Since the metrics in HDP dataset are heterogeneous, so the selected metrics from both the dataset does not have a single common feature between them.

3.3 Metric Matching

Metric matching is the core part of heterogeneous defect prediction. It solves the problem of heterogeneity in cross-project defect prediction. The basic idea behind metric matching is to find the best similar metrics which have higher values and have higher tendency for a module being defective. There are various techniques available for metric matching. Jaechang Nam et al. conducted an experiment on various techniques and concluded that Kolmogorov–Smirnov test-based matching technique performed better than the other.

Kolmogorov–Smirnov test: It is a nonparametric test of equality which can be used to compare two samples based on the correlation between them. To find the best-matched metrics, multiple pairs of samples are taken, and after applying KS-test, p-value (probability value) from the KS-test statistics has been taken to find the best-matched metrics. p-value interpretation is similar to another hypothesis test, which is generated with the help of the test statistics of KS-test which is calculated by:

$$D_{n,m} = \sup \left| F_{1,n}(x) - F_{2,m}(x) \right| \tag{1}$$

where $F_{1,n}$ and $F_{2,m}$ are the distributions of two samples and n and m are the size of both the samples undertaken and sup is supremum function.

With the help of test statistic, the p-values are looked up from the KS-test p-value table.

3.4 Maximum Weighted Bipartite Matching

Result from KS-test is in the form of matrix where each value represents the p-value for every pair of metrics. Maximum weighted bipartite matching is a way to find the best-matched metrics from the p-value matrix. We have used Ford Fulkerson algorithm which uses the flow network to find the maximum flow. The idea to find the metrics pair came from the very famous problem of finding the best applicant for a job which was solved using the above-said algorithm.

3.5 Feature Transformation

Before building the prediction model, feature transformation technique can be used to reduce some features by combining possibly correlated features to save the computation cost. After conducting an experiment between linear discriminant analysis (LDA) and principal component analysis (PCA), it was found that LDA did not prove to be helpful maybe because it is applicable to linear datasets, i.e., datasets where data points can be distinguished by a linear decision boundary. In this paper, PCA is used for the feature transformation.

3.6 Ensemble Learning Technique

In machine learning, sometimes different classifiers give different results for a single instance based on their learning capability. Ensemble learning is a method to solve that problem. It is one of the most powerful ways to boost the performance and accuracy of a model. Voting classifier is one of the ways of ensemble learning where votes of different classifier are taken and on the basis of votes, class label is decided for that instance. In this approach, 11 different classifiers are used to decide the majority, they are as follows:

- Logistic regression
- Support vector machine (both Linear and RBF kernel were used separately)
- K-nearest neighbours
- Naïve Bayes
- Decision tree

- Random forest classifier
- Ada-boost
- XG-boost
- Multi-layer perceptron
- Probabilistic neural network

4 Implementation

Proposing an ensemble learning algorithm, the aim is to ensure a robust method to ameliorate the calibre of the software product. The implementation details of the algorithm using flowchart are discussed below:

4.1 Pre-processing the Dataset and Model Training

For training the prediction model, the number of metrics between source and target dataset must match, but in HDP dataset, number of metrics available is not uniform. To make it uniform and to select the most significant metrics, feature selection technique is applied.

After feature selection, feature scaling is applied to normalize the dataset. Feature scaling is one out of many essential pre-processing steps if the metric values in dataset are not in the standard range. Like in the dataset which we are using, if we look at the cyclomatic complexity metric and LOC, they differ greatly in range.

Now KS-test is applied to the top 30% metrics from both the source and target dataset to find the best-matched metrics with the help of maximum weighted bipartite matching. After applying, the above techniques, the dataset for training and testing is ready. The ensemble voting classifier is trained with the source dataset and tested on the target dataset (Fig. 2).

4.2 Evaluation Criteria

The defect prediction models are assessed by the values of the area under the curve (AUC) to compare with the previous existing technique. Models are also judged by mean squared error (MSE), precision, recall and $f1$-score to show the errors in the prediction models. The formula for the calculation of the above measures is given as follows:

Fig. 2 HDP approach
overview

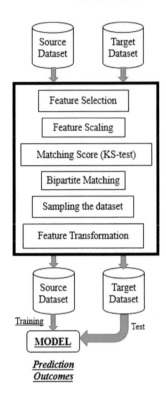

$$\text{MSE} = \frac{1}{n}\sum_{i=1}^{n}(\text{Actual}_i - \text{Predicted}_i)^2 \tag{2}$$

To understand below performance measure, first understand various prediction outcomes:

- True positive (TP): buggy instances predicted as buggy
- False positive (FP): clean instances predicted as buggy
- True negative (TN): clean instances predicted as clean
- False negative (FN): buggy instances predicted as clean

$$\text{Precision} = \frac{\text{TP}}{\text{TP + FP}} \tag{3}$$

$$\text{Recall} = \frac{\text{TP}}{\text{TP + FN}} \tag{4}$$

$$F1 - \text{score} = \frac{2 * (\text{Precision} * \text{Recall})}{\text{Precision + Recall}} \tag{5}$$

5 Result and Discussion

Around 67 combinations of source and target dataset were made and whose AUC value along with other performance measures is higher are displayed in Table 3. The table rows consist of the source dataset by which the prediction model is trained and the target dataset on which the predicted model is tested along with values of performance measures like AUC, MSE, precision, recall and $F1$-score.

In our study as shown in the table, it is found that EQ dataset from AEEEM group as a source dataset works best with almost half of the dataset. The results shown in the table is 1–15% higher than existing method for HDP.

Table 3 Prediction result of source and targets in terms of best AUC values

Source	Target	AUC values (baseline methods) [14]	AUC	Mean squared error	Precision	Recall	$F1$-score
KC3	EQ	0.776	0.79	0.29	0.71	0.71	0.71
JM1	JDT	0.767	0.80	0.19	0.82	0.81	0.81
JM1	ML	0.692	0.74	0.31	0.84	0.68	0.73
EQ	Apache	0.720	0.71	0.29	0.71	0.71	0.71
EQ	Safe	0.837	0.80	0.23	0.77	0.77	0.77
JDT	ZXing	0.650	0.67	0.32	0.67	0.68	0.67
ML	JM1	0.688	0.73	0.27	0.76	0.73	0.74
ML	MC2	0.682	0.70	0.29	0.69	0.70	0.69
EQ	KC3	0.678	0.74	0.39	0.91	0.61	0.69
EQ	ar4	0.805	0.83	0.36	0.86	0.64	0.67
ML	ar5	0.911	0.93	0.13	0.89	0.86	0.87
EQ	ar6	0.676	0.69	0.44	0.85	0.55	0.61

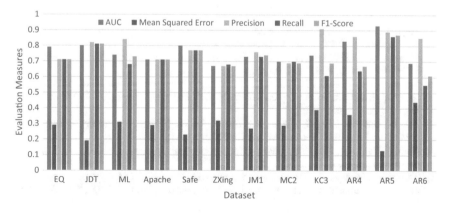

Fig. 3 Graphical representation of Table 3

Figure 3 represents the bar chart representation of Table 3. The representation for a dataset is grouped in terms of the target dataset. For example, the AUC, MSE, precision, recall and $F1$-score of EQ is shown at one place.

The datasets are highly complex in nature, it has the highest degree of nonlinearity and the datasets also has the class imbalance problem. Heterogeneous defect prediction after applying various pre-processing techniques with ensemble voting classifier gives promising results.

6 Conclusion and Future Work

By using various classifiers in an ensemble voting technique for heterogeneous defect prediction with effective pre-processing techniques gives the promising result, which will encourage the researchers to use this technique in various machine learning problems.

In this era, it has become important to sort and manage the test analysis work as it enhances the possibilities of the forecasting the errors as early as possible which in turn benefits the organization as well as helps in building liaisons with the customers. The simple aim of identifying the faults to lessen the tedious work of the developers propels the researchers into the onus of ensuring that all the faults are identified and sorted out by the programmers within the tight deadline put by the organization. In this paper, we have put an effort in analyzing the ensemble learning technique which provides us with the outcomes which are comparable with the baseline method. To improve the accuracy of the prediction model, our future work will be based on some deep learning techniques for metric matching and we will also be trying some other feature selection techniques.

References

1. Shen, V.Y., Yu, T.J., Thebaut, S.M., Paulsen, L.R.: Identifying error-prone software—an empirical study. IEEE Trans. Software Eng. 4, 317–324 (1985)
2. Munson, J.C., Khoshgoftaar, T.M.: The detection of fault-prone programs. IEEE Trans. Softw. Eng. 18(5), 423–433 (1992)
3. Lessmann, S., Baesens, B., Mues, C., Pietsch, S.: Benchmarking classification models for software defect prediction: a proposed framework and novel findings. IEEE Trans. Softw. Eng. 34(4), 485–496 (2008)
4. Ma, Y., Luo, G., Zeng, X., Chen, A.: Transfer learning for cross-company software defect prediction. Inf. Softw. Technol. (Elsevier) 54(3), 248–256 (2012)
5. Nam, J., Pan, S.J., Kim, S.: Transfer defect learning. In: 35th IEEE international conference on software engineering (ICSE), pp. 382–391 (2013)
6. Turhan, B., Menzies, T., Bener, A.B., Stefano, J.D.: On the relative value of cross-company and within-company data for defect prediction. Empirical Softw. Eng. 14(5), 540–578 (2009). Springer

7. Shepperd, M., Bowes, D., Hall, T.: Researcher bias: the use of machine learning in software defect prediction. IEEE Trans. Softw. Eng. **40**(6), 603–616 (2014)
8. Challagulla, V., Bastani, F.B., Yen, I.L.: Empirical assessment of machine learning based software defect prediction techniques. Int. J. Artif. Intell. Tools World Sci. **17**(2), 389–400 (2008)
9. Gan, Y., Zhang, C.: Research of software defect prediction based on GRA-SVM. AIP Conf. Proc. **1890**(1), 040116 (2017)
10. Menzies, T., et al.: The promise repository of empirical software engineering data. June 2012 [Online]. Available http://promisedata.googlecode.com
11. D'Ambros, M., Lanza, M., Robbes, R.: Evaluating defect prediction approaches: a benchmark and an extensive comparison. Empirical Softw. Eng. **17**(4/5), 531–577 (2012)
12. R. Wu, H. Zhang, S. Kim, and S. Cheung, "ReLink: Recovering links between bugs and changes," in Proc. 16th ACM SIGSOFT Int. Symp. Found. Softw. Eng., 2011, pp. 15–25
13. Turhan, B., Menzies, T., Bener, A.B., Di Stefano, J.: On the relative value of cross-company and within-company data for defect prediction. Empirical Softw. Eng. **14**, 540–578 (2009)
14. Chang, J., Fu, W., Kim, S., Menzies, T., Tan, L.: Heterogeneous defect prediction. IEEE Trans. Softw. Eng. **44**(9), 874–896 (2018)

Fuzzy Lattice-Based Orthogonal Image Transformation Technique for Natural Image Analysis

S. Jagatheswari and R. Viswanathan

Abstract Fuzzy lattice theory have been widely used in image processing as it allows map functions residing in an original space to functions in a transformed space, resulting in powerful knowledge extraction and image pattern recognition. Despite recognition efficiency and best means of knowledge extraction, the computational complexity and the noise rate involved have been an open problem to be addressed. In this paper, to reduce the computational complexity by optimizing the number of granules between pixels and improving the PSNR through linear fuzzy transform, a method called Euclidean Fuzzy Lattice Orthogonal Image Transform (EFL-OIT) has been presented.

Keywords Fuzzy lattice · Orthogonal normalization · Linear triangular membership · Euclidean similarity

1 Introduction

Fuzzy transforms play a significant role similar to Fourier, Laplace, Hilbert, Mellin, Z-, and wavelet transforms. The objective of these transforms is to initially map functions present in the original space to transformed space before mapping them back into original space, retaining the image quality. The first transformation being the direct transform and the second transformation called the inverse transformation. Many researchers have designed fuzzy transforms using lattice function. Mathematical Morphology was concentrated in [1] with lattice-based versions improving the image quality and scalability. Orthogonal moments based on higher level representations of numerical data for efficient recognition of image patterns

S. Jagatheswari (✉)
Vellore Institute of Technology, Vellore, India
e-mail: jaga.sripa@gmail.com

R. Viswanathan
Kongu Engineering College, Erode, India
e-mail: hod_maths@kongu.ac.in

© Springer Nature Singapore Pte Ltd. 2020
S. S. Dash et al. (eds.), *Artificial Intelligence and Evolutionary Computations in Engineering Systems*, Advances in Intelligent Systems and Computing 1056, https://doi.org/10.1007/978-981-15-0199-9_26

were presented in [2]. This orthogonal moment in turn resulted in the improvement of data processing speed. With the objective of measuring the image quality, measurement through Mean Square Error (MSE) and Penalty Function (PF) was presented in [3] based on the concept of fuzzy transform. Yet another image reconstruction technique using approximation properties of F-transform was provided in [4] ensuring less computational complexity. The advancement in image processing mechanisms has become a major interest for many researchers. In [5], the detected images were evaluated using Pratt's Figure of Merit (FOM), Jaccard's Index (JI), and Dice's Coefficient (DC), respectively, resulting in optimal results. Fuzzy Logic-based Edge Detection [6] offered enhanced solution to edge detection from smooth and noisy clinical images, offering also scalability. An integration of morphological and fuzzy algebraic systems was introduced in [7] that resulted in better processing through the use of lattice fuzzy image operators. Retrieval through two-stage performance was done in [8] by ranking by secondary medium in the first stage whereas, performing content-based image retrieval on the top-k items, ensuring image retrieval efficiency. A short overview of models and methods based on fuzzy sets was provided in [9]. However, in all the above said methods, robustness was not addressed. To address issues related to robustness, fuzzy solutions of partial differential equations were presented in [10]. One of the most popular mathematical framework is the 'Lattice Theory' which is used in different domains namely, mathematical morphology, fuzzy sets, formal concept analysis, and so on.

In [11], pareto and lexicographic orderings was presented, therefore improving the image processing efficiency. In [12], lattice library for 3D data was presented using less resource and fewer data points. A review of lattice theory for image transformation was provided in [13]. In [14], fuzzy aggregation based on neighborhood supported model was designed with the objective of improving robustness and object localization. Yet another pattern matching and classification model was designed in [15] using Hausdorff distance and sum of minimal distances resulting in the reduction of error rate. Removal of noise present in video was provided in [16] with the help of fuzzy rules. Image transformations are significantly employed for image analysis, image compression and decompression, image encryption, decryption, and so on. In [17], interval-valued fuzzy transform was presented to reduce the noise level in images. Lattice-based fuzzy description logics were presented in [18] to ensure tight complexity bounds. An algorithm for generating bipolar fuzzy formal concepts was constructed in [19] for handling bipolar information. The different transforms like Fourier and Radon Transform [20] is planned to use to get more accuracy. The remaining article is organized in the following sections. Section 2 presents the developed methodology for image transformation followed by experimental setup in Sect. 3. Finally, conclusions are drawn in Sect. 4.

2 Euclidean Fuzzy Latticed Orthogonal Image Transform

2.1 Orthogonal Function Normalization

In this section, the design of orthogonal function normalization (OFN) is constructed. Algorithm 1 shows the process of normalizing through orthogonal function. As shown in Fig. 1, three monochromatic components corresponding to the three channels 'R,' 'G,' and 'B' are compressed. The set of RGB pixel values from the corresponding RGB channels extracted is initially subjected to a spectrum (i.e., separate spectra for R pixel values, separate spectra for G pixel values and separate spectra for B pixel values, respectively) and fed as input to orthogonal function. Let 'P' be an image and 'S_{ij}' represents the spectra input pixel position value of pixels of the image 'P' with $0 \leq \mu_p(S_{ij}) \leq 1$, where $0 < \mu_p(S_{ij}) \leq 1$ denotes membership and $\mu_p(S_{ij}) = 0$ denotes non-membership.

The membership value is considered to represent information and is used for further processing, whereas the non-membership value does not contain information and are discarded. Throughout this paper, let us assume that 'L' is a lattice denoted as '$(L, \vee, \wedge)$' with membership information with unit element '1,' whereas non-membership information with zero element '0,' respectively.

The membership valued image is then represented as

$$FL = \sum_{i=1}^{M} \sum_{j=1}^{N} S_{ij} \mu_p(S_{ij}) \quad \text{where } S_{ij} \in P \tag{1}$$

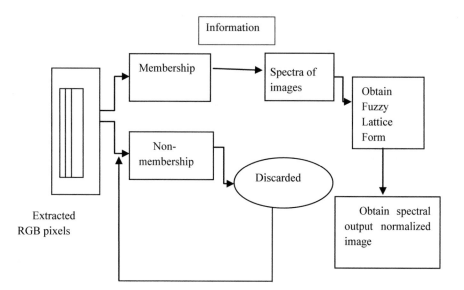

Fig. 1 Illustration of orthogonal function normalization

Here, a band of input color image, corresponding to three channels are subjected to a spectrum 'S_{ij}', of sizes '$M \times N$' interpreted as fuzzy lattice 'FL(a, b).' The fuzzy latticed image is obtained by normalizing the spectral input image 'S_{ij}' with respect to the total image size 'I_{size}' using an orthogonal function. Let us consider a spectra input image 'S_{ij}' with a specific amount of spectra noise 'S_n' is subjected to normalization using orthogonal function. Then, the spectral output image 'S_0' (normalized value) is expressed as given below.

$$S_0 = \frac{1}{I_{size}} * \text{Max} \left[0, \frac{(S_{ij})^2 - (S_n)}{(S_{ij})^2} \right] \tag{2}$$

Algorithm 1 Orthogonal Normalization Algorithm

Input: Input image '$I_{image}=P,Q...$', channels '$C=R,G,\text{B}$', image size '$M*N$', spectra input pixel 'S_{ij}', spectra output pixel 'S_o', spectra noise 'S_n',
Output: Optimized number of granules
1: **Begin**
2:　　**For** each Input image 'I_{image}'
3:　　　**If** $\mu_p(S_{ij}) \le 1$
4:　　　　Denotes membership
5:　　　　Membership value represent information
6:　　　　Measure resultant membership valued image using (1)
7:　　　　Measure normalized image using (2)
8:　　　**End if**
9:　　　**If** $\mu_p(S_{ij}) = 0$
10:　　　　Denotes non-membership
11:　　　　Membership value do not represent information, discarded
12:　　　**End if**
13:　　**End for**
14: **End**

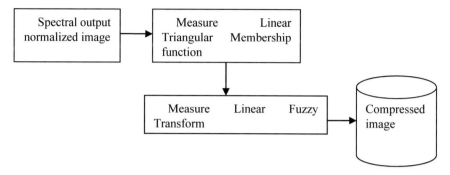

Fig. 2 Illustration of linear fuzzy transform

2.2 Linear Fuzzy Transform

The spectral output image coefficients 'S_0' are subjected, one by one, to optimal linear fuzzy transform that improves the PSNR. To perform image reconstruction from the compressed bit stream, it is subjected to corresponding reverse transformations. Figure 2 shows the illustration of linear fuzzy transform model.

In the explanation given below, we will speak about the linear fuzzy transform of a spectral output image function '$S_0 \rightarrow f$' which is a discrete function '$f: A \rightarrow B$' of two variables, defined over the set of pixels $P = \{(i,j)/ i = 1, 2 \ldots M; j = 1, 2 \ldots N\}$, the linear fuzzy triangular membership function is as given below.

$$A_1(p) = \begin{cases} \frac{(p-p_1)}{I_{\text{size}}}, & p \in [p_1, p_2] \\ 0, & \text{otherwise} \end{cases} \tag{3}$$

$$A_m(p) = \begin{cases} \frac{(p-p_{m-1})}{I_{\text{size}}}, & p \in [p_{m-1}, p_m] \\ 0, & \text{otherwise} \end{cases} \tag{4}$$

$$A_l(p) = \begin{cases} \frac{(p-p_{l-1})}{I_{\text{size}}}, & p \in [p_{l-1}, p_l] \\ 0, & \text{otherwise} \end{cases} \tag{5}$$

With the above linear fuzzy triangular membership function (3), (4), and (5), let us consider two fuzzy sets 'A_i' and 'B_j,' where '$i = 1, 2, \ldots, m$' and '$j = 1, 2, \ldots, n$' establish fuzzy partition of '$[1, u] * [1, v]$.' Then, the linear fuzzy transform for spectral output image coefficients 'S_0' is an image of the map

$$f(s_0) : \{A_1, A_2, A_3, \ldots, A_m\} * \{B_1, B_2, B_3, \ldots, B_n\}$$

Algorithm 2 shows the process of transformation of normalized image to a compressed form by applying linear fuzzy model.

Algorithm 2 Linear Fuzzy Transform Algorithm

As illustrated in Algorithm 2, for each spectral output image coefficient, the

Input: spectral output image coefficients 'S$_0$'
Output: Improved PSNR rate
1: **Begin**
2:      **For** each spectral output image coefficients 'S$_0$'
3:                    Measure Linear Fuzzy Triangular Membership function using (3), (4) and (5)
4:                    Measure Linear Fuzzy Transform
5:      **End for**
6: **End**

linear fuzzy transform performs two important functions. The first function, being the measurement of triangular membership function, helps in minimizing the complexity (i.e., memory and time used). The second function being the measurement of linear fuzzy transform performed using a mapping function. This in turn increases the PSNR rate.

2.3   Fuzzy Transform Euclidean Similarity Image Matching

Finally, the similarity image matching is performed in this section using a fuzzy transform Euclidean similarity model. Figure 3 illustrates the fuzzy transform euclidean similarity image matching with the help of a diagram. The compressed sample image is fed as input. The three compressed channels 'R,' 'G,' and 'B' of each image obtained is applied with inverse linear fuzzy transform to perform similarity matching between the images. Finally, a reconstructed image is obtained. With this obtained image, the transformed testing image is verified. The inverse

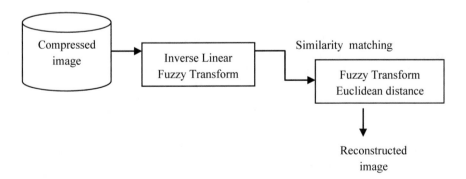

Fig. 3 Fuzzy transform euclidean similarity image matching

function of 'IF' with respect to '$\{A_1, A_2, \ldots, A_m\}$ and $\{B_1, B_2, \ldots, B_n\}$' is expressed as given below.

$$\text{IF}_{MN}\left(A_i, B_j\right)) = \sum_{i=1}^{m}\sum_{j=1}^{n}(S_0)A_i(u, v)B_j(u, v) \tag{7}$$

With the obtained resultant inverse function value, similarity matching between compressed and testing image is measured. The EFL-OIT method uses fuzzy transform euclidean distance between two points. The fuzzy transform euclidean distance is expressed as given below.

$$\text{FTED}\left(A_{x_1y_1}, B_{x_2y_2}\right) = \sqrt{(x_2 - x_1)^2 + (y_2 - y_1)^2} \tag{8}$$

With the obtained fuzzy transformed distance measure from (8), linear fuzzy triangular membership function (3), (4), and (5) is used as a basis in the EFL-OIT to measure the similarity level. The compressed fuzzy transform of an image is then compared with the inverse fuzzy transform image to measure the grade of similarity between the two images. It is expressed as given below.

$$\mu_A(p) = \frac{p}{\text{MAX}(A)} \tag{9}$$

As illustrated in Algorithm 3, compressed image and testing image are taken as input. To these two sets of images, inverse linear fuzzy transform is applied to obtain transformed portions. Next, distance between these two compressed and testing images is measured using fuzzy transform euclidean distance. With this, the grade of similarity between two images helps in measuring the similarity level.

Algorithm 3 Fuzzy Transform Euclidean Similarity Matching

Input: Compressed image '$[S_o](A_i . B_j)$', Testing Image 'TI'
Output: Improved image matching efficiency or accuracy
1: **Begin**
2:     **For** each compressed image '$[S_o](A_i . B_j)$' and testing image 'TI'
3:         Apply Inverse Linear Fuzzy Transform using (7)
4:         Measure Fuzzy Transform Euclidean distance using (8)
5:         Measure grade of similarity between the two images using (9)
6:     **End for**
7: **End**

3   Experimental Setup

The experiment is conducted on factors such as computational complexity, accuracy, and noise (PSNR) rate. The result analysis of Euclidean Fuzzy Latticed Orthogonal Image Transform (EFL-OIT) method is compared with existing lattice fuzzy mathematical morphology (L-fuzzy MM) [1] and fuzzy lattice reasoning fuzzy associative measure (flrFAM) [2], respectively. Computational complexity (i.e., time) involved during the normalization process is the time taken to obtain the spectral output image 'S_0' with respect to the image size 'I_{size}' used for testing.

$$CT = I_{size} * \text{Time}(S_0) \tag{10}$$

Accuracy on the other hand, measures the rate of image matching with respect to different images with varying sizes provided as input. Table 1 is provided with the computational complexity and accuracy rate involved using EFL-OIT, L-fuzzy MM and flrFAM, respectively. From Table 1, we demonstrate that the computational complexity (i.e., measured by the execution time) and accuracy of the proposed EFL-OIT method is better than their counterparts namely L-fuzzy MM [1] and flrFAM [2], respectively. Six different images are used for conducting the experiments namely child, old man, chip, girl, owl, and butterfly with varying sizes. As illustrated in table, the computational time and accuracy are observed to be better using EFL-OIT when compared to L-fuzzy MM [1] and flrFAM [2], respectively. Figure 4 illustrates the computational complexity comparisons for image transformation averaged over six different images. It can be observed that the proposed measurement outperforms the others, indicating that it best describes the statistical distortion. With each images having different size, the computational complexity involved as varies and comparatively observed to be lower using EFL-OIT method.

From Table 1, the image size of child was 42 KB; therefore, the computational complexity using EFL-OIT method was observed to be 2.1 ms, 2.47 ms using L-fuzzy MM whereas 3.066 ms using flrFAM. Therefore, the computational complexity using EFL-OIT method when the image given as input being child was reduced by 16% compared to L-fuzzy MM and 46% reduced by applying flrFAM, respectively. The results depicted in Table 1 shows that the accuracy provided by

Table 1 Computational complexity and accuracy using EFL-OIT, L-fuzzy MM and flrFAM

Images	Computational time (ms)			Accuracy (%)		
	EFL-OIT	L-fuzzy MM	flrFAM	EFL-OIT	L-fuzzy MM	flrFAM
Child	2.15	2.49	3.13	85.14	73.11	68.22
Old man	4.85	5.05	6.25	89.32	77.29	72.31
Chip	5.14	5.44	6.64	80.85	62.82	57.93
Girl	4.35	4.65	5.85	85.12	73.09	68.12
Owl	9.25	9.55	10.75	82.35	70.32	65.43
Butterfly	11.48	11.88	12.98	89.01	77.03	72.14

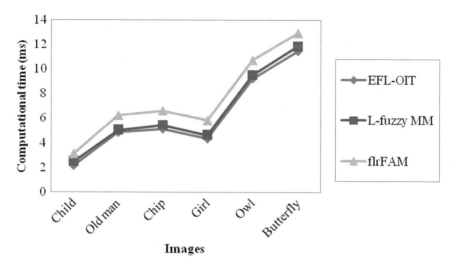

Fig. 4 Measure of computational complexity

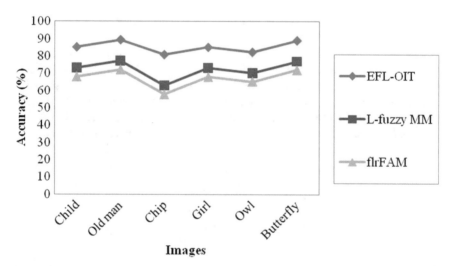

Fig. 5 Measure of accuracy

the EFL-OIT method is found to be competitive enough as compared to the solution provided by the L-fuzzy MM [1] and flrFAM [2]. Figure 5 shows the measure of accuracy or the image matching efficiency rate with respect to six testing images provided as input. The dual factor, euclidean distance and fuzzy measure applied in EFL-OIT method helps in improving the accuracy rate of EFL-OIT method by 15% compared to L-fuzzy MM. Moreover, the compressed image is not considered to image matching in EFL-OIT method, but an inverse linear fuzzy transform is

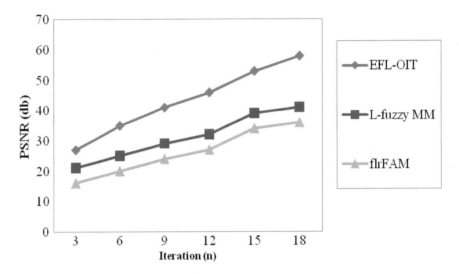

Fig. 6 Measure of PSNR

applied to the compressed image and then the distance measure is performed to analyze the matching efficiency.

The results therefore confirm that the image matching efficiency of EFL-OIT method is significantly higher by 21% compared to flrFAM. Finally, the PSNR is used as measure to obtain the error rate with minor changes in the statistical behavior of the image being processed after image compression. The PSNR measures the statistical difference between the training and testing images in decibels (dB), and it is calculated using (11). Higher PSNR value indicates better quality.

$$\text{MSE} = \frac{1}{U * V} \sum_{p=0}^{N} \sum_{q=0}^{M} I(p,q) - I'(p,q) \qquad (10)$$

$$\text{PSNR} = (10 \log 10) * \left[\frac{R^2}{\text{MSE}} \right] \qquad (11)$$

'$I(p, q)$' corresponds to the training image whereas '$I'(p, q)$' corresponds to the testing image, respectively, with 'N' number of rows and 'M' number of columns in the image. Lower value of 'MSE' provides with the efficiency of the method. The PSNR is measured through the peak signal level 'R,' whereas the mean square error 'MSE' obtained from (11). The MSE measures the average of the squares of the errors, and it quantifies the difference between the values implied by an estimator and the true values of the quantity being estimated. Figure 6 shows the PSNR values with respect to different iterations, with single iteration being performed using one testing image and as an average of about 18 testing images used for conducting experiments and changes being observed by applying EFL-OIT method,

L-fuzzy MM, and flrFAM, respectively. The proposed EFL-OIT method yields improved PSNR in all the experiments.

In addition, by applying the linear fuzzy transform in the EFL-OIT method to the spectral output image coefficients further enhances the PSNR rate by 28% compared to L-fuzzy MM and 40% compared to flrFAM, respectively.

4 Conclusion

In this paper, we have designed a Euclidean Fuzzy Latticed Orthogonal Image Transform (EFL-OIT) method for image transformation to reduce computational complexity through optimization of granules between pixels and improve PSNR through linear fuzzy transform model. Euclidean is performed to measure the rate of accuracy. Performance evaluation result shows that the algorithm extends the matching efficiency and PSNR rate compared with similar algorithms for different number of images and iterations and has a good performance in terms of computational complexity. The different transforms like Fourier and Radon Transform is planed to use to get more accuracy.

References

1. Sussner, P.: Lattice fuzzy transforms from the perspective of mathematical morphology. Fuzzy Sets Syst. **288**, 115–128 (2016)
2. Kaburlasos, V.G., Papakostas, G.A.: Learning Distributions of image features by interactive fuzzy lattice reasoning in pattern recognition applications. IEEE Comput. Intell. Mag. **10**(3), 42–51 (2015)
3. Di Martino, Ferdinando, Hurtik, Petr, Perfilieva, Irina, Sessa, Salvatore: A color image reduction based on fuzzy transforms. Inf. Sci. **266**, 101–111 (2014)
4. Perfiljeva, I., Vlasanek, P.: Image reconstruction by means of F-transform. Knowl.-Based Syst. **70**, 55–63 (2014)
5. Perez-Ornelas, F., Mendoza, O., Melin, P., Castro, J.R., Rodriguez-Diaz, A., Castillo, O.: Fuzzy index to evaluate edge detection in digital image. Plos One 1–19 (2015)
6. Haq, I., Anwar, S., Shah, K., Khan, M.T., Shah, S.A.: Fuzzy logic based edge detection in smooth and noisy clinical images. Plos One 1–17 (2015)
7. Maragos, Petros: Lattice image processing: a unification of morphological and fuzzy algebraic systems. J. Math. Imaging Vis. **22**(2), 333–353 (2005)
8. Arampatzis, A., Zagoris, K., Chatzichristofis, S.A.: Dynamic two-stage image retrieval from large multimedia databases. Inf. Process. Manag. **49**(1), 274–285 (2013)
9. Bloch, I.: Fuzzy sets for image processing and understanding. Fuzzy Sets Syst. **281**, 280–291 (2015)
10. Zeng, Y., Lan, J., Zou, J., Wu, C., Li, J.: A fast and robust method for image segmentation using fuzzy solutions of partial differential equations. Int. J. Signal Process., Image Process. Pattern Recognit. **8**(10), 389–400 (2015)
11. Bloch, I.: Lattices of fuzzy sets and bipolar fuzzy sets, and mathematical morphology. Inf. Sci. **181**(10), 2002–2015 (2011)

12. Linner, E.S., Moren, M., Smed, K.-O., Nysjo, J., Strand, R.: LatticeLibrary and BccFccRaycaster: software for processing and viewing 3D data on optimal sampling lattices. SoftwareX 1–9 (2016)
13. Grana, M.: Lattice computing: lattice theory based computational intelligence. Lattice Comput. 1–9 (2008)
14. Chiranjeevi, P., Sengupta, S.: Neighborhood supported model level fuzzy aggregation for moving object segmentation. IEEE Trans. Image Process. 23(2), 645–657 (2014)
15. Lindblad, J., Sladoj, N.: Linear time distances between fuzzy sets with applications to pattern matching and classification. IEEE Trans. Image Process. 23(1), 126–136 (2014)
16. Mélange, T., Nachtegael, M., Kerre, E.E.: Fuzzy random impulse noise removal from color image sequences. IEEE Trans. Image Process. 20(4), 959–970 (2011)
17. Strauss, O.: Non-additive interval-valued F-transform. Fuzzy Sets Syst. 270, 1–24 (2015)
18. Borgwardt, S., Penaloza, R.: Consistency reasoning in lattice-based fuzzy description logics. Int. J. Approx. Reason. 55(9), 1917–1938 (2014)
19. Singh, P.K., Aswani Kumar, C.: Bipolar fuzzy graph representation of concept lattice. Inf. Sci. 288, 437–448 (2014)
20. Karur, S.P.: Contributions of mathematical model in bio medical sciences-an overview. Int. J. Appl. Sci.-Res. Rev. 33–39 (2016)

DC Smart Grid System for EV Charging Station

T. K. Krishna, D. Susitra and S. Dinesh Kumar

Abstract Increasing the use of electric vehicles (EVs) is regarded as the right step to overcome the air pollution and carbon emission. Economic Times report, Government of India, is spending Rs. 5.65 lakhs crores for importing crude oil in 2017–2018. This huge amount of import affects our economic status and increases dependence on Arab and Eastern countries. In order to overcome this fossil fuel dependence, electrical vehicle is the ideal solution. But major difficulties to implement EV are stepping up the charging station across the country and to reduce transmission losses and distortion across smart grid network. In AC smart grid system, distortion and harmonic losses frequently occur; moreover, most of storage batteries and EV are DC. Hence, an AC/DC converter is required for an AC smart grid system. In DC smart grid system, distortion and harmonic losses are nullified. In India, we mostly use solar energy and fuel generator, and in case of wind energy AC source can be converted into DC using AC/DC converter. This paper models an integrated electric vehicle charging battery storage system and motor load operating in the presence of DC source smart grid. The major constraints are voltage quality maintenance and reduction of EV charging time. Both of these constraints are achieved using SEPIC converter. Using this converter, higher output voltage is achieved to drive a motor load.

Keywords EV—electrical vehicle · DC smart grid · SEPIC · FOPID · Fuzzy logic controller

T. K. Krishna (✉) · D. Susitra · S. Dinesh Kumar
Department of EEE, Sathyabama Institute of Science and Technology (Deemed to be University), Chennai, Tamil Nadu 600119, India
e-mail: tkkoushick22@gmail.com

D. Susitra
e-mail: susitra.eee@sathyabama.ac.in

S. Dinesh Kumar
e-mail: dineshselvaraj05@gmail.com

© Springer Nature Singapore Pte Ltd. 2020
S. S. Dash et al. (eds.), *Artificial Intelligence and Evolutionary Computations in Engineering Systems*, Advances in Intelligent Systems and Computing 1056,
https://doi.org/10.1007/978-981-15-0199-9_27

1 Introduction

The paper briefly explains usage of electric vehicles (EVs) to reduce the usage of fossil fuel, air pollution and carbon emission. This huge amount of import affects our economic status and also increases our dependence on Arab and Eastern countries. The immense necessity for renewable energy resources is increasing day by day due to the depletion of available resources and greenhouse gas emission. The challenge in execution of such energy sources is voltage fluctuation in DC grid.

The right solution to overcome the all above-mentioned problem is electrical vehicle. Several published papers from researchers belonging to different countries on their point of view are presented. In [1], electrical vehicle charging station is designed in such a way the voltage quality is maintained and the charging time is reduced using 'buck–boost' converter but its future scope is achieved in our proposed method. In [2], M. Shenghad formulated problem in hybrid energy supply in smart grid using non-deterministic polynomial method.

In [3], Eduardo provides a mathematical algorithm for DC smart grid system easily adopted for AC systems also. Research in [4] shows low-voltage distribution network which provides fast charging for six electrical vehicles only.

These disadvantages are also overcome in the proposed method.

In [5], two main techniques are analyzed: quadratic and dynamic programming. In [6], the issue in battery charging is discussed. Coordinated charging technique is implemented for power loss minimization, load factor maximization in grid and fixed start delay. Quadratic and dynamic programming techniques are used and inferred that the later one has produced fuel economy. In [7], Fang He suggests the principles of equilibrium to overcome the economic and distribution difficulties in PHEV. In [8], S. Masoum has designed a system with two management layers to improve distribution side efficiency with reduced peak demand, voltage regulation and system losses.

In [9], mathematical calculations are formulated to reduce the capital and operating charges in EV charging station. In [10], Author has implemented an optimized charging and discharging for batteries using swapping stations. A balanced demand and supply are obtained. In [11], Shaoyun Ge has suggested a planning scheme for selection of station and grid partition. In [10, 12] an analysis of economic status of the battery is provided and all other concerns are considered to provide a superior grid regulation.

In [13], Zhipeng Liu has formulated both screening and algorithm. In [14], fast-charging station is developed using the respective compensation scheme and by overcoming the physical and distribution distraction. This method is also used to remove the grid voltage fluctuation due to source fluctuation. Bracalea et al. [15] discusses the significance of low-voltage hybrid DC smart grid. In [16], protection of DC smart grid is done using DCCB relay, in order to overcome the overcurrent and transmission loss characteristics. In [17], DC smart system is designed as such to maintain the power quality and uses maximum power tracking techniques and storage equipment cost also reduced. In [18], Robert has introduced two control techniques for achieving localized control in DC distribution system and introduces small stabilization by injecting the power back to added bus which is generated due to severe loss.

In addition to all the above-listed constraints, [15, 17–19] deal with energy management with continuous data transfer based on system's performance, economic status and production forecasting. This layer plays a major role in real-time application of the smart EV charging station. In [20], specific application areas of DC smart grid such as electrification transportation, PHEV/PEV battery/charging facility, intelligent energy management and V2G and communication requirements are discussed. In [21], importance of solid-state transformer (SST) having key characteristics such as VAR compensation, voltage regulation, source disturbance rejection and microgrid integration is presented.

Based on this extensive survey, it is inferred that the major constraints in EV charging system are not addressed completely in any of these papers. This research attempts to address and rectify these issues. Simulations are carried out using SEPIC/FLC-based control systems satisfying the major constraints at all real-time operating conditions, and a constant output voltage is obtained even with pulsating source.

2 Open-Loop Simulation

Open-loop simulation is divided into two parts: First part provides the comparison between buck–boost converter and SEPIC based on their characteristics like output voltage, output current and output power.

The second part consists of comparison between the $\prod$-filter, T-filter and C-filter of SEPIC in the proposed system whereas buck–boost converter in the existing system.

The block diagram, circuit diagram and the simulation results of existing and proposed system are given below.

2.1 Comparison Between Buck–Boost and SEPIC Converter

See Figs. 1 and 2.

Simulation Results

See Figs. 3, 4, 5, 6, 7, 8 and 9.

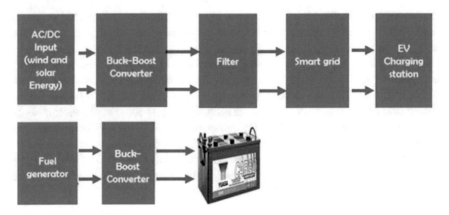

Fig. 1 Open-loop block diagram with buck–boost converter

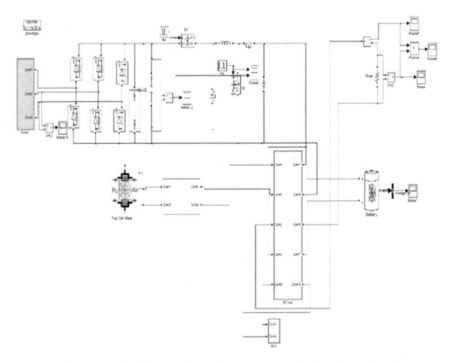

Fig. 2 Circuit diagram of DC smart grid EV charging system with buck–boost converter

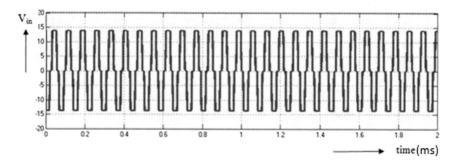

Fig. 3 Input voltage

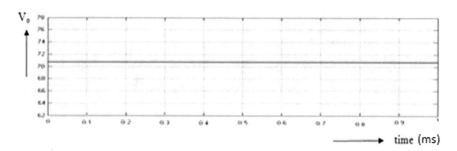

Fig. 4 Voltage across rectifier

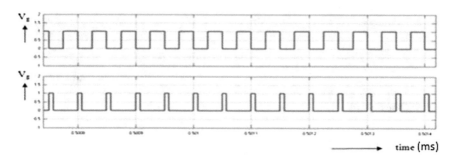

Fig. 5 Switching pulse for buck–boost converter S1, S2

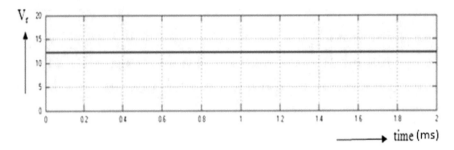

Fig. 6 Output voltage across R-load

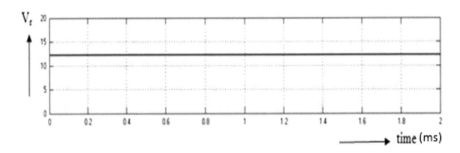

Fig. 7 Output current through load

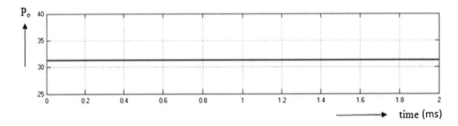

Fig. 8 Output power

2.2 Single-Ended Primary Inductor Converter

The SEPIC is a type of DC/DC chopper and operates based on the gate pulse from control circuit. It produces a non-inverting and low repulsive output voltage. It is similar to the operation of flyback converter. In this research, SEPIC operates with the higher output rating of 1020 KVA (Fig. 10).

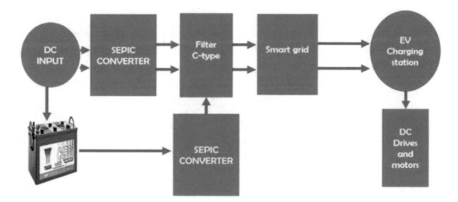

Fig. 9 Open-loop block diagram with SEPIC

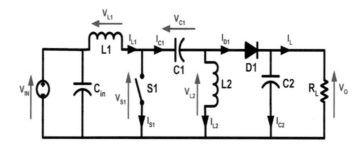

Fig. 10 SEPIC

2.3 Mode of Operation

The mode of operation of SEPIC is based on rapid and high-frequency switching of MOSFET. The switch is ON for a high pulse. At this moment, L1 is charged by input and L2 is charged by C1. The diode is under reverse-biased condition, and C2 maintains the output. The switch is OFF when the pulse is low. The stored energy in L1 and L2 flows through the diode and load. The capacitors are charged. The output is high when the pulse is low. This is due to the fact that when the charging time of inductors is high, the stored energy is high. But if the low pulse lasts for a long time, the capacitors fail to charge and result in failure of converter operation (Fig. 11).

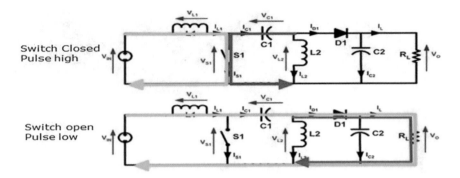

Fig. 11 Modes of operation

2.4 DC Smart Grid EV Charging System with SEPIC

See Fig. 12.

Simulation Results

See Figs. 13, 14, 15, 16, 17, 18 and 19.

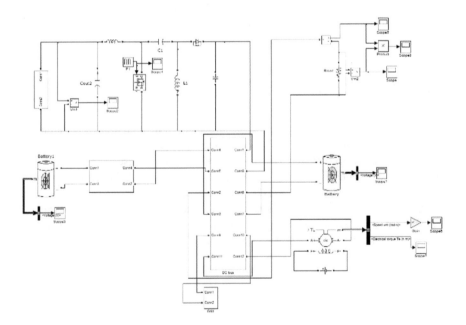

Fig. 12 DC smart grid EV charging system with SEPIC (C-filter)

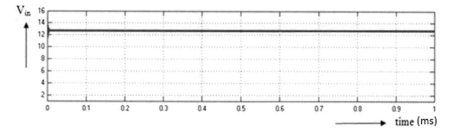

Fig. 13 Input voltage

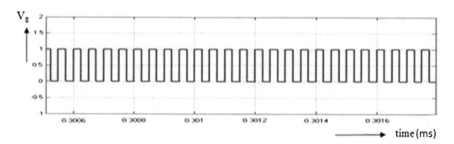

Fig. 14 Switching pulses of SEPICS1

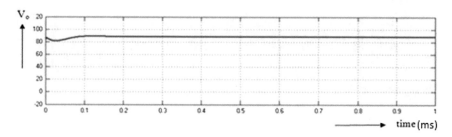

Fig. 15 Output voltage across R-load

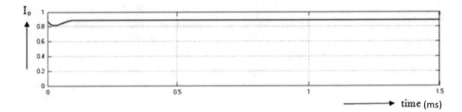

Fig. 16 Output current across load

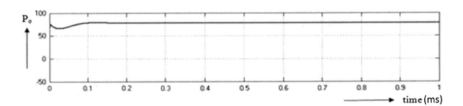

Fig. 17 Output power

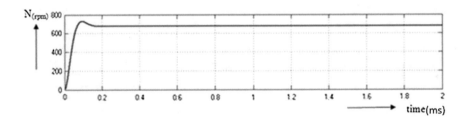

Fig. 18 Motor speed

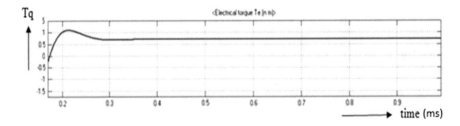

Fig. 19 Motor torque

2.5 Comparison Between ∏-Filter, T-Filter and C-Filter

See Figs. 20 and 21.

$$\text{Voltage Ripple Factor} = \frac{(\text{Vmax} - \text{Vmin})}{\text{Vaverage}}$$

$$\text{Current Ripple Factor} = \frac{(\text{Imin} - \text{Imax})}{\text{Iaverage}}$$

From the results obtained from Tables 1 and 2, it is concluded that SEPIC with C-filter is more efficient and provides lesser voltage and current ripple compared to other filters.

3 Closed-Loop Simulation

Closed-loop simulation provides a comparison between PI converter, FOPID converter and FL converter of the system based on their characteristics like input voltage, output voltage, output current, motor speed, motor torque. Based on these characteristics, the time domain parameters of Tr, Ts, Tp, Ess are determined; on comparing these values, the converter with lower value of Ess is taken for the proposed system. FL converter is taken as a suitable converter for the proposed system, since it has comparatively lower value of Ess.

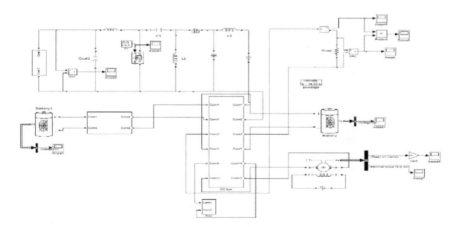

Fig. 20 DC smart grid EV charging system with SEPIC (∏-filter)

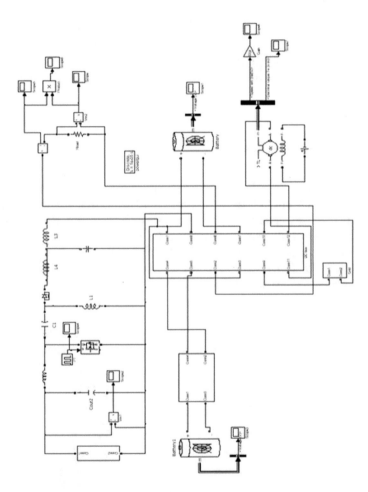

Fig. 21 DC smart grid EV charging system with SEPIC (T-filter)

Table 1 Comparison of output voltage, output current, output power

Converters	Vin (V)	Vout (V)	Io (A)	Po (W)
Buck–boost	12	70	0.44	32
SEPIC	12	88	0.89	78

Table 2 Comparision of output voltage and current ripple

Filter	Vor (V)	Ior (A)
C-Filter	0.011	0.005
T-Filter	0.002	0.003
∏-Filter	0.003	0.001

3.1 Comparison Between PI Converter, FOPID Converter and FL Converter

See Fig. 22.

Simulation Results

See Figs. 23, 24, 25, 26, 27, 28, 29, 30, 31, 32, 33, 34 and Table 3.

Simulation Results

See Figs. 35, 36, 37, 38, 39 and 40.

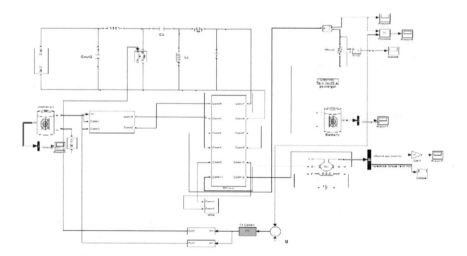

Fig. 22 Circuit diagram of DC smart grid with PI controller

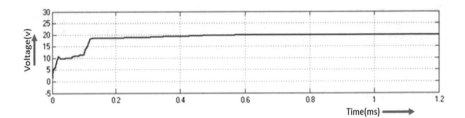

Fig. 23 Input voltage

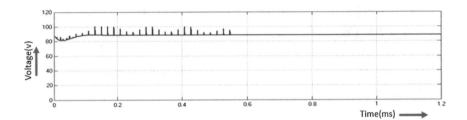

Fig. 24 Output voltage

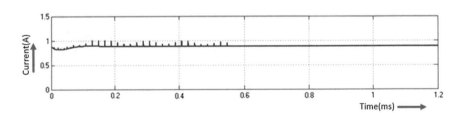

Fig. 25 Output current

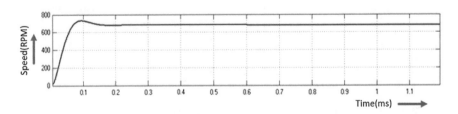

Fig. 26 Motor speed

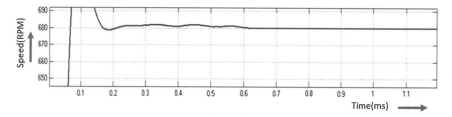

Fig. 27 Motor speed zoom out

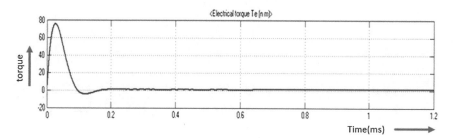

Fig. 28 Motor torque

3.2 Fuzzy Logic Controller

See Fig. 41.

3.3 Overview of Fuzzy Logic Controller

See Fig. 42 and Table 4.

Simulation Results

See Figs. 43, 44, 45, 46 and 47.

From the results obtained from Table 5, it is concluded that FL controller is more efficient and provides lesser steady-state error (Ess) compared to other controllers.

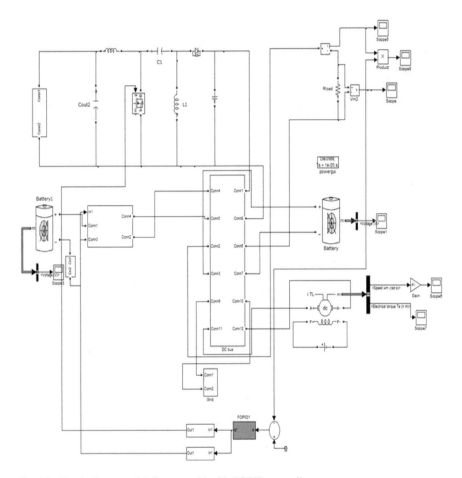

Fig. 29 Circuit diagram of DC smart grid with FOPID controller

Fig. 30 KD (saturation)

Fig. 31 KI (integrator)

Fig. 32 Kp (PID controller)

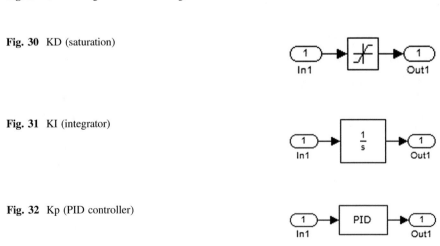

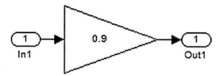

Fig. 33 Ks (gain)

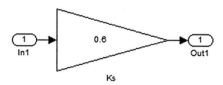

Fig. 34 Gain

Table 3 Functional block parameters of PI and FOPID

Parameters	PI controller	FOPID controller
Kp	0.16	0.16
KI	0.6	0.5
Ks	0.6	0.9
KD	–	0.000009

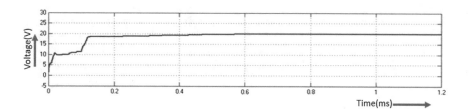

Fig. 35 Input voltage

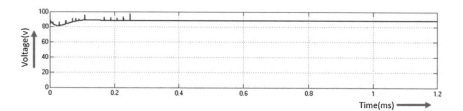

Fig. 36 Output voltage

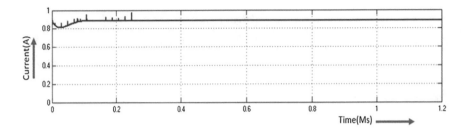

Fig. 37 Output current

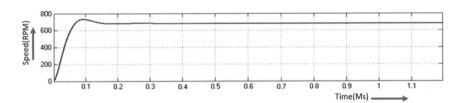

Fig. 38 Motor speed

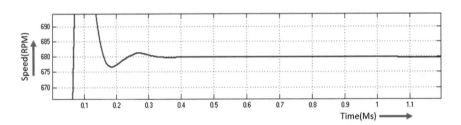

Fig. 39 Motor speed zoom out

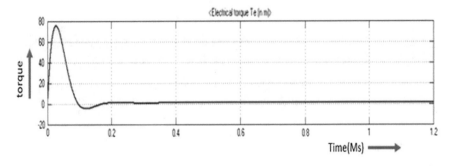

Fig. 40 Motor Torque

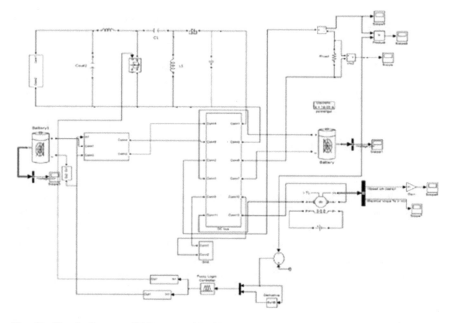

Fig. 41 Circuit diagram of DC smart grid with FL controller

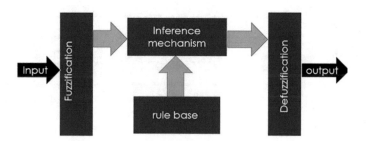

Fig. 42 FLC block diagram

Table 4 Rule base table of FLC	e/Δe	P1	P2	P3
	P1	P3	P2	P1
	P2	P1	Z	P3
	P3	Z	P2	P3

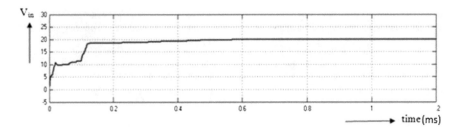

Fig. 43 Input voltage

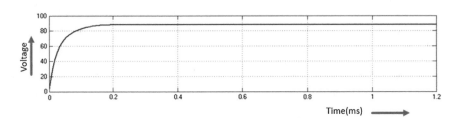

Fig. 44 Output voltage

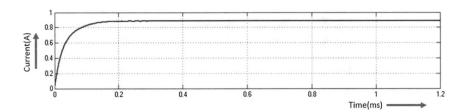

Fig. 45 Output current

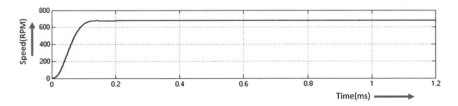

Fig. 46 Motor speed

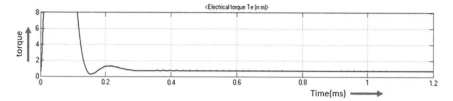

Fig. 47 Motor torque

Table 5 Comparison of time domain parameter

Controllers	Tr (ms)	Ts (ms)	Tp (ms)	Ess (V)
PI	0.11	0.60	0.30	3.3
FOPID	0.09	0.26	0.23	1.9
FL	0.05	0.19	0.16	0.6

References

1. Khan, A., Memon, S., Sattar, T.P.: Analysing integrated renewable energy and smart-grid systems to improve voltage quality and harmonic distortion losses at electric-vehicle charging stations. IEEE Access **6**, 404–415 (2017)
2. Sheng, M., Zhai, D., Wang, X., Li, Y., Shi, Y., Li, J.: Intelligent energy and traffic coordination for Green cellular network switch hybrid energy supply. IEEE Trans. Veh. Technol. **66**, 631–646 (2016)
3. Jiménez, E., Carrizosa, M.J., Benchaib, A., Damm, G., Lamnabhi Lagarrigue, F.: A new generalized power flow method for multi connected DC grids. Proc. Int. J. Elect. Power Energy Syst. **74**, 329–337 (2016)
4. Yong, J.Y., Ramachandaramurthy, V.K., Tan, K.M., Mithulananthan, N.: Bi-directional electric vehicle fast charging station with novel reactive power compensation for voltage regulation. Int. J. Electr. Power Energy Syst. **64**, 300–310 (2015)
5. Clement-Nyns, K., Haesen, E., Driesen, J.: The impact of charging plug-in hybrid electric vehicles on a residential distribution grid. IEEE Trans. Power Syst. **25**, 371–380 (2009)
6. Lin, C.-C., Filipi, Z., Wang, Y., Louca, L., Peng, H., Assanis, D., Stein, J.: Integrated, feed-forward hybrid electric vehicle simulation in SIMULINK and its use for power management studies. SAE Technica Paper 2001-01-1334 (2001)
7. He, F., Wu, D., Yin, Y., Guan, Y.: Optimal deployment of public charging stations for plug-in hybrid electric vehicles. Transp. Res. Part B: Methodol. **47**, 87–101 (2013)
8. Masoum, A.S., Deilami, S., Moses, P.S., Masoum, M.A.S., Abu-Siada, A.: Smart load management of plug-in electric vehicles in distribution and residential networks with charging stations for peak shaving and loss minimisation considering voltage regulation. IET Gener. Transm. Distrib. **5**, 877–888 (2011)
9. Jia, L., Hu, Z., Song, Y., Luo, Z.: Optimal siting and sizing of electric vehicle charging stations. In: 2012 IEEE International Electric Vehicle Conference (2012)
10. Zheng, Y., Dong, Z.Y., Xu, Y., Meng, K., Zhao, J.H., Qiu, J.: Electric vehicle battery charging/swap stations in distribution systems: comparison study and optimal planning. IEEE Trans. Power Syst. **29**, 221–229 (2013)
11. Ge, S., Feng, L., Liu, H.: The planning of electric vehicle charging station based on grid partition method. In: 2011 International Conference on Electrical and Control Engineering (2011)

12. Tomic, J., Kempton, W.: Using fleets of electric-drive vehicles for grid support. J. Power Sources **168**, 459–468 (2007)
13. Liu, Z., Wen, F., Ledwich, G.: Optimal planning of electric-vehicle charging stations in distribution systems. IEEE Trans. Power Deliv. **28**, 102–110 (2013)
14. Etezadi-Amoli, M., Choma, K., Stefani, J.: Rapid-charge electric-vehicle stations. IEEE Trans. Power Deliv. **25**, 1883–1887 (2010)
15. Bracale, A., Caramia, P., Carpinelli, G., Mottola, F., Proto, D.: A hybrid AC/DC smart grid to improve power quality and reliability. In: 2012 IEEE International Energy Conference and Exhibition (ENERGYCON) (2012)
16. Yamauchi, H., Kina, M., Kurohane, K., Yona, A., Senjyu, T.: Protection design of DC smart grid. J. Int. Counc. Electr. Eng. **2**, 242–249 (2012)
17. Kurohane, K., Senjyu, T., Yona, A., Urasaki, N., Muhando, E.B., Funabashi, T.: A high quality power supply system with DC smart grid. IEEE PES T&D 2010 (2010)
18. Balog, R.S., Weaver, W.W., Krein, P.T.: The load as an energy asset in a distributed DC smart grid architecture. IEEE Trans. Smart Grid **3**, 253–260 (2012)
19. Wang, B., Sechilariu, M., Locment, F.: Intelligent DC microgrid with smart grid communications: control strategy consideration and design. In: 2013 IEEE Power & Energy Society General Meeting (2012)
20. Su, W., Eichi, H., Zeng, W., Chow, M.-Y.: A survey on the electrification of transportation in a smart grid environment. IEEE Trans. Ind. Inf. **8**, 1–10 (2012)
21. She, X., Yu, X., Wang, F., Huang, A.Q.: Design and demonstration of a 3.6-kV–120-V/ 10-kVA solid-state transformer for smart grid application. IEEE Trans. Power Electron. **29**, 3982–3996 (2014)

PVW: An Efficient Dynamic Symmetric Cipher to Enable End-to-End Secure Communication

V. Panchami, Varghese Paul and Amitabh Wahi

Abstract There is a global move towards the usage of mobile devices due to the tremendous growth in mobile technologies. These mobile devices are utilized by more than billions of users for commercial and social purposes. In many scenarios, sensitive information should be sent from our mobile phones through the Internet, which is considered as an open and insecure network. Therefore, we have to secure our sensitive data before sending from our mobile phones. These mobile phones have limitations in resources such as memory space, processing and battery capacity. In this paper, a secure, fast and efficient cryptographic algorithm for mobile and lightweight cryptographic application is proposed. PVW cipher is a block symmetric cipher, which follows a Feistel structure. PVW cipher works on two basic principles, confusion and diffusion. The basic operations used are permutation, substitution, mixing, addition and XOR (exclusive OR). In PVW cipher, the basic operations are dynamic in nature. The inputs to PVW cipher are 128-bit plain text and 160-bit key. In PVW cipher, there are 8–16 rounds. Finally, we show an implementation of PVW cipher in eight rounds. The performance of PVW cipher is measured in terms of time complexity, space complexity, security. The results show that PVW cipher is very efficient, fast and secure against brute force attack, differential attack and linear attack.

Keywords Substitution · Transposition · Confusion · Diffusion · Differential cryptanalysis · Linear cryptanalysis

V. Panchami (✉)
Computer Science and Engineering, Indian Institute of Information Technology,
Kottayam, India
e-mail: panchamam036@iiitkottayam.ac.in

V. Paul
Rajagiri School of Engineering and Technology, Kochi, Kerala, India
e-mail: vp.itcusat@gmail.com

A. Wahi
Information Technology, Bannari Amman Institute of Technology, Coimbatore,
Tamil Nadu, India
e-mail: awahi@rediffmail.com

© Springer Nature Singapore Pte Ltd. 2020   329
S. S. Dash et al. (eds.), *Artificial Intelligence and Evolutionary Computations
in Engineering Systems*, Advances in Intelligent Systems and Computing 1056,
https://doi.org/10.1007/978-981-15-0199-9_28

1 Introduction

Mobile phones are used for a wide range of applications such as for making phone calls, voice or video call, chatting, sending messages, mobile commerce and mobile banking and to access social networking sites. Now, there are many mobile applications and mobile versions of these commercial and social purposes, and sites are freely available. A major portion of these mobile devices, which are used to communicate between users and other purposes, are of the smartphones. In the present scenario, when considering the global Internet traffic, the mobile traffic accounts for more than half of the global Internet traffic. If a user uses these smartphones for commercial purposes, he has to enter his credential to the Internet or if he communicates with other users, his data should be kept confidential. Thus, some actions must be taken to ensure data confidentiality. Encryption [4] is a best method to achieve data confidentiality. Therefore, the aim of this research work is to apply the area, cryptography in an extremely different area called mobile computing for secure mobile communication.

2 Literature Survey

The various symmetric ciphers analysed are DES, 3-DES, AES, RC6, Blowfish, TEA, IDEA, Skipjack and GOST. In DES, hardware implementation of DES is very fast but software implementation is slow. Broken in Brute force Attack. It is vulnerable to linear and differential cryptanalysis. 3-DES is more secure than DES and easy to implement. It runs 3 times slower than DES. AES is faster in both software and hardware. AES [1] provides a strong resistance against both linear and differential cryptanalyses. Increasing the key size of AES leads to the increase in power consumption, while reducing the number of rounds leads to power savings but it makes AES insecure. RC6 is a secure compact and simple block cipher giving high performance in small devices. In RC6, brute force attack [2] is possible and it is also vulnerable to differential cryptanalysis. Blowfish is one of the fastest block ciphers except when changing the keys. Blowfish [3] is not subjected to any patents. Use Feistel cipher using large key-dependent S-boxes. Each new key requires pre-processing equivalent to encrypting about 4 kilobytes of text, which is very slow compared to other ciphers. TEA [4] is simple to implement and faster. It is not subjected to any patents. TEA is also susceptible to related key attack, which requires 2^{32} chosen plain text under a related key pair with 2^{32} time complexity. IDEA is a linear, differential and algebraic cryptanalysis which is difficult. The key size of Skipjack [5] is only 80 bits. It can be broken through differential cryptanalysis. GOST, since S-box is secret, Linear, differential and algebraic cryptanalysis is difficult. GOST [6] is vulnerable to generic attacks based on its short (64-bit) block size.

3 Proposed System: PVW Cipher

3.1 Key Whitening

In this paper, we propose a block symmetric cipher for secure mobile applications. The plain text is 128 bits, and the key size is 160 bits. The 160-bit key is treated as the master key K, and it is subdivided as multiples of 8 into k_1 as 40 bits, k_2 as 24 bits, k_3 as 32 bits and k_4 as 64 bits. In Fig. 1, PVW cipher follows the Feistel structure. The plain text of 128 bits is divided into two equal parts, left-hand side (LE) and right-hand side (RE) of 64 bits each. The two halves have undergone a whitening process. The plain text whitening process [7] decreases the difficulty of key search attacks against the remainder of the cipher. The left-hand side is XORed with the concatenation of the subkeys k_1 and k_2, while the other half (RE) is again subdivided into two equal halves of 32 bits each; the 32-bit left-hand side is XORed

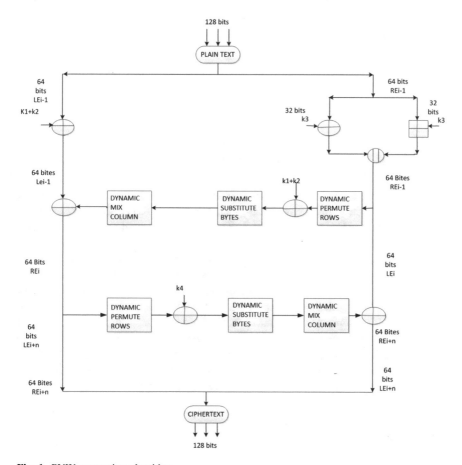

Fig. 1 PVW encryption algorithm

with the subkey k_3, and the right-hand side is added with the k_3. Here, adding is done with addition modulo 2^{32}, since our field is 32 bits.

The output of both sides each having 32 bits is again concatenated to form 64 bits. Next starts the rounds, there are 8–16 rounds for PVW cipher, and in this paper 8 rounds are used. Each round has a round function. There are two inputs for the round function, the right-hand side and the subkeys. In the first round, the right-hand side and the concatenated k_1 and k_2 are used as keys, while in the second round right-hand side and the subkey k_4 are used as the input of the round function. Each round function has four basic operations, and it is a combination of permutation, XOR, substitution and mixing to ensure confusion and diffusion [8] in our PVW cipher.

For the first round:

$$\mathrm{RE}_i = [\mathrm{LE}_{i-1} \oplus F(\mathrm{RE}_{i-1}, (k_1 + k_2))] \tag{1}$$

$$\mathrm{LE}_i = \mathrm{RE}_{i-1} \tag{2}$$

For the second round:

$$\mathrm{LE}_{i+1} = [\mathrm{RE}_i \oplus F(\mathrm{LE}_i, (k_4))] \tag{3}$$

$$\mathrm{RE}_{i+1} = \mathrm{LE}_i \tag{4}$$

4 Subkey Generation Algorithm

The aim of the step is to generate the subkeys k_1, k_2, k_3, k_4 and four seeds, which used to create entries in PVW dynamic colour table (DCT). The four seeds are SEEDA, SEEDR, SEEDG and SEEDB. First, we have to read the master key K, a number having 40 decimal digits (160 bits). From the master key K, read first 10 digits and store it as the first subkey k_1. Read the next 5 digits, the next 9 digits and the last 16 digits from K and store it as k_2, k_3 and k_4, respectively. The subkey values k_1, k_2, k_3, k_4 are applied to the elliptic curve equation in the Step 13 of Algorithm 2 to get the master seed w. The master seed w ranges from 0 to 4294967291, by applying mod function to the equation with the closest prime number of 2^{32} (4294967296). This master seed w acts as the initial values to generate the four seeds. The master seed w is passed through congruent residue function to generate four unique values which act as four seeds. Then, the master key K is considered to be accepted; otherwise, again enter a valid key.

5 Dynamic Permute Bytes

The first step of the round function is permute bytes, and the 64-bit half of the plain text in each round is stored in state array S. The state array is an 8×8 matrix. The permutation is based on subkeys k_3, k_3 which is 32 bits. It can be represented as 8-digit decimal number. The first row is permuted with respect to the first decimal in k_3. For example, if the k_3 is 15632637, then the first row of the state array S is made one-time circular left shift, and the second row is permuted circularly 5 times. The third, fourth, fifth, sixth, seventh and eighth row is circularly permuted 6 times, 3 times, 2 times, six times, three times and seven times, respectively. The output of each row is copied to the output array O.

6 Dynamic Round Function

The next step of the round function is the XOR operation. In this step, for the first round the values in the output array O are XORed with the subkeys, which is concatenation of two subkeys k_1 and k_2 ($k_1 + k_2$). In the second round, the output array O is XORed with the subkey k_4 of 64 bits. In PVW cipher, 8 rounds are used. The first subkeys ($k_1 + k_2$) and k_3 are used in the first round. Then, a circular left shift is done for the third round, 2 shifts for the fourth round, 3 shifts for the fifth round, 4 shifts for the sixth round, 5 shifts for the seventh round and 6 shifts for the last eighth round. The output of the XOR operation is copied to the output array O.

7 Dynamic Substitute Bytes

In substitute bytes, each row (8 bits) in output array is substituted with another 8 bits. Here, substitution is done with a dynamic S-box called dynamic colour table (DCT) [9], which depends on the subkeys K_1, K_2 and K_3. The dynamic colour table (DCT) is a 64×4 S-box. So, there are 256 entries in the S-box. Sixty-four entries in the S-box are unique. In PVW cipher, the dynamic colour table is a key-dependent S-box. The 256 values in S-box are generated based on the subkeys, k_1, k_2 and k_3. Each 8-bit output array is replaced by the 8-bit value intersecting by the 6-bit row and the 2-bit column. To ensure the basic property of a symmetric cipher, confusion is achieved through this step substitute bytes.

8   Dynamic Mix Bytes

In mix bites, we use a lightweight maximum distance separable (MDS) matrix, to enable diffusion in PVW cipher. The MDS [10] increases the avalanche characteristics of the block cipher. The results in the output array and converted into bytes. Those bytes are stored in a 4×2 matrix, which is called the mix matrix. The 4×4 MDS matrix is multiplied by the 4×2 mix matrix to get 4×2 output, which is 64 bits. Here, addition and multiplication are done in GF (2^8), and XOR operation is used for addition. This output is again copied to the output matrix O.

In the first round, the following operation is done:

$$\text{MDS} \, X \, A = \begin{pmatrix} 01 & 01 & 01 & 02 \\ 01 & 01 & 02 & 01 \\ 01 & 02 & 01 & 01 \\ 02 & 01 & 01 & 01 \end{pmatrix} \times \begin{pmatrix} 1B & 09 \\ B8 & A6 \\ 24 & 2D \\ 34 & AD \end{pmatrix} \Rightarrow \begin{pmatrix} A0 & A1 \\ A2 & A3 \\ A4 & A5 \\ A6 & A7 \end{pmatrix}$$

In the second round:

$$\text{MDS} \, X \, A = \begin{pmatrix} 01 & 02 & 01 & 03 \\ 02 & 01 & 03 & 01 \\ 01 & 03 & 01 & 02 \\ 03 & 01 & 02 & 01 \end{pmatrix} \times \begin{pmatrix} 0A & 27 \\ 1A & C5 \\ D8 & 30 \\ BA & 82 \end{pmatrix} \Rightarrow \begin{pmatrix} B0 & B1 \\ B2 & B3 \\ B4 & B5 \\ B6 & B7 \end{pmatrix}$$

9   Implementation

The PVW cipher is implemented in Java and Android [11]. The IDE used is Eclipse ADT; operating system used is Windows 7 and Android. The smartphone specifications where PVW application is installed: Android version used is 4.4.2, kernel version is 3.4.67, smartphone model number is 8x-1000, and the baseband version is Modem_XOLO_8x-1000_S059_13042015. The programming languages used are Android and Java. Figure 2 shows the screenshot of PVW mobile application.

10   Analysis of PVW Cipher

10.1   Theoretical Correctness

Theorem *The encryption algorithm is the inverse of the decryption algorithm. Encryption function at each round*:

Fig. 2 User interface of
PVW cipher mobile
application

$$RE_i = LE_{i-1} \oplus F[DM(S(k_1 k_2 \oplus DP(RE_{i-1})))] \tag{5}$$

$$LE_i = RE_{i-1} \tag{6}$$

Decryption function at each round:

$$LE_{i-1} = RE_i \oplus F[DM(S(k_1 k_2 \oplus DP(LE_i)))] \tag{7}$$

$$RE_{i-1} = LE_i \tag{8}$$

Proof Let

$$RE_i \oplus F[DM(S(k_1 k_2 \oplus DP(LE_i)))] = LE_{i-1}$$

Applying commutative law in Eq. 7

$$LHS = RE_i \oplus F[DM(S(k_1 k_2 \oplus DP(LE_i)))]$$
$$= (LE_{i-1} + F[DM(S(k_1 k_2 \oplus DP(RE_{i-1})))]) \oplus F[DM(S(k_1 k_2 \oplus DP(LE_{i-1})))]$$

Applying Eq. 5

$$= (LE_{i-1} + F\lceil DM(S(k_1 k_2 \oplus DP(LE_{i-1})))\rceil) \oplus F\lceil DM(S(k_1 k_2 \oplus DP(LE_{i-1})))\rceil$$

Applying Eq. 12

$$= LE_{i-1} = RHS$$

10.2 PVW Throughput

The encryption time for PVW is computed by taking the time to produce the ciphertext from a plain text block. The throughput of the encryption scheme is calculated as the total plain text in bytes encrypted divided by the encryption time in seconds [12]. From Table 1, Figs. 3 and 4, it is very clear that our PVW cipher has high throughput for encryption and decryption when compared with other ciphers. It shows that PVW is very efficient cipher. Efficiency is expressed in terms of throughput. Throughput of the encryption algorithm,

$$T = Tp/Et \tag{9}$$

where T_p is total plain text in bytes and E_t is average time of the various encryptions.

Table 1 Analysis of throughput (Kb/sec) of different symmetric encryption algorithms

File size in KB	DES	3-DES	AES	Blowfish	PVW
29	20	33	39	16	12
58	29	61	60	17	18
116	39	51	42	26	21
242	49	82	99	51	37
323	48	120	145	36	48
789	83	168	145	56	58
953	145	238	223	115	106
5783	258	286	263	212	196
7176	269	317	211	232	228
7197	1286	1479	1489	1074	983
22,467	1727	1807	1589	1459	1167
41,759	2100	2301	1537	1894	1678
98,645	2610	2753	1956	2158	1857
Average time	666.3846	745.846	599.846	565.08	493
Throughput (T)	21.41	19.13	23.79	25.25	28.95

Fig. 3 Throughput analysis of algorithm

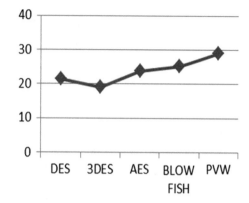

Fig. 4 Throughput analysis of Decryption encryption algorithm

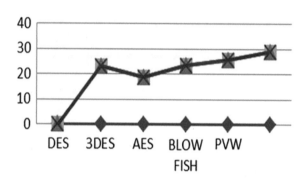

10.3 Security: Brute Force Attack

In PVW cipher, the key used is 160 bits. If the key size is n bits, then the possible combination of key is 2^n. Therefore, the possible combination of 160-bit key is 1.46×10^{48}. In one year, there are 31536000 s.

$$1 \text{ year} = 365 \times 24 \times 60 \times 60 = 31536000 \, s \qquad (10)$$

A fast supercomputer is capable of generating 33.86 quadrillion keys per second. Therefore, in one year the supercomputers are able to generate 1.067×10^{24} keys.

$$33.86 \, \text{PFLOP/S} = 33.86 \times 10^{15} \, \text{FLOP/S} \qquad (11)$$

$$31536000 * (3.386 \times 10^{16}) = 1.067 \times 10^{24} \qquad (12)$$

The time, T, to play a brute force attack to break 160-bit key is 1.369×10^{24}. So using brute force attack, it will take 1.369×10^{24} years to break the PVW cipher. We analyse the security in terms of brute force attack on the key K of size 160 bits produces a higher value collision resistance that is impractical.

$$T = \frac{1.461 \times 10^{48}}{1.067 \times 10^{24}} = 1.369 \times 10^{24} \qquad (13)$$

11 Conclusion

In this paper, we developed a new symmetric cipher for mobile devices to ensure secure end-to-end mobile communication. The above result shows that the computational cost of PVW cipher is very less, good security and high performance. PVW cipher is very efficient for small as well as big messages. In PVW cipher, efficiency is calculated in terms of throughput. The throughput rate is very high for PVW encryption and decryption algorithm when compared to all other symmetric block ciphers. It is proven that PVW cipher is very fast and efficient cipher for mobile platforms. It is also proven that our PVW cipher is resistant against brute force attacks and linear and differential attacks. Thus, PVW cipher is secure too.

References

1. Daemen, J., Rijmen, V.: AES Proposal: Rijndael. National Institute of Standards and Technology (2013)
2. Kofahi, N.A., Al-Somani, T., Al-Zamil, K.: Performance evaluation of three encryption/decryption algorithms. In: Proceedings of the 46th IEEE International Midwest Symposium on Circuits and Systems, MWSCAS '03, pp. 27–30 (2013)
3. Mousa, A.: Data encryption performance based on blowfish., 47th International Symposium ELMAR (2005)
4. Schneier, B., Kelsey, J., Whiting, D., Wagner, D., Hall, C., Ferguson, N.: Twofish: A 128-Bit block Cipher, available: http://www.counterpane.com/twofish.html
5. Hussain, I., Shah, T., Gondal, M.A., Wang, Y.: Analyses of SKIPJACK S-box. World Appl. Sci. J. **13**(11), 2385–2388 (2011). ISSN 1818-4952
6. Schneier, B.: The GOST encryption algorithm. Dobb's J **20**(1) 123–124 (1995)
7. Potlapallyt, N.R., Ravit, S., Raghunathant, A., Niraj, K.: A study of the energy consumption characteristic of cryptograhic algorithms and security protocols. IEEE Trans. Mobile Comput. [Online]. **5**(2), 128–143 (2006)
8. Shannon, C.E.: Communication theory of secrecy systems. Bell Syst. Tech. J. **27**(4), 656–715 (1948)
9. Vijayan, P., Paul, V., Wahi, A.: Dynamic colour table: a novel S-box for cryptographic applications. Int. J. Commun. Syst. e3318 (2017)
10. Gupta, K.C., Ray, I.G.: On constructions of MDS matrices from companion matrices for lightweight cryptography. In International Conference on Availability, Reliability, and Security, pp. 29–43. Springer, Berlin, Heidelberg (2013)
11. Lee, S.: Creating and using databases for android applications. Int. J. Database Theory Appl. [Online]. **5**(1), 99–106 (2012)
12. Elminaam, D.S.A., Abdul Kader, H.M.: Performance evaluation of symmetric encryption algorithms. Commun. IBIMA, [Online] **8**(8), 58–64 (2005)

Smart Monitoring System for Football Players

P. Sagarika, V. Arjun, Shiv Narain and Sabitha Gauni

Abstract By implementing an IoT-based smart monitoring system, the aim is to facilitate and aid the process of analyzing different aspects of football players which are otherwise difficult to identify during practice sessions. The major applications designed in this system are force imparted on the ball by the player by using a 3-axis accelerometer, angular movements of the player during game time by using a gyroscope, stamina of the player and tackle detection system to indicate any type of foul committed on the player and the nature of the foul. This is done by implementing the system on the foot of a player and collecting real-time values based on the actions produced by the player. By doing this, we gather accurate numerical values and graphical models of the player, thus enabling the coaches and managers to take decisions for enhancing quality of players. Player comparison is one of the most vital aspects for a manager to determine which player fits into his tactics whether or not a particular player is suited to his style of play and philosophical approach, thus potentially increasing his team's chances of winning trophies. A tackle detection system is also implemented to aid the referee in making just and fair decisions to mitigate any unethical actions such as diving, which dampen the quality of the beautiful game. This system helps the referee decide if a player has involved himself in a diving simulation or was genuinely fouled by the opponent. Also, based on the intensity of the foul, this system helps the referee decide whether the player should be given a warning (yellow card) or should be sent off (red card).

Keywords Football analysis · Monitoring system · Real-time values · Node MCU · MPU 6050

P. Sagarika (✉) · V. Arjun · S. Narain · S. Gauni
SRM Institute of Science and Technology, Chennai, India
e-mail: sagarika2905@gmail.com

V. Arjun
e-mail: ronarjun@gmail.com

S. Narain
e-mail: shivnarain_peeyushanand@srmuniv.edu.in

S. Gauni
e-mail: sabithag@srmist.edu.in

© Springer Nature Singapore Pte Ltd. 2020
S. S. Dash et al. (eds.), *Artificial Intelligence and Evolutionary Computations in Engineering Systems*, Advances in Intelligent Systems and Computing 1056,
https://doi.org/10.1007/978-981-15-0199-9_29

1   Introduction

Football is one of the oldest and most popular sports in the world. The fan base
around this sport is huge. With advancement in technology, football has become all
the more interesting in terms of tactical and technical analysis. In olden days, there
was not any major technology which could be used to study opponents' tactics. But
modern-day football is different. It allows coaches to do a thorough study of his
rival or opponents using visual analysis and set up his team accordingly [1]. This
can be both an asset and a liability in that it allows our team to study the opponents
and allow the opponents to study us. But there is a need for a machine to analyze
and evaluate the live performances of players, something which has been of urgent
need to the managers to help them study players more effectively and quickly. This
machinery [2] provides an easy platform for coaches to learn players and thus
determine how to develop him. Every manager has a unique view to the game of
football, due to their own philosophical approach and in-game tactics. A player
having less acceleration than another need not mean he is less preferred. Managers
who focus more on possession-based and passing game need not require his player
to have a high acceleration as opposed to another manager who prefers defensive

Fig. 1 Former archetype of
football machinery

football and launch his attackers quickly on the counterattack. Hence, this system helps coaches, managers and trainers to determine which player will be suited to his methodology and understanding of the sport. Moreover, a tackle detection system is built which helps the referees in giving the right decision whether a player engaged himself in a dive or was genuinely fouled by an opponent. Thus, the chances of a game to be spoilt because of an incorrect decision by the referee are ameliorated (Fig. 1).

2 Prototype

The entire football machinery is split into three different systems which are integrated into our soccer analysis application (Fig. 2).

2.1 Sensor System

A node MCU interfaced with MPU 6050 is used in this system for collection of values.

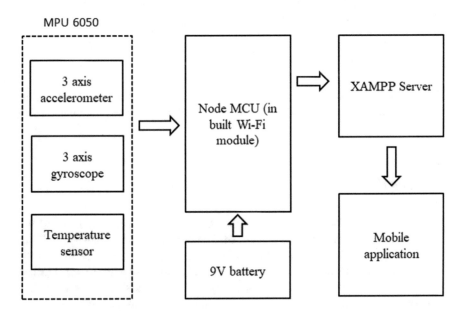

Fig. 2 Architecture design

Fig. 3 MPU 6050

2.1.1 MPU 6050

MPU 6050 has 8-pin module with in-built MEMS technology. It produces 6-axis output values. The DMP is used for synchronization and detection. MPU 6050 is embedded with a temperature sensor. It can take VDD supply voltage range of 2.33.46 V. The gyro operating current is 3.6 mA, and gyro + acceleration operating current is 3.8 mA. It uses an I2C bus interface (Fig. 3).

2.1.2 Node MCU

The node MCU uses an open-source platform. It has an in-built Wi-Fi chip. The serial communication protocols used are UART, SPI, I2C. It is programmable using Arduino IDE (Fig. 4).

2.2 Database System

The input values are taken from MPU 6050. Acceleration in *XYZ*-axes is found out, and roll, pitch and yaw values are obtained. The parameters are calculated in DMP of node MCU. Application programming interface (API) testing is done with the help of Postman. Creation of database and table is done in XAMPP server. Further, we develop a mobile application displaying various parameters consisting of a user-friendly GUI. Then, the sensor values are sent to the mobile application and are viewed on the device.

Fig. 4 Node MCU

2.3 Analytics System

This football machinery is placed on the foot of a player, and it collects real-time values and presents graphical models of the player. An analytical system has been devised which aids the coaches and managers to take different types of decisions such as buying and selling the player for a particular fee, trading the player for a possible upgrade depending on the quality of the player and training and upgrading the existing player to enhance the quality of the team (Fig. 5).

Fig. 5 Mobile application

3 Methodology

This system comprises of two hardware components, namely MPU 6050, which is the sensor, and node MCU, which is the microcontroller. The MPU 6050 has an in-built 3-axis accelerometer and a 3-axis gyroscope which helps us in calculating the sensor values of acceleration and angular movements by the corresponding football player. The raw values obtained in the three axes are converted into a mean value. Further, the force, velocity and the stamina of the player are calculated. These values are then sent to the database which is explained below.

XAMPP is a cross-platform for Apache, MySQL, Pearl and PHP programming languages. MySQL is used primarily to create a database. Additionally, a table is created inside the corresponding database, and columns are added to the table for each parameter. The values from sensor are stored in this table.

A mobile application is developed using ionic framework 3, where a profile for the player is created highlighting his personal details. He gets an evaluation chart based on his performance.

4 Theoretical Analysis

4.1 Force Imparted by the Player

Force is a very important parameter to be measured. A goal can be achieved only with the right amount of force. A high or low amount of force will not help in conceding a goal. The force can be calculated using the acceleration value obtained.

Accelerometer in the MPU 6050 provides 3-axis output data so that the RMS acceleration can be obtained.

$$\text{Acc(rms)} = \sqrt{\left(\text{Acc}(x)^2 + \text{Acc}(y)^2 + \text{Acc}(z)^2\right)}$$

The equation for force exerted by the player is

$$F = m * a$$

where mass is the weight of player during play.

4.2 Angular Movements of the Player

The node MCU also has a built-in gyroscope which measures the angular movements. The angle of the foot is important to make a perfect shot toward the goal.

$$\text{Gyro(rms)} = \sqrt{\left(\text{Gyro}(x)^2 + \text{Gyro}(v)^2 + \text{Gyro}(z)^2 \right)}$$

4.3 Stamina of the Player

Stamina can be determined for a period of time by calculating the average of the acceleration over that period of time divided by the time interval under which this parameter is calculated.

$$\text{Stamina} = \text{Acc(rms)}/\text{time}$$

5 Tackle Detection System

Football is known as the beautiful game, yet there are many unethical tactics which players often resort to, to get decisions in their favor. The tackle detection system is devised in order to remove any possible corrupt and ugly methods which destroy the game. This mainly aids the referee to make just and fair decisions, thus not disrupting the quality of the game.

There can be two ways a player can commit himself to the floor, either by being fouled by the opponent or by engaging in a diving simulation without any harm committed by the opponent. If a contact by the opponent is made on the player, it is a genuine punishable offense; else if there is no contact, it is a dive. A threshold for force is set to indicate whether the foul is worthy of a yellow card or a red card with the value of a red card being set to a higher limit.

6 Results

The parameters are calculated, stored in the database and reflected in the mobile application developed in both numerical and graphical ways.

The data from the database sent to the mobile application will help the coaches and managers in making tactical decisions by monitoring the performance of the players during both practice and game play. It can aid the coaches and managers to make important decisions such as buying/selling and exchanging players suited to his tactics. The numerical values are plotted in a graphical method for easier analysis and comparison of the different approaches by each player (Figs. 6, 7 and 8).

```
◎ COM4
│
16:25:04.156 -> Angular movement = 2810.32
16:25:04.204 -> Force = 8.47
16:25:04.204 -> Acceleration = 18.81
16:25:04.204 -> Angular movement = 2810.32
16:25:04.204 -> Force = 8.47
16:25:04.204 -> Acceleration = 18.81
16:25:04.204 -> Angular movement = 2810.32
16:25:04.204 -> Force = 8.47
16:25:04.204 -> Acceleration = 18.81
16:25:04.204 -> Angular movement = 2810.32
16:25:04.204 -> Force = 8.47
16:25:04.204 -> Acceleration = 18.81
16:25:04.204 -> Angular movement = 2810.32
16:25:04.204 -> Force = 8.47
16:25:04.204 -> Acceleration = 18.81
16:25:04.204 -> Angular movement = 2810.32
16:25:04.204 -> Force = 8.47
16:25:04.204 -> Acceleration = 18.81
16:25:04.204 -> Angular movement = 2810.32
16:25:04.204 -> Force = 8.47
16:25:04.204 -> Acceleration = 18.81
16:25:04.204 -> Angular movement = 2810.32
16:25:04.204 -> Force = 8.47
16:25:04.204 -> Acceleration = 18.81
16:25:04.204 -> Angular movement = 2810.32
16:25:04.204 -> Force = 8.47
16:25:04.204 -> Acceleration = 18.81
16:25:04.204 -> Angular movement = 2810.32
16:25:04.252 -> Force = 8.47
16:25:04.252 -> Stamina = 1.51
```

Fig. 6 Serial monitor output

7 Discussion

This model is very compact in size so it can be used as a wearable during training to analyze the performance of different football players. The players wear this machinery under the socks on the calf [2], and his quality is determined by the actions he produces on the ball when his performance is reflected in a mobile application designed particularly for this product. The major parameters included in this model are player acceleration, force imparted by the player, tackle detection system, stamina of the player and his foot orientation. The major contribution of this work, is to ease, the role of managers and coaches in comparing and analyzing different aspects of different players and thus make a decision which player fits into their system or style of play.

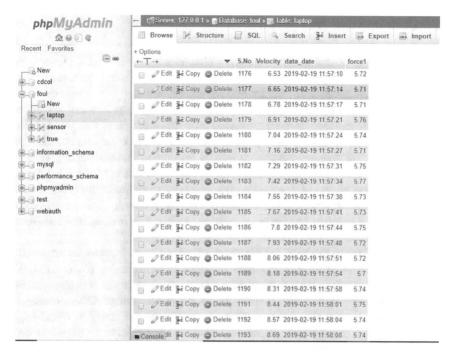

Fig. 7 Database

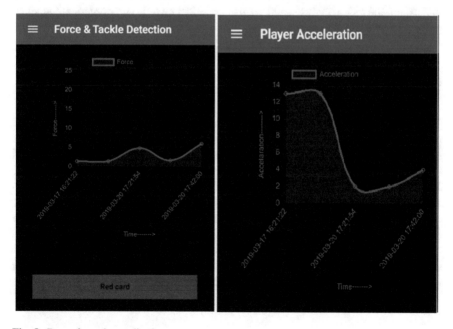

Fig. 8 Pages from the application

8 Comparative Analysis

The Zepp Play Soccer is one such device which uses a wearable track player stats, captures a real-time game report and uses auto-clip technology to create video highlights providing overall insights into motion picture analysis. While the Zepp Play Soccer highlights the importance of technical skills predominantly such as goal conversion, distance covered, preferred foot, this machinery, a slight variation albeit significant, focuses on more physical attributes such as stamina, acceleration with few technical characteristics such as kick speed and tackle detection system. This model is also an extension to the existing models in that not only coaches and managers are benefitted. Referees also find it easier to evaluate the type of foul, whether it is a diving simulation or a genuine punishable offense. This model is also very economical compared to the other existing models, thus making it affordable to the common man.

"Soccer: Who has the ball" [1] generates visual analytics from video arrays [3], produces player statistics [4] and thus provides a platform for player analysis. Self-analysis one of the most fundamental aspects of a sportsman and often failing to analyze themselves has been detrimental to their development. Often, managers tend to overlook what they have in their own academy and prefer to spend huge money on high-profile players. Hence, this model is also useful to scouts who travel from country to country to identify the players best suited for their team and get players for minor fee for higher resale value if the coach develops them into world-beaters. Thus, this model can be used for talent identification as well, only that except the video analysis, and we use graphical models and numerical values of players to evaluate player stats.

9 Conclusion

The smart monitoring system facilitates and enhances the quality of the beautiful game. It can be extended to be a part of football analytics, which analyzes players based on their performance. This system helps a player identify his strengths and weakness, draw a parallel between the two and aid his own development. Self-analysis one of the most important aspects of any sport and often failing to analyze himself has been detrimental to a player's development. It is also used by managers to selectively decide which player is best suited to his tactics. Also, the job of referees becomes easier with this system as it helps eradicate any ill-practices of the game such as diving. This is particularly important as referees are often made scapegoat and given scrutiny due to bad decisions none of which is their fault because not everything can be observed with the naked eye.

References

1. Theagarajan, R., Pala, F., Zhang, X., Bhanu, B.: Soccer: who has the ball? generating visual analytics and player statistics. In: IEEE Conference on Computer Vision and Pattern Recognition Workshops (CVPRW) (2018)
2. Sajjad Hossain, H.M., Khan, M.A.A.H., Roy, N.: SoccerMate: a personal soccer attribute profiler using wearables. In: 2017 IEEE Conference on Pervasive Computing and Communications (2017)
3. Stein, M., Janetzko, H., Breitkreutz, T., Seebacher, D., Schreck, T., Grossniklaus, M., Couzin, I.D., Keim, D.A.: Director's cut: analysis and annotation of soccer matches. IEEE Comput. Graph. Appl. **36**(5) (2016)
4. Stein, M., Janetzko, H., Lamprecht, A., Breitkreutz, T., Zimmermann, P., Goldlucke, B., Schreck, T., Andrienko, Grossniklaus, M., Keim, D.A.: Bring it to the pitch: combining video and movement data to enhance team sport analysis. IEEE Trans. Vis. Comput. Graph. **24**, 13–22 (2017)

Artificial Intelligence and Cybersecurity: Past, Presence, and Future

Thanh Cong Truong, Ivan Zelinka, Jan Plucar, Marek Čandík and Vladimír Šulc

Abstract The rapid development of Internet services also led to a significant increase in cyber-attacks. Cyber threats are becoming more sophisticated and automation, make the protections ineffective. Conventional cybersecurity approaches have a limited effect on fighting new cyber threats. Therefore, we need new approaches, and artificial intelligence can aid to counter cybercrime. In this paper, we present the capability of adopting artificial intelligence techniques in cybersecurity and present some of those intelligent-based approaches already in place in practice. Furthermore, we highlight the limitations of AI-based approaches in cybersecurity as well as suggest some directions for research in the future.

Keywords AI · Cybersecurity · Malware detection · Intrusion detection · Phishing detection · Advanced persistent threat

1 Introduction

Ever since the concept of artificial intelligence (AI) was proposed in 1956, many areas take advantage of AI, such as gaming, natural language processing, health care, manufacturing, education as well as cybersecurity. Today, cybersecurity is now a top concern in the digital age, where the role of AI is more crucial than ever

T. C. Truong (✉) · I. Zelinka · J. Plucar
Department of Computer Science, Faculty of Electrical Engineering and Computer Science, VSB-Technical University of Ostrava, University of Ostrava, Ostrava, Czech Republic
e-mail: cong.thanh.truong.st@vsb.cz

I. Zelinka
e-mail: ivan.zelinka@vsb.cz

J. Plucar
e-mail: jan.plucar@vsb.cz

M. Čandík · V. Šulc
Department of Management and Informatics, Faculty of Security Management, Police Academy of the Czech Republic in Prague, Prague, Czech Republic

© Springer Nature Singapore Pte Ltd. 2020
S. S. Dash et al. (eds.), *Artificial Intelligence and Evolutionary Computations in Engineering Systems*, Advances in Intelligent Systems and Computing 1056,
https://doi.org/10.1007/978-981-15-0199-9_30

before. Based on its powerful automation and data analysis capabilities, AI can be used to analyze large amounts of data with efficiency, accuracy, and speed. An AI system can take advantage of what it knows and understand the past threats to identify similar attacks in the future, even if their patterns change. For this reason, the use of AI to combat security threats is an unavoidable trend.

Contemporary with the development of technology, cyber threats are becoming sophisticated and stealthy. Cybercriminals are always changing their methods, making attacks challenging to predict and prevent. Consequently, conventional security systems, which use rules and signatures, have become ineffective against flexible, continually evolving cyber-attacks. Thus, we need advanced approaches such as applying AI-based techniques that provide flexibility and learning capability to assist humans in combating cybercrimes.

The main aim of this study is to present the capability of adopting AI techniques for combating cybercrimes, to demonstrate how these techniques can be an effective tool for combating cyber-attacks. Furthermore, this study highlights the limitations of AI-based approaches in cybersecurity as well as suggests some directions for research in the future.

The rest of this paper proceeds as follows. Section 2 introduces the advantages of AI for combating cybercrimes. It is followed by Sect. 3, which give a brief overview of machine learning (ML) and deep learning (DL). Then, Sect. 4 lists out trendy applications of ML and DL in system security and defense. Section 5 discusses the limitations of existing approaches as well as points out the scope for future work. Finally, Sect. 6 concludes the paper.

2 General Idea About Applying AI on Cybersecurity Issues

Generally, AI has several advantages when it comes to cybersecurity in the following aspects:

- *AI can discover new and sophisticated changes in attack flexibility*
  Conventional technology is focused on the past and relies heavily on known attackers and attacks, leaving room for blind spots when detecting unusual behavior in new attacks. For example, privileged activity in an intranet can be monitored, and any significant mutation in privileged access operations can denote a potential internal threat. If the detection is successful, the machine will reinforce the validity of the actions and become more sensitive to detect similar patterns in the future.
  Furthermore, the machine can learn and adapt better to detect anomalous, faster and more accurate operations. This is especially useful when cyber-attacks are becoming more sophisticated, and hackers are creating new and innovative approaches.

- *AI can handle the volume of data*
  AI can enhance network security by developing autonomous security systems to detect attacks and respond to breaches. The volume of security alerts that appears daily can be very overwhelming for security groups. Automatically detecting and responding to threats have helped to reduce the work of experts and can assist in detecting threats more effectively than other methods.
  When a large amount of security data is created and transmitted over the network every day, network security experts will gradually have difficulty tracking and identifying attack factors quickly and reliably. This is where AI can help by expanding the monitoring and detection of suspicious behavior. This can help network security personnel react to situations that they have not encountered before, replacing the time-consuming analysis of people.
- *AI security system can learn over time to respond better to threats*
  AI helps detect threats based on application behavior and the entire network activity. Over time, AI security system learns about regular network traffic and behavior and makes a baseline of what is the normal pattern. From there, any deviations from the norm can be spotted to detect attacks.

As discussed above, AI techniques can enhance cyberspace security more effectively. Many AI techniques are used to deal with threats, including neural networks, fuzzy logic, expert system, machine learning, and deep learning. However, among these techniques, machine learning and deep learning bring the most achievements.

3 Machine Learning and Deep Learning in Cybersecurity Domain

Machine learning is a subset of AI which focuses on teaching machines how to learn new things and make decisions by applying algorithms to data. Moreover, ML has strong ties to mathematical techniques that enable a process of extracting information, discovering patterns, and drawing conclusions from data. Two most concentrated methods of ML are classification and regression.

There are different types of ML algorithm, but in general, they can be divided into several categories: Supervised Learning, Unsupervised Learning, Semi-supervised Learning, and Reinforcement Learning.

Deep learning is a subfield of machine learning, and it uses data to teach computers how to do things only humans were capable of before. This is done by imitating the human brain mechanism to interpret the data. The core of DL is that if we construct larger neural networks and train them with as much data as possible, their performance continues to increase.

Machine learning and deep learning are very effective in wide-ranging problems in cybersecurity. Machine learning methods are applied in various aspect of computer security, for instance: spam filtering, botnet detection, network anomalies detection, and user behavior anomalies detection. Likewise, deep learning has shown significant improvements in malware detection and network intrusion detection to the existing solutions.

4 Practical Applications of AI on Cybersecurity

4.1 Malware Detection

Traditional methods to detect malware are using signature-based method. However, this method cannot detect newly generated malware. Furthermore, malware authors use obfuscation techniques to change the characteristics of the malware to avoid detection. For these reasons, the blacklists method is not enough for dealing with modern malware. Therefore, intelligent-based ML and DL have been exploited to keep pace with malware evolution.

Technically, a classification malware session typically proceeds through four phases:

- Gathering: collecting a dataset of malicious and benign executables. The dataset is then split into training and testing sets; the first one is used for training the classification model, while the second one is used for model evaluation.
- Feature Generation:

  - Feature Extraction: Feature extraction is the process of collecting discriminative information from a set of samples. The main task of the feature extraction is to extract various types of features. The extraction processes can be performed through static, dynamic, or hybrid analysis. In general, researchers mostly focus on the following feature types in Windows environment: byte sequences, opcodes, application programming interface (API) and system calls, network activity, file system, central processing unit (CPU) registers, portable executable (PE) file characteristics, strings. Likewise, in the Android environment, the malware analysis pays attention to these features: permissions, system calls, and API calls.
  - Feature Selection: The process of selecting only the most important and relevant features for building a machine learning model. The feature selection process can be done manually or automatically using statistical methods and algorithms.

- Training and Testing: Applying a classifier to a set of training data and tuning the hyperparameter. After that, the model's accuracy will be tested with the test data. If the results qualify for the requirements, then it will proceed to the deployment phase. On the contrary, the model will be rebuilt.
- Deployment: In this phase, the model is applied to classify the malware.

In the ML approaches, typically a classification algorithm is applied to the collected features for malware detection. There are many classifiers used for the classification task such as Naive Bayes method, Support Vector Machines (SVM), and Decision Trees.

DL approaches applied to malware classification involve feature extraction of malware or using DL algorithms for the malware classification task.

Tables 1 and 2 list some recent and noteworthy contributions works using machine learning and deep learning for malware detection, respectively.

4.2 Intrusion Detection

Intrusion detection is a protection technique in cybersecurity that can intercept and respond to intrusions, which are defined as "attempts to compromise the confidentiality, integrity, or availability of a computer or network, or to bypass the security mechanisms of a computer or network" [11]. An intrusion detection system (IDS) is a system that is supposed to detect when the system has signs of possible incidents, violations, or imminent threats. Intrusion detection systems can be classified into various types. IDS can either be divided as host-based intrusion detection systems (HIDS) or network-based intrusion detection systems (NIDS). The first one monitors activity on a single system, or host, while the latter observes the traffic on a network.

Table 1 Malware detection using ML methods

References	Algorithm	Features
[1]	Decision tree, random forest	PE file characteristics
[2]	Graph matching	APIs calls, system calls
[3]	SVM	Network
[4]	Decision tree, random forest, Naïve Bayes, SVM	Byte sequences and APIs/system calls
[5]	Random forest	Network
[6]	Neural networks	PE file characteristics, strings
[7]	Naïve Bayes, random forest, SVM	Byte sequences, APIs/system calls, file system, and windows registry

Table 2 Malware detection using DL methods

References	Algorithm	Features
[8]	Neural networks	API/system call sequences
[9]	Deep neural networks	API call sequences
[10]	Deep belief network	Log files

An intrusion detection session typically proceeds through four phases:

- Data Collection: collecting data for feature extraction, many tools are used such as Argus, BRO-IDS, Netflow, Tcptrace, and Netmate. For training and validating IDS, several datasets are applied for instance: DARPA [12], KDD'99 [13], DEFCON [14], CAIDA [15], LBNL [16], CDX [17], Kyoto [18], Twente [19], UMASS [20], ISCX2012 [21], ADFA [22], and CICIDS2017 [23].
- Feature Engineering: Choose the appropriate features for building IDS; this highly affects accuracy. This phase may include feature extraction, feature selection.
- Intrusion Detection: analysis engine is used for discovering existing attacks from the features. Various types of techniques have been applied to detect the intrusion, such as statistical techniques, knowledge-based techniques, and artificial intelligence-based.
- Respond: initiates actions when an intrusion is detected.

Traditional IDSs are limited to detect only known vulnerabilities. The most commonly detect known attacks based on defined rules or detect anomalous behavior by baselining the network.

AI-based techniques are appropriate for developing IDS and outperform other techniques because of their flexibility, adaptability, fast calculation, and learning ability. Hence, many researchers studied AI-based methods to improve the performance of IDS [24–34]. The focus is on developing features generation and selection and improving classifiers to reduce the false alarm.

Many AI-based techniques have been applied to intrusion detection such as fuzzy logic, expert systems, genetic algorithm, machine learning, and deep learning. A current trend in IDS development is the implementation of ML and DL methods. Some recent notable studies are listed as follows.

Farnaaz and Jabbar [35] used Random Forest algorithm to build a model for intrusion detection system and evaluated the model with NSL-KDD dataset. The result of their experiments showed that they obtained the average accuracy of 99.67% without feature selection.

Ambusaidi et al. [36] presented the Least Square Support Vector Machine for designing IDS by selecting the important features. Their experiments showed that they obtained an accuracy rate of 99.79%, 99.91%, and 99.77% for KDD Cup 99, NSL-KDD, and Kyoto 2006 + datasets, respectively.

Kim et al. [37] suggested applying long short-term memory (LSTM) architecture to a Recurrent Neural Network using KDD Cup 1999 dataset. As a result of this approach, the attack detection rate is 98.8%, and the false alarm rate is about 10%.

Shone et al. [38] proposed a novel deep learning-based intrusion detection method called NDAE. The authors used TensorFlow and evaluated their method by using KDD Cup '99 and NSL-KDD datasets. They have claimed that their model has achieved an accuracy of 97.85%.

4.3 Phishing Detection

Phishing is a deceptive way to forge reputable organizations like banks, online trading sites, and credit card companies to trick users into sharing financial information such as login names, passwords, their transactions, or their sensitive information. This attack can also install malware on a user's device. They are a big concern if the user has no knowledge of this type of attack or is not aware of it. Phishing attacks often rely on social networking techniques to deceive users, such as e-mail spoofing, website spoofing. It is usually done via e-mail or text messages containing forms with an embedded hyperlink to a fake website.

There are numerous approaches to detect and prevent phishing. One of those solutions is to use e-mail security gateway. However, this approach just identifies the phishing e-mail but is unable to prevent the user from clicking on a malicious link embedded within the e-mail that may threat to the system. Another method is using intrusion detection system or intrusion prevention system to identify and deter phishing e-mails, yet they are unable to read encrypted e-mail. Further, they depend on the response time so that the system may be compromised before detection alerts. Also, the other approach is embedding a security toolbar within a web browser to alert end-user about phishing attacks, but their detection rate is low.

AI techniques appear appropriate for detecting and preventing phishing since this problem can be transformed into a typical classification task. For example, AI can automatically classify e-mail as phishing and even detect unusual activity in the compromised account. Besides, it also can prevent domain spoofing and alerts when users encounter impersonated websites. There have been several studies to detect and prevent phishing attack using a variety of AI techniques in [39–50]. Some remarkable studies are listed as follows.

Gowtham and Krishnamurthi [40] developed a classification model by using Support Vector Machine (SVM) to combat phishing. The result of their experiments showed that they achieved a rationale performance with 99.6% of True Positive Rate and 0.44% of False Positive Rate.

Thabtah et al. [41] applied the dynamic Neural Networks model to detect phishing attacks. The results showed that their approach revealed promising predictions when compared to another algorithm like Bayesian networks and decision trees.

Yi et al. [42] proposed a model based on Deep Belief Networks for web phishing detection. The model was then evaluated on real IP flows data from the Internet Service Provider. According to their experiments, the model accomplished 89.2% detection rate and the False Positive Rate of 0.6%.

4.4 Advanced Persistent Threat

Advanced persistent threat (APT) is a term used to describe an attacking campaign, usually by a group of attackers, using advanced attack techniques to exploit sensitive data and remain undetected. Because APT attacks are usually carried out with many resources, the attackers usually aim at valuable targets, such as large corporation's security agencies and government organizations, with the ultimate goal of long-term information stealing. The consequences of these attacks are enormous, and the target is to steal intellectual property, compromise sensitive information, destroy the important infrastructure of the organization, or acquire the entire domain name of the organization.

The typical APT attack process consists of the following steps: define the target, establish a foothold, escalate privileges, internal reconnaissance, move laterally, maintain presence, and complete mission.

There is no one-in-all solution that provides effective protection because of the complexity and concealment ability of APT. To defend against APT attacks, multiple layers of security working together are required, such as intrusion detection and prevention, firewall, endpoint protection, data loss prevention technologies, and others. However, these traditional techniques based on feature matching are not enough to deal with APT. Hence, the defense systems need to be enhanced with new technologies such as intelligent techniques to work in the APT context. Researchers in [51–55] proposed a variety of AI techniques for countering APT.

Siddiqui et al. [51] proposed a fractal-based APT detection scheme uses the feature vector of TCP/IP to reduce the APT damage. The authors have claimed that their method performed better than the traditional machine learning algorithms k-NN.

Moon et al. [52] applied a decision tree to build intrusion detection system to detect APT attacks. It can detect intrusion from the beginning and quickly react to APT to minimize damage. The results show that the proposed system has a high rate of APT detection.

Burnap et al. [53] used machine activity metrics and self-organizing feature map approach to distinguish legitimate and malicious software. The authors have claimed that their approach showed promising for APT detection.

Ghafir et al. [54] proposed a machine learning-based approach called MLAPT to detect and predict APT. According to the authors, their system is capable of early prediction of APT attacks. The experiments show that MLAPT has a true positive rate and the false positive rate with 81.8% and 4.5%, respectively.

Yu et al. [55] suggested a network intrusion detection model using deep learning approach, called dilated convolutional autoencoders (DCAEs) for malicious network traffic classification such as a botnet, web-based malware, exploits, scans, and APT. According to their experiments, their model can detect complex attacks from lots of unlabeled data.

5 Discussion

AI methods have played a crucial role in cybersecurity applications and will continue a promising direction that attracts investigations. However, some issues must be considered when applying AI-based techniques in cybersecurity.

- Adversarial attacks: adversarial is a rising trend of attacking AI model, where malicious actors design the inputs to make models predict erroneously. If AI can learn to detect potential threats, another AI should be capable to learn from observing the detector make its decisions and use that knowledge to surpass the defense system. Furthermore, recent advances in DL led to the development of generative adversarial networks, in which the DL network is capable of automatically producing adversarial samples against a target ML system.
- Poisoning data: AI-based security system learns from given data. So, the output results are inaccuracy if its inputs are poisoned, and cybercriminals are already trying to do this. The attacker could pollute the training data from which the algorithm is learning in such a way that the system misbehaves. Different domains are vulnerable to poisoning attacks, for example, network intrusion, spam filtering, or malware analysis.
- Model stealing: These techniques are used to thieve (for example, duplicate) models or recover training data via backbox examining. In this occasion, the attacker learns how ML algorithms work by reversing techniques. From this knowledge, the malicious authors know what the defense engines are looking for and how to avoid it. For example, malicious actors could steal spam filtering models or malware prediction to be able to surpass such models.
- False positives and false negatives: The misclassification in the cybersecurity domain is a significant problem. False positive annoys the security staffs, making it difficult for remediation in case of an actual attack. Notably, in a large enterprise, the false positive alerts account for hundreds to thousands daily will overload the security operators. In contrast, failing to detect malware, a network intrusion or a phishing e-mail could cause a disaster for the organization.

Although the AI-based techniques have limitations, however, they also open a lot of new research ideas and suggestions. First, in-depth research needs to be done to counter against adversarial attacks, poisoning data, and model stealing as well as reduce false positives and false negatives rate. Second, we must study how to defend against malicious actors that using AI attack. Finally, the security of smart devices and is also worth attention when AI is embedded in those devices. These are promising directions to research in the future.

6 Conclusion

Artificial intelligence, as a technique widely investigated in recent years, has been applied practically in many fields. Discussions and studies on the intersection of artificial intelligence and cybersecurity have become an urgent need. In this paper, we have provided a brief and comprehensive review about combining AI and cybersecurity. Specifically, for the application of AI on cybersecurity, we have introduced malware detection, intrusion detection, phishing detection, and APT issues. In the future, based on the historical and contemporary development of the malware, it can be seen that new malware is coming, as predicted in the [56]. This paper is to outline a possible dynamics, structure, and behavior of a hypothetical (up to now) swarm malware as a background for a future antimalware system. The research shows how to capture and visualize the behavior of such malware when it walks through the operating system. The swarm virus prototype, designed here, mimics a swarm system behavior and thus follows the main idea of swarm algorithms. The information of its behavior is stored and visualized in the form of a complex network, reacting virus communication and swarm behavior. As the paper shows, the swarm behavior pattern could also be incorporated into an antimalware system and analyzed for a future computer system protection. AI plays a more and more important role in the cybersecurity as the intelligent scanners, firewalls, antimalware, but also as the intelligent espionage tools and weapons.

Acknowledgements The following grants are acknowledged for the financial support provided for this research: Grant of SGS No. SP2019/137, VSB Technical University of Ostrava.

References

1. Bai, J., et al.: A malware detection scheme based on mining format information. Sci. World J. (2014)
2. Elhadi, A.A.E., et al.: Improving the detection of malware behaviour using simplified data dependent API call graph. Int. J. Secur. Appl. **7**(5), 29–42 (2013)
3. Kruczkowski, M., Szynkiewicz, E.N.: Support vector machine for malware analysis and classification. In: Proceedings of the 2014 IEEE/WIC/ACM International Joint Conferences on Web Intelligence (WI) and Intelligent Agent Technologies (IAT), vol. 2. IEEE Computer Society (2014)
4. Uppal, D., et al.: Malware detection and classification based on the extraction of API sequences. In: 2014 International Conference on Advances in Computing, Communications and Informatics (ICACCI). IEEE (2014)
5. Kwon, B.J., et al.: The dropper effect: insights into malware distribution with downloader graph analytics. In: Proceedings of the 22nd ACM SIGSAC Conference on Computer and Communications Security. ACM (2015)
6. Saxe, J., Berlin, K.: Deep neural network-based malware detection using two-dimensional binary program features. In: 10th International Conference on Malicious and Unwanted Software (MALWARE). IEEE (2015)

7. Wüchner, T., Ochoa, M., Pretschner, A.: Robust and effective malware detection through quantitative data flow graph metrics. In: International Conference on Detection of Intrusions and Malware, and Vulnerability Assessment. Springer, Cham (2015)
8. Kolosnjaji, B., et al.: Deep learning for classification of malware system call sequences. In: Australasian Joint Conference on Artificial Intelligence. Springer, Cham (2016)
9. Tobiyama, S., et al.: Malware detection with the deep neural network using process behaviour. In: Computer Software and Applications Conference (COMPSAC), 2016 IEEE 40th Annual, vol. 2. IEEE (2016)
10. David, O.E., Netanyahu, N.S.: Design: deep learning for automatic malware signature generation and classification. In: 2015 International Joint Conference on Neural Networks (IJCNN). IEEE (2015)
11. Bace, R., Mell, P.: NIST special publication on intrusion detection systems. Booz-Allen and Hamilton Inc., Mclean (2001)
12. Lincoln Laboratory: MIT Lincoln Laboratory: DARPA Intrusion Detection Evaluation (n.d.). https://www.ll.mit.edu/ideval/data
13. Hettich, S., Bay, S.D.: The UCI KDD Archive (1999). http://kdd.ics.uci.edu
14. The Shmoo Group: DEFCON 8, 10 and 11 (2000). http://cctf.shmoo.com/
15. Center for Applied Internet Data Analysis: CAIDA Data (n.d.). http://www.caida.org/data/index.xml
16. Lawrence Berkeley National Laboratory (LBNL) and International Computer Science Institute (ICSI): LBNL/ICSI Enterprise Tracing Project (2005). http://www.icir.org/enterprise-tracing/Overview.html
17. Sangster, B., et al.: Toward Instrumenting Network Warfare Competitions to Generate Labeled Datasets. CSET (2009)
18. Song, J., et al.: Statistical analysis of honeypot data and building of Kyoto 2006+ dataset for NIDS evaluation. In: Proceedings of the First Workshop on Building Analysis Datasets and Gathering Experience Returns for Security. ACM (2011)
19. Sperotto, A., et al.: A labelled data set for flow-based intrusion detection. In: International Workshop on IP Operations and Management. Springer, Berlin (2009)
20. Prusty, S., Levine, B.N., Liberatore, M.: Forensic investigation of the OneSwarm anonymous filesharing system. In: Proceedings of the 18th ACM Conference on Computer and Communications Security. ACM (2011)
21. Canadian Institute for Cybersecurity: Intrusion detection evaluation dataset (ISCXIDS2012) (n.d.). http://www.unb.ca/cic/datasets/ids.html
22. Creech, G., Hu, J.: Generation of a new IDS test dataset: time to retire the KDD collection. In: 2013 IEEE Wireless Communications and Networking Conference (WCNC). IEEE (2013)
23. UNB Canadian Cyber Security, Intrusion Detection Evaluation Dataset (CICIDS2017). http://www.unb.ca/cic/datasets/ids-2017.html
24. Xiang, C., Yong, P.C., Meng, L.S.: Design of multiple-level hybrid classifier for intrusion detection system using Bayesian clustering and decision trees. Pattern Recognit. Lett. **29**, 918–924 (2008)
25. Shafi, K., Abbass, H.A.: An adaptive genetic-based signature learning system for intrusion detection. Expert Syst. Appl. **36**(10), 12036–12043 (2009)
26. Tong, X., Wang, Z., Haining, Y.: A research using hybrid RBF/Elman neural networks for intrusion detection system secure model. Comput. Phys. Commun. **180**(10), 1795–1801 (2009)
27. Wang, G., et al.: A new approach to intrusion detection using Artificial Neural Networks and fuzzy clustering. Expert. Syst. Appl. **37**(9), 6225–6232 (2010)
28. Wagner, C., François, J., Engel, T.: Machine learning approach for IP-flow record anomaly detection. In: International Conference on Research in Networking. Springer, Berlin (2011)
29. Lin, S.-W., et al.: An intelligent algorithm with feature selection and decision rules applied to anomaly intrusion detection. Appl. Soft Comput. **12**(10), 3285–3290 (2012)

30. Yassin, W., et al.: Anomaly-based intrusion detection through k-means clustering and naive Bayes classification. In: Proceedings of 4th International Conference on Computer Informatics, ICOCI, vol. 49 (2013)
31. Shrivas, A.K., Dewangan, A.K.: An ensemble model for classification of attacks with feature selection based on KDD99 and NSL-KDD data set. Int. J. Comput. Appl. **99**(15), 8–13 (2014)
32. Lin, W.-C., Ke, S.-W., Tsai, C.-F.: CANN: an intrusion detection system based on combining cluster centres and nearest neighbours. Knowl.-Based Syst. **78**, 13–21 (2015)
33. Hodo, E., et al.: Threat analysis of IoT networks using artificial neural network intrusion detection system. In: 2016 International Symposium on Networks, Computers and Communications (ISNCC). IEEE (2016)
34. Subba, B., Biswas, S., Karmakar, S.: Enhancing performance of anomaly-based intrusion detection systems through dimensionality reduction using principal component analysis. In: 2016 IEEE International Conference on Advanced Networks and Telecommunications Systems (ANTS). IEEE (2016)
35. Farnaaz, N., Jabbar, M.A.: Random forest modelling for network intrusion detection system. Procedia Comput. Sci. **89**, 213–217 (2016)
36. Ambusaidi, M.A., et al.: Building an intrusion detection system using a filter-based feature selection algorithm. IEEE Trans. Comput. **65**(10), 2986–2998 (2016)
37. Kim, J., et al.: Long short term memory recurrent neural network classifier for intrusion detection. In: 2016 International Conference on Platform Technology and Service (PlatCon). IEEE (2016)
38. Shone, N., et al.: A deep learning approach to network intrusion detection. IEEE Trans. Emerg. Top. Comput. Intell. **2**(1), 41–50 (2018)
39. Xiang, G., et al.: Cantina+: a feature-rich machine learning framework for detecting phishing web sites. ACM Trans. Inf. Syst. Secur. (TISSEC) **14**(2), 21 (2011)
40. Gowtham, R., Krishnamurthi, I.: A comprehensive and efficacious architecture for detecting phishing webpages. Comput. Secur. **40**, 23–37 (2014)
41. Thabtah, F., Mohammad, R.M., McCluskey, L.: A dynamic self-structuring neural network model to combat phishing. In: 2016 International Joint Conference on Neural Networks (IJCNN). IEEE (2016)
42. Yi, P., et al.: Web phishing detection using a deep learning framework. Wirel. Commun. Mob. Comput. **2018** (2018)
43. Jain, A.K., Gupta, B.B.: Towards detection of phishing websites on client-side using machine learning based approach. Telecommun. Syst. **68**(4), 687–700 (2018)
44. Tyagi, I., et al.: A novel machine learning approach to detect phishing websites. In: 2018 5th International Conference on Signal Processing and Integrated Networks (SPIN). IEEE (2018)
45. Zuhair, H., Selamat, A.: Phishing hybrid feature-based classifier by using recursive features subset selection and machine learning algorithms. In: International Conference of Reliable Information and Communication Technology. Springer, Cham (2018)
46. Li, Y., et al.: A stacking model using URL and HTML features for phishing webpage detection. Futur. Gener. Comput. Syst. **94**, 27–39 (2019)
47. Qabajeh, Issa, Thabtah, Fadi, Chiclana, Francisco: A recent review of conventional vs. automated cybersecurity anti-phishing techniques. Comput. Sci. Rev. **29**, 44–55 (2018)
48. Adebowale, M.A., et al.: Intelligent web-phishing detection and protection scheme using integrated features of Images, frames and text. Expert Syst. Appl. (2018)
49. Shirazi, H., Bezawada, B., Ray, I.: Know thy domain name: unbiased phishing detection using domain name based features. In: Proceedings of the 23rd ACM on Symposium on Access Control Models and Technologies. ACM (2018)
50. Sahingoz, O.K., et al.: Machine learning based phishing detection from URLs. Expert Syst. Appl. **117**, 345–357 (2019)
51. Siddiqui, S., et al.: Detecting advanced persistent threats using fractal dimension based machine learning classification. In: Proceedings of the 2016 ACM on the International Workshop on Security and Privacy Analytics. ACM (2016)

52. Moon, D., et al.: DTB-IDS: an intrusion detection system based on decision tree using behaviour analysis for preventing APT attacks. J. Supercomput. **73**(7), 2881–2895 (2017)
53. Burnap, P., et al.: Malware classification using self-organising feature maps and machine activity data. Comput. Secur. **73**, 399–410 (2018)
54. Ghafir, I., et al.: Detection of advanced persistent threat using machine-learning correlation analysis. Future Gener. Comput. Syst. **89**, 349–359 (2018)
55. Yu, Y., Long, J., Cai, Z.: Network intrusion detection through stacking dilated convolutional autoencoders. Secur. Commun. Netw. **2017** (2017)
56. Zelinka, I., Das, S., Sikora, L., Šenkeřík, R.: Swarm virus-next-generation virus and antivirus paradigm? Swarm Evol. Comput. **43**, 207–224 (2018)

CNN and Sound Processing-Based Audio Classifier for Alarm Sound Detection

Babu Durai C. Ramesh and Ram S. Vishnu

Abstract Artificial neural networks (ANN) has evolved through many stages in the last three decades with many researchers contributing in this challenging field. With the power of math, complex problems can also be solved by ANNs. ANNs like convolutional neural network (CNN), deep neural network, generative adversarial network (GAN), long short-term memory (LSTM) network, recurrent neural network (RNN), ordinary differential network, etc., are playing promising roles in many MNCs and IT industries for their predictions and accuracy. In this paper, convolutional neural network is used for prediction of abnormal hospital instrumental beep sounds in high noise levels. Based on supervised learning, the research has developed the novel CNN architecture for beep sound recognition in noisy situations. The proposed method gives better results with an accuracy of 96%. The prototype is tested with various architecture models for the training and test data, out of which two-layer CNN classifier predictions were the best.

Keywords CNN · Deep neural network · Adam optimization · Sound processing · Backpropagation · Peak detection

1 Introduction

With the rapid development of technology and research, Artificial intelligence has its major plays in many industries like search engines, computer vision, image retrieval, business analytics and a lot more. Artificial intelligence brought out astonishing results like self-driving cars, image recognition, speech processing, cancer cells detection and intuition and much more to come in the mere future. Among the popular machine learning algorithms, neural networks are widely used

B. D. C. Ramesh · R. S. Vishnu (✉)
Sri Sairam Engineering College, Chennai, Tamil Nadu, India
e-mail: vishnurameeengg@gmail.com

B. D. C. Ramesh
e-mail: crbdurai.eee@sairam.edu.in

© Springer Nature Singapore Pte Ltd. 2020
S. S. Dash et al. (eds.), *Artificial Intelligence and Evolutionary Computations in Engineering Systems*, Advances in Intelligent Systems and Computing 1056, https://doi.org/10.1007/978-981-15-0199-9_31

by major data scientist and research engineers around the world. As of now, majority of research works are based on supervised learning and semi-supervised learning. In the upcoming years, it is targeted to produce high productivity using reinforced learning and unsupervised learning algorithms. Other popular algorithms used in machine learning include support vector machines (SVM), K-nearest neighbors (KNN), logistic regressions, decision trees, genetic algorithm, random forest, Naïve Bayes, etc. In this proposal, it is focused on supervised learning using convolutional deep neural network.

2 Deep Neural Network (DNN)

The mathematical theory of artificial neural network was first proposed by Frank Rosenblatt in 1958. He assumed that the coordination of body activity and thoughts is the results of interactions among neurons in brain. Neural networks commonly known as artificial neural network (ANN) are the mathematical matrix operations consisting of input, hidden, and output layers. It uses the concept of human neurons in brains that are gradually optimized over the course of time by interaction of humans with the environment. Human brain is estimated to have neurons in the order of billions. In the available firmware, it is not possible to incorporate the power of such counts of perceptron. Researches are undertaken to improve hardware performance of Machine Learning algorithms, especially ANNs. There are lot of DNN models in the industries that are mainly involved in cloud-based image recognition, better search results, business analytics, etc. One of the examples is Google's Deep Mind.

3 Data Preprocessing

Supervised Learning is used for our classification model. The training and testing data consists of three hundred samples of noise and alarm (beep) sound and are labeled accordingly. One hot encoding is employed for output labels. In the case of alarm sound detection, it has two classes as noise and beeps (alarms). Neural networks are designed to deal with numbers, not with strings or characters. So, they must be encoded into numbers matching the output shape of the neural network. For, e.g., if there are two classes as dogs and cats, labels for the network must be (0, 1) for cats and (1, 0) for dogs and so on. If the model is trained on more data, the accuracy gets better, and it is tested experimentally.

The sampled sound is converted into a Mel spectrogram image of data usually as rectangular array of data. The Mel-frequency scale is a quasi-logarithmic spacing roughly resembling the resolution of the human auditory system. The matrix data has various patterns for various sounds like human voices, air-conditioners, fans, environmental noises, alarms, etc. Normal Spectrogram is different from a Mel spectrogram data. The Mel-scale is closely related to biological hearing and has proven to be more successful in speech processing, segregation, and recognition. It produces unique patterns and textures in sound data by which convolutional neural networks are good at detecting patterns in 2D or 3D matrix data. Lot of other sound features like Chroma, Pitch, Tonnetz, and Mel-frequency cepstral coefficient (MCFC) may also be used for feature extraction for complex multi-feature multi-class classification problems.

3.1 Mel Spectrogram Calculation

In a normal spectrogram computation, the output spectrum is based on time-series information of the frequency bins of each window of raw audio data. Fast Fourier transform is applied to each window of audio data of defined length, and the audio frequencies are extracted mathematically and plotted to the vertical axis with value in the horizontal axis as time intervals which contribute to the spectrogram. Spectrograms play an important role in frequency analysis of audio signals [1]. The following figures are the plot of Mel spectrogram versus normal spectrogram on the right side (Figs. 1 and 2).

If a time-series input is provided, then its magnitude spectrogram S is first computed and then mapped onto the Mel-scale by mel_f.dot (S^{power}). The Mel spectrogram is the input data to the model. Python language provides inbuilt functions for Mel spectrogram calculations in Librosa library package. The data is passed through standard scalar where each data points is scaled between two points such that it well suits for the neural network.

$$\text{Standard Scalar} = \frac{x_i - \text{mean}(x)}{\text{stdev}(x)} \tag{1}$$

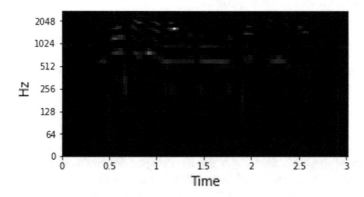

Fig. 1 Mel spectrogram image

Fig. 2 Normal spectrogram image

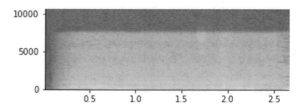

3.2 Data Augmentation

Data augmentation plays a major role in improving the accuracy of the model [2]. The prototype model is trained on a balanced data. When the data is unbalanced, it is better practice to analyze the metrics using "f1score". The audio data is augmented using the following methods.

- Time Stretching.
- Pitch Shifting.
- Rolling series data.

The 300-sample dataset has enlarged to a massive 900 sample dataset by application of the above methods in each sample data. If the training data has many diverse datasets, the model becomes capable of learning and classifying the required outputs in noisy environments. The training dataset includes input data coupled with respective output labels. Some better practices include having diverse data features instead of repeating the same data and its feature in every iteration. The training is done in a powerful machine-like personal computer. The model trained and optimized can be deployed into the dedicated environments, for example, raspberry pi, voice bonnet, and other embedded applications to deal with real-time noisy data.

4   CNN Model Development

The model can be subdivided into two layers of convolutional layers with pooling layers, a hidden perceptron layer (optionally be increased based on training data), and an output layer. The first convolutional layer faces the input data, applies filters to extract features, i.e., curves, lines, edges, etc. The output shape of the first convolutional layer is defined by Eq. 2.

$$\text{Out1} = ((N + 2p - f)/s + 1) \times ((N + 2p - f)/s + 1) \tag{2}$$

for the input shape of order $N \times N$ and filter shape of order $f \times f$. The notations "p" and "s" represent padding and strides in the convolution operation, and they are chosen optionally during optimization and training. It is followed by a MaxPooling

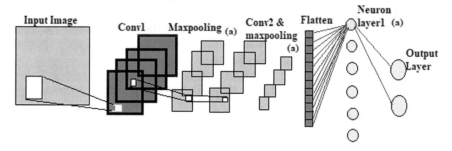

Fig. 3 Proposed CNN architecture

layer for further extraction. Further, the following convolutional layer involves the same operation and MaxPooling layer. The output shape consists of 2D array of data. The 2D matrix is flattened into 1D array of data for feeding it into a deep neural network through synapses, which has perceptron or neuron units having rectified linear unit (ReLU) activations $f(x) = \max(0, x)$. The output of this layer is passed through the synapses layer and other activation layer having "Sigmoid" activation. The two perceptron layers are densely connected with output of neuron. The output is categorized as either alarm or noise sounds based on the output data, i.e., maximum(out1, out2). This is a basic primitive model for predicting beep sound patterns. For multi-level audio classification, high-level CNN classifiers with huge datasets must be used [2]. Model is tested under different testing conditions in real time and provided a better accuracy in results. Figure 3 represents the CNN architecture for the prototype. Here, the neuron connections for a single neuron are visualized.

4.1 Optimization of the Model

The "Backpropagation" algorithm is deployed to optimize hyper parameters of the model at the training stage [3]. Backpropagation uses differential calculus to update weights in the model depending on the error value for each input, i.e., how far away the predictions are from the actual value. The weights in the network are randomly initialized before training. During training process, the error in weight values is gradually updated by Backpropagation process using chain rule. On iteratively reducing the weight's error, a series of weights in the network that produce good predictions may be obtained. Software giants like Google trained their deep learning models with some millions of data. The rate at which the model learns depends on the value of learning rate fixed during training. Generally, it is in the order of 0.001 for Adam optimization. Higher the learning rate, lower the accuracy and vice versa. Each layer weights is updated based on the error obtained in the next layer during forward propagation of the input. The model gets optimized over the time during training, and for each batches of training data, weights are updated so that the model fits to the input and output labels. Backpropagation is done by applying "Chain rule" to the model.

Given a forward propagation function:

$$f(x) = A(B(C(x))) \tag{3}$$

Chain rule can be used to find $f(x)$ with respect to x as,

$$f'(x) = f'(A).A'(B).B'(C).C'(x) \tag{4}$$

Assuming $B(C(x))$ to be a constant B and differentiated normally with respect to B. This technique is applied throughout the network, and weights are modified or updated based on the cost function for the purpose of making good predictions as the input data propagates through the model. Adaptive moment estimation (Adam) is the state-of-the-art optimization algorithm that integrates the ideas from RMSProp and momentum. It is preferred comparing with the classical stochastic gradient descent procedure. It estimates adaptive learning rates for each hyper parameter as follows.

- First, the exponentially weighted average of past gradients is computed (ϑ_{dW}).
- Second, the exponentially weighted average of the squares of past gradients is computed (s_{dW}).
- Third, these averages have a bias toward zero, and to counteract this, a bias correction is applied ($\vartheta_{dW}^{corrected}, s_{dW}^{corrected}$).
- Finally, the hyper parameters are updated using the data from the averages computed (Eq. 5).

$$
\begin{aligned}
\vartheta_{dW} &= \beta_1 \vartheta_{dW} + (1 - \beta_1)\frac{\partial J}{\partial W} \\
s_{dW} &= \beta_2 s_{dW} + (1 - \beta_2)\left(\frac{\partial J}{\partial W}\right)^2 \\
\vartheta_{dW}^{corrected} &= \frac{\vartheta_{dW}}{1-(\beta_1)^t} \\
s_{dW}^{corrected} &= \frac{s_{dW}}{1-(\beta_1)^t} \\
W &= W - \alpha \frac{\vartheta_{dW}^{corrected}}{\sqrt{s_{dW}^{corrected}} + \varepsilon}
\end{aligned}
\tag{5}
$$

The concept that the model is not able to classify input in test data but be able to classify input in training data is called overfitting. The model is unable to generalize and accurately predict the output. Overfitting of the model can be avoided by having diverse data in training input by data augmentation, adding dropout and avoiding too much hidden layers in the model. In underfitting, the model is unable to predict desired output even in the training data itself. This can be reduced by reducing the dropout rate in the model, increasing hidden layers, adding more features to the input data. Fine-tuning is done by changing the trainable parameters of the model and training it accordingly to obtain the best results. Better predictions can be obtained in real time by having training data similar as that of real-time data. When the number of classes for classification increases, i.e., more features, complexity of the network should be arbitrarily high and vice versa. Adding too much parameter (layers) for small class predictions can result in bad predictions as some weights in the network may not be optimized for minimal features. The detailed documentation is available in Python keras library for Adam optimization of the model. Expected accuracy may be obtained by training the CNN model with new data every time (fine-tuning). The rest of the prototype has sound processing for peak sound detection in beep sounds for classifying whether the sound is abnormal or normal sound. The first part of the prototype is the CNN, and the second part is sound processing. In our use case, again classifying Hospital instrumental beep sounds based on the interval between the peaks for abnormal and normal beep sounds. A heuristic method of peak detection is used to detect peaks in alarm sound pattern. A sample n at a time is selected as peak if it satisfies the following conditions.

$$x(n) == \text{maximum}(x[n - \text{pre\_max}(\text{upto})\, n + \text{post\_max}]),$$
$$x(n) > = \text{mean}(x[n - \text{pre\_avg}(\text{upto})\, n + \text{post\_avg}]) + \text{delta},$$
$$n - \text{previous\_}n > \text{wait}.$$

where,

x—array [shape = (n,)] input signal to peak picks from
pre_max—integer >= 0, number of samples before n over which max is computed
post_max—integer >= 1, number of samples after n over which max is computed
pre_avg—integer >= 0, number of samples before n over which mean is computed
post_avg—integer >= 1, number of samples after n over which mean is computed
delta—threshold offset for mean
wait—number of samples to wait after picking a peak.

This method is used to detect peaks in beep sounds. Detailed explanation is depicted in librosa.util.peak_pick function in librosa package python. This is fast, and the parameters can be varied according to the environmental situations, i.e., high and low noise levels. If the estimated number of peaks is more than the threshold for a time interval, it is considered as an abnormal alarm, and an indication is set by general means.

5   Experimental Investigation

The prototype uses raspberry pi 3B+ board as the working board for machine learning model to run. Sound is recorded and sampled at same frequency as that of training data for a period of three seconds and converted into a linear array of data. The sampling rate at test conditions was 22100 Hz which is the recommended

Fig. 4 Raspberry pi 3B+
model

sampling rate for raspberry pi [4]. The linear array of data is transformed into a Mel spectrogram matrix. It is set on its way toward the model. The matrix data must be reshaped as per the model's input shape. The model classified the data, made some predictions, and outputted probabilistic values. The maximum probability value is chosen as output prediction for that particular input. The data flow path, i.e., forward propagation of input data for a particular class is defined during training. It has python language support, and all its dependencies can be installed using pip package installer.

Different sounds produce various Mel-scale patterns. A raspberry pi B+ model and a test alarm sound's Mel spectrogram after standard scaling are shown in Figs. 4 and 5.

The CNN model showed a mean accuracy of 76% without data augmentation which again improved steadily to a mean accuracy of 96% during training and is robust to noisy real-time environment. The accuracy and loss history during training and testing are depicted in Figs. 6 and 7.

Fig. 5 Mel spectrogram for alarm

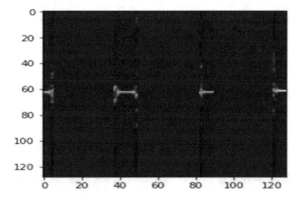

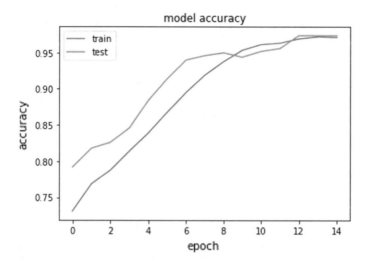

Fig. 6 Model accuracy during training

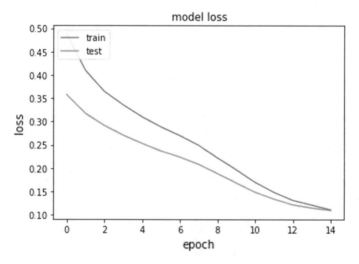

Fig. 7 Model test and train loss

6 Results and Discussion

The prototype acquired an accuracy of 96.5% as average in classifying alarm sounds from environmental sounds. Without data augmentation, the prototype encountered an accuracy of around 76%. In real-time noisy test, the model failed to predict some of test alarm sounds before data augmentation. After augmentation, the accuracy is raised above 93%. Other classifiers like SVM, KNN, random forest produced relatively less accuracy and requiring more time in real-time predictions compared to CNN. The architecture can also be trained with new datasets each time (fine tunning) in the cloud which is an additional advantage compared to other classifiers. The model as of now is in its miniature level, and it produces robust classification of alarm sounds. In the future, it is aimed to implement reinforced learning where the model trains itself to the new alarm signals instead of tedious supervised learning for each new signals.

7 Summary

In this work, in order to predict alarm sounds, the implementation of a novel convolutional neural network has been proposed. Mel spectrogram is considered as the input feature matrix for the model. The computation speed was also good, and real-time predictions were accurate for hospital situations. High accuracy is obtained only by data augmentation process as the model fits and gets optimized for the input features. For multi-class predictions, various features like Chroma, Pitch, Tonnetz, Mel-frequency cepstral coefficients (MFCCs) ought to be used as discussed already.

8 Conclusion

The proposed model predicts the alarms well and is fine-tuned for different kind of alarm and beep sounds. Convolutional neural network detects pattern in both image and audio matrix data with high speed in around 10 ms time that includes pre-processing and prediction. Feature engineering plays an important role in optimizing the weights in the network improving the predictive analyzer algorithm outputs.

Acknowledgements I wish to submit a new manuscript entitled CNN and sound processing-based audio classifier for alarm sound detection. I declare that this work is not currently under consideration for publication elsewhere or under implementation by the public.

References

1. Huzaifah, M.: Comparison of Time-Frequency Representations for Environmental Sound Classification Using Convolutional Neural Networks. arXiv preprint arXiv:1706.07156 (2017)
2. Salamon, J., Bello, J.P.: Deep convolutional neural networks and data augmentation for environmental sound classification. IEEE Signal Process. Lett. **24**(3) (2017)
3. Wang, L., Zeng, Y., Chen, T.: Back propagation neural network with adaptive differential evolution algorithm for time series forecasting. Expert. Syst. Appl. **42**, 855–863 (2015)
4. Peng, C.Y., Chen, R.C.: Voice recognition by Google Home and Raspberry Pi for smart socket control. In: Tenth International Conference on Advanced Computational Intelligence (ICACI) (2018)

LoRa—Pollution and Weather Monitoring System

Yeddula Aravind Kumar Reddy, K. Suganthi, Paluru Mohan Kumar, Afjal Ali Khan and K. Sai Pavan

Abstract A decentralized system is created to store the data collected from the IOT sensors using Arduino module with the transformation of sensor data from analog to digital and then hook up with LoRa transmitter, from here all data forwarded to LoRa receiver which coupled with NodeMCU. The Things Network is used to decode the encrypted data to cloud storage every time the transaction is received from which the data reserved in Cloud. Here we use thingsspeak.com with the visualizing channels for each sensor result has taken out with suitable graphs along with certain date and time. As keeping security, distance in mind we compact with different types of architectures based on LoRa modules and device to device security and cloud storing methodologies.

1 Introduction

Population is increasing and broadening everyday, as population increases pollution in atmosphere also increases [1]. Mostly lot of people living in cities and pollution is also increasing in cities rapidly [2]. The main motivation to this paper tends to make eco-friendly cities. Activity of human being causes adulteration to atmosphere pollution causes lot of health issues and making a large impact on our community. Pollution is caused more by transportation and even with cooking, urbanization in remote areas also.

Y. A. K. Reddy (✉) · K. Suganthi (✉) · P. M. Kumar · A. A. Khan · K. S. Pavan
Department of Electronics and Communication Engineering, SRM Institute of Science and Technology, Chennai, Tamil Nadu, India
e-mail: yaravind951@gmail.com

K. Suganthi
e-mail: suganthk@srmist.edu.in

© Springer Nature Singapore Pte Ltd. 2020
S. S. Dash et al. (eds.), *Artificial Intelligence and Evolutionary Computations in Engineering Systems*, Advances in Intelligent Systems and Computing 1056,
https://doi.org/10.1007/978-981-15-0199-9_32

As long as technologies increasing more number of solutions came up with different ways, in metropolitan cities all data can be monitored and stored [3]. With these types of smart equipment even health related issues also sorting.

Communication plays a prominent role in every aspects of life, as of now only short range communication placed via Bluetooth, ZigBee modules [4], with the implementing of LoRa module we can extend the range coverage up to certain kilometres. By using storing devices, it can help to evaluate data of being concentrated from initial state to final state of result from output, it can backup files, securely maintain the data through end-to-end communication with encryption before transferring to cloud storage devices, it is accessible from every corner of the world by using Internet.

Now-a-days, there are many policies to maintain data securely from the management tools; lots of services are available based on cost deployment.

This paper brings you to determine how much pollutant happening around our atmosphere by means of human activities and weather report from daily access can be handled. The data collected from sensors are transferred to Arduino Uno via analog to digital converters with the mean time sources that can be sent to LoRa transmitter. With the wireless spreading of data from LoRa transmitter to LoRa receiver engaged with NodeMCU to upload the files by using Internet.

LoRa [5] module is used to long range coverage of wireless transmission mechanism, By the recent priority with Low power technologies can progress new solutions in wireless communication systems, with the different IOT appliances LoRa technology can get upper hand in Long ranges. The major part of this work is to bring new methodologies regarding LoRa module and its technologies and evaluating different results with real-time constraints.

This paper consists of different sections as follows. Section 2 reviews literature survey, Sect. 3 has block diagram, Sect. 4 have proposed frame work, Sects. 5 and 6 has components used and the results of the model followed by conclusion.

In the last years, several AI-based forecasting systems were presented in the literature for the prediction of different air pollutants. The potential benefits of using various artificial intelligence techniques and in particular, machine learning (ML) techniques for solving different environmental problems were emphasized by various research efforts that have been reported in the literature and by new research opportunities such as those mentioned.

2   Block Diagram

See Fig. 1.

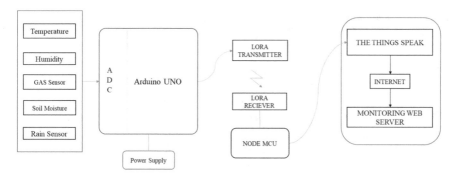

Fig. 1 Block configuration

3 Frame Work of Proposed Model

This proposed model overcomes the drawbacks of the previous models which are present. This model has LoRa transmitter and receiver which is used for the transfer of data after collecting from Arduino Uno and NodeMCU. The data which is given from the sensors are send to the Arduino and from there LoRa transmitter transfer to the receiver through wireless medium.

After the collection of data which is collected will be displayed on the computer screen, and the data are stored in the things speak cloud storage; the user can take the necessary operation or precaution to control the pollution [6]. LoRa is a long-range module which can transfer the data in the range of 1 km. The LoRa is a low power device for better lifespan of the device. NodeMCU is used for connecting LoRa to the cloud to store the data. The data collected in different days will be given in the graphical way in things speak cloud storage. The LoRa collects the pollutant level and packs them in frame with the physical location.

4 Comparison with Other Technologies

See Table 1.

5 Components

5.1 Arduino Uno

The Arduino is based on ATmega328 microcontroller board. Arduino Uno consists of 6 analog inputs, 14 digital I/O (Input/Output) pins, 16 MHz crystal oscillator, power jack, USB connection, and reset button. It is powered by either USB

Table 1 Comparison

Feature	LoRaWAN	SigFox	LTE-M	WiFi (802.11n)
Data rate	290–50 kbps	100 bps	200 kbps	450 Mbps max
Range	10 kms	30–50 kms	2–5 kms	50 m
Battery life time	96–120 (8–10) Months	72–84 (6–7) Months	12–24 (1–2) Months	84–96 (7–8) Months
Gateway (Yes/No)	Yes	Yes	No	No
Interference immunity	Vey high	Low	Medium	High

Fig. 2 Arduino Uno

connection or external power supply. External power may be from AC to DC adapter or battery source. It can be operated at an external supply of 6–20 V. The Ardunio has 32 KB of storage and also have 2 KB of SRAM and 1 KB of EEPROM.

Each 14 digital pins is used as I/O, using pinMode(), digitalwrite(), and DigitalRead() functions (Fig. 2).

5.2   Temperature and Humidity Sensor (DHT11)

Digital signal output is generated by the DHT11 temperature and humidity sensor. By using the digital-signal-acquisition technique and temperature and humidity sensing technology, it ensures high reliability and excellent long-term stability. This sensor is mostly used due to great performance and less cost. The sampling rate of DHT11 sensor is 1 Hz and has smaller body size [7].

The sensor has three ports; pin 1 is used for vcc, pin 2 for data for sending to arduino, and pin 3 for ground. This sensor can gauge the estimations of dampness at the scope of 20–90% of relative humidity (RH) and a temperature in the scope of 0–50 °C (Fig. 3).

Fig. 3 Temperature and humidity sensor

5.3   Gas Sensor (MQ-2)

The MQ-2 is a gas sensor which detects the combustible gas present in the air and shows the output values in analog voltage. The MQ-2 sensor is sensitive to LPG, Propane, Hydrogen, NH_3, CO, CO_2 and smoke [8–10]. These gas sensors are used in industry, used in gas leakage detection, houses, etc.

Its main features are wide detection scope, fast response, stable and long life, high sensitivity. MQ-2 sensor composed by ceramic tube, SnO2 sensitive layers, measuring electrode and heater are fixed into a crust made by stainless steel and plastic [11]. The MQ-2 sensor contains 3 pins; pin 1 Vcc is connected to power supply, pin 2 to the ground, and pin 3 works as signal carrier to Arduino (Fig. 4).

Fig. 4 Gas sensor (MQ-2)

Fig. 5 Rain sensor

5.4 Rain Sensor

This sensor is used to detect the rain. It is also used as a switch when rain drop falls on the board and used to measure rain intensity. This sensor consists of two parts for more convenience they are rain board and control board. The analog output is used in detection of drops in the amount of rainfall. It uses a wide range of comparator.

Its features are Anti-oxidation, anti-conductivity, with long use time. It consists of four pins vcc, gnd, input and output pins (Fig. 5).

5.5 Soil Moisture Sensor

Soil moisture sensor is used to test the moisture of soil, when the lack of water or shortage of water the sensor output is at high level else the output is at low level. It is mostly used in agriculture, irrigation purpose, nurseries, and green houses.

The main features of this sensor is adjustable sensitivity, convenient installation, we can configure threshold level output is more accurate.

Soil moisture content may be determined due to its effect on the dielectric constant to measure the capacitance between the present in the soil. The probe gives a frequency to measure the dielectric constant. This sensor consist of six pins; pin one is output pin collects the data and send to Arduino, pin two is power supply Vcc, pin 3 and pin 6 are ground, pin 4 and 5 are transmitter and receiver ports (Fig. 6).

Fig. 6 Soil moisture sensor

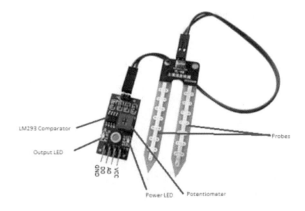

5.6 NodeMCU (ESP8566EX)

NodeMCU offers a complete and self-contained WiFi network which is used to host the application or to offload Wifi network from another application processor. It is a 32-bit processor with on-chip SRAM. It is designed for wearable electronics, mobile purpose and Internet of things application for achieving low power consumption of several techniques [12].

Its features are open-source, interactive, programmable, low cost, simple, smart, Wi-Fi enabled. The NodeMCU is used for connecting the LoRa receiver to the cloud for collecting the data of different time and date.

It consists of 17 general purpose input/output pins and they are assigned with various functions by the hard ware. Each GPIO can be configured with internal pull-up, input available for sampling by a software register (Fig. 7).

5.7 LoRa Module SX1278

The term LoRa stands for long range. It is a wireless radio frequency technology introduced by a company called Semtech. This LoRa technology can be used to transmit bidirectional information to long distance without consuming much power.

Fig. 7 NodeMCU

Fig. 8 LoRa module
SX1278

This property can be used by remote sensors which have to transmit its data by just operating on a small battery.

In order to achieve high distance with low power LoRa [13] compromises on bandwidth, it operates on very low bandwidth. The maximum bandwidth for Lora is around 5.5 kbps, this means that you will be able to send only small amount of data through LoRa. So, you cannot send audio or video through this technology, it works great only for transmitting less information like sensor values. Comparison of LoRa with Wi-Fi or Bluetooth, but these two do not stand anywhere near LoRa. Bluetooth is used to transfer information between two Bluetooth devices and Wi-Fi is used to transfer information between an access point (Router) and station (Mobile). But LoRa technology was primarily not invented to transmit data between two LoRa modules. LoRa modules do come in different frequency ranges, the most common being the 433, 915 and 868 MHz (Fig. 8).

6 Results

After collecting the data in cloud [6] for different days, the outputs of the sensors are showed in different graphical ways (Figs. 9, 10, 11, 12 and 13).

The data collected from the sensors are present in the graphical representation of each sensor in analog reading.

The hardware kit of pollution and weather monitoring system using LoRa (Figs. 14 and 15).

Fig. 9 Temperature

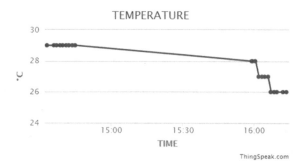

Fig. 10 Humidity

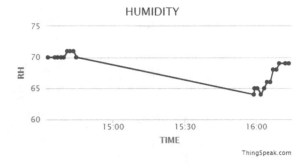

Fig. 11 Gas sensor

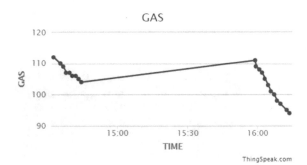

Fig. 12 Rain sensor

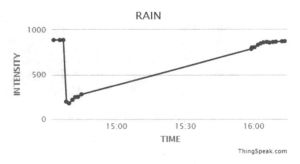

Fig. 13 Soil moisture

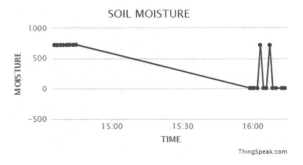

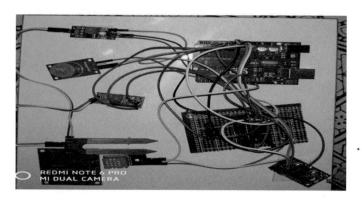

Fig. 14 Prototype of transmitter model using LoRa

Fig. 15 Prototype of receiver model using LoRa

7 Conclusion

This smart equipment has pinned as major eventual for IoT approach and it cut down the composite situations, extravagant things etc. The arranged module pollution and weather monitoring using LoRa defeated the difficulty of conventional system. As long as considering the expensive of cost, different locations with mentioning particular date towards the data has collected on web server using cloud storage.

Key asset towards this work to bring to take place in both remote and metropolitan city areas. Communication via device to device wirelessly has to get more efficiency in mutual authentication in cloud storage elements. We classify the carrying of consuming energy and range coverage up to certain kilometres in overall process of monitoring.

References

1. Dunwell, J.M.: Global Population Growth, Food Security and Food and Farming for the Future, pp. 23–38. Cambridge University Press. Cambridge (2013)
2. Air Quality Monitoring. https://www.qld.gov.au/environment/pollution/monitoring/air-monitoring (2017)
3. Semtech, T.: Air Pollution Monitoring. https://www.semtech.com/wireless-rf/internet-of-things/downloads/Semtech_Enviro_AirPollution_AppBrief-FINAL.pdf (2017)
4. World, R. W.: LoRa vs Zigbee| Difference between LoRa and Zigbee. www.rfwireless-world.com/Terminology/LoRa-vs-Zigbee.html
5. Raju, V., Varama, A.S.N, Raju, Y.S.: An environmental pollution monitoring system using LoRa. In: International Conference on Energy, Communication, Data Analytics and Soft Computing (ICECDS-2017) (IEEE 2017)
6. Sarathkumar, M., Barathiselvaraj, M., Chikkandhar, V.: Advanced Pollution Monitoring and Controlling System Using Cloud Interface, 02 (2016). ISSN (Online): 2321–0613
7. Mouser Electronics. DHT11 Temperature and Humidity sensor. https://www.mouser.com/ds/2/758/DHT11-Technical-Data-Sheet-Translated-Version-1143054.pdf
8. Engineeringtoolbox.com. Carbon Dioxide Concentration—Comfort Levels. https://www.engineeringtoolbox.com/co2-comfort-level-d1024.html
9. Kills, C.M.: Permissible levels of Carbon Monoxide—Carbon Monoxide Kills. http://www.carbonmonoxidekills.com/are-you-at-risk/carbon-monoxide-levels/
10. K. I. Ltd.: What Are Safe Levels of CO and CO_2 in Rooms? https://www.kane.co.uk/knowledge-centre/what-are-safe-levels-of-coand-co2-in-rooms
11. Pololu. MQ-2 SENSOR. https://www.pololu.com/file/0J309/MQ2.pdf
12. Adafruid Industries. ESP8266EX Datasheet. https://cdn-shop.adafruit.com/product-files/2471/0A-ESP8266__Datasheet__EN_v4.3.pdf
13. de Carvalho Silva, J., Rodrigues, J.J., Alberti, A.M., Solic, P., Aquino, A.L.: LoRaWAN—a low power WAN protocol for Internet of things: a review and opportunities. In: National Institute of Telecommunication (IEEE 2017)

Secure Storage and Accessing of Organ Donor Details

A. Geetha, R. M. Ishwarya and R. Karthik

Abstract The area cloud computing itself is a very vast domain with increasing new technologies each day. Here, an online organ donation system is developed with the help of the aforementioned technologies. The main aim of the project is to make sure that the organs of the people, who have come forward to donate, reach the respective individual in need of the particular organ. Since everything is stored in the cloud, anybody can access it anytime, which in turn puts the data at risk. The secured sharing of donor details is necessary. Encryption is done at two levels to provide security, one while the data are entered and the other done by a third party providing proxy re-encryption. Now, when a patient in the hospital has been declared dead after thorough verification for any complications, based on the deciding parameters such as the organs that can be transplanted, blood group, HIV status, and location, a comparison is done with the cloud database using query filtering, which results in the nearest recipient available meeting the requirements. The performance enhancement of cloud data using encryption schemes is of high potential.

Keywords Cloud computing · Proxy re-encryption · Query filter · Cloud database · Encryption

A. Geetha (✉) · R. M. Ishwarya · R. Karthik
Department of Computer Science and Engineering, Easwari Engineering College, Chennai 600089, India
e-mail: geetha.vinodh1@gmail.com

R. M. Ishwarya
e-mail: rmishu97@gmail.com

R. Karthik
e-mail: karth2109@gmail.com

© Springer Nature Singapore Pte Ltd. 2020
S. S. Dash et al. (eds.), *Artificial Intelligence and Evolutionary Computations in Engineering Systems*, Advances in Intelligent Systems and Computing 1056,
https://doi.org/10.1007/978-981-15-0199-9_34

1 Introduction

Cloud computing in today's world has enormous usage, and it is becoming a trend to make everything available to all users at any time, now that the internet is available at every town and city. Another reason cloud computing is growing is that of the easy and cheap solutions to a lot of services provided by cloud services. Most of these services are the pay-for-use basis, and so there is no requirement for payment of a huge sum for services that people do not even use. Despite the ubiquitous and agile services offered by cloud, the main disadvantage in availing cloud services and why most people fear to use it is because of the violent threats and attacks that hackers do to the data on the cloud. Even though encryption of data can be done, it has become an easy task for the intruders and hackers to decode the encryption to retrieve the data. So, it is necessary to deploy encryption at newer levels and different algorithms to maintain the secrecy of information on the cloud. It is the main reason many companies do not reveal their security measures to the outside world. In order to extensively make use of cloud services, we must concentrate on providing high-security measures with high-speed data store. Organ donation can be done more extensively in India but there comes a need for a more agile system to easily locate the availability of organs and to find the corresponding donors before the organs get deceased. There are several difficulties in this process, such as knowing whether the patient is an organ donor, whether he/she is HIV positive, what blood group he belongs to, and the organs that can be donated. These are the parameters that are considered for the effective way to notify the hospitals about the organs that are currently available for transplantation.

1.1 Encryption Techniques

The process of encrypting the data by a proxy, after it has been encrypted at user side, i.e., proxy re-encryption is being implemented to ensure the security of data by keeping the decryption key with the proxy itself. So, whenever anyone wishes to access the encrypted data, they must request the proxy for the key to decrypt. The access list must be set at proxy in order to issue to keys to only those who are in that list. But as the proxy cannot be trusted with the data, there comes a need to encrypt the data initially and send it to the proxy. By encrypting the data at system side, we can avoid the attack from others to some level. With the help of proxy re-encryption system, we can avoid both insider and outsider attacks to an appreciable level. Insider attacks can be prevented because the encryption being applied at proxy is unknown to anybody but the proxy itself. This type of encryption is being implemented in the below architecture.

In Fig. 1, the Personal Health Record (PHR) of users is being encrypted with El-Gamal encryption and sent to a Setup and Re-encryption Server (SRS) for the proxy encryption. After encryption, it is stored in the cloud. Whenever the user

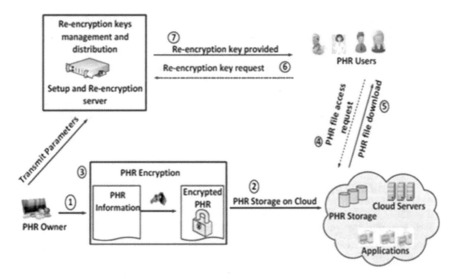

Fig. 1 Encryption and proxy re-encryption

wants to access the PHR, they first obtain the file from the cloud by using the public key available to all users in the system. To decrypt the second level of encryption, users must request the key from the SRS. Now, the SRS will check for the key that belongs to the particular user. After confirming the user with the access list, the corresponding key will be provided, with which the decryption is done. By performing two levels of encryption, we can provide security to the system to a certain level. To access this encrypted data, we must need two keys. One is the public key that is available to all users in the system and the other is the private key which is available only at the proxy. So to obtain the data, we must get the key from SRS and then decrypt it. This is the encryption technique used in our system.

1.2 Access Control

Access control is defined as the process of assigning the accessibility rights to certain specific people, who have permission to read or make changes to that file. There are three types of access control available. They are:

- Discretionary Access Control
- Mandatory Access Control
- Role-Based Access Control.

Out of these, Role-based access control (RBAC) is widely used in cloud computing for restricting those who can access particular data. Anybody else not mentioned in the access control list will be denied permission to enter the system or whatever activity that has been restricted. The main advantage of using this access control system is that nobody else can cause damage to the system. It maintains the privacy and confidentiality of user data. By the efficient implementation of the access control, we can easily protect it from attackers. But there is one exception when the attacker pretends to be someone with the access permission and tries to gain access into the system. This can be avoided when several parameters are taken into account such as the location of the user, with which we can make sure whether it is the user that seeks permission or somebody else.

1.3   Cloud Platform

There are various providers of cloud such as Amazon Web Services (AWS), Microsoft Azure, Google Cloud Platform, etc. Since the website uses C#.NET with Visual Studio, we are moving with Microsoft Azure for the cloud services for our system. In this paper, we discuss a model that makes use of SQL database cloud service for the storage of highly sensitive information about the organ donors in India and also the organ requirement list collected from various hospitals, which contain the information about those people having deceased or malfunctioning organs. As sensitive data come into use, the security measures applied must be highly effective and efficient. The usage of cloud ensures the availability of details at all respective hospitals, and so, the effective distribution of available organs becomes feasible. As cloud platforms are pay-per-use basis, any platform which meets the user needs can be used.

2   Related Work

There are several task scheduling approaches that have already been proposed in cloud computing environments. Some of them are studied and the important results are being briefed for reference. In the following, we explain some of the previous studies.

Liu et al. [1] 2013 proposed a multi-owner data sharing scheme known as Mona to overcome the hindrances while sharing the data in a multi-owner environment such as preserving the data and identity privacy. By using this multi-owner sharing scheme, we can make the storage overheads independent of the number of revoked users. The computational cost of Mona is very less when compared to other schemes. It is significantly reduced stating the efficiency of this multi-owner data sharing scheme. Also, the computational cost increases as the number of users revoked increases but the computational cost for the basic operations being done is almost the same for smaller and larger files. The total storage of each user in the system is 572 bytes. He clearly explains the operations such as file generation, access, and deletion. Along with this, the file updating option is also available. Mona, a secure multi-owner data sharing system is cost-efficient and with very less memory overhead.

Dutta et al. [2] 2014 proposed C-Cloud, a cost-efficient cloud for surplus cloud computing resources, which is a democratic infrastructure of the cloud for renting and sharing cloud resources. It also includes the non-cloud equipment such as PCs, laptops, etc. With the help of C-Cloud, large amounts of cloud resources can be shared thereby making the owners of the machines earn a lot of cash by subletting their idle system for the use of others. An incentive mechanism is being used in this system for exceptionally standing out of other systems. The main disadvantage of this model is that the interaction between the C-Cloud and those who periodically shared their resources to check the status. In this, an open-source distribution for volunteer computing has been used for interaction purposes.

Li et al. [3] 2015 did a security analysis on the one-to-many order-preserving encryption-based cloud data search. In order to perform the ranked search in encrypted cloud data, a tool called order-preserving encryption (OPE) is used. But while using deterministic OPE, the ciphertexts started to show relevance, which is not encouraged in any encrypted data. Hence, a probabilistic approach called one-to-many OPE to overcome the flaw in deterministic OPE. This scheme employs a random probabilistic method to assign the ciphertexts. A bucket m is picked and then randomly any value in the bucket is chosen as ciphertext. This one-to-many OPE can be developed in many ways including dividing the plaintext of the same value into several sets and dividing the corresponding set into sub-buckets. By doing so, we improve the security of the data and pose a challenge for the attacker to decrypt it.

Chen et al. [4] 2016 proposed a secured processor architecture, Cyber DB, in order to securely store the encrypted data and to execute queries over the encrypted data. This architecture uses the AES encryption scheme but in two different modes: AES-CRT and AES-OFB. The former is the counter mode of AES and the latter is the feedback mode of AES. The usage of these modes is to offload the encryption and decryption work and to transform block cipher into a stream cipher. The Cyber DB provides the most common operations of any database, which are the creation of table, insertion, deletion, and updating of records. The performance is evaluated with DBmBench, TPC-H queries. During all the evaluation process, the overhead was only minimal and the security was highly ensured. The major issue with the AES modes is that seeds must be unique for each datum. Otherwise, the system is vulnerable to "two-time" pad attack. It can also be done by sending the wrong seed by the adversary to the seed register. It introduces on an average of 10% performance overhead, which can be further reduced by using register sharing of attribute seed and program execution.

Zhou et al. [5] 2016 put forth a role-based encryption technique to develop a large-scale secured Electronic Health Record (EHR) system, which is called Personally Controlled Electronic Health Record (PCEHR). It was designed by the Australian government but with several security issues. The issues were met by the proposed PCEHR system as it captures practical access policies based on roles in a flexible manner and provides the security of data storage. This patient-centric health record system uses ISAAC symmetric encryption system and 160-bit elliptic curve.

It uses JPBC and PBC as pairing-based crypto library. JPBC wraps the PBC library to generate MNT curve. With this system, it can be seen that the ciphertext size remains the same when the number of ancestor role changes. This system has an access list which has the organizations/companies that the users have granted permission to access his/her records. It provides two levels of access: Basic access control and advanced access control. With the former, the user need not provide explicit access to organizations but the system itself will assign with all healthcare organizations. The latter provides a Provider Access Consent Code (PACC) which any organization must possess in order to access the records of the user. The code is set and controlled by the users.

Sallam et al. [6] 2017 introduced a High-Efficiency Video Coding (HEVC) selective encryption system that encrypts the highly sensitive part of the video with Rivest Cipher 6 block cipher technique. It provides low complexity overhead and faster encoding time for real-time applications. This technique should maintain the video format compliance, same bit rate, and real-time constraints. In this paper, they also compared RC6 with every mode of AES to prove the efficiency of using RC6 over AES. In order to avoid a brute force attack, a larger key space must be employed and by using RC6 in this technique, which is a 128-bit key length encryption, we obtain 2^{128} key space size. It provides immunity against brute force attack. The histogram analysis of the plain video and encrypted video shows that both of them are very different, which supports the security feature. The RC6 encryption enhances faster encoding as it is less complex and simple when compared to the AES. With the help of HEVC SE, we can securely transmit a video file very quickly and with high level of security to receiver end.

Ali et al. [7] 2018 proposed a system where the Personal Health Records (PHRs) of individuals are securely stored in the cloud by performing two levels of encryption. The cloud storage used for this system was Amazon Simple Storage Services (Amazon s3). In SeSPHR, initially, encryption is done and this encrypted data are sent to a setup and proxy re-encryption server (SRS), which is a trusted third party. Here, the encrypted file is again encrypted with a key, which will be unknown to the users in the system. In case anybody wants to access the data from the system, they need two keys out of which one is available to all the users, i.e., the public key and the key is present only at SRS. In order to obtain it, one must request the setup and proxy re-encryption server for the key. The main technique used is the proxy re-encryption for issuing the keys to authenticated users. Java Paring-Based Cryptography (JPBC) was used to perform encryption of PHRs. The encryption algorithm used in this system is the El-Gamal encryption scheme. The turnaround time for encryption is given as $TT\text{-}up = tEnc + tup$ and the turnaround time for decryption is given as $TT\text{-}down = tDec + tdown$. The major factor affecting the turnaround time is the encryption and decryption part as the uploading and downloading time are minimal effect. The formal analysis and verification part was done with High-Level Net (HLPN), Z3 solver and SMT-lib. The system ensures the confidentiality of and provides access control of records that even some part of the data are not even accessible by those in the system. The performance analysis is done based on time consumed. It was found that the decryption phase was 24.38% lesser than that of encryption.

Lin and Kumar [8] 2018 improvised the existing robust thin-plate spline (RTPS), which was used to match the legacy contact-based fingerprints to contactless fingerprint databases. The problem this RTPS had was some deformations and accuracy issues. Moving to the contactless fingerprint is a need because we want to provide much secure, hygiene, and accurate fingerprint detection. In order to remove any deformations that occurred in the conversion of a robust thin-plate spline-based, deformation correction model (DCM) has been proposed by Chenhao Lin. By the usage of DCM, we can achieve results in accurate alignment of key minutiae features that were observed in both the methods. The results suggest that the contact-based fingerprint matching using the deformation correction model and the robust thin-plate spline provides more accurate findings with equal error rate (EER) as 4.46%, whereas the thin-plate spline from other researches had significantly higher EERs.

3 Proposed System

Our paper proposes a system which securely stores the organ donor details and allows accessing of the same only at a particular time. The system and functional architecture of the model are described below.

3.1 System Architecture

In order to perfectly keep track of the organ donors, we store the details of the donor over the cloud, so that it is available everywhere. As the data are highly sensitive, we perform encryption on them before storing it in the cloud. It is done because in case of an accident if the patient goes into the brain dead situation, there is no way to know if he is a pledged donor. To know this efficiently, we built a new system, which will read the fingerprint of the brain dead patient and cross-check with the fingerprints in the database. If a match is obtained, then the particular patient is an organ donor. Then the donor's details are extracted, once they are successfully decrypted. For the successful implementation of the system, we need the requirement list of organs that are currently needed, which will be collected from various hospitals. Now, the extracted donor details are run against this list to find the matching donors. The parameters that are used for finding the matching donors are blood group, HIV status, organs, and finally location. The requirement from the hospitals that satisfy all the above parameters is shortlisted. With the obtained list, we send notifications to the respective hospital to indicate the availability of organs.

The major components of our model (Fig. 2) include the mobile application, website, Setup and Re-encryption Server, cloud database. The mobile application is

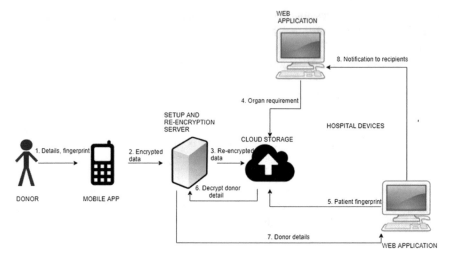

Fig. 2 System architecture of the system

used to get the details from the pledged organ donors; website is used at hospitals in order for them to enter the organ requirement details and to find the details of the organ donor; Setup and Re-encryption Sever to provide the decryption keys to the hospitals while accessing the organ donor details; cloud database to store all the details on the cloud for easy access to the data anywhere anytime. At the end of the entire process, hospitals will obtain notification about the availability of organ as per the requirement provided by that hospital.

3.2   Functional Architecture

The functional architecture (Fig. 4) gives the entire functional modules of our system. The system can be briefly classified into three modules, each of them performing some task to reach the final output or goal. The modules are:

- Donor data encryption
- Donor authentication
- Notification generation.

3.2.1   Donor Data Encryption

The donor data encryption module mainly encrypts the data and stores in the cloud database. Initially, the organ requirement list must be collected from the hospitals. The requirement list data are encrypted and then stored securely in the database.

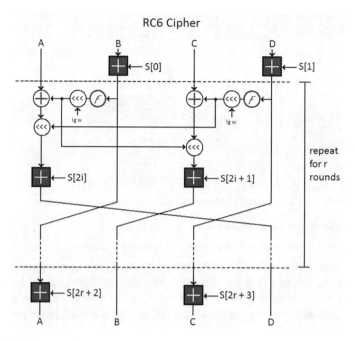

Fig. 3 RC6 implementation

Any hospital in the system can access this list as they are allowed to do so. Now, an Android application is created to read the data from the pledged organ donors. In order to connect the mobile application with the web, a service API is being used. It is where the actual encryption of data is done. The encryption algorithm used in this system is Rivest Cipher 6 (RC6) encryption (Fig. 3), which is derived from RC5. RC6 is a 128-bit block cipher with key sizes varying between 128, 192, and 256. RC6 requires more CPU and memory resources for encryption but it results in higher efficiency when the encryption speed is taken into account. After encryption, the encrypted data are sent to a proxy to perform the proxy re-encryption. When the second level of encryption is also done, the data are stored in the cloud database. Once the encryption phase is over, all the encrypted details along with fingerprints and requirement list are stored in the cloud (Fig. 4).

3.2.2 Donor Authentication

In case an accident occurs and the patient is diagnosed to be brain dead or any person who wants to donate an organ, in order to verify whether they have signed up as an organ donor, we must read their fingerprint. Once it is done, it is cross-checked with the fingerprints stored in our cloud database of pledged organ donors. In case of a match, the details of the patient are extracted after decrypting

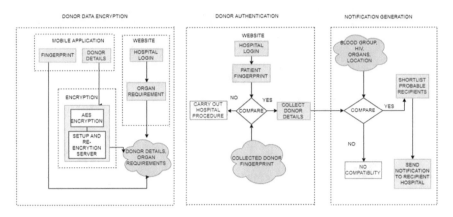

Fig. 4 Functional architecture of the system

the two levels of encryption. Now, to decrypt the first level of encryption, we must request the key from Setup and Re-encryption Server (SRS). Once the SRS has authenticated the request as positive, it will issue the key. With the public key which is available to all hospitals in the database, we can decrypt the second level of encryption. Now, we can obtain the details of the organ donor in a matter of time.

3.2.3 Notification Generation

With the collected details of the donor, we perform SQL query filtering on the cloud database containing the organ requirement list. The major fields to be considered are blood group, HIV positive or negative, organs that are needed, and location. If all the fields match, then the concerned hospital providing that detail is notified the availability of organs. All the notifications are sent directly to the website that the hospital uses. Then the hospital contacts the particular individual and carries out the procedures. By this way, we may increase the chances of survival as there is no time wasted trying to contact the patient's family to check if he is an organ donor.

4   System Implementation

The proposed system has been implemented for certain hospitals from which the requirements are gathered and to which notifications are sent when matching donors are found when an organ donor who died being in the hospital. The modules defined are implemented to support our project.

firstname	lastname	dob	gender
1	YgeNqHfp3C9eoRB3CYh0kBehwT9kYW3mpUk9giaq8/Ckzw...	Mt6yalaH7KKtMdOWyjiqovNBPfBpUn5X2fiU59AolB/K3rv67J...	vcz51yXRe6a7RyB9tHc2I1z9lBCuo8GWeyO0ybHwsST6PGq...	NjWP2gXqShcSgL3rFv9NyClDa4FeWW3OCp0RGK\
2	jY42H2R5eFhkGM+8Tg9iNJP5cdne/7g1fV7MRyqqDrkoWnN...	vrPoNqy51zsnzmJLB+oa1KN5DZly5/MF8VVJ32FTGJ78M6ZGJ...	H9VYe9EbNVbWf6uX8bkCZPNG/rb01bxi4tDaOgAuM7KdxQy...	7L+sm91347Y7LeSCeC/Givhs5NKZjY9Emmeoy Vbxd#
3	8XX4u8DVSLxTB8fThy3B4g+=	qMFkvwciQiTNOnk/ezC4+w==	GNUtyzGOhEKl34zEse1hvqM97zBNEuKQpqsleyL31fQ=	2+fvwq0MLcdHZO4d9ZZhbQ==
4	8dM7xBJy+TSxBkaxUmYiY7bxTwr21X+L/KiJFwsaDE=	LU/RCHGoDHBloxgeuVvRpA==	aNuX4TXjku/3QWl8ww7/GKZ+XTz41EwKUKcK4QLx=	HkOB1oeSzA3fOOFvFBV3Sw==

Image 1 Encrypted data stored in cloud database

4.1 Donor Data Encryption

The output of this donor data encryption module is the successful storage of organ donor details and the requirement list from the hospital over the cloud. The main part is the encryption of the details carefully to avoid attackers to view the original data of the donors. The organ donor details are collected using an Android application from the pledged donors. The fingerprint is read using a fingerprint reader and the threshold is set to maintain the quality of the fingerprint. In case it does not meet the specified threshold, then it is asked to scan the fingerprint again. The requirement list is collected from the hospitals using a website. Images 1, 2, and 3 show the result of this module.

4.2 Donor Authentication

The result of this module is the collection of the organ donor details from the database if the patient admitted is an organ donor. It is done by comparing the fingerprint of the patient with those in the database and if a match is found, then the hospital requests the Setup and Re-encryption Server for the decryption key. Once

Image 2 Requirement list from the hospitals

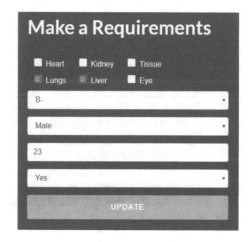

Image 3 Collection of donor details

localhost:1948 says

Finger matched

OK

Image 4 Authentication of patient

the key is issued to the hospital, it can decrypt the outer encryption and with the public key available to hospitals in the database, the inner encryption is decrypted. Now, we can obtain the organ donor's details. The output is displayed in Images 4 and 5.

ID	104
firstname	9CFsneFieAubY+243kmO9AI9xMowPod3O6U4Qb8bW50NI3eSA6wbXLIS+ThWgLa1
lastname	XLBvfx3f3IfsmXI2Q2zD+gwkIxZWjbS4Ak1x/G/79Kfa51JcVfcl2eqddTul9T41mxX
dob	ZK3j72/aEkUsPich1Um35q7NItCQxyWsFZr/iat+L7PIWqzjcUzYVSvZKUwmn/1TbajVkWh15NRd3T8PvfWV+o+
gender	ckfa+h4WKiBcN0clF3UFjBGWy52SO/r/yfVnST4mbmbClonfuaZdNgzKM2hxVTTuPf
bloodgroup	9xL9HhovO37P5Vj4EdB38hL0UJT17U+jEhrDQgIQXNfAmcVT8fW1ILnPayqVLs0T
email	IP/86W64J/sgqpzDFgpxn+Mjh+vyeyoTntblwyNP3aeXsf56yu+ibkeJhakMAP0Y7aIRGSCJcDF4Ico9Z68wv9S1Po9pi7P8+AUHC1zxfAYqY
phone	1MDeiXJBguBYiTdQIZIKvpoUq9P0Gk1BUM936RkvX45R9tP1r6Ns/IIII5+R97meOjWmhVz6yNydj7dfkUvZljMeZv
address	C5MPSk3v0PJ8rgMXYHy1DRjU/RNO2HrbNwkknYAAvHhNS0pZzjFfQpkytgGRCVVzF2oGUAXFT1SWFXtAScazT
choice	3WY4v0+j8JS6QktGcXHyIpsCpZnE3iSo3jY1rTC9GrmTbXfFyyFHrFvfX79YbvEL7F1vlhsRDhWfPokegwnsg+GbJRkUuOoHJZTaxU+FWxOv5

Request for Re-encryption Key

1045 Request

Image 5 Requisition of decryption key

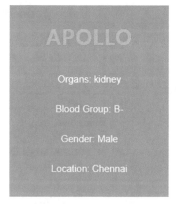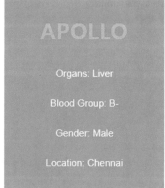

Image 6 Notification sent to the hospitals

4.3   Notification Generation

The final module produces the notification to the hospitals following some parameters. First, the collected details of the organ donor are compared with the requirement list from the cloud. The parameters taken into account are blood group, matching organs, HIV positive or not, and finally the location. Once the parameters are satisfied, the notification is sent to the corresponding hospital which provided the matching requirement. The result of this module is the display of notification in the hospital website. The output is displayed as in Image 6.

A notification is like will be displayed in the respective hospital's website login itself. After that, the hospital carries out further procedures like calling the patient in and arranging for the organ transplantation.

5   Performance Analysis

In this section, we are going to study about the system's performance and accuracy that has been achieved. The dataset for the system is given manually since all the organ donors are asked to register themselves in the Android application. The organ requirement list must be obtained from the hospitals from the website itself. We are ensuring the privacy and confidentiality of the data provided with our encryption techniques. The RC6 encryption is proved to be one of the most secure encryption algorithms, and the equal error rate (EER) of this algorithm is only 4.46% which is comparatively lower than most other algorithms. With the usage of Setup and Re-encryption Server, we provide added security to the organ donor details.

The encryption time required for the system is a bit longer than other models since our model uses two levels of highly secure encryption algorithms—Advanced

Graph 1 Encryption time
required for users

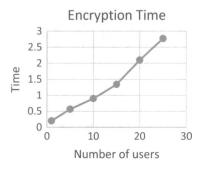

Graph 2 Decryption time
required for users

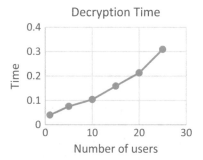

Encryption Scheme (AES) and Rivest Cipher 6 (RC6). But on the other hand, decryption time is shorter because these algorithms we used are block ciphers and only during encryption, the operation needs to be done block by block but during decryption, there is no need for such mechanism. Parallelism can be implemented depending upon the processor we are using to decrease the decryption time. Graph 1 shows the encryption time for several numbers of users while Graph 2 shows the decryption time required for the same number of users.

The above graphs prove that even though the time taken for encryption is higher than many systems, the decryption is comparatively very low because once the key has been issued by the Setup and Re-encryption Server (SRS), it can be decrypted very quickly. Therefore, the performance of the system can be increased as the rate at which decryption occurs is 26% faster than earlier systems.

The system that we use has an Intel i5 processor running at 1.7 GHz with 8 GB of RAM and is a 64-bit operating system. The fingerprint reader that has been used is Mantra MFS100 V54, which is used in both Android application and the website. The cloud storage that has been used for the system is Microsoft Azure. These are the system specifications for the project.

6 Conclusion and Future Work

Although many existing systems have implemented the system, they were not able to successfully find the match and availability of organs at the needed time. Our proposed system successfully finds the available organs and sends the notification to the hospitals where there are people in need of organs. It can be extended in many directions. One of them is to increase the location accuracy to find the nearest hospital midway between the donated organ and the recipient in real time. By doing so, we increase the success of organ transplant as the time available for the transplantation process can be increased.

References

1. Liu, X., Zhang, Y., Wang, B., Yan, J.: Mona: secure multi-owner data sharing for dynamic groups in the cloud. IEEE Trans. Parallel Distrib. Syst. **24**(1), 1182–1191 (2013)
2. Dutta, P., Mukherjee, T., Hegde, V.G., Guhar, S.: C-Cloud: a cost-efficient reliable cloud for surplus computing resources. In: IEEE 7th International Conference on Cloud Computing, Sept., pp. 986–987 (2014)
3. Li, K., Zhang, W., Yang, C., Yu, N.: Security analysis on one-to-many order preserving encryption based cloud data search. IEEE Trans. Inf. Forensics Secur. **10**(9), 1918–1926 (2015)
4. Chen, B.H.K., Cheung, P.Y.S., Cheung, P.Y.K., Kwok, Y.-K.: Cyber DB: a novel architecture for outsourcing secure database processing. IEEE Trans. Cloud Comput. **6**(2) (2016)
5. Zhou, L., Varadharajan, V., Gopinath, K.: A secure role-based cloud storage system for encrypted patient-centric health records. IEEE J. Comput. **59**(11), 1245–1257 (2016)
6. Sallam, A.I., Faragallah, O.S., El-Rabaie, E.-S.M.: HEVC selective encryption using RC6 block cipher technique. IEEE Trans. Multimed. **20**(7), 1636–1644 (2017)
7. Ali, M., Abbas, A., Khan, M.U.S., Khan, S.U.: SeSPHR: a methodology for secure sharing of personal health records on cloud. IEEE Trans. Cloud Comput. **14**(8), 1–1 (2018)
8. Lin, C., Kumar, A.: Matching contactless and contact-based conventional fingerprint images for biometrics identification. IEEE Trans. Image Process. **27**(12), 2008–2021 (2018)

The Requirements of the Technique of Communication from Machine to Machine Applied in Smart Grids

Mehmet Rida Tur and Ramazan Bayindir

Abstract As a result of increasing population and energy demands of individuals, the demand for energy is increasing day by day. It will be difficult and inefficient to meet this demand with the current network structure. Problems such as inefficient transmission caused by high transmission losses, problems in the integration of renewable resources into the network, problems such as inelastic demand and compensation, meeting the increasing demand with traditional network structures are inefficient and unsustainable. Expectations of existing networks to be efficient, flexible, and reliable in terms of supply–demand balance bring up the applications of Smart Grids. Smart Grid applications aim to ensure that every stage of the production, transmission, distribution, and consumption processes of energy is observable and controllable, and therefore, effectively manageable. In this study, wide area management in Smart Grids and distributed predictions are presented with distributed control applications. In this way, the applicability of Machine-to-Machine communications on machine networks was studied. For this purpose, Machine-to-Machine systems and Smart Grid systems are compared in separate layers. In addition, using Machine-to-Machine system architectures, Machine-to-Machine system architectures were developed. The proposed architecture is based on a Smart City platform and the European Telecommunications Standards Institute Compatible Machine-to-Machine communication frame.

Keywords Smart grid · Machine-to-machine component · Information and Communication Technologies · Smart cities · European Telecommunications Standards Institute

M. R. Tur
Department of Electricity and Energy, University of Batman, Batman, Turkey
e-mail: mrida.tur@batman.edu.tr

R. Bayindir (✉)
Department of Electrical and Electronic Engineering, Faculty of Engineering, Gazi University, Besevler, 06500 Ankara, Turkey
e-mail: bayindir@gazi.edu.tr

© Springer Nature Singapore Pte Ltd. 2020
S. S. Dash et al. (eds.), *Artificial Intelligence and Evolutionary Computations in Engineering Systems*, Advances in Intelligent Systems and Computing 1056,
https://doi.org/10.1007/978-981-15-0199-9_35

1 Introduction

Smart Grids and Cities, the concept of using Information and Communication Technologies (ICT) to improve the quality of life, is a necessary research area that is on the agenda both in academia and industry. The research carried out in this area is mainly to create effective communication platforms that can integrate various field systems into an urban-scale system. This new communication platform paradigm, known as Machine-to-Machine (M2M) communication, provides reliable information flow to decision-making systems from connected objects (sensors and actuators). Such platforms support new interaction models that are not in human control, often produce small amounts of data and have limited resources, i.e., objects connected to objects that are limited to memory, energy, and computing power. M2M refers to the technology that enables wireless or wireless communication of different devices. It is not right to think about M2M concept for a single area. In parallel with the developments in communication systems, internet, and sensor technologies, M2M concept is increasing its importance. The M2M idea offers systems that are interconnected with low-cost, scalable, and reliable technologies and can also be controlled remotely. Many applications have been developed in order to increase productivity and reduce costs by communicating between M2M and devices. Some of these applications are developed in areas such as public safety, transportation, health, and energy management. The development of M2M systems leads to new projects and technologies. These are the concept of Internet of Things (IoT) and Smart Grids.

When we look at this day, it is seen that the structure of the existing electricity grids is not very different from the structure of the networks that were put into operation a hundred years ago [1–7]. The developments that can be considered as important are generally in the materials and designs of transformer, line, mast, and insulation systems. In terms of measurement, monitoring, protection, and control equipment, interconnected systems with electromechanical systems dominate a shift toward the use of microprocessor-equipped systems in the last two decades [8–10]. With this orientation, the introduction of improved algorithms and technologies in the implementation of these functions led to improvements in the quality and reliability of networks.

The standardization process of M2M communication continues with the main institutions ETSI and 3GPP. The main focus is to promote effective cooperation between shareholders to improve interoperability, develop business perspectives and address potential problems and propose new and better technological trends. D. Niyato et al. studied energy management in Smart Grids [11]. Sood et al. [12] focused on intelligent network infrastructure, application, and communication requirements. M. E. Kantarci et al. proposed low-cost energy management for residential areas using wireless sensor networks [13]. Smart Grids offer significant advantages in order to meet the increasing energy demands of the countries continuously, to intervene more quickly and to control the harmful emissions of gases emitted to the environment more rapidly [14]. At the same time, it is very important to make energy cheaper and use efficiently. For this purpose, in addition to the energy-saving devices used in the networks, it is located in daily life in the systems that control the flow in the network and aim to prevent the leaks. The contribution of Smart Grids to the economies of the country is at a great extent.

Using renewable energy sources is getting more and more energy and this means increasing energy suppliers. In addition to the correct and uninterrupted transmission of energy, there is a need for new energy markets and platforms that enable energy trade. Older generation, transmission, distribution, and onsite billing systems are now being replaced by new generation systems. In 2005, the European Technology Platform (ETP) was established in order to accelerate this transition and create a common vision for the Smart Grids in Europe. Accordingly, it was decided to ensure the sustainability, competition, and security of supply for every energy-consuming organization [15]. Transition to Smart Grids use in the energy distribution sector has become crucial for Turkey. The first reason is the privatization of energy distribution companies, the need for legal separation, increasing competition, and technological requirements. The second reason is to prevent the use of illegal energy. Although the use of illicit energy varies according to regions, it may be up to 76 and 79% in some regions [16, 17]. With the development of sensor technologies, humidity, temperature, pressure, image, speed and so on. Such systems can easily measure values. For example, today, with the development of micro-electronic structures, infrared cameras have entered our mobile phones [18]. While technology has progressed rapidly, it has become inevitable to use M2M systems in industry [19]. In this study, it is tried to explain how M2M and Smart Grid systems can be constructed on M2M system architectures in the literature and in the world.

2 The Structure of Communication Technology

In parallel with the technological progress, there are many applications developed on M2M. Both the multiplicity of applications and the desire to use different architectures together made M2M work as a standard inevitable. One global cooperation group called oneM2M on M2M and IoT issues work to develop standards. To this end, the world's first global M2M standard was published by oneM2M on February 4, 2015 [20].

2.1 Communication Technology of Machine to Machine

M2M architecture consists of three basic layers as main structure [21]. The first layer contains sensors that provide data for the M2M system. These sensors transmit the data required for the M2M system. For example, heat, humidity, gas detectors, cameras, smartphones, etc. devices are some examples of this layer.

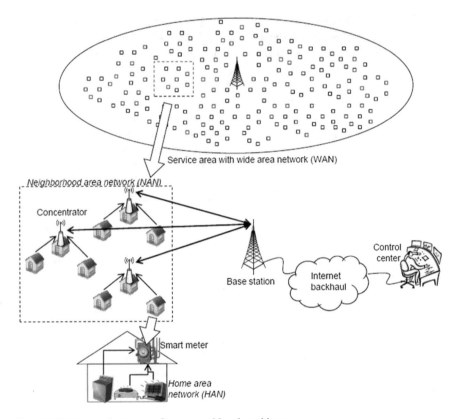

Fig. 1 M2M network structure for smart grid and partitions

When the data is transferred to the system, the second layer, communication network, is transferred to the desired area. This field may be human interacting systems or machines capable of direct communication. In Fig. 1, the above-listed layers are shown schematically.

Data should be presented to the end user as meaningful information via the network. At this point, the third layer, the application layer, is activated. Here, users can analyze data and, if necessary, give command to the M2M system.

Basic components of M2M technologies are technologies such as sensors, RFID and communication technologies such as Wi-Fi, Ethernet, GSM, RF to transmit this information to other machines. One of the fundamentals of M2M technology is the ability of machines to transmit the information they produce or collect from their surroundings to other machines in remote locations without any human intervention. For M2M technology, standards have also been created in some vertical areas, but there is no general standard for all M2M applications. For example, in smart home applications, KNX, Z-Wave, Zigbee, Modbus are the standards that determine which format to send the information they receive. Similarly, standards such as OSOS (Automatic Meter Reading Standard) have been published in smart meter

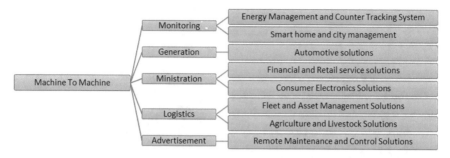

Fig. 2 Classification of communication between machines

Fig. 3 The percentages of sectors in order of importance

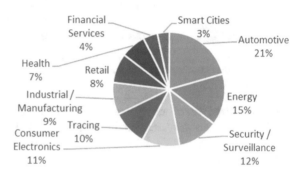

applications. It is not possible to categorize these technologies when it comes to this wide concept, which is based on the communication of machines with one another, but we can consider the classification study as a general approach in Fig. 2.

For the importance of M2M services, Analysis Mason is the result of the survey conducted among mobile operators with the highest market volume. The percentages of the sectors in order of importance are shown in Fig. 3.

According to the statement of ETSI in 2012, M2M infrastructure in the world has the capacity to connect 50 billion devices and this number is increasing day by day. According to the statement made by the OECD, the M2M market could reach 700 million in 2020. According to data from another source (Analyses Mason, 2013), it is estimated that in 2021, there will be a total of 2 billion mobile-connected devices in the world.

2.2 Smart Grid and Communication Technology

The idea of Smart Grids is based on the principle of being flexible, accessible, reliable, and economical. The solutions developed within this framework offer a bidirectional communication along the line of production and consumption in order to reduce energy losses, facilitate energy transfers, and increase efficiency and

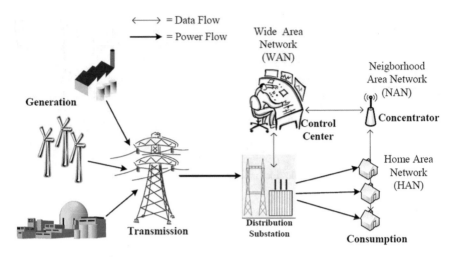

Fig. 4 A general building model of the smart grid

security. Smart Grids are systems that require continuity in terms of domestic market, supply security, and environment. This is shown in detail in Fig. 4.

ETSI divides Smart Grid systems into three main layers as shown in Fig. 5 [22]. These layers are:

1. Energy Layer: It deals with energy (production, distribution, transmission, and consumption), sensors, storage, connection, distribution, and power systems.
2. Control and Link Layer: It provides access technology, management functions, and energy controls such as transformer automation, monitoring and diagnostics, control automation and protection, time synchronization, measurement, routing.
3. Service Layer: It covers all services related to Smart Grid use, billing, electronic commerce, data modeling, subscription management, activation, implementation, and business processes.

As seen in Fig. 5, M2M system architecture and Smart Grid system architecture are similar. Based on this similarity, the two systems with different architectures can be combined under one roof and the Smart Grid applications can be developed. In this case, by adding smart network information onto M2M information, system can be developed or ready systems can be examined easily. The following items attempt to explain the changes to M2M's layers.

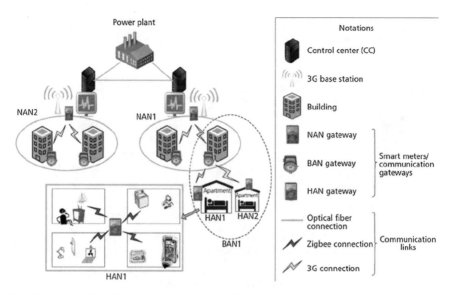

Fig. 5 A general building model of the smart grid

2.3 Implementing Smart Grids on Machine to Machine in the Perspective of Standards

ETSI's M2M standards include an easy-to-deploy scenario for Smart Grid systems. Control centers can be used within M2M standards while energy is transferred from point to point. The connections shown in dashed lines in Fig. 6 can be performed with high bandwidth fiber networks. The connection indicated at the bottom of the figure shows that through the M2M gateways, each stage can be monitored from the control centers, while the energy flow takes place from the supplier to the demander [23]. Based on this notation, M2M standards were introduced at certain points and a healthy system was developed. OneM2M organization serves in different areas in terms of standards, one of which is Smart Grid systems.

The two-way control of M2M systems and the two-way transmission of Smart Grids are similar. For this purpose, systems with gateways placed at the right points and service-oriented architectures that provide device management support ensure that service classes can be operated more easily.

2.4 Smart Grid Applications in M2M Device Area Perspective

Smart Grid systems consist of many devices such as sensors, storage systems, smart meters, and energy production control. It can also be added to control systems and

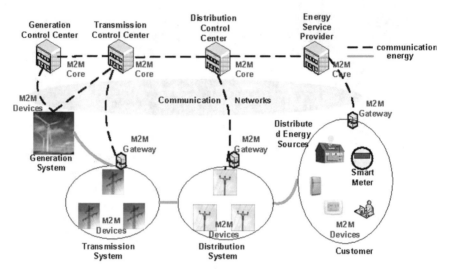

Fig. 6 Implementation of the smart grid system on ETSI M2M

different devices that vary from user to user. The device concept is the existing components on the basis of M2M applications. The ETSI and oneM2M organization have been developing standards for these components for years. With the proper application of these standards, M2M device management can be integrated into Smart Grid systems. Some of the devices that serve the end user are working with batteries. In this case, it is possible to check the battery's energy level and in such operations, the device management operations that M2M has been offering for years can be used in Smart Grids. This will facilitate the development of Smart Grids.

2.5 Smart Grid Applications in M2M Communication Perspective

Smart Grid systems primarily include applications developed to save electricity, improve service quality, reduce energy production, and distribution costs. Previously, the systems that were left to the user's savings with the Household Energy Management Applications have now left their place to a complete network of Smart Grids in order to create new world markets and to facilitate the traceability of the energy produced from different plants (heat, wind, hydro, etc.). Thus, applications have been developed in favor of both sides in the supply–demand relationship.

Applications made with a short distance wireless, long-distance cable, or hybrid communication model in simple M2M applications should be maximized in Smart

Grid systems. Existing power line systems cannot provide the cellular data required by Smart Grid systems. For this reason, the existing infrastructure should be changed in accordance with the system so that energy demanders can control the energy use in the house even from their mobile phone and see the usage details and also the manufacturer can make the measurements correctly for the correct billing of the energy.

Some problems may be encountered because of the different standards of local and wide area networks when the abovementioned infrastructure is changed. Therefore, this transition process can be shortened by using M2M structures. In M2M applications, TCP/IP-based communication is performed. It is not recommended since this communication is made in the energy transmission line and high energy is consumed for low data [11], but communication can be performed with this protocol at the main control points. In this way, the transition process needed to renew the infrastructure can be shortened.

2.6 Smart Grid Applications in M2M Application Layer Perspective

The most common application area among M2M applications is smart measurement systems. ETSI conducted a study on M2M's use of intelligent measurement in its technical report ETSI TR 102 691 V1.1.1 published in June 2010 [24]. OneM2M shows its use in M2M's intelligent measurement systems in its technical report OneM2M-TR-0001-UseCase on September 23, 2013. In addition, these two organizations also worked for Smart Home and Home Energy Applications. Figure 7 illustrates the relationships between Smart Grid, intelligent measurement, and home automation systems [25].

The measurement infrastructure is one of the key areas of Smart Grid systems. Accurate measurement of energy supply and demand is very important on both sides. It may be desirable to check the different operations such as making the correct invoicing and energy switching on from a single center. This works in the same way as M2M smart home applications. By correctly configuring M2M's intelligent metering systems, the service layer of Smart Grid systems can be easily designed. Intelligent grid systems can be successfully implemented over an M2M platform with the use of service-based architectures and accurate storage of network data transferred to the central database for billing, analysis, troubleshooting, and investment planning. With the development of cloud-based applications, data was collected from a single point and analyzed from a single point. This facilitated the development of M2M platforms on the cloud and Smart Grid applications on the same platform.

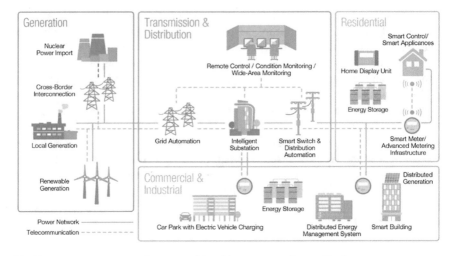

Fig. 7 Smart meter systems for smart grid and home automation applications

3 Required Research Subjects in Machine-to-Machine Communication Used in Smart Grid

- **Security**

One of the main issues in M2M communication is security; according to 3GPP's features, it still needs a lot of work to make such communications more reliable. Remote provisioning and subscription change is the most important issue that needs to be addressed.

- **Bandwidth Requirement**

In the M2M communication, a large number of MTC devices regularly send data updates. In the future, the expected bandwidth requirement for a single device is 2.5 Mbps. Due to the limited bandwidth availability in the main carrier network, there is a need to prioritize and allocate sufficient bandwidth to user traffic and traffic generated by M2M devices.

- **Low Cost and Optimal Design**

Smart devices and network components must be cost-effective (hardware and maintenance) and optimally designed for M2M applications. Optimized designs are required to improve service quality (QoS).

- **Traffic Characterization**

Human–Human (H2H) communication traffic differs from M2M communication. In M2M communication, there are three types of communication; as discussed before, each application type has different traffic requirements. Traffic characterization is important for designing optimized components that will improve the QoS required for M2M applications.

- **Delay**

Some applications are not delayed, others are not. Real-time communication is required for some M2M applications. In wired networks, the latency in optical networks is quite low, but for wireless networks, a significant part is a trivial matter, considering the general trend in the increased bandwidth requirements of end users received by real-time delay-sensitive applications. In future hybrid wired wireless networks, maintaining an end-to-end tolerable delay level is something to work with.

- **Special Protocols**

Special protocols are required for M2M communication to make data transmission more efficient and to minimize protocol overhead. Existing Internet protocols for TCP/UDP M2M communication are inadequate and new special protocols need to be investigated. The necessity is also due to the need for very little human intervention in M2M applications.

3.1 Recommendations for a Road Map for Integrated Construction in Smart Networks

In order to realize the possibilities and expectations mentioned in the previous section, the following points should be taken into consideration. What should be done to ensure the most efficient communication between machines and infrastructure development and current communication technologies:

- Smart meters must have microcomputer, memory, and communication units as a minimum. These devices must have access to wide area communication (GSM/GPRS, PLC, satellite communication) as well as in-house power systems and have narrow field communication modules (ZigBee, Wi-Fi, PLC, etc.) for Machine-to-Machine communication.
- It should be ensured that renewable energy systems (solar and wind) are disseminated in households and manageable with smart meters. In this way, applications such as in-house load management and management of energy supply to the grid should be realized by programmable smart meters.
- The power unit of household appliances shall be capable of communicating with and managing smart meters. Traditional smart plugs should be designed to provide this compatibility. Smart sockets should be manageable with smart meters.
- Electric vehicle charging systems must be installed in garages or parking spaces.

What to do on a global scale:

- Global energy control and management systems should be established to evaluate the energy situation and demands of all countries and manage energy traffic on a global scale.
- Independent agencies should be established for Smart Grid research and standardization. These organizations should be authorized to take measures and implement sanctions to ensure the development and integration of Smart Grids in accordance with certain standards.

What to do on a country scale:

- A national energy control and management system should be established to evaluate the energy situation and demands of all cities and to manage energy traffic.
- State support programs and renewable energy resources that have the advantages of developing technology and local production and local consumption measures should be taken and mechanisms should be established.

What to do in the city scale:

- Efforts should be made to bring the cities to produce their own electricity. Renewable energy (solar and wind) production areas should be established in the nearby regions of each city. In addition, solid waste disposal facilities should be transformed into power generation stations and integrated into the Smart Grid.
- Extensive field management practices should be developed for intra-city energy management so that the energy map of the city should be monitored in real time and the energy balance should be maintained. It should be able to carry out tasks such as local remuneration, commissioning or deactivating renewable resources, managing energy import and export operations.

What to do in the neighborhood scale:

- It should be ensured that the local distribution network provides communication (PLC) from the power line. It should be ensured that power line communication components can be monitored and managed remotely.
- Renewable energy and storage systems should be established to support local production and consumption and the energy obtained from it should be able to support local consumption.
- Charges and filling stations on a local basis should be extended to service electric vehicles.

4 Conclusion

In this study, communication and control applications and technologies that can provide distributed control and wide area network management of Smart Grids are examined. With the help of these applications and technologies, smart network architecture has been projected and smart network application opportunities are discussed. It has been emphasized that the intelligent network architecture, which includes communication and control systems for smarter management and efficient use of energy systems, must have a functionally intertwined three-layer structure (power–control–communication).

Development of communication between machines and applications developed in different areas and basic energy and natural resources are managed effectively and at the same time, illegal usage is tried to be prevented. Thus, it is aimed to shorten and expand the project development process by eliminating the need for additional IT infrastructure for Smart Grids. Although M2M communication has numerous application scenarios, the possible implementation in Smart Grids is discussed in the article. Research shows that the number of smart meters increased accordingly, resulting in a significant increase in delays. However, delays may be tolerated to some extent, as the traffic generated by smart meters does not have real-time restrictions in most cases.

The capacity of the master carrier network and the number of smart meters in a cluster covered by a single-base station/access point also determine the amount of delay to wait. If smart meters are required to perform real-time updates in certain scenarios or in specific clusters, scheduling algorithms may be required to prioritize and determine the high-priority data from these smart meters. In addition, an important issue that needs to be addressed is to allocate sufficient bandwidth to user traffic as well as to traffic generated from smart meters or other M2M traffic in the mainstream network.

References

1. Lo, C.H., Ansari, N.: The progressive smart grid system from both power and communications aspects. IEEE Commun. Surv. Tutor. **14**(3) (2012)
2. Friedman, T.L., Crowded, H.F.: Why we need a green revolution and how it can renew America. Farrar, Straus and Giroux, New York (2008)
3. Collier, S.: Ten steps to a smarter grid. IEEE Ind. Appl. Mag. **16**(2), 62–68 (2010)
4. Freris, L., Infield, D.: Renewable Energy in Power Systems. Wiley, UK (2008)
5. The Smart Grid: An Introduction, U.S. Department of Energy, DOE (2008)
6. Lima, C.: Smart grid communications: enabling a smarter grid. IEEE SCV Commun. Soc. Monthly Meet. (2010)
7. See, J., Carr, W., Collier, S.: Real time distribution analysis for electric utilities. In: Proceedings of IEEE Rural Electric Power Conference, pp. B5–B5–8. North Charleston, SC (2018)

8. Usta, O., Sonsuz, K., Eksi, S.: Akıllı Sayaç Okunma Sistemleri için Alternatif İletişim Ağlarının Değerlendirilmesi, 13. Ulusal Elektrik-Elektronik-Bilgisayar Kongresi, Ankara (2008)
9. Usta, O., Sonsuz, K., Ozmen, Y., Eksi, S.: Elektrik Dağıtım Şirketleri İle Tüketiciler Arasında İki Yönlü Bilgi İletişimi, 12. Ulusal Elektrik-Elektronik-Bilgisayar Kong, Eskişehir (2007)
10. Usta, O., et. al.: Data communications for power system relaying, IEEE 7803-3879/98 (2012)
11. Niyato, D., Lu, X., Ping, W.: Machine-to-machine communications for home energy management system in smart grid. Commun. Mag. IEEE 49(4), 53–59 (2011)
12. Sood, V.K., Fischer, D., Eklund, J.M., Brown, T.: Developing a Communication Infrastructure for the Smart Grid, Electrical Power & Energy Conference (EPEC), 2009, pp. 1–7. IEEE, Montreal, QC (2012)
13. Erol, K., Mouftah, H.T.: Wireless sensor networks for cost-efficient residential energy management in the smart grid. IEEE Trans. Smart Grid 2(2), 314–325 (2011)
14. Hilty, L.M., Aebischer, B., Rizzoli, A.E.: Modeling and evaluating the sustainability of smart solutions. Environ. Model Softw. 56, 1–5 (2014)
15. Final report of the CEN/CENELEC/ETSI. Joint Working Group on Standards for Smart Grids (2011)
16. Bayındır, R., Demirtaş, K.: Akıllı Şebekeler: Elektronik Sayaç Uygulamaları, Politeknik Dergisi Cilt:17 Sayı:2, pp. 75–82 (2014)
17. Deloitte: Akıllı Sayaç, Akıllı Şebeke ve İleri Ölçüm Altyapısının Kurulması, Etkileri ve Yönetilmesi, Deloitte Türkiye (2012)
18. METU MEMS: Microbolometer type uncooled infrared detectors in standard CMOS technology. METU MEMS
19. Uluslararası Yatırımcılar Derneği. 2023 Hedefleri yolunda Bilgi ve İletişim Teknolojileri. İstanbul: Uluslararası Yatırımcılar Derneği, pp. 18–24 (2012)
20. Internet: OneM2M. M2M İçin Dünyanın İlk Küresel Standartları. OneM2M (2015)
21. OneM2M: Technical Report, Document Number; OneM2M-TR-0002-Architecture_Analysis_Part_1-V-0.2.0. OneM2M (2013)
22. ETSI: Standards for smart grids. ETSI. http://www.etsi.org/technologies-clusters/technologies/575-smart-grids
23. Lu, G., Seed, D., Starsinic, M., Wang, C., Russell, P.: Enabling smart grid with ETSI M2M standards. In: Wireless Communications and Networking Conference Workshops, WCNCW, pp. 148–153. IEEE (2012)
24. ETSI Technical Report. Document Number: ETSI TR 102 691 V1.1.1, 05 (2010)
25. ETSI Technical Report. Document Number: ETSI TR 102 935 V2.1.1, 09 (2012)

On Generative Power of Rewriting Graph P System

Meena Parvathy Sankar and N. G. David

Abstract In the field of research, the system "membrane computing" was originated and motivated, by the organization and working of a living cell and particularly, by the role of membranes into compartments of living cells into "protected reactors." The computing model in this area is called P system named after the scientist Paun. In the computer science field, diagram shows a major role to represent information and they are found to be more suitable than text. In such cases, it is easy to manipulate higher dimensional structures like graphs in information processing other than linear structure such as strings. This novel idea forms a great motivation for developing graph grammars and graph languages. The proposed objective based on the notions of membrane computing and edNCE graph grammars generating directed graphs with labeled edges and nodes and encouraged by the structure and working of a living cell, and also the mechanism of graph transformations paved the way to introduce rewriting graph P system. In this research article, the set of all *L*-shaped graphs with equal arm lengths using this device was generated.

Keywords Rewriting graph P system · Graph grammars · Rewriting P system · Conditional communication · edNCE graph grammars

AMS Subject Classification 68Q42 · 68Q45

M. P. Sankar (✉)
Department of Mathematics, SRM Institute of Science and Technology,
Kattankulathur, Chennai 603203, India
e-mail: meenaparvathysankar@gmail.com

N. G. David
Department of Mathematics, Madras Christian College, Chennai 600059, India
e-mail: ngdmcc@gmail.com

© Springer Nature Singapore Pte Ltd. 2020
S. S. Dash et al. (eds.), *Artificial Intelligence and Evolutionary Computations in Engineering Systems*, Advances in Intelligent Systems and Computing 1056,
https://doi.org/10.1007/978-981-15-0199-9_36

1 Introduction

The structure and working of living cells inspired Paun [1] and he introduced a new field called membrane computing which is now a "fast emerging front" of computer science. In this field, membrane systems or P systems are progressive models of computation rooted in formal language theory. Inspired by the biological process, the objects can pass through membranes or regions. In each of the regions, the objects are subjected to the rewriting rules that have to be applied in a parallel manner which is chosen in a nondeterministic manner. The membranes are compartments which can contain other membranes as well as objects and it represents all the facts in the system. They react among themselves and they are formed by means of the rules that are considered as generalizations of biochemical reactions occurring within the cells.

On the other hand, the local conversions on graphs can be modeled in a mathematical way using the tool provided by graph grammars [2]. A graph is rewritten by using the embedding process. There are two types of embedding in graph grammar: gluing and connecting approach. In this, the production consists of a mother graph, daughter graph, and the embedding mechanism. In the gluing approach, in the daughter graph, the nodes and edges are identified with certain parts of the remainder part of the host graph, whereas in the second type of the embedding process, new edges are used to connect the daughter graph with the host graph. In the connecting approach, to apply the production rules to the graphs, the context-free rules are used. Node replacement mechanism is considered more powerful since there are no limitations on the right-hand sides of the productions to apply the rules. edNCE graph grammar is the generalization of NLC mechanism. The connection instructions are applied directly to the nodes in the daughter graph rather than through their labels.

The novelty idea that parallel processing mechanism used in P systems with atomic objects and each graph is rewritten by only one rule and it is checked with the communication conditions; a new generative device rewriting graph P system for edNCE graph grammars is constructed. Using this device, the set of all L-shaped graphs has been generated with three membranes.

2 Preliminaries

In this section, some basic definitions regarding P systems and examples of P systems [1, 3] and graph grammars [2, 4] are discussed.

2.1 P System

A P system is a construct $\Pi = (V, T, C, \mu, w_1, \ldots, w_m, (R_1, \rho_1), (R_2, \rho_2), \ldots, (R_m, \rho_m))$, where the symbols are represented as V and its components are called as objects. The output symbol is denoted as T and $T \subseteq V$. The catalysts are represented as C and $C \subseteq V\text{--}T$. The structure of the membrane is represented as μ containing m membranes; the strings wi, $1 \le i \le m$, denote the multi-sets over the symbol V connected with the regions 1, 2, ..., m of μ. The strings are rewritten by using the rule $u \rightarrow v$, where u specifies the string over the symbol V, and $v = v'$ or $v = v'\delta$, where v' is the string over $\{a_{here}, a_{out}, a_{ij}/a \in V$, here denotes the string to stay in the current membrane, out denotes the string to go out of that membrane, a_{ij} $1 \le j \le m$ denotes to which region the string has to go}, and δ is a special symbol that is not in V, and when it is used, it dissolves the region and its contents will be contents of the preceding membrane. The evolution rules over V are represented as R_i, $1 \le i \le m$ associated with region; the partial order relation is denoted as ρ_i over R_i.

Example A P system that generates $L(\Pi) = \{n^2 : n \ge 1\}$.
   $\Pi = (V, T, C, \mu, w_1, w_2, w_3, (R_1, \rho_1), (R_2, \rho_2), (R_3, \rho_3))$ where
   V represents the symbols namely $\{a, b, d, e, f\}$, T is the output alphabet namely $\{e\}$,
   c denotes the catalysts, and in this system, it is empty represented as ϕ
   μ is the membrane structure consisting of three membranes represented diagrammatically as

   In the region one, there is no string present initially $w_1 = \lambda$, and the rewriting rule is denoted by $R_1 = \{e \rightarrow e_{out}\}$, and the partial ordering is also empty $\rho = \phi$.
   In the region two, there is no initial string which is present in the membrane $w_2 = \lambda$, and the rewriting rule is $R_2 = \{b \rightarrow d, d \rightarrow de, r_1: ff \rightarrow f, r_2: f \rightarrow \delta\}$, priority relation is $\rho_2 = \{r_1 > r_2\}$, r_1 has priority over r_2.
   In the region three, the initial string is present, $w_3 = af$, and the rewriting rule is $R_3 = \{a \rightarrow ab, a \rightarrow b\delta, f \rightarrow ff\}$, and the priority relation is $\rho_3 = \phi$.
   The schematic diagram is given in Fig. 1.
   The working of the P system generating $L(\Pi) = \{n^2 : n \ge 1\}$ is given diagrammatically is as in Fig. 2.

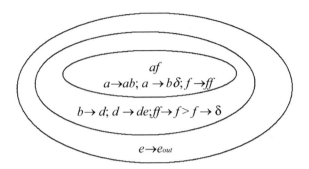

Fig. 1 Membrane structure

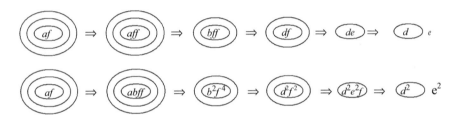

Fig. 2 Derivation

2.2 *Rewriting P System [5]*

The rules to process the string objects are context-free rule and along with target specified as the target indicator, which represents whether the string should stay in the same region denoted by here, and whether the string should go out of the region denoted by out and in that allows the string to go into the inner membrane after rewriting step.

The additional observation in rewriting P system [6, 7] is that the rules are applied to the strings in parallel and each single string is rewritten with only one rewriting rule. In this case, the strings are exposed to maximal parallelism. The organization and the functioning remain the same as in P system. The computation is considered as successful for the strings which are a finite set in each region and gathers the strings over the terminal symbols that are generated after rewriting rules are used in that region.

Fig. 3 Derivation

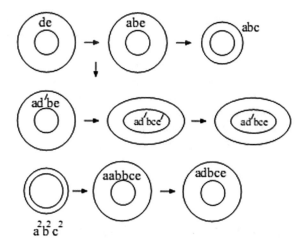

Example Consider the rewriting P system which generates $L(\Pi) = \{a^n b^n c^n : n \geq 1\}$
It consists of two membranes.

$$\Pi = (V, \bar{T}, \mu, M_1, M_2, (R_1, \rho_1), (R_2, \rho_2))$$

Where V is a finite set of symbols $\{a, b, c, d, d', e, e'\}$ and the terminal symbols
T are $\{a, b, c\}$, consists of two membranes which is depicted diagrammatically as

The two initial strings present in the two regions are $M_1 = \{de\}$ located in the
first membrane and the second membrane does not contain any initial string,
$M_2 = \phi$.

The rewriting rules in the region one are $R_1 = \{(d \rightarrow ad'b, \text{here}), (e \rightarrow ce', \text{in}),$
$(d \rightarrow ab, \text{here}), (e \rightarrow c, \text{out})\}$, and the partial ordering is empty, $\rho_1 = \phi$.

The rewriting rules in the region two are $R_2 = \{(d' \rightarrow d, \text{out}), (e' \rightarrow e, \text{here})\}$, and
the partial ordering is empty, $\rho_2 = \phi$. The evolution of first string abc is as in Fig. 3.

Thus, $L(\Pi) = \{a^n b^n c^n : n \geq 1\}$ is generated.

2.3 Node Replacement and the NLC Methodology [2]

The procedure of constructing a graph by applying node replacement tool is to
change the subgraph M (mother graph) of the host graph H by a daughter graph
D and using the inserting mechanism to joint D into the remaining of the resulting
graph H.

Node Label Controlled (NLC) mechanism is a good sample for the mechanism in
which the nodes are replaced using this methodology. In the replacement process,

the daughter graph is connected to host nodes that are close to the mother node and the mother node consists of one node only. The main idea of graph grammars based on node rewriting consists of series of iterations on this rewriting leads to a transformation considered as global into a graph that is based on local transformation. In such grammar, the productions and the connection instructions are finite. The NLC approach described can be extended and modified in various ways. The generalization is edNCE graph grammars which is more dominant than the NLC grammar.

edNCE graph grammar generates graphs where the nodes and edges are labeled and the edges are directed. edNCE grammars are graph grammars with neighborhood controlled embedding, the "d" denotes that the graphs are directed and "e" denotes that the edges of the graph are labeled.

2.4 edNCE Graph Grammars [4]

- An edNCE grammar is $G = (\Sigma, \Delta, \Gamma, \Omega, P, S)$ where the symbols of node labels are denoted as Σ, the symbols of the terminal node labels as $\Delta \subseteq \Sigma$, the symbols of the edge labels are denoted as Γ, the symbols of final edge labels are denoted as $\Omega \subseteq \Gamma$ and P is a finite set of productions, and the initial nonterminal is represented as $S \in \Sigma - \delta$.
- In the edNCE graph grammar, a construction rule is of the form $X \rightarrow (D, C)$. $X \in \Sigma - \Delta$ and $(D, C) \in \mathrm{GRE}_{\Sigma,\Gamma}$. $\Sigma - \Delta$ contains the labels of nonterminal nodes, and $\Gamma - \Omega$ contains the nonterminal edge labels. C is a set of joining commands. A graph with embedding over Σ and Γ is a pair (H, C) with $H \in \mathrm{GR}_{\Sigma,\Gamma}$, and C is a connection relation of (H, C). $\mathrm{GR}_{\Sigma,\Gamma}$ denotes the set of all graphs over node and the edge labels.

3 Rewriting Graph P System

Definition A rewriting graph P system (RGPS) is a construct $G = (\Sigma, \Delta, \Gamma, \Omega, \mu, M_i, R_i)$ where the node labels are represented a Σ, terminal node labels are represented as $\Delta \subseteq \Sigma$ edge labels are denoted as Γ, terminal edge labels are denoted as $\Omega \subseteq \Gamma$.

The structure of the membrane is represented as μ for the regions. The set of graphs available in the region is denoted as M_i, $i = 1, 2, ..., m$ and R_i, $i = 1, 2, ...,$ m indicates the set of graph rewriting rule in the ith region.

R_i contains the rule $X \rightarrow (D, C; tar)$ where X, D, and C are as explained in edNCE graph grammar, tar is a target indicator, the indicator consists of here, out, in$i$ and it specifies the region, where the resultant graph should go.

L_n (RGPS) represents the family of graph languages generated by RGPS.

Example Consider an example G_1 generating the set of all streets represented by L (G) where the unlabeled edges have label * and unlabeled nodes have label #.

$$G_1 = (\Sigma, \Delta, \Gamma, \Omega, \mu, M_1, M_2, R_1, R_2)$$

with

- $\Sigma = \{S, X, \#\}$
- $\Delta = \{x_1, x_2\}$
- $\Gamma = \{h, r, a, b, *\}$
- $\Omega = \{a, b\}$
- $\mu = $ ⬭ 2 ⟩ 1
- $M_1 = \phi$, $M_2 = s$
- $R_1 = (S \rightarrow P_2, C_2 = ((\#, h/b, x_1, \text{in}), (\#, h/b, x_2, \text{out})), \text{here})$
  $\quad\quad (S \rightarrow P_3, C_3 = ((\#, h/b, x_1, \text{in}), (\#, h/b, x_2, \text{out})), \text{out})$
- $R_2 = (S \rightarrow P_1, = C_1 = \phi, \text{out})$

where P_1, P_2, P_3 are the three productions as shown in Fig. 4.

The construction procedure with parallelism mode establishes the replacement of all the occurrences of one node by applying only one rule, which is nondeterministically selected between all that are available in the region and rules that can be applied to that node. The communication conditions are represented as P_i's and

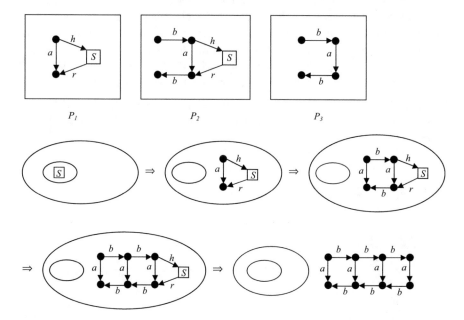

Fig. 4 Derivation

F_i's, the permitting and forbidding conditions of the region i, $i = 1, 2, ..., n$. It is of the form vacant or checking of the symbol as follows.

Rule 1: Empty: The empty permitting condition is (true, in), (true, out), and an empty forbidding condition is (false, not in) (false, not out). This rule is applied to all graphs. The resultant graph thus obtained after the application of rewriting rules can enter into any of the inner membrane or they can exit the existing membrane.

Rule 2: Symbols checking: For $A \in V$, the permitting condition P_i is denoted as a set of pairs (A, α), where α denotes $\{in, out\}$ and for $B \in V$, the forbidding condition F_i is denoted as a set of pairs $(B, not \alpha)$, α denotes $\{in, out\}$. A graph W can enter to a lower membrane if the permitting condition contains a symbol and the condition "in" (i.e.) $(A, in) \in P_i$ with $A \in$ lab (W) and the forbidding condition is $(B, not in) \in F_i$, $B \notin$ lab (W). Similarly, the graph goes out of a membrane i, if the condition contains the symbol and the condition "out" (i.e.) $(A, out) \in P_i$ and $(B, not out) \in F_i$ for all $B \notin$ lab (W).

The working of the system is defined in the following way. The rules are chosen nondeterministically in each region; the graph for which the rule has to be applied is rewritten by a rule from that region. Then the graph is checked with the communication conditions in that region. If the resultant graph satisfies the requested conditions, it either enters into the inner membrane if it exists or it leaves the current membrane. If the resultant graph is satisfied by both in and out conditions, then the graph produced is directed out of the membrane or to a membrane that is present inside which is nondeterministically selected. When a graph completes the rewriting rule and if it is sent to any other another membrane is "consumed" and no duplicate of it is available for the next step in the same membrane. When a rewriting step cannot be applied to a graph, then the graph is subjected to communication conditions, and the graph that has been produced leaves the membrane or it remains inside forever depending on the result of the checking of the communication conditions. Hence, rewriting has precedence over communication. After a series of transitions referred as computation and the computation is halted when the system generates a set of all simple graphs that are collected in the skin membrane which contributes to the language produced by the system denoted by $L(\pi)$.

Theorem 1 *The set of all L-shaped graphs with equal arm lengths can be produced using rewriting graph P system with conditional communication using three membranes.*

$\Pi = (\Sigma, \Delta, \Gamma, \Omega, \mu, M_1, M_2, M_3, (R_1, P_1, F_1), (R_2, P_2, F_2), (R_3, P_3, F_3))$ with $\Sigma = \{S, S_1, S_2, \#\}$
$\Delta = \{x_1, x_2\}$
$\Gamma = \{h, r, a, b, *\} \, \Omega = \{a, b\}$

$M_1 = S_\square, M_2 = \phi, M_3 = \phi,$

$R_1 = \{(S \rightarrow g_1, C_1 = \phi, \text{in }); (S_1 \rightarrow g_3, C_3 = ((\#, h/a, x_1, \text{in}), (\#, r/a, x_2, \text{out})), \text{in})\}$

$P_1 = (\text{true}, \text{in}), F_1 = \{(\text{false}, \text{not in}), (\text{false}, \text{not out})\}$

$R_2 = (S_1 \rightarrow g_4, C_4 = ((\#, h/a, x_1, \text{in}), (\#, r/a, x_2, \text{out})), \text{out}),$

$\quad (S_2 \rightarrow g_5, C_5 = ((\#, h/b, x_1, \text{in}), (\#, r/b, x_2, \text{out})), \text{out}),$

$P_2 = (\text{true}, \text{out}), F_2 = (S_1, S_2, \text{not out})$

$R_3 = (S_2 \rightarrow g_2, C_2 = ((\#, h/b, x_1, \text{in}), (\#, r/b, x_2, \text{out})), \text{out})$

$P_3 = (\text{true}, \text{out}), F_3 = (S_2, \text{not out})$ (Fig. 5).

The rewriting P system works in the below-mentioned way. It consists of three membranes. In the RGPS with the conditional communication after applying the rewriting rule to the nonterminal, the resultant graph will check the permitting and forbidding condition before leaving the membrane. Initially, a node labeled with S is in the skin membrane. After applying the production, it is checked with permitting and forbidding conditions. After checking with both the conditions, it enters into the second or the third membrane. If the resultant graph enters into the second membrane, the terminal rules are applied and the first member of the graph language is generated and it is collected in the skin membrane. If the resultant graph enters into the third membrane, the upper arm is increased by one unit after applying the production g_2. The resultant graph comes to the first membrane and thereafter applying the production g_3, the right side of the graph is increased by one unit and the resultant graph either enters into the second or the third membrane. If it enters into the second membrane, the second member of the graph language is generated and it is collected in the skin membrane. The process can repeat.

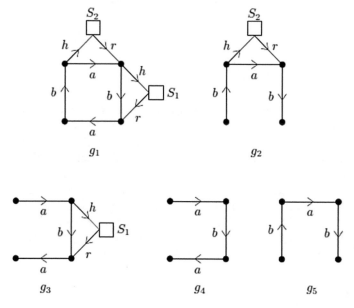

Fig. 5 Rules

Theorem 2 *The set of all L-shaped graphs with arm length equal in size can be generated using rewriting graph P system using two membranes.*

Proof

$$\Pi = (\Sigma, \Delta, \Gamma, \Omega, \mu, M_1, M_2, R_1, R_2) \text{ with}$$
$$\Sigma = \{S, S_1, S_2, \#\} \Delta = \{x_1, x_2\}$$
$$\Gamma = \{h, r, a, b, *\} \Omega = \{a, b\}$$

$$\mu = \quad \boxed{\bigcirc \quad 2} \quad 1$$

$M_1 = S_\square,$
$M_2 = \phi$
$R_1 = \{(S \rightarrow g_1, C_1 = \phi, \text{in } 2), (S_1 \rightarrow g_3, C_3 = ((\#, h/a, x_1, \text{in}), (\#, r/a, x_2, \text{out})), \text{in}_2),$
$\quad (S_2 \rightarrow g_5, C_5 = ((\#, h/b, x_1, \text{in}), (\#, r/b, x_2, \text{out})))\}$
$R_2 = \{(S_1 \rightarrow g_4, C_4 = ((\#, h/a, x_1, \text{in}), (\#, r/a, x_2, \text{out})), \text{out}),$
$\quad (S_2 \rightarrow g_2, C_2 = ((\#, h/b, x_1, \text{in}), (\#, r/b, x_2, \text{out}), \text{out})\}$

The system works in the following way. It consists of two membranes. The first membrane consists of three productions and the second membrane consists of two productions. Initially, the node labeled with S is in the first membrane. After applying the first production, the resultant graph is checked with the communication condition and so, it enters into the second membrane since the target is in$_2$. In the second membrane, the graph is rewritten by applying the production g_2 and since the target is out, the resultant graph is sent out of the membrane. The process repeats and the generated graph language is collected in the first membrane. Hence, the theorem is proved.

4   Conclusion

P systems can be effectively used as a framework to exhibit the simulation of biological systems dynamics, since it has a direct communication within cells and it is a suitable way to describe these systems. Conditional communication is an interesting variant in the study of rewriting graph P system. In this system, the graph is considered as an object. Graph grammars use the NLC methodology to generate graphs for rewriting. This novel idea paved the way to introduce the system. Here, the set of all L-shaped graphs with arms of equal length is generated. Our future work is to study their properties and to generate graphs using other variants.

References

1. Paun, G., Rozenberg, G.: A guide to membrane computing. Theoret. Comput. Sci. **287**, 73–100 (2002)
2. Rozenberg, G. (ed.): Handbook of Graph Grammars and Computing by Graph Transformation, vol. 1. World Scientific, Hong Kong (1997)
3. Paun, G.: Computing with membranes. J. Comput. Syst. Sci. **61**, 108–143 (2000)
4. Habel, A.: Hyperedge replacement grammars and languages. In: Lecture Notes in Computer Science, p. 643. Springer, New York (1992)
5. Bottoni, P., Labella, A., Martin-vide, C., Paun, Gh: Rewriting P systems with conditional communication. Lect. Notes Comput. Sci. **2300**, 325–353 (2002)
6. Subramanian, K.G., Hemalatha, S., SrihariNagore, C., Margenstern, M.: On the power of P systems with parallel rewriting and conditional communication. Rom. J. Inf. Sci. Technol. **10** (2), 137–144 (2007)
7. Paun, G., Rozenberg, G., Salomaa, A.: Handbook of Membrane Computing. Oxford University Press Inc., New York (2010)

Study on Synthesis Methods for Real-Time Control of Car-Like Mobile Robot

Igor Prokopyev and Elena Sofronova

Abstract This paper presents a comparative study of synthesis methods for real-time control of an unmanned car-like mobile robot while moving along a given trajectory. Three methods are considered: a feedback control using the PID controller, a model predictive control, and artificial neural networks. A comparison is based on computational experiments. A complex trajectory with crossing paths is considered for path tracking.

Keywords Control synthesis · Unmanned mobile robot · Path tracking

1 Introduction

Numerous studies in the field of improving the autonomous movement systems of unmanned vehicles are aimed at developing a nonlinear control law for tracking trajectories in real time.

Traditionally, real-time control of unmanned vehicles is divided into three subtasks: trajectory generation, position estimation, and path tracking.

At present, a feedback control is still widely used to track trajectories because of its simplicity and ease of implementation [1], but due to the increased power of computers and microprocessors, a model predictive control (MPC) is also used to solve it [2, 3], as well as fuzzy controllers, artificial neural networks (ANNs) [4], and adaptive methods. Along with these methods, there is a class of modern methods of symbolic regression, genetic programming, grammatical evolution, network operator method. Based on evolutionary algorithms, these methods allow

I. Prokopyev · E. Sofronova (✉)
Federal Research Center "Computer Science and Control" of Russian Academy of Sciences, Moscow, Russia
e-mail: sofronova_ea@mail.ru

I. Prokopyev
e-mail: fvi2014@list.ru

© Springer Nature Singapore Pte Ltd. 2020
S. S. Dash et al. (eds.), *Artificial Intelligence and Evolutionary Computations in Engineering Systems*, Advances in Intelligent Systems and Computing 1056,
https://doi.org/10.1007/978-981-15-0199-9_37

to create numerical methods for the control synthesis [5–7], but they are not usually used for the control synthesis in real time.

This paper covers the results of an experimental comparison of the control synthesis methods for the path tracking of an unmanned vehicle along a spatial trajectory in real time. In order to assess the methods more accurately, we included uncertainty in the model and complicated the trajectory of motion. In the work, a finite difference model of the control object constructed using Ackermann geometry is considered [8].

2   Control Synthesis Problem Statement

Consider the formal statement of the control synthesis problem for path tracking. A mathematical model of the control object is given as

$$\dot{x} = f(x, u), \tag{1}$$

where x is a state vector, u is a control vector, $x \in R^n$, $u \in U \subseteq R^m$, U is a closed limited set, $m \leq n$.

Initial conditions are

$$x(0) \in R^n. \tag{2}$$

A spatial trajectory of motion is given in the form of $n-r$ dimensional manifolds

$$\varphi_i(x) = 0, \; i = \overline{1, r}, \; r < n. \tag{3}$$

The trajectory (3) is called spatial, since it does not depend on time.

A time of control process termination is given t_f.

It is necessary to find control in the form of a function of state-space coordinates

$$u = h(x), \tag{4}$$

where $h(x)$ is a synthesizing function, $h(x) : R^n \to R^m$, that has the following property $\forall(x) \in R^n$, $h(x) \in U$.

The function should provide a minimum of the function describing the accuracy of tracking

$$J = \frac{1}{t_f r} \int_0^{t_f} \sqrt{\sum_{i=1}^{r} \varphi_i((x(t))} dt \to \min. \tag{5}$$

To solve the control synthesis problem (1)–(5) for the motion of an object along a spatial trajectory (3), we use a feedback controller, MPC controller, and a controller with ANN. To study the proposed approaches, an unmanned car-like mobile robot has been created.

3 A Model of Car-Like Mobile Robot

A control object is an unmanned car-like mobile robot, see Fig. 1a, built using a 1/10 size automobile chassis. The robot weighs 1.5 kg, the distance between the axes is $H = 0.4$ m, and the maximum speed is up to 10 m/s.

Calculations are performed using the NVIDIA Jetson TX2 Kit. Sensors, video camera, and the controller are attached to Jetson. Jetson regulates the speed of the engine of the chassis as well as the rotation angle of wheels. Raspberry Pi 3 with a camera module is used as a system for orientation and navigation of the robot by ArUco markers.

Two front wheels are used for steering and two rear wheels are used for tracking. The kinematic scheme of robot motion control is shown schematically in Fig. 1b.

Figure 1b shows the steering angle α, the orientation angle of the robot θ relative to the axis x, the distance between the front and rear axes of the robot H, the reference point located in the middle of the rear axis of the robot R, and its coordinates x, y.

The finite difference equations of a mobile robot model are as follows

$$\mathbf{x}(k+1) = \begin{bmatrix} x(k+1) \\ y(k+1) \\ \theta(k+1) \end{bmatrix} = \begin{bmatrix} x(k) + v(k)\Delta t \cos(\theta(k))\cos(\alpha(k)) \\ y(k) + v(k)\Delta t \sin(\theta(k))\cos(\alpha(k)) \\ \theta(k) + v(k)\Delta t \sin(\alpha(k))/H \end{bmatrix}, \qquad (6)$$

(a) **(b)**

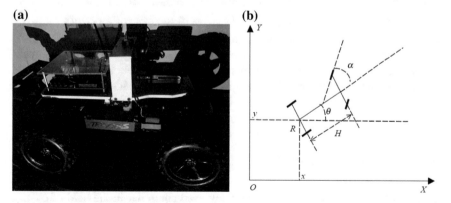

Fig. 1 A car-like mobile robot [8]

where $\mathbf{x}(k)$ is a state vector,

$$\mathbf{x}(k) = [x(k),\, y(k),\, \theta(k)]^T, \tag{7}$$

$v(k)$ is a linear velocity, Δt is a time interval, an integration step k is an index of time interval

$$\mathbf{x}(k) = \mathbf{x}(t_k) = \mathbf{x}(k\Delta t),\ t_k \in \left[0; t_f\right],\ k = \overline{0, L},\ L = \left\lfloor \frac{t_f}{\Delta t} \right\rfloor.$$

A control vector is

$$\mathbf{u}(k) = [v(k),\, \alpha(k)]^T. \tag{8}$$

Such a nonlinear system is open-loop controllable, which can be linearized in order to use traditional linear feedback control to control the robot. But if the robot moves along complex trajectories, turns around corners, etc., the linearization may lead to the loss of controllability. The movement of such models is difficult to simulate since there are many physical sources of uncertainty, e.g., friction, load, etc. Adaptive control should be used in this case.

4 Path Tracking Scheme

To track the trajectory, information on the desired, generated trajectory and the measured results of movement of a real control object are needed. Figure 2 shows a generalized scheme of a control system using feedback control, MPC and ANN.

In Fig. 2, $\mathbf{x}_r(k) = [x_r(k),\, y_r(k),\, \theta_r(k)]^T$ is a generated trajectory, $\mathbf{x}_m(k) = [x_m(k),\, y_m(k),\, \theta_m(k)]^T$ is a measured trajectory of real mobile robot movement.

As a trajectory of movement, we use a complex trajectory with crossing, one of the figures of Lissajous. Position estimation plays a key role in path tracking. To assess the position of the robot, ArUco markers located on the ceiling and a vision system connected to the robot through the Raspberry Pi 3 board are used.

4.1 Feedback control

Consider a feedback as the most commonly used controller in the field of control. The feedback controller receives an error on the deviation of the real movement from the desired trajectory and passes it through three links. In general, the PID controller has the form

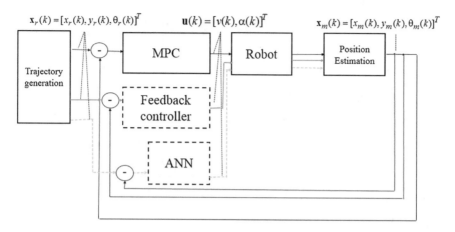

Fig. 2 A generalized scheme of control system

$$y = k_1 e(t) + k_2 \int\limits_0^t e(t)\mathrm{dt} + k_3 \dot{e}(t), \tag{12}$$

where $e(t)$ is the error of deviation of the real movement from the desired trajectory, k_1, k_2, k_3 are scalar coefficients of links that are searched in order to improve the quality of control, y is a control signal. In most cases, the quality of control is determined by the transition process time, overshooting, and static error.

When the feedback controller is used in the control loop of car-like mobile robot [9], the following control is transmitted to the object

$$\alpha = \arctan \frac{Hw}{v}, \tag{13}$$

where H is a distance between the front and rear axes of the robot, v is a linear velocity, w is an angular velocity,

$$w = \frac{vk(s)\cos(\theta_e)}{1 - k(s)e} - k_\theta |v|\theta_e - (k_e v \frac{\sin(\theta_e)}{\theta_e})e, \tag{14}$$

$k(s)$ is a curvature of the path at s, θ_e is an error of heading movement, e is the error of deviation of the real movement from the desired trajectory, k_θ, k_e are coefficients of controller. Linear velocity of object v is considered to be constant.

Figure 3 shows how a feedback controller automatically applies an accurate and responsive correction to the control function, which allows tracking the desired path with the minimum mean square error (MSE) (see Table 1). The desired, generated, trajectory is red marked, the traced trajectory is green.

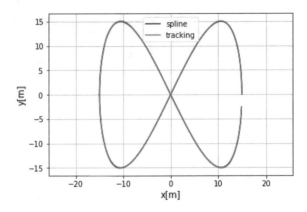

Fig. 3 Path tracking using feedback controller

Table 1 Comparison of MSE for feedback control, MPC and ANN for three experiments

Control	Feedback			MPC			ANN		
Exp. No	1	2	3	1	2	3	1	2	3
Position x	0.012	1.146	0.105	0.002	0.76	0.039	0.007	1.201	0.107
Position y	0.033	0.825	0.133	0.006	0.419	0.08	0.009	0.794	0.263
Orientation θ	0.002	0.236	0.036	0.037	0.157	0.141	0.053	0.226	0.25

4.2 Model Predictive Control

Model predictive control defines the task of path tracking as an optimization problem. The solution to the optimization problem is the optimal trajectory. MPC includes simulation of various actuator inputs, predicting the final trajectory and choosing this trajectory with minimal costs. The current state and the reference path are shown in Fig. 4. At each time step, the control vector is optimized to minimize the cost of the predicted trajectory.

Fig. 4 Path tracking using MPC

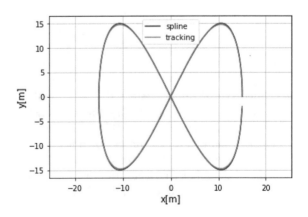

Fig. 5 Path tracking using ANN controller

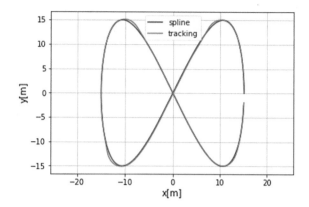

4.3 Artificial Neural Networks

To implement the controller based on the ANNs, at the first stage, we collected data for training using MPC for path tracking (see Fig. 2). For training, Gaussian noise with a zero average value was used, which was added to the state of the robot to increase the representative capacity of the model.

The ANN controller model was implemented in Keras; the hyperparameters were obtained using a genetic algorithm. As a result, a model with two hidden layers and 768 neurons in a layer was obtained, with an optimizer—adagrad with an activation function—tan$h$.

Figure 5 shows that the ANN controller tracks the desired trajectory no worse than the controller based on the feedback control.

5 Computational Experiments

To make a comparative analysis of the methods, three types of experiments were carried out. In the first experiment, the desired trajectory was used, along which the object could move exactly. The resulting modeling plots for the three methods considered are shown in Figs. 3, 4 and 5. Mean time of control calculation at each step was $T_{\text{feedback}} = 0.0022$ s, $T_{\text{MPC}} = 1.12$ s, $T_{\text{ANN}} = 0.0022$ s. The ratio of computation time between the methods for all subsequent experiments remained unchanged.

In the second experiment, the model of the control object was subjected to random effects

$$
\mathbf{x}(k+1) = \begin{bmatrix} x(k+1) \\ y(k+1) \\ \theta(k+1) \end{bmatrix} = \begin{bmatrix} x(k) + v(k)\Delta t \cos(\theta(k)) \cos(\alpha(k)) + \xi(k) \\ y(k) + v(k)\Delta t \sin(\theta(k)) \cos(\alpha(k)) + \xi(k) \\ \theta(k) + v(k)\Delta t \sin(\alpha(k))/H \end{bmatrix},
$$

Fig. 6 Path tracking with
uncertainty using feedback
controller

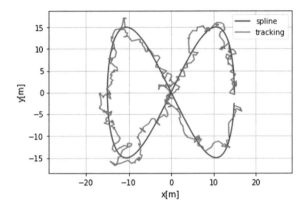

Fig. 7 Path tracking with
uncertainty using MPC

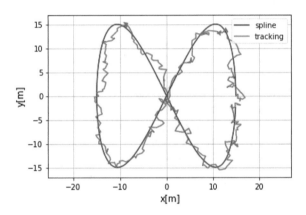

where $\xi(k)$ is a random variable with a normal distribution and a standard deviation of 0.7. The results of the experiment with random effects are shown in Figs. 6, 7 and 8. In the third experiment, the geometry of the trajectory was changed so that

Fig. 8 Path tracking with
uncertainty using ANN
controller

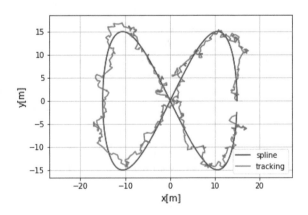

the object obviously could not move along the trajectory precisely because of the existing restrictions on control. The results of experiments with movement along a complex trajectory are shown in Figs. 9, 10 and 11.

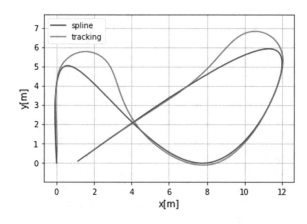

Fig. 9 Complex path tracking using feedback controller

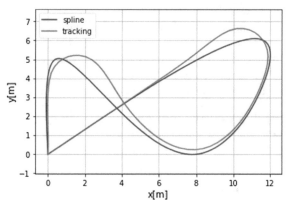

Fig. 10 Complex path tracking using MPC

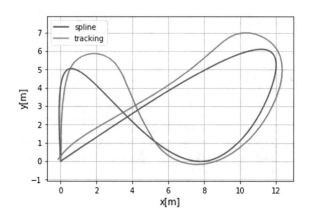

Fig. 11 Complex path tracking using ANN controller

6 Conclusion

The results of an experimental comparison of the three methods of real-time control synthesis for path tracking are presented. The most popular methods were considered: feedback control, MPC, and ANNs. To obtain more reliable results, three types of experiments were carried out. In the first experiment, the reference trajectory was used, along which the object could move exactly. In the second experiment, the model of the control object was subjected to random effects. In the third experiment, the geometry of the trajectory was changed so that the object obviously could not move along the trajectory exactly.

It is shown that all three methods provide accurate path tracking for complex trajectory. At the same time, the duration of the calculation of the control for the MPC was 500 times longer than for other methods, which imposes certain requirements on the onboard computer. In addition, in a real robot, the control command is not executed instantly—there is a delay, since the command is distributed throughout the system. The actual delay may be about 100 ms. The contributing factor to the delay is the dynamics of the drive; for example, the time elapsed from the moment when the steering angle was set to the moment when this angle was actually reached.

In all three experiments, model predictive control has shown the best results in terms of MSE. Controllers based on ANNs and feedback showed comparable results, but about two times worse than the controller based on MPC. An ANN-based controller trained on MPC using a vehicle model that takes into account the dynamics of the drive can be more accurate.

Acknowledgements Research was partially supported by the Russian Foundation for Basic Research, project No. 18-29-03061-mk.

References

1. Gu, D., Hu, H., Brady, M., Li, F.: Navigation system for autonomous mobile robots. In: Proceedings of International Workshop on Recent Advances in Mobile Robots, pp. 24–33. Leicester, UK (1997)
2. Clarke, D., Mohtadi, C., Tuffs, P.: Generalised predictive control-part I. Basic Alg. Autom. **23** (2), 137–148 (1987)
3. Rawlings, J.B.: Tutorial overview of model predictive control. IEEE Control Syst. Mag. **20**(3), 38–52 (2000)
4. Haykin, S.: Neural Networks. A Comprehensive Foundation, Prentice Hall, 842 pp. (1999)
5. Koza, J.R.: Genetic Programming: on the Programming of Computers by Means of Natural Selection, 819 pp. MIT Press (1992)
6. Diveev, A.I., Kazaryan, D.E., Sofronova, E.A.: Symbolic regression methods for control system synthesis. In: 22nd Mediterranean Conference on Control and Automation (MED '14), pp. 587–592 (2014)

7. Diveev, A.I.: A numerical method for network operator for synthesis of a control system with uncertain initial values. J. Comput. Syst. Sci. Int. **51**(2), 228–243 (2012)
8. Gu, D., Hu, H.: Neural predictive control for a car-like mobile robot. Int. J. Robot. Auton. Syst. **39**(2–3) (2002)
9. Paden, B., Cáp, M., Yong, S.Z., Yershov, D., Frazzoli, E.: A survey of motion planning and control techniques for self-driving urban vehicles. IEEE Trans. Intell. Veh. **1**(1) (2016)

A Brief Review of the IoT-Based Energy Management System in the Smart Industry

Amam Hossain Bagdadee, Li Zhang and Md. Saddam Hossain Remus

Abstract Nowadays, manufacturing circumstances, rising energy costs, increased concern for the environment, and changes in the practice of consumers promoting green manufacturing organizations. The remarkable increase in modernization in recent years requires an economical, productive, and smart solution such as transportation, management, the state, personal satisfaction. Energy demand for IoT applications is growing. In this particular situation, energy management is an important issue because of high energy savings and efficient energy crisis reasons. In this paper, IoT energy management tends to achieve green energy response and communication from supply and demand. As a result, smart industrial planning must be able to use energy productively to overcome related difficulties. At that time, we will provide a structure for combining reservations and increasing energy efficiency in smart industries based on IoT. Also, I will explain the energy generation in smart industries. This is a solution to extend the low power of equipment with control-related problems too. In this paper has discussed case studies. The smart factory energy planning and the second include wireless power transfer from IoT implements in the smart industry. The situation of smart industrial facilities has shown as contextual analysis, including progress considering the balance of energy supply and demand, unsolved issues also discussed in the field of IoT-based energy management system.

Keywords IoT · Smart industry · Energy management · Power system · The industrial sector

A. H. Bagdadee (✉) · L. Zhang
College of Energy and Electrical Engineering, Hohai University, Nanjing, China
e-mail: a.bagdadee@hhu.edu.cn

Md. Saddam Hossain Remus
Jahangirnagar University, Dhaka, Bangladesh

© Springer Nature Singapore Pte Ltd. 2020
S. S. Dash et al. (eds.), *Artificial Intelligence and Evolutionary Computations in Engineering Systems*, Advances in Intelligent Systems and Computing 1056,
https://doi.org/10.1007/978-981-15-0199-9_38

1 Introduction

Smart industry settings use communication and network technology to manage problems accelerated by the automation system. Nowadays, energy efficiency of manufacturing method provides some advantages to manufacturing affiliates, such as cost savings despite unexpected increases in energy costs, industrial building a good reputation with the government, and satisfying the world's natural regulation with the green products. Internet of Things (IoT) is a manager who offers smart industries, sensing devices and actuators are a real segment in addition to communication and system devices as well [1, 2]. Newly developed self-sufficiency technologies such as the Internet of Things (IoT) gradually increase the monitoring of industrial production in real time. The zone where IoT plays extraordinary works is an observation of energy consumption. This industry is an extraordinary electric consumer. As shown in the approaching of the universal energy organization, the industrial sector reported 41.96% of world electricity consumption in 2018 [3]. The range is quite significant when developing modern countries that usually represent a more significant economic level. In line with this policy, realizing the demand response (DR) in the field of the industrial sector is very important and fundamental. These modern sensing devices do not only increase the reliability of the power framework but also reducing the energy costs of industrial facilities. Sensing devices are used to identify and observe industrial operations in different situations continuously [4, 5]. As quickly as possible, it is expected that industrial, personal, office, manufacturing devices, machines will have the ability to detect, communicate, and process data widely [6, 7]. Perhaps, it is trying to build a fully enhanced system for interrelated ideas between smart industries and different modern technologies. Also, smart industrial planning must produce energy from a consumer and environmental perspective.

This article will discuss IoT energy management in smart industries. This contribution can explain the improved structure for considering smart industry by using IoT. This study shows energy management goals, types of operations, and methods of regulation. This paper also explains energy-productive solutions for smart industries using IoT. A case study analysis was introduced to show the benefits obtained by planning the progress of smart factory networks. Furthermore, context analysis is carried out to investigate the performance benefits achieved by planning unique energy sources.

2 The Structure of Industrial Energy Management by IoT

The energy management in smart industries is divided into two main types, such as approved energy arrangements and energy storage activities. In addition to several research points, this arrangement is shown in Fig. 1. Energy-saving responses for smart industries utilize IoT, optimization of reservations, pre-model of energy use,

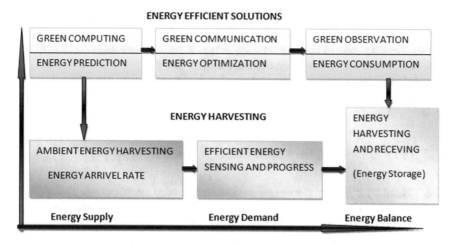

Fig. 1 Energy-saving and harvesting model

and so on [8]. A methodology is based on low-power energy management and cognitive energy management structures. With energy harvesting, IoT devices can harvest energy from a comprehensive source as well as unique renewable energy sources. The essence of an environmental power plant is to extend of IoT equipment. Exploration points included in two types of energy storage are the plan for receiving energy storage, the level of energy input, the status of the necessary number of committed energy sources, the condition of particular energy sources, and the multidirectional energy direction [9, 10]. In the IoT configuration, the concept of smart grids is to transform conventional electricity networks into different energy worldviews and to build energy availability, automation, and coordination with all providers that are more ecologically consistent. Consumers in one measurement, the energy management problems identified in the sensor-actuator layer, the organization layer, and the application layer are introduced. In other measurements, energy management problems are classified according to considerations identified on the supply side, demand side, and supply–demand balance [11]. The primary purpose of green identification and green correspondence is the focus of each layer in a row from the top base. These two measurements reproduce the problems identified in IoT energy management in interrelated mapping systems from the perspective of understanding energy coordination and data communication related to IoT energy management.

2.1 The Internet of Things (IoT)

Internet of Things (IoT) views about the world is characterized as an intelligent thing and transactions from intelligent communication systems. From a certain point of view, smart objects are items that have several essential functions besides some discretionary items. By other smart objects focused security framework (mandatory) that collaboration with inclusion conditions such as metering, monitoring, sensing, and in making data collection to extort data for increasingly rich information and communication management. In this way, the smart of objects can range from RF (Radio-frequency) tags that are involved in the wireless sensor systems.

2.2 IoT-Based Energy Management Technology

The demand for energy consumption concerns several organizations that provide creative monitoring responses in the industrial sector, such as Epi-Sensor, Wi-Lem, Wattsup, SATEC, Change Electric, Energy Metering Innovation LTD, General Electricity, Mitsubishi, Siemens, and Schneider. Therefore, several organizations provide Emergency Energy Management (EEM) software applications to break down collected information, such as Resource Kraft, Google, e-sight energy, EFT-energy, and so on [12, 13]. By summarizing supplier-approved procedures that can decide on the design of a general energy monitoring framework by utilizing Internet technology innovations, as outlined in Fig. 2. This can be attributed via a wired or wireless system. Available energy meters can obtain several parameters (for example, power consumption, power factor, and maximum/minimum voltage) so that they can adapt in monitoring and analyzing energy consumption indicating abnormal energy crisis conditions. Monitoring of the new intelligent meter energy

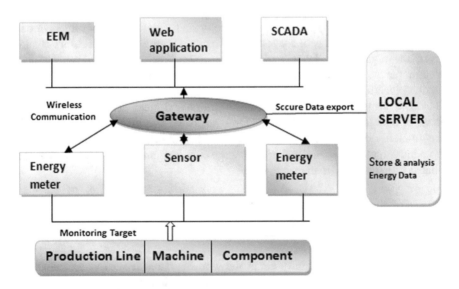

Fig. 2 IoT-based wireless network

assembly carried out that meter can be used for various monitoring targets, such as all power plant lines, one phase machine.

At the middle level, information collected is sent to the gateway and then communicated with local PC or the Internet with standard exchange rules such as ZigBee Wireless Innovation [14, 15]. When a wireless system is used, sensors can be set up to be significantly more adaptive throughout the manufacturing floor. In this aspect, this principle contrasts with the use of smart systems characterized by anterior segments (standard, open, multifunctional, direct-addressed articles) and the use of established web authorizations. Finally, as in Fig. 2, information can be transferred to EEM software to the analysis of energy frameworks, such as industrial building management frameworks (IMS), recommended assembly. The Manufacturing Execution Framework (MEF) that primarily for Industrial Resource Arrangement (IRA). The data from the smart measurement framework can also be incorporated into the supervisory control and data acquisition system (SCADA).

2.3 Energy Harvesting

In the sensor-actuator layer, the supplier management system must overcome the heterogeneity that results from assembling assets. Control the distribution of hybrid energy and energy among different species, which changes into energy supply patterns [16]. Harvesting energy means capturing temporary energy lost in several forms, based on the type of heat, light, solids, vibrations, electromagnetic waves, and so on. This energy is obtained from one or more sources that often occur, and the energy collected is stored and monitored for the future. The energy interactions identified as environmental power plants mainly include various resources that collect energy.

1. Renewable energy: The renewable energy is a vital power supply for sensors and actuators and is often monitored by control circuits and very low power management to support detection and stimulation operations. Wind and solar power plants are unique renewable energy sources, contributing 62% and 13% of the total capacity to collect renewable energy produced without hydro.

    (a) Solar energy: The solar energy is an unlimited resource that can be converted into heat intensity under the influence of solar power plants. Solar energy abuse is perfect for versatile independent wireless electronics. Given the high power density of outdoor applications, this is the most reliable renewable energy source that can be accessed for the required asset framework. Semiconductor polymer solar cells are attractive devices with minimal effort to obtain solar energy, depending on composite materials with hetero-mass structures [17]. Regardless, less productive polymer solar cells, claiming that power changes can swing when exposed to light, nanotechnology serving photovoltaic and photoelectrochemical applications in harvesting non-material solar energy can be a better way to overcome the increasing generation of positive energy solar generation identified by solar cell parameters and current green energy conditions. This can be formalized in terms of trademark voltages.

(b) Wind energy: Devices that change wind energy can be classified into devices with variable speed and devices at a fixed speed. The first uses an electronic interface for associations with the grid, and the latter depends on a list generator of limited participation for the direct association framework. Wireless sensor networks (WSN) in wind energy can be provided for continuous activities through very low power implementation circuits, where wind-driven sensor nodes estimate equivalent electrical parameters detect weather conditions [18, 19]. Power plants from wind fields can be displayed as wind speed elements, and time checks for information about power and wind speed are very valuable for predicting wind energy.

2. Thermal Energy: The thermal energy is achieved during thermoelectric and piezoelectric property where thermal energy is converted to electric power. Thermoelectric generator (TEG) converts to heat that is not used but can be accessed from the surrounding heat source to electricity. The heat source is present in various structures (for example, hot surfaces, the framework of heating, ventilation, and cooling (VC), and the human body). In particular, the pyroelectric property provides a way to handle distributed sensors and actuator systems that are encouraged for wireless energy sustain and continuation.

3. Energy storage: This is cultivated by using physical media equipment (e.g., collection equipment) to store energy obtained for direct use or indirect use. Also, a power storage framework is being developed for shaving up the power and integrating sustainable resources in smart electricity network applications. Energy communication is determined explicitly in energy storage, including electricity and thermal energy storage technology.

(a) Electrical energy storage: Electrical energy storage provides electrical energy through a specific external electrical interface. In IoT (biofuels, hydrogen, liquid nitrogen, photocatalyst storage), electrochemical energy storage (batteries, flow cells), mechanical energy storage (e.g., flywheel, compressed air, wind power plants, hydropower collectors, micromechanical devices), and electromagnetic energy storage.

(b) Thermal energy storage: The thermal energy storage technology can utilize moderate latent heating by utilizing wasted heat and cooling air atmosphere. This technology is a framework that relies on nitrogen or cryogenic air fluids, ice storage, eutectic frameworks, non-flammable trains, liquid salts, occasional storage, steam collectors, and various conventional storage media. Multilevel material is an attractive source for developing thermal energy storage technologies of energy storage density and isothermal conduction in the form of phase changes.

3   Challenges for IoT-Based Smart Industry

Demand for customers continues to increase, so demands for energy management requirements occur throughout the world. Dangerous atmospheric aberrations and air pollution do not harm dangerous people, and what will happen from now on. This is due to increased volume emissions with the expansion of energy demand. Then in 2020, there will be more than 50 billion IoT devices related to the internet, as indicated by measurements made by Cisco [20]. In line with this policy, it is vital to oversee the energy of the IoT. This will make it easier to recognize great industrial ideas. Following are some models that can reduce energy use with proper energy management. Factory equipment is a vital source of energy use. Demand management is the key to changing energy use by overcoming the manufacturing framework in machinery. Also, enthusiastic, smart operations can also promote optimal energy management and operations.

3.1   Information Altering

One attacker can change traded information such as effective costs sent preceding pinnacle periods, making them the most minimal costs. As an outcome, this could make the factory expanding their energy consumption rather than dropping them, accordingly about an overloaded power system.

3.2   Approve and Control Access Problems

Science a few devices monitor and control remotely, such as a smart meter installed in a power distribution substation or sensor and actuator in place, even the attacker and representatives could attempt to pick up an unapproved get to appropriate, to control them, in this way harming physical resources (ex, transformers) or prompting power outages.

3.3   Cyberattack

Smart networks can be considered as the largest Physical Cyber Network (PCN), such as physical structure and ICT frameworks that communicate with smart network physical resources (transformers, circuit breakers, smart meters, links, etc.). Where the ICT component manages or oversees physical substances. Nowadays, cyberattacks, such as the Stuxnet attack, can damage physical resources, which make it difficult for conventional electricity networks.

4 Energy Management Solutions for Smart Industries

Along with the expansion of IoT applications for smart industries, regulation of energy production is also ongoing for low-power equipment. There is different planning that benefits energy that can reduce energy consumption or accelerate asset use [21, 22]. Following are some basic research patterns regarding the regulation of productivity of the energy industry in smart industries that utilize IoT.

4.1 Light Rules/Convention

Light means overheads are reduced by rules. Smart industries, which have been allowed for IoT, require using different conventions for communication. There are several conventions in this document such as Message Line Telemetry Transfers (MLTT), Mandated Application Contracts (MAC) extendable notifications and proximity protocols, forwarding message line conventions, 6lowPAN, and various IoT attachments and I will play. MLTT and MAP are the most common rules. MLTT is a light contract that collects data from IoT equipment and sends it to the server. MAP targets mandatory devices and systems for internet exchanges. All of these conventions cover situations and specific applications that function properly. However, changing provisions will hamper industrial buildings that are important for IoT. Therefore, IoT equipment comes from various manufacturers or need to use a different convention.

4.2 Scheduling Optimization

Optimizing scheduling for smart industries using IoT implies asset optimization because of its capacity for energy consumption and reduces power usage. In this way, the applicant's demand-side management (DSM) is the most important. This implies the management of industrial power use by adjusting the form of the load of the framework, thereby reducing costs. The DSM includes two main initiatives: load transfer and energy protection. Loads mobilization encourages the exchange of client deposits from high to low. Receiving this allows to monitor power and provide liberty for various consumers.

4.3 An Ancient Model for Energy Consumption

There is no distrust that there is a fundamental significance in the old model for energy use in IoT-centered smart industries. They imply a variety of uses in the smart industry, including an ancient model for traffic and travel, a prescience model to control temperature and power, and so on. For example, different expectation models such as the neural system and Markov selection procedures can be integrated here. The exploitation of the perceptive model reduces not only large amounts of energy use but also has many social benefits.

4.4 Low-Power Equipment

IoT devices from smart industrial applications operate with limited battery life; low energy management plan architecture or operational structure is essential to orient, smart industry executives who use IoT for energy management. In most cases, IoT application contracts are currently not seen from the standpoint of energy productivity. More specifically, the radio duty cycle of IoT devices is an essential element of energy efficiency, and specialists have reduced the radio duty cycle of IoT devices and thus investigated the approach to achieving that energy production architecture.

4.5 Intelligent Management System

IoT equipment is inherently different, and related management is not reliable. After that, it is essential to explore the intelligent management structure that accepts subjective knowledge and methodologies through smart industries that utilize all IoT. The system must incorporate thinking and learning to improve the choice of the IoT system. According to related infrastructure, an intelligent management system that consciously decided to choose about IoT devices was demonstrated.

5 Case Study: Smart Factory

In this segment, as shown in Fig. 3, the executive introduces the smart factory situation as an IoT energy context analysis from the perspective of the supply side, demand side, and demand–supply balance. Management systems that feature power plants, smart network ideas, virtualization, programming, advanced metrology infrastructure, demand reaction, and innovation from vehicles to networks are not dependent on efforts to achieve high ability and energy protection.

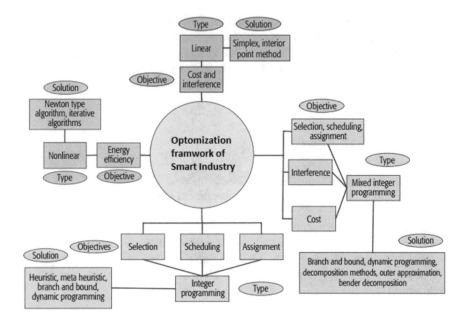

Fig. 3 IoT-based smart industry construction

5.1 Supply Side

The problem identified by smart factory executives on supply-side energy tends to be through solar power and smart grid-based microgrid concepts.

5.2 Photovoltaic Power Generation

Solar energy exploitation specifically applies to independent wireless electronic equipment. Solar energy is the highest source that can be accessed for a framework that requires assets, given the high power density of external applications. Semiconductor polymer solar cells are an attractive tool for minimizing the acquisition of solar energy depending on combined with hetero-mass structures. As is possible, polymer solar cells have a low power change effect and may be unstable under the coverage of light that cannot be connected in cruel situations. Again, nanotechnology can be advantageous for photovoltaic and photoelectrochemical applications in solar energy harvesting, and non-material ones overcome the possibility of third-generation solar advances.

Photovoltaic technology is widely used to power generation plant facilities and promote reasonable progress in green IoT. The quasi-Z Staggered Inverter Source Course (qZS-CMI) provides clear points of interest for the photovoltaic framework, including appropriate DC connection voltage and voltage assistance functions. However, qZS-CMI cannot overcome discontinuous and probabilistic fluctuations of solar power injected into the grid. For this reason, batteries for storing energy can compare stochastic fluctuations in photovoltaic power plants.

MPPT algorithm calculations can target solar power plants to make full use of solar-oriented energy, execution, and swift and unique reactions. However, if partial shading occurs in a solar photovoltaic frame, the supply voltage change indicates many vertices. The stable voltage and consistent current attributes of a photovoltaic power plant are possible to evaluate maximum power points with high accuracy using linear cycle strategies. Then artificial brain strategies provide a productive solution to communicate the maximum strength of the photovoltaic framework with the independent operation.

5.3 Smart Microgrid

Microgrids provide a very valuable position that can adapt to operate regardless of whether it is connected to the electricity network or separated in autonomous mode. The incorporation of distributed energy resources (DER) into microgrid network design strengthens the foundation of virtual power plants and motivates self-utilization of personally created electricity [23, 24]. Such a DER is connected to a grid through an electronic power processor that facilitates the reduction of misery and decentralized voltage regulation. In addition to the continuous power supply entrance, regression-based power transmission in a distributed generator is needed so that this generator functions as a conventional synchronous generator.

As a possibility, frequency control movements on electricity networks using frequency-based power-sharing because the frequency is controlled like a typical variable across the network. Modern renewable power plants in microgrid depend on allowing these power levels to change according to the type of local frequencies along this path, which reduces the type of power available from the network. With smart factory, the operation of a smart microgrid can handle difficult problems faced with standard energy [25]. On the other hand, the only operation triggered by an error can cause voltage and frequency to drop, causing the release of distributed resources, thus affecting the reliability and accessibility of the electricity network. The standard approach to the introduction of a single microgrid combines a strategy of correspondence-based, dynamic, and autonomous and plots half a variety of single placements that voltage deviations and high-frequency impedances that are regulated in a photovoltaic-based microgrid, dynamic power generation can reduce the power factor for regular coupling with the electricity network. The photovoltaic voltage source inverters are connected for power control identified in dynamic and reactive power operation.

5.4 Demand Side

Problems identified in the energy demand management side at smart factory are often obtained through virtualization, network software characteristics, and advanced measurement infrastructure ideas.

5.5 Virtualization

Virtualization means to form virtual substances that converse physical substance to redesign and optimize assets and can be categorized into three main types, such as full virtualization, standard virtualization, and partial virtualization. The movement of virtual machines is developed as a primary means of server farms, and as a capacity framework for solidification considers the provision of energy management. System virtualization is an interesting way to deal with combining available assets into the system by dividing the available transmission capacity into free channels. This will increase energy productivity. In this particular situation, coherent units (such as legitimate switches, sensible interfaces, reliable switches, intelligent firewalls, and coherent load balancers) can be replicated in virtual system topologies to combine energy controls.

5.6 Network Demand Software (NDS)

Property of Network Demand Software (NDS) is a world perspective of system infrastructure that allows generating system excellence by reflecting low-level functions. This is prepared by separating the control and information fields. Approaching and state of the system are smartly focused, and the application occupies the necessary infrastructure of the system. The information field encapsulates physical system segments, including switches, virtual switches, and various wireless points. Control panels are released from the system nodes. The controller promotes the expansion, updating, and cancelation of flows and connected ways, and utilizes restrictions and data related to requests obtained from the system through traffic. Data transmission at the equipment level depends on the flow and controls the information. Energy-conscious systems are required to realize high-energy, energy-efficient networks and NDS support different system capacity and smart applications with low operating costs through relocation of equipment, software, and management having certain constructive situations of the rapid expansion of agricultural servers frustrated the energy use involved, which affected the sustainable development of cloud benefits at smart factory. The NDS procedure can be connected to the flow reservation that considers energy. This means planning flow with time measurements and characterizing the particular guidance method of each stream. Restrictive control of connection assets often leads to increased connection utilization of each stream and as a result, resolves connection transfer rate problems approach achieves higher energy efficiency contrast and traditional direction rules. I suggest that SDN is different from system virtualization (SV) and system capacity virtualization (SCV). System virtualization creates virtual transits and capacities identified in the physical system. Paradoxically, in NDS, the physical system at a fundamental level is changed to achieve the homogeneity of equipment parts using the separate transmission control panel in addition to transmission of traffic or exchange panel.

6 Point of View on IoT-Based Energy Management

IoT-based energy management has several attractive qualities and perspectives, including green identification goals, green communication with green registration.

6.1 Supply Side

IoT in the prospect understands interconnection everywhere using physical, digital social space and has characteristics related to energy-related problems.

6.2 Spatiotemporal Consistency

The IoT-based energy management considers spatial measurement (e.g., alignment between energy supply and demand in various fields) and direct measurement (e.g., energy protection that is suitable for future use). If possible, IoT energy management eliminates some actual physical barriers, because there are limits to their existence to survive.

6.3 Heterogeneity

Heterogeneity refers to various devices, systems, and management that are firmly related to energy management. The heterogeneity enhances the ability to achieve a balance between energy use and energy productivity and promotes the progress of various hybrid sources that strengthen agreements.

6.4 Dynamic

Dynamic topology, management, and power planning are experienced in the context of discovery and remote networks. The energy management, together with the use of sustainable electricity supplies, is needed to overcome the dramatic nature of power quality instability in this situation. Nowadays, energy prediction is a well-known way to achieve sufficient energy management and trends in potential situations in the future. The energy use and accessibility methodologies must be motivated by legitimate human thinking and must create a multipurpose expectation model that can adapt to modify in energy use design.

6.5 Social Attributes

Social characteristics are based on work that is indispensable in IoT, and it is necessary to build a multiasset energy management model that considers the right ownership and relevance of substances. The social obligation of the electricity business is primarily to understand the electricity sustainability framework.

6.6 Self-operation Activity

Self-management complimentary individuals from their energy management obligations and regulates intelligent energy productivity through the use of autonomous automatic frameworks. Within the framework of independent power, energy resources are considered as intellectual elements that are equipped for self-management such as placement, optimization, improvement, learning, and so on.

6.7 Green Observation

In the energy harvesting/storage, effective processing/ sensing, and energy buffering, there is a need for systematic solutions to adapt equipment that effectively uses resources to the system fully. For example, in wireless sensor applications and actuator systems (WSAS), sensors and actuators are classified using substantial multitopology techniques.

6.8 Green Communication

The IP-based communications set of rules can be used for all IoT layers. IPv6 through individual wireless area control systems are intelligent for the sensor actuator; navigation over a low control and loss system is appropriate for the system.

7 Discussion of Internet Energy Management

Internet of energy management (IEM) is considered a web-based smart grid, or an Internet-based energy framework, as suggesting changes to traditional power grids. Smart networks tell the world view that it allows the generation, transmission, distribution, and utilization of productive energy throughout the cycle of action for favorite energy management scheme organization. Following eco-friendly, IoT patterns need to improve technology to overcome the failure of relevant physical equipment characteristics in the IEM configuration. Between real energy management devices and virtual internet parts, there is intelligent mapping by combining them through stacks of standard interoperable Internet conventions. The development of renewable energy sources and the development of distributed power plants have brought new difficulties about the use of the Internet as a special tool to support energy management. IEM, with two critical effects, is needed to strengthen rigorous monitoring, management, and the direction of the power framework. In part, a separate and isolated energy framework will almost certainly build interconnection through the Internet, provide assistance, and optimize energy management. Communication frameworks and energy infrastructure are interconnected through data connections and power connections. The increase in IEM is a team on design framework, interoperability, legal structure, and market development. About the second most crucial part of IEM that effects of increasing reliability, security, costs, implementation, versatility, similarity and adaptability of the energy management framework based on the supply side, demand side, and supply and demand general energy approaches has been made with balance considerations.

8 Conclusion

In this study, demonstrated the scientific structure of IoT-based energy management, including sensor-actuator layers, organizational layers, and application layers. We consider energy issues related to exclusively considering the supply and demand side of IoT energy management, demand side, and energy balance. The sensor-actuator layer is responsible for the acquisition and capacity of energy, capable of detection and management, and energy support to achieve green identification. The system layer is focused on energy penetration and energy load adjustments to achieve an environmentally friendly response. Moreover, application layers tend to be used to estimate energy allocations, productive frameworks, energy measurements, and green processing. Analysis of the context of smart factory situations is taken into relation to describe empowerment technology for energy management. Also, a new point of view of IoT energy management is the combination of digital physical-social aspect and combined communication with Internet energy management in this topic.

Acknowledgements This work was supported by Hohai University under (CSC) No. 2017GXZ019296.

References

1. Vlacheas, P., et al.: Enabling smart cities through a cognitive management framework for the internet of things. IEEE Commun. Mag. **51**(6), 102–111 (2013)
2. Qayyum, F., et al.: Appliance scheduling optimization in smart home networks. In: IEEE Access, vol. 3, pp. 2176–2190 (2015)
3. Mishra, D., et al.: Smart RF energy harvesting communications: challenges and opportunities. IEEE Commun. Mag. **53**(4), 70–78 (2015)
4. Ozawa, T., et al.: Smart cities and energy management. Fujitsu Sci. Tech. J. **50**(2), 49–57 (2015)
5. Pohls, H.C., et al.: RERUM: building a reliable IoT upon privacy-and security-enabled smart objects. In: IEEE Wireless Communications and Networking Conference Workshop, pp. 122–127. Istanbul, Turkey (2014)
6. Bagdadee, A.H.: Power quality analysis by the ripple technique. Phoenix Res. Publ. J. Appl. Adv. Res. **2**(4), 227–234. https://doi.org/10.21839/jaarv2i4.74 (2017)
7. Bagdadee, A.H.: Imitation intellect techniques Implement for improving power quality in supply network. In: IEEE International Conference on Signal Processing, Communication, Power and Embedded System (SCOPES). https://doi.org/10.1109/scopes.2016.7955611 (2016)
8. Bagdadee, A.H., Bayezid Islam, Md.: To improve power failure and protect sustainability of the environment in Bangladesh by renewable energy. Int. J. Energy Environ. Res. **3**(1), 29–42. ISSN: 2055-0200 (UK) (2015)
9. Bagdadee, A.H., Siobhan, N.: Developing model of control stratagem with variable speed drive by the synchronous speed in micro-hydro plant. Int. J. Power Renew. Energy Syst. (IJPRES) **2**, 88–100. ISSN 2374-376X (The USA) (2015)
10. Chen, L.R., Chu, N.Y., Wang, C.S., Liang, R.H.: Design of a reflex-based bidirectional converter with the energy recovery function. IEEE Trans. Ind. Electron. **55**(8), 3022–3029 (2008)
11. Paudyal, S., Cañizares, C.A., Bhattacharya, K.: Optimal operation of distribution feeders in smart grids. IEEE Trans. Ind. Electron. **58**(10), 4495–4503 (2011)
12. Chiang, H.C., Ma, T.T., Cheng, Y.H., Chang, J.M., Chang, W.N.: Design and implementation of a hybrid regenerative power system combining grid-tie and uninterruptible power supply functions. IET Renew. Power Gener. **4**(1), 85–99 (2009)
13. Lee, Y.J., Khaligh, A., Emadi, A.: Advanced integrated bidirectional AC/DC and DC/DC converter for plug-in hybrid electric vehicles. IEEE Trans. Veh. Technol. **58**(8), 3970–3980 (2016)

14. Bagdadee, A.H.: Assessment of PV operation in Bangladesh. Int. J. Energy Environ. Res. **2** (1). ISSN: 2055-0200 (UK) (2014)
15. Bagdadee, A.H.: Status and reform towards development energy sector of Bangladesh. Eur. J. Adv. Eng. Technol. **2**(2), 24–28. ISSN: 2394-658X (India) (2015)
16. Lasseter, R.H.: Smart distribution: coupled microgrids. Proc. IEEE **99**(6), 1074–1082 (2011)
17. Li, D., Wang, S., Zhan, J., Zhao, Y.: A self-healing reconfiguration technique for smart distribution networks with DGs. In: Electrical and Control Engineering (ICECE) International Conference, pp. 4318–4321. Yichang, China (2011)
18. Bonino, D., et al.: ALMANAC: internet of things for smart cities. In: Proceedings of the International Conference in Future Internet of Things and Cloud, pp. 309–316. Rome, Italy (2015)
19. Evans, D.: The internet of things: how the next evolution of the internet is changing everything. Cisco Tech. Rep. (2011)
20. Bagdadee, A.H.: Rural electrification through micro-grid in Bangladesh. Eng. Sci. Technol. J. (ESTJ) **10**(5). ISSN: 1465-2382 (UK) (2015)
21. Erol-Kantarci, M., Mouftah, H.T.: Wireless sensor networks for cost-efficient residential energy management in the smart grid. IEEE Trans. Smart Grid **2**(2), 314–325 (2011)
22. Tsui, K.M., Chan, S.-C.: Demand response optimization for smart home scheduling under real-time pricing. IEEE Trans. Smart Grid **3**(4), 1812–1821 (2012)
23. Khanouche, M.E., et al.: Energy-centered and QoS-aware services selection for the internet of things. IEEE Trans. Autom. Sci. Eng. **13**(3), 1256–1269 (2016)
24. Al-Fuqaha, A., et al.: Internet of things: a survey on enabling technologies, protocols, and applications. IEEE Commun. Surv. Tutor. **17**(4), 2347–2376 (2015)
25. Bagdadee, A.H.: To reduce the impact of the variation of power from renewable energy by using supercapacitor in smart grid. WSEAS Trans. Power Syst. **11** (USA) (2016)

Innovative Automated Shopping Trolley with RFID and IoT Technologies

C. N. Yogalakshmi and Vivek Maik

Abstract The innovation across the globe has gone through dramatic changes and has made lives even simpler. Having or showing extreme avaricious nature towards the time being spent unwantedly, humans have always worked on to find an effective way to quench the purpose. IoT being the easiest way in connecting things together and solving the problems makes living simpler. In this proposed system, we have introduced a trolley that is connected with RFID reader and eradicates counter systems at shopping areas. It also helps in transmitting the items that have been purchased from the store to the destination of the customer's choice inside the shopping mall. Introducing the QR lock makes the journey of the trolley till it makes it to the destination safe and secure. Theft alarm is also been placed over the seal, if the lock that's been sealed with the QR is been tried opening manually without a proper QR code.

Keywords Smart trolley · Automated trolley · RFID trolley · RFID · QR lock · IoT smart trolley

1 Introduction

At present trend, it seems hectic when a person has got to shop his own needs in a shopping market. The complication rises when the customer is supposed to wait in a queue for getting items billed. Billing section creates misunderstandings as well as problems between individuals when one overtakes the other in a queue. The muddle develops along with arguments when one seems to be either taking time for billing or is interrupted by any other while standing in the queue in front of others. This

C. N. Yogalakshmi (✉) · V. Maik
Department of Electronics and Communication Engineering, SRM Institute
of Science and Technology, Chennai, India
e-mail: Kalayoga2@gmail.com

V. Maik
e-mail: maik.vivek@gmail.com

© Springer Nature Singapore Pte Ltd. 2020
S. S. Dash et al. (eds.), *Artificial Intelligence and Evolutionary Computations
in Engineering Systems*, Advances in Intelligent Systems and Computing 1056,
https://doi.org/10.1007/978-981-15-0199-9_39

given proposal has a remedy for the given social issue. RFID reader is used in this as an innovative approach to solve the issue. Barcode are usually printed over a paper or on any adhesive layer. They are not approximate and are pretty much tedious when it is being billed. And when it is been rough handled by people, it loses its prints over it and leads to defective piece, causing trouble while billing. Whereas RFID tags are read easily and it's near filed technology produces much quicker output than compared to the barcodes. This is further attached with the IoT where the product details are stored in the cloud and also connecting the trolley with the server. The transmitter is connected with the trolley to the server in order to update the items that have been taken by the customer. Once after the billing process is done, customers needn't carry the products whereas they can just direct the trolley to the destination where their respective vehicle is parked. The trolley then follows it way after getting the sign of paid receipt which will be reflected over the smartphone of the customer right after the amount is paid. When once the trolley reaches the destination there security guard manually is let to check the cart which opens only to the particular QR code which is generated after the amount paid from the smart phone of the respective owner of the cart. Once the amount is being paid a normal QR code is generated allowing the customer to lock the cart for safety purpose when it travels down to its respective destination being assigned. The lock is set such way that it opens only when the QR code generated after the payment matches the lock in the cart. Through this way the safeness of the cart is been ensured. Even if any malicious person tries to harm the cart by any manual attempts, antitheft alarm is placed above the lock to ensure it starts glowing and will alert the customer by popping on their gadget as a notification which is been connected.

2 Literature Review

As far as the innovations we have come across in day–to-day life, to solve the issues dealing with time consumption and converting it into time efficiency, the model with RFID along with IoT will produce appropriate results in reducing the time consumption. It will also help the shop dealer to have a proper way to deal with rush at the odd hours or at the peak time. In this proposed system, we have introduced RFID which is interfaced with IoT is attached with the trolley. Later the trolley is been handled by the customers to pick the objects they are interested in. There is a LED screen placed at the handle of the trolley for the customer to get a glance of the items they have picked. The amount is been calculated in the trolley. This happens with the process by the reader sensing the products and that gets reflected on the screen. Once after the shopping is done by the customer, they will be provided with the options to proceed with billing or to delete the items which they think are not. When some malicious customer tries to delete the items in the cart by just operating it in the trolley LED screen provided but haven't actually taken the product out, the RFID senses it and will let the customer to proceed for billing only if the product is been taken out of the trolley manually. After finalizing the products, the customer is set to proceed for the payment. Once after reading through the list the customer confirms the list and proceeds for payment. Once the customer goes for the payment section, the customer is given various options like Google pay, Paytm, debit/credit card for getting their payment done. In favouring digitalization, we here are providing such options to encourage in building online platform. The payment is done through their respective gadgets like mobile phones of the customer is connected to the arduino and is also connected with the Bluetooth v5 model which the customer can access through their device. Once after the payment is done, the customer is allowed to walk through the exit where another RFID reader is placed in order to read the items which checks whether the items in the trolley are paid and that is also updated with the server connected in the shop. The details on what has been bought by the customer, gets immediately updated through the zigbee transmitter and it is further connected with cloud. The trolley is also connected with the IoT which keeps updating on the details on the products that the customers have bought. Once the payment is done, a unique QR lock is been generated in their gadgets and that can be accessed when the lid of the trolley is been closed and reaches its destination. It actually works as a system which protects the items that has been purchased by the customer while travelling from its point of start to its destination. An antitheft alarm is also been placed over it to protect when it's been manually mishandled by any other person and sends a notification to the respective owner as a notification. This novelty in this system helps in making the work of the customer in easy lifting the bags of heavy items that is been purchased.

Once the customer is out of the shop they can assign instructions to the trolley by updating it to report to the parking lot where they have parked their vehicle. As soon as the trolley is set to start from the place to the destination there is an automatic lid which covers the trolley which when can be opened at the parking lot only when the QR lock code is been scanned from the respective customer who got the code after having done with their payment. The automated trolley reaches there without any delay as soon it's been assigned the instructions. The customers can now go to their lot and check their trolley. There is a security guard checking if the trolley really belonged to them with the help of the receipt in the phone and items in the cart. This automation happens since the trolley is with IoT. The area of the parking is been placed in the format of a grid and the customer is allowed to choose the grid which his/her vehicle belongs to and the trolley then picks up the shortest route and reaches the destination assigned to it. It more or less is the same concept on how does the Google maps work in suggesting a less reduced traffic pathway for the users. Here the grid is been suggested and the program to set the area limit is been done through the MATLAB coding. This has to be installed in the mall prior with the construction and the shop owners should have it installed. The design and allotment of the grid is done prior I the malls and the route map of it installed in every carts belonging to the respective malls/shops. Through the algorithm set up which is done through MATLAB coding, the shortest route to reach the destination is calculated through programs and then the trolley starts its way to the destination.

3 Proposed System

In this system, a developed technology like RFID which requires less time than the barcode with an IoT attached with an arduino and buzzer in the automatic shopping trolley is being introduced. The mechanism behind is that the components like zigbee transmitter, Bluetooth module, RFID reader, power supply, LED screen and a GPRS is connected with an arduino. The RFID reader reads the RFID cards in the trolley and gives the customer the list of products in the trolley through the LED screen which is placed at the handle. Further selecting the items, the customer is given various options in choosing the mode of payment among which he is let to choose any given options say for instance Paytm, Google pay, credit/debit card etc. Once the payment is done, it gets the server updated on the products which got sold. The customer will get the paid receipt in their mobile for future reference. Through zigbee transmitter the server gets updated on the products which got sold. Then the customer assigns the trolley to its destination where it should go and it passes through another RFID reader which is placed at the exit to make sure that the items inside the trolley is been paid. After it crosses the customer then covers the trolley with an inbuilt QR lock attached lid which only opens when the particular QR scanned belonging to the owner of the items in the trolley. The trolley uses the grid mechanism which should be created by the shop dealers on the route to reach the parking lot. It almost is equivalent to the technology proposed in the Google maps. The algorithm in letting the trolley chooses it smallest route is been done in the MATLAB first externally when connected with the system. Then it is converted into the embedded format which can be attached with the trolley further on. Once the trolley reaches the destination there is a security guard placed at each destination to check the respective carts reaches its owner and the owner uses the QR code which appears after the payment they make. When scanned with the respective QR code the customer can open the lid get their products. The innovation in using the QR lock based lid as well as RFID with IoT makes the shopping work very simple and is time consuming.

4 System Architecture

System is constructed having the arduino as the integral part along with the RFID reader and the transmitter. It is then connected with input and output devices like LED screen, server Bluetooth and also the power supply. The block diagram of the circuit is given in the figure below.

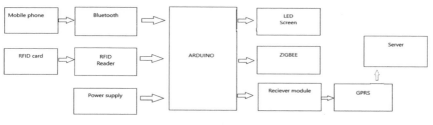

(1) BLOCK DIAGRAM

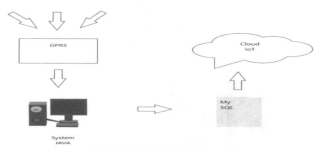

(2) SERVER BASED CONNECTION

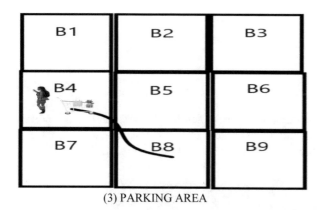

(3) PARKING AREA

The system inbuilt in the trolley with respect to the direction to choose the shortest distance in the parking lot is done in the MATLAB coding and is converted into one embedded form of coding which can then be installed in every of the trolleys in the shopping mall. The important components used in the system are RFID reader, arduino, Zigbee, server connection. Let's see the components in detailed.

4.1 RFID

Microchips are embedded into the RFID tags which when can be read using radio signals. The cost of each tags are too lower when compared to the barcodes. Barcodes comes on either paper or on an adhesive layer which when if roughly handled by the customer gets damaged. Using an antenna the radio signals are sent from the reader to which the tag responds by sending back the information stored in the microchip. There are two types of RFID reader, active and passive reader.

4.2 Arduino UNO

It is an ATmega-328 based microcontroller which has 14 input and output portals along with USB connection and a power jack. This is further connected to the rest of the two important components like Zigbee, RFID and to the LED screen.

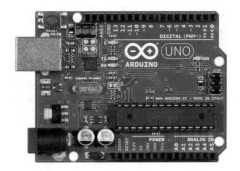

4.3 Zigbee

It is a standards-based wireless technology developed for machine to machine connection and in internet of things in networks. It is enabled for low cost. It consumes less battery as well as uses small packets compared with Wi-Fi and Bluetooth.

4.4 LED Screen

An LED screen is used to display pictures or videos with the help of light emitting diode. In here it helps the customer in making them aware of the items inside their trolley. It is also further attached with the arduino and keeps the server indirectly updated.

4.5 Bluetooth

HC-05 is a SPP serial protocol port which connects arduino serially. It is used here to connect the mobile phones of the respective customer to proceed for their payment. This mode makes the payment easier as well as generates the QR code after the payment is done through which the QR lid can be opened at the parking lot for the customer.

5   Simulation Code

Code used for finding the shortest distance in making the trolley reach the destination is been done in MATLAB. And also the code in which RFID can be read is also done through MATLAB.

Code used to figure the shortest distance by the trolley to reach the destination:-:
clearvars

```
W = [1 1 1 1 1 1 1 1 1 1 1 1 2 2 2 2 2 2 2 2 2 2 3 3 3 3 3 3 3 3 3 3
     4 4 4 4 4 4 4 4 4 4 4 4 4 4 5 5 5 5 5 5 5 5 5 5 6 6 6 7 7 7 7 7 7 7 7 7
     1.41 1.41 1.41 1.41 1.41 7 7 6 13 22 22 22 22 22 22 22 22 22 22 22 22 22 22
     11 11 11 11 11 11 11 11 10 10 10 10 10 10 10 10 10];

S = [1 2 2 3 3 4 4 5 5 6 6 7 7 8 8 9 9 10 10 11 11 12 12 13 13 14 14 15
     15 16 16 17 17 18 18 19 19 20 20 1 2 19 3 18 8 13 19
     12 3 8 4 7 18 13 17 14 1 19 2 20 2 18 3 19 7 3 4 8 5 7 4 6 18 14 13 17 14 16 17
     15 13 9 8 12 3 13 8 18 9 11 10 12 21 8 21 18 21 13 21
     3 22 9 22 12 22 10 22 11];
```

$$T = [2\ 1\ 3\ 2\ 4\ 3\ 5\ 4\ 6\ 5\ 7\ 6\ 8\ 7\ 9\ 8\ 10\ 9\ 11\ 10\ 12\ 11\ 13\ 12\ 14\ 13$$
$$15\ 14\ 16\ 15\ 17\ 16\ 18\ 17\ 19\ 18\ 20\ 19\ 1\ 20\ 19\ 2\ 18\ 3\ 13\ 8\ 12\ 9$$
$$8\ 3\ 7\ 4\ 13\ 18\ 14\ 17\ 19\ 1\ 20\ 2\ 18\ 2\ 19\ 3\ 3\ 7\ 8\ 4\ 7\ 5\ 6\ 4\ 14\ 16$$
$$17\ 13\ 16\ 14\ 15\ 17\ 9\ 13\ 12\ 8\ 13\ 3\ 18\ 8\ 11\ 9\ 12\ 10\ 8\ 21\ 18\ 21\ 13\ 21\ 3\ 21$$
$$9\ 22\ 12\ 22\ 10\ 22\ 11\ 12];$$

$$DG = \text{sparse}(S, T, W, 22, 22);$$

Names =
  {`Node1', `Node2', `Node3', `Node4', `Node5', `Node6', `Node7', `Node8',
  `Node9', `Node10', `Node11', `Node12', `Node13', `Node14', `Node15',
  `Node16', `Node17', `Node18', `Node19', `Node20', `Point1', `Point2'};

```
x = [1 3 5 5 5 7 7 7 9 11 11 9 7 7 7 5 5 5 3 1 6 10];
%y-coordinates%
y = [2 2 2 6 10 10 6 2 2 2 -2 -2 -2 -6 -10 -10 -6 -2 -2 -2 0 0];
G = digraph(DG,Names);
plot(G,'XData', x,'YData', y);
```

6 Simulation Results

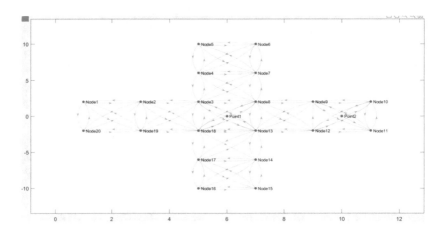

7 Result and Conclusion

The proposed system maintains a record of the items which got sold, and if the item is in less stock, it also generates a notification to the shop to restore the items. The automated trolley helps the customer not to carry such heavy bags. The RFID along with IoT makes the model look much simpler and eradicates the billing system as well as the counter section at the shopping area. This in turn saves the paper which is used for billing and also is not time consuming. The consumer is here unaided except for his trolley and does his shopping peacefully without any misunderstanding with the other customer which happened in the case of queue. This also lets the crowd jam to happen less and makes the shopping area much more pleasant which also provides profit to the shop dealers.

8 Future Scopes

We need to make sure that the devices shouldn't be roughly handled by the customers. Also we need to make sure that the memory of the device is updated from time to time. If any new technology comes up in the future we can also add that along with the existing trolley much simpler than it is. This can be installed in malls in metro cities or in any grocery shop. Further applications of it can be even done in the area of pharmacy where usually standing in queue to get tablets shouldn't happen in case of emergency.

Bibliography

1. Xia, F., Yang, L.T., Wang, L., Vinel, A.: Internet of things. Int. J. Commun. Syst. **25**(9), 1101 (2012)
2. Castillejo, P., Martinez, J.F., Rodriguez-Molina, J., Cuerva, A.: Integration of wearable devices in a wireless sensor network for an e-health application. IEEE Wirel. Commun. **20**(4), 38–49 (2013)
3. Mitton, N., Papavassiliou, S., Puliafito, A., Trivedi, K.S.: Combining cloud and sensors in a smart city environment. EURASIP J. Wirel. Commun. Networking **2012**(1), 1 (2012)
4. Song, T., Li, R., Xing, X., Yu, J., Cheng, X.: A privacy preserving communication protocol for IoT applications in smart homes. In: International Conference on Identification, Information and Knowledge in the Internet of Things (IIKI) 2016 (2016)
5. Maini, E., Sheltar, J.: Wireless intelligent billing trolley for malls. Inter. J. Sci. Eng. Technol. **3**(9), 1175–1178 (2014)

Fuzzy Logic: Adding Personal Attributes to Estimate Effect of Accident Frequencies

Arya Vijayan, Anupama S. Kumar and Jayasree Narayanan

Abstract In the field of risk assessment, one of the main aspects is the frequency of accident circumstances. Generally, the risk assessment evaluation is done using generic frequency approach. In case of accident, human factors associated with assigning the impact and cause of the hazard is not considered. Rather more focus is given to generic frequency approach which does not explicitly use the human factors. The main reason behind ignoring this factor primarily lies in the fact that its inclusion makes the entire procedure more complex. However, the proposed fuzzy logic approach helps in achieving the associated complexities in risk computation when we include human factors.

Keywords Risk assessment · Fuzzy logic · Accident frequency · Generic frequency approach · Frequency modifier

1 Introduction

In industries for ensuring safety the commonly used parameter is risk. For calculating risk, the factors considered are frequency and magnitude of its probable consequences. In the present scenario, generic frequency approach is used for evaluating the risk frequency. That is, the human factor is not more considered, so that employee safety and organization safety are measured only based on securities provided for the machines, etc. For example, if an accident had occurred in an industry, then to provide insurance and all, the insurance providers will consider

A. Vijayan (✉) · A. S. Kumar · J. Narayanan
Department of Computer Science and Engineering, Amrita Vishwa Vidyapeetham,
Amritapuri, India
e-mail: aryavijayan5925@gmail.com

A. S. Kumar
e-mail: anupamask91@gmail.com

J. Narayanan
e-mail: jayasreen@am.amrita.edu

© Springer Nature Singapore Pte Ltd. 2020
S. S. Dash et al. (eds.), *Artificial Intelligence and Evolutionary Computations
in Engineering Systems*, Advances in Intelligent Systems and Computing 1056,
https://doi.org/10.1007/978-981-15-0199-9_40

general safety measures provided by the company and according to that the insurance is provided without considering the involvement of human factors. Sometimes the accident may occur in an industry due to some human factors. But in generic frequency, the human factors are not explicitly considered, so we are not able to accurately find out the amount of influence of humans to the industry safety and the amount of safety assured to employees by the company cannot be determined properly. Thus, the ample scope of applications and the huge challenges encountered presently in this area provide enough opportunities for future work and research.

So here we are aiming to involve a human factor aspect within the industrial risk analysis field through creating a coefficient that revises the value of generic failure frequency using fuzzy methodology (fuzzy frequency modifier). To evaluate failure frequency, the input variables must be identified using fuzzy set, linguistic variables, and membership function. This leads the input to be fuzzified, they are connected using fuzzy inference using the fuzzy rules. Finally, defuzzification must be done to get the final crisp numbers which represent the ultimate fuzzy output.

This modifier varies between the range of 1–1.5. This choice has been done taking into consideration the HSE (Health and Safety Executive) explanation, which ascertains that within the petrochemical industry the mishaps caused due to human factors account up to 50%.

2   Related Work

The technical aspects involved in human safety measures are a matter of in-depth research which is dealt with, in the Safety Science. This field of study covers almost all the fields which have the risk of encountering hazards due to activities involving human. Safety Science will empower academic analysts, specialists and decision-makers in organizations, government agencies, and worldwide bodies, to increase their data level on the most recent trends in the field, from policy makers and administration researchers to transport engineers. There is an ample scope of research in the field of Safety Science.

In [1] Safety assessments predicted based on probability are often faced with the issue of limited availability of analytic data. Analysis done based on these data collected is not even subjected to any prior check.

To predict the impact of fetching the data from different sources without having any background check, it is doing a comparative study based on the sample data collected from some reliable source. As a part of analysis, they are doing a case study based on a reactor cooling module. Comparison based on frequency and probability values is done for our process. They saw a match in data when compared to the data collected from corner conditions. It also observes that the safety parameters obtained are not affected by the quality of collected data.

In the research work [2], they included the human factors in risk analysis which alters the generic failure frequencies utilizing fuzzy logic. This hypothesis allows incorporating qualitative factors that are dismissed by conventional strategies and bargaining the uncertainties involved. This strategy is apt for using it as an appropriate device for coordinating the human factor in risk assessment as it is uncommonly situated in rationalizing the vulnerability associated with imprecision or ambiguity. By including the human factor element in the risk estimation technique, a fuzzy modifier has been created. For testing purpose, the model was used in two case studies of chemical plants and new frequency values were obtained and along with the risk evaluation. Not much works have been implemented in this field/area. So, we have valid scope in extending this research work by including some extra features to get the accident frequency more accurately. After a detailed study, it is found that, there are some input parameters that can highly influence the final output, i.e., risk frequency. For that, we introduced hiring factor as a new human factor and the included parameters are having high weight, and the variation of these parameters will highly affect the final frequency.

3 Frequency Calculation

In risk assessment field to ensure safety, estimation of frequency is very important. Because risk can be measured on multiplying the frequency in which an event occurred by its probable consequences. But the existing risk frequency calculation is based on the generic frequency approach which does not explicitly include human factors. So here we are introducing a new fuzzy logic to control the frequency of an event which also includes human factors, so that the overall risk will change.

4 Our Approach

4.1 Work-Flow Diagram

The work-flow diagram of the system is shown in Fig. 1; the flow goes in the order; enter the crisp values as input, fuzzification process, assign rules, weight assignment, implication method, aggregation, and at last do the defuzzification method to get the frequency as crisp value.

4.2 Inputs/Identification of Variables

We are classifying the data in different ways. Depending on the classification, variables were selected to create the model. A straightforward approach to see human components is to consider three viewpoints: the activity, the people furthermore, the association, and how they influence individual's well-being and

Fig. 1 Work-flow diagram

security-related conduct. In view of this characterization, variables were chosen to create the model for this study. This determination regards that the general human factor is formed of three distinct elements illustrative of three fundamental classifications: organizational factor, job characteristic factor and personal characteristic factor. Every factor is further characterized by the influence of the basic variables. The organizational factor takes mainly contracting, training and communication and reporting under consideration. Same to this, job characteristic factors include workload, environmental condition, and workplace design. Finally, the personal characteristic factors are skills and knowledge, personal behavior.

After the detailed study, it is found that some other parameters are also giving more contribution in finding the frequency in safety scenario than the existing inputs considered. So, to make the estimation of risk frequency more accurate, we can include hiring factor also as a human factor. In the hiring factor, we can include age of the person, experience of the person, etc. The employees whose is so old, sometimes having less memory power and low energy, which can cause inaccurate way of handling the machines and it will lead to accidents in the plant. Secondly, while considering the experience, for example, a task performed by highly skilled operator will cause low error probability for that task and less accident scenarios. So, we are including the hiring factor for estimation of risk frequency, the system can give more accurate result for risk factor.

4.3 Fuzzification Process

Fuzzification is the process of transforming definite data into linguistic variables by computing degree of membership of the data to each fuzzy set. We are using Mamdani model for fuzzification. In this process, input uses three linguistic variables (poor, medium, Excellent) shown in Fig. 2a, the input score value ranges from 0 to 100 and the output uses five linguistic variables (very low, low, medium, high, very high) shown in Fig. 2b, the score value ranges from 0 to 10.

The fuzzy set theory and membership function theory are correlated. Membership function defines membership values in the range of 0–1 that indicates the degree of membership of an element to that set. Membership functions could be categorized as Z-shape, S-shape and Pi-shape which is used for lower (poor), upper (excellent) and intermediate (medium) fuzzy sets, respectively.

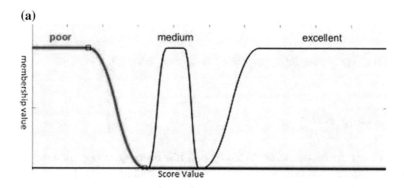

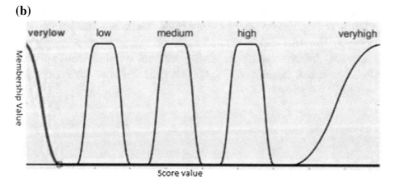

Fig. 2 **a** Membership function for input. **b** Membership function for output

Table 1 Linguistic variable, their score range, and associated numerical value

Linguistic variable	Score range	Numerical value
Poor (16–32)	16–19	0
	20–23	1
	24–27	2
	28–32	3
Medium (33–47)	33–37	4
	38–42	5
	43–47	6
Excellent (48–64)	48–52	7
	53–56	8
	57–60	9
	61–64	10

The linguistic variable, their score range, and associated numerical value are shown in Table 1 [2].

The Z-shape membership function is like as in Eq. (1).

$$f(x, y, z) = \begin{cases} 1, & x < a \\ 1 - 2\left(\frac{x-a}{b-a}\right)^2, & a \le x \le \frac{a-b}{2} \\ 2\left(\frac{x-a}{b-a}\right)^2, & \frac{a-b}{2} \le x \le b \\ 0, & x \ge b \end{cases} \tag{1}$$

For that we set the input parameter as 'a' and 'b'.

4.4 Assign Rules

Fuzzy rules connect the inputs to the output. It is based on proposition, i.e., if then. It has two parts: antecedent part and consequent part (same to if then). Once the rules are established, the antecedents are connected using logical AND operator. Formulating these rules is necessary for fuzzy inference process which uses the aggregation with implication of the output rules. Therefore, this results in a fuzzy number output. The rules can be created based on weight of each input parameters. In our approach, different inputs are having different weight according to its effect and importance in risk calculation. So, we can create rules to our system according to the weights.

Table 2 Weights of the variables of the system	Group 1	Age	0.50
		Experience	0.50
	Group 2	Contracting	0.20
		Training	0.60
		Communication and reporting	0.20
	Group 3	Workload management	0.33
		Environmental condition	0.33
		Safety equipment	0.33
	Group 4	Personal behavior	0.50
		Skills knowledge	0.50
	Group 5	Organizational factors	0.33
		Job characteristic factors	0.33
		Personal characteristic factors	0.33

4.5 Weight Assignment

The presentation of weights in this technique is pertinent as this may altogether influence the failure frequency esteem computed, furthermore because of the way that not every one of the factors may have the same significance. The scientific technique utilized for this reason is the Analytical Hierarchical Process (AHP). Also, for the new input factor, i.e., hiring factor also has its own significance in the estimation of risk frequency. So, in Group 1 we can include the weights for age and experience of the employees. The weights for different input parameters are shown in Table 2 [2].

4.6 Implication

It is a graphical process where for every rule the membership degree of the subsequent part is transformed in an area value. Here, it is using minimum implication method.

4.7 Aggregation

It is an inference process, in which the area obtained by implication process are combined into a single fuzzy set. Here, we are using summation method for inference. Aggregation is the last step of the inference process.

The previous step, i.e., the implication process, is resulting a set of areas in the fuzzy membership function according to the input score that we had given. For each of the input parameter, it will map to an area in the fuzzy membership function. But we need an output as a combination of the input that we had given. For example, we want to combine the area of the age and experience of a person into a single area, as hiring factor. Similarly, to obtain the final single fuzzy risk frequency modifier, we need to combine the areas of hiring factor, organizational factor, job characteristic factor and personal factor. For that, we are using the fuzzy tool 'Aggregation'. The order of execution of rules for the fuzzy system will not affect the aggregation process.

Like the previous fuzzification steps, the aggregation also has different types. The mainly used methods are: summation and maximization method. In summation method, it sums up the areas of different inputs. But in the case of maximization method, it takes the maximum area form different inputs under consideration. Since we are assigning some weights for different input parameters, we need to consider the cause of all the inputs in one way or another, so that here we are using the summation method for aggregation.

4.8 Defuzzification

We aim to involve the human factor in analyzing industrial risk, by creating a coefficient that revises the value of the generic failure frequency, using fuzzy logic. This modifier varies between the range of 1–1.5.

To obtain the end result of fuzzy logic methodology, the final step is defuzzification that yields a final crisp number that represents a fuzzy output. Here, we are using centroid method for defuzzification.

$$Z^* = \frac{\int \sigma(c) * Z * dz}{\int \sigma * (z) * dz} : \text{Defuzzification} \tag{2}$$

5 Experiments and Results

We use the data set of two chemical industries, which is used in [2] and our own data that we have collected by conducting a questioner with experts. After we applied it to the proposed work, it is found that as the value for different human factors (organizational factor, job characteristic factor, personal factor, and hiring factor) is increased, then the corresponding fuzzy frequency value for risk low. Also, it is analyzed that a small change in the input parameter which is having highest weight as shown in Table 2 can affect largely on the final output frequency. The total result evaluation is shown in Table 3a and b and the variation of hiring factor using different score values is shown in Table 3c.

Table 3 (a) Pool result for considered company. (b) Evaluation of hiring factor using different input scores. (c) Values of the modifier

(a)

		Company A total score	Company B total score
Organizational factors	Contracting	19	64
	Training	25	64
	Communication and reporting	34	56
Job characteristic factors	Workload management	34	64
	Environmental condition	43	64
	Safety equipment	22	64
Personal characteristic factor	Personal behavior	31	51
	Skills knowledge	29	51
Hiring factors	Age	50	25
	Experience	29	39

(b)

Age	Experience	Hiring factor
50	15	4.68
40	10	4.63
35	10	4.6
35	10	4.9
40	5	5.0

(c)

Variables	Company A value with weight	Company B with weight
Organizational factors	1.3	8.5
Job characteristic factors	2.5	8.2
Personal characteristic factor	3.4	5.8
Hiring factors	1.43	4.68
Organizational factors	1.45	1.1

For any new combination of human factor's score value, we can apply the fuzzy logic and can efficiently find the fuzzy risk frequency for a set of human score values. This should be ideally compared with expert from the field of safety assurance to check the decision obtained.

6 Conclusion

Generally, risk in an industry is calculated by generic frequency approach, which does not explicitly use human factors. So, with the aim of taking human factors for finding risk, a fuzzy frequency modifier was created using fuzzy set theory. For that a new fuzzy rule base is generated according to the weights of input parameters, which is effectively working with varying ranges of different input parameters. Mainly four kinds of human factors are considered as the input parameters and they are organizational factor, job characteristic factor, personal factor, and hiring factor. If all these factors are good (higher value) for a company, then its risk frequency will be less.

For getting more accurate risk value we introduced hiring factor as a new parameter that includes age and experience of the person (employee), which will be considered while recruiting a new employee. And thus, more accurate human effect on an accident scenario is determined. And, the fuzzy toolbox that we have developed can be used to determine value of safety contribution by each individual factor.

We evaluated our system with two data sets. And the results show that the fuzzy logic can be effectively used for finding risk frequency in safety measurement. And thus, this result can be used for different application. Our proposed method can be used to verify that whether a newly establishing company is following the rules and regulations put forward by HSE (Health and Safety Executive) for safety and different insurance providers, investigators can find the real cause of an accident scenario in different industries and the influence of human factors on it, etc.

The proposed method is developed for finding the safety of chemical industries by including human factors. The same can also be used in different domain for risk analysis.

References

1. Health and Safety Executive (HSE). Failure rate and event data for use within risk assessments. http://www.hse.gov.uk/landuseplanning/failure-rates.pdf(03.12.12) (2012)
2. Dan, J.G., Arnaldos, J., Darbra, R.M.: Introduction of the human factors in the estimation of accident frequencies through fuzzy logic. Saf. Sci. J. (2017)
3. Health and Safety Executive (HSE). Managing Contractors: A Guide for Employers. http://www.hse.gov.uk/pubns/priced/hsg159.pdf(16.11.12) (2011)
4. Jang, J.: Fuzzy Inference Systems. Prentice-Hall, NJ (1997)
5. Health and Safety Executive (HSE). Human factors in the management of major accident hazards. http://www.hse.gov.uk/humanfactors/topics/toolkitintro.pdf(17.11.12) (2005)
6. Klir, G.J., Yuan, B.: Fuzzy Sets and Fuzzy Logic: Theory and Applications. Prentice-Hall, NJ (1995)
7. Nait-Said, R., Zidani, F., Ouzraoui, N.: Fuzzy risk graph model for determining safety integrity level. Int. J. Qual. Stat. Reliab. **112** (2008)
8. https://www.wikipedia.org

9. Nair, J.J., Govindan, V.K.: Automatic segmentation employing fuzzy connectedness. Inter. Rev. Comput. Softw. **7** (2012)
10. Mathew, R., Kaimal, M.R.: A fuzzy approach to the 2×2 games and an analysis of the game of chicken. Inter. J. Knowl. Based Intell. Eng. Syst. **8**(4), 181–188 (2004)
11. Mohanlal, P.P., Kaimal, M.R.: Design of optimal fuzzy observer based on TS fuzzy model. In: Fuzz IEEE (2004)

Preconditions of GPA-ES Algorithm Application to Big Data

Tomas Brandejsky

Abstract Herein presented contribution speaks about preconditions of genetic programming algorithm (GPA) used to work with so-called big data. In the paper, the different ways of decrease of number of (big) training data evaluations as well as parallelization of this process are presented. The paper also discusses using a floating window to provide evolution steps on small data subsets. The presented approach is based on reduction of the number of evaluations to allow fitness function evaluation in acceptable computational time. The original GPA-ES algorithm is described too. This algorithm was developed and tested for many years for application in area of accurate symbolic regression. Now the interest is to apply it to the application domain of big data, especially to data-based modeling and knowledge discovering fields. Then, there are described experiments verifying these hypotheses and the results are discussed. These experiments are based on symbolic regression (discovering of differential equations) describing a training data set representing movement of the Lorenz attractor deterministic chaos system.

Keywords Big data · Evolutionary algorithm · Evaluation scheme · Evaluation reduction · Floating window · Data subset

1 Introduction

Increasing capabilities of nowadays computers, servers, and especially server clusters and supercomputers allows collecting, storing, and analyzing unbelievable amounts of data. Such amounts of data (and related methods of their processing) are called big data, a term which has no precise definition. This term is deeply bound

T. Brandejsky (✉)
University of Pardubice, 532 10 Pardubice, Czech Republic
e-mail: tomas.brandejsky@upce.cz

© Springer Nature Singapore Pte Ltd. 2020
S. S. Dash et al. (eds.), *Artificial Intelligence and Evolutionary Computations in Engineering Systems*, Advances in Intelligent Systems and Computing 1056,
https://doi.org/10.1007/978-981-15-0199-9_41

with everyday increase of produced amount of data by the human population. For example, Technopedia [1] defines big data following way:

Big Data Definition—What does Big Data mean?

Big data refers to a process that is used when traditional data mining and handling techniques cannot uncover the insights and meaning of the underlying data. Data that is unstructured or time sensitive or simply very large cannot be processed by relational database engines. This type of data requires a different processing approach called big data, which uses massive parallelism on readily-available hardware.

This and many similar techniques point to the most significant quality of big data—an inability to apply standard analytic and processing tools and techniques developed for "small" or "standard size" data like relational databases.

Frequently used big data processing framework Hadoop uses a MapReduce evaluation scheme to deliver related data to suitable nodes of large computer systems.

When studying publications about research in the big data field, it is possible to obtain the meaning that genetic programming belongs to above the mentioned tools suitable only for "small" data and evolutionary systems. In common, these systems are applicable rather to optimization of big data tool parameters than for direct big data processing. To underline this meaning, it is possible to introduce such works as [2] demonstrating application of Genetic Algorithm for Data Clustering. In this work, there are also used applications of parallel GA working in Hadoop and MapReduce environment to provide big data clustering. Clustering is divided into two phases and operated on many communicating nodes.

On the opposite side, only few publications describe applications of genetic programming in the big data area. Project "Automatic Programming for Optimization Problems with Big Data" at School of Computing, University of Portsmouth listed suitable big data problems for GPA application as Financial Forecasting, Traffic Optimization, Cloud Optimization and Scheduling [3]. It is possible to mention also any other potential application domains of genetic programming in the field of big data as symbolic regression (search of model describing stored data), discovering game-playing strategies (and not only game-playing but also business ones), forecasting, induction of decision trees, etc. The work [4] describes interesting application of GPA in mining of big data obtained from the Large Hadron Collider. This work describes large set of sampling methods as random sampling, weighted sampling, incremental data selection, topology-based subset selection, balanced sampling and many multilevel sampling and hierarchical sampling methods. Some authors applied genetic programming in big data domain indirectly in automatic synthesis of automatic classifiers of data, as [5].

2   Genetic Programming Algorithms

Genetic programming algorithms are successful in many areas of automatic solving of constructive tasks as building models, programs, technical systems, etc. Thus, there is a legitimate question why they are not popular in the field of big data. The main problem is in their nature. A typical evolutionary algorithm (not only GPA) works with a population of individuals to be evolved. In each evolutionary cycle, these individuals are evaluated, e.g., by training data. Thus, the computational complexity of evolutionary algorithms can be expressed as a multiple of population size, number of evolutionary steps (populations), and data samples to be processed in each step in a very simplified way. Big data means a potentially big number of data samples to be processed, because population size and number of evolutionary steps are limited by solved problem nature, besides other things, and thus success rate of their reduction is limited.

With respect to capabilities of GPA, the search of a way to apply them to big data is interesting. There are a few ways as parallelization of GPA work, decrease of GPA evolutionary step number, or data reduction in the sense of MapReduce application.

2.1   GPA Parallelization

There are well-known parallel implementations of GPA, one of the first was developed by pioneer of GPA research, Prof. Koza [6]. The earlier work was work [7]. Nowadays there are thousands of works about parallel and distributed genetic programming algorithms, of which a large collection is cited, e.g., in the work [8]. This work also explains the difference between parallel and distributed approach. Not looking at this terminological problem (many so-called parallel GPAs are by [8] rather distributed), these algorithms make parallel the computing of fitness function from the viewpoint of dividing of population into smaller groups.

Big data brings different a problem. Not the population but the training data set is too large to compute. Big data proposes a different approach. We cannot divide population of individuals and neither can we compute different sub-populations on each node sharing the same training data set with others. We can apply the opposite solution which is to divide data set and to equip each node with separate part of training data.

Such change of approach opens many ways of big data GPA evolution. At first, we can reason if all data are needed. Really big data collections typically contain a number of redundancies and not all redundancies can be simply reduced in the standard clustering process. First experiment described in the next chapter studies influence of training data set reduction on symbolic regression by GPA process.

Also, it is possible to expect that if parts of training data set distributed to particular computing nodes are similar, and discovered solutions will be similar too.

This hypothesis is implied from the previously described reason and experiment results. This idea can be also expressed where the solution across all data is an intersection of all particular solutions.

If the results produced by particular computing nodes on the base of their subset of training data are not similar, it will be faster to migrate sub-populations between computing nodes than to share big training data. There is only one additional requirement—genome representation must enable serialization of individuals. Movement of individuals on another node allows inclusion of additional knowledge in different data subsets. GPAs are not fast in forgetting of previously learned knowledge as it is illustrated by the second experiment comparing speed of learning using standard work with a data set and using floating window, as it will be discussed later. There exist only few exceptions from this rule—GPAs with improved long term adaptability like haploid and diploid algorithms [9]. These nature-based algorithms preserve genetic diversity but they do not reach optima as fast as standard GPAs and thus they will not be reasoned more in this paper.

Another way to conform with the previous approach is to decrease the number of GPA evolutionary steps. It is possible to do this by two different ways. Straightforward one is to use extremely large populations, but this way brings difficult problems. Especially, large populations typically contain a lot of duplicit genetic material and thus computing efficiency is small. The mechanism of duplicit gene reduction must be added. The second problem is that very small populations by some observations [10] are significantly more computationally efficient but the remittance is in the form of an unasked big number of evolutionary cycles.

On the base of the data redundancy prerequisite, an alternative approach might be formulated. Because each small subset of training data can contain enough useful information, in each cycle of evolutionary algorithm work, a different subset can be used. If the training data set is divided into a sufficient number of subsets, each of them can be used only once.

Floating window can also be implemented differently. For example, it can move about one data point after each evolutionary cycle of GPA. In such case, each data point is read at least once. Repeated reading is possible if the developed model reasons longer time history and higher-order model.

2.2 Data Reduction

Standard big data style approach based on data reduction, especially MapReduce one, is not discussed more because its application is not contradictory to other ones discussed herein.

The most significant property of any big data manipulating GPA is the number of fitness function evaluations without respect to population size influence. It should be also useful to study the relation between number of samples in data window and total number of required iterations of evolutionary algorithm, as it will be examined by the following experiments.

Currently under development are arrhythmia monitors for ambulatory patients which analyze the ECG in real time [1–3]. Software QRS detectors typically include one or more of three different types of processing steps: linear digital filtering, nonlinear transformation, and decision rule algorithms [4].

3 Experiments

Experiments map features of GPA with respect to its applicability to big data problems because they were provided before acquisition of new HW oriented to big data processing. Thus, used training data were representing model, very simplified problem. It tended also to small number of iterations.

The first experiment uses different size of training set to demonstrate dependencies between number of samples, number of iterations, and number of fitness function evaluations. Data are taken from beginning of the same vector representing Lorenz attractor system movement (1 and 2).

$$
\begin{aligned}
x'(t) &= -y(t) - z(t) \\
y'(t) &= x(t) + ay(t) \\
z'(t) &= b + z(t)(x(t) - c)
\end{aligned}
\tag{1}
$$

$$
a = 0.1, \ b = 0.1, \ c = 5.7 \tag{2}
$$

The second experiment does the same work, but the data are taken by a floating window moving with each iteration of evolutionary algorithm. This experiment studies influence of floating window application in comparison with static data window studied in the previous experiment.

3.1 GPA-ES Algorithm

As subject of exploration, GPA-ES algorithm was used. This algorithm optimizes capabilities of standard GPA to area of highly precise symbolic regression (symbolic regression producing results which are not over-complicated and are comparable to results produced by human analysists [11]). GPA-ES algorithm was described many times, e.g., in [12].

Computational complexity of GPA-ES is expressed as (3), as it was analyzed in [12]:

$$
O(\mathrm{GPAES}) = pqnm \log(m) + pqn \log(n) \tag{3}
$$

where

n is number of GPA individuals
m is number of ES individuals
l is complexity of structures created by GPA
k is average number of constants in GPA genes, where
p is number of GPA populations
q is number of ES populations.

Experiments stopped when sum of error squares was less than 10^{-7}. Experiments were repeated 1000 times for different initial seed magnitudes of used pRNG (standard rand() function of C++ stdlib function). Used GPA population was very small. It contained 10 individuals to reach higher resolution in evolutionary cycle number.

3.2 Experiment Results

The first experiment studied influence of data window size on efficiency of evolutionary algorithm. In period of 512 to 128 samples in training data set, the efficiency was increasing (number of evolutionary cycles was decreasing) and start of expected opposite behavior was observed after 64 samples (Fig. 1). This fact points to sensitivity of small data sets to local properties of training data, but also it concludes that even very small data sets can contain enough data for model construction.

The second experiment was performed to analyze how can experiment results be influenced by use of a floating window, which is close to the big data approach. In this case, the speed of evolution decreases (number of evolutionary steps increases)

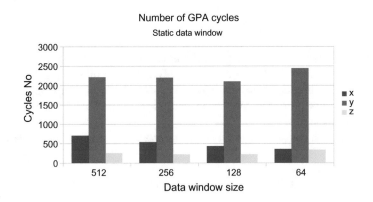

Fig. 1 Number of cycles of GPA in Lorenz attractor system model regression for variables x, y, and z using static data set

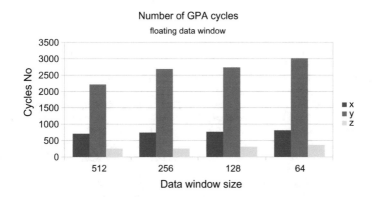

Fig. 2 Number of cycles of GPA in Lorenz attractor system model regression for variables x, y, and z using floating window in large data set

with decreasing size of data window, as expected (Fig. 2). This fact underlines influence of local property of data to the first experiment. We can also observe that the number of evolutionary cycles is the same or higher than in the first experiment. This effect is probably caused by lack of straight direction of evolution of GPA individuals. It can be also observed that the loss of evolution efficiency by decrease of training data set size is significantly smaller than increase of required evolutionary cycles and thus number of fitness function evaluations.

4 Conclusions

Herein presented initial study of research of GPA applicability in big data analysis discusses reasons about limits and needed provisions to use GPA on big data. Especially, problem of efficiency given by large number of evaluations is studied. There are suggested ways of efficient evaluation organization. Their effect is demonstrated on simple (not big data) problem of symbolic regression of Lorenz attractor system from vector of 512 data sets. Thus, herein presented evaluation scheme opens way to GPA applications in the big data field.

Acknowledgements This work was supported by The Ministry of Education, Youth and Sports from the Large Infrastructures for Research, Experimental Development and Innovations project "IT4Innovations National Super-computing Center—LM2015070".

References

1. Technopedia: Big data. https://www.techopedia.com/definition/27745/big-data (2019). Accessed 11 March 2019
2. Nirvanshu, H., Sana, M., Omkar, S.N.: Big data clustering using genetic algorithm on hadoop mapreduce. Inter. J. Sci. Technol. Res. **4**(4), (2015). ISSN 2277-8616
3. Bader-EL-Den, M., Adda, M.: Automatic programming for optimization problems with big data. https://www.findaphd.com/phds/project/automatic-programming-for-optimization-problems-with-big-data/?p62613 (2019). Accessed 12 March 2019
4. Hmida, H., Hamida, S., B., Borgi, A., Rukoz, M.: Scale genetic programming for large data sets: case of Higgs bosons classification. Procedia Comput. Sci. **126**, 302–311 (2018). https://doi.org/10.1016/j.procs.2018.07.264
5. Kojecký, L., Zelinka, I., Šaloun, P.: Evolutionary synthesis of automatic classification on astroinformatic big data. Int. J. Parallel Emergent Distrib. Syst. **32**(5), 429–447 (2017). https://doi.org/10.1080/17445760.2016.1194984
6. Andre, D., Koza, J.R.: Parallel genetic programming: A scalable implementation using the transputer network architecture. In: Angeline, J.P., Kinnear Jr, K.E. (eds) Advances in Genetic Programming 2, The MIT Press, (1996)
7. Hugues, J., Pollack, J., B.: Pararllel genetic programming and fine-grained SIMD architecture. In: Siegel, E.V., Koza, J.R. (eds.) Working Notes for the AAAI Symposium on Genetic Programming, pp. 31–37. MIT, Cambridge (1995)
8. Poli, R., Langdon, W.B., McPhee N.F., (With contributions by Koza, J.R): A field guide to genetic programming. Published via http://lulu.com and freely available at http://www.gp-field-guide.org.uk (2008)
9. Vekaria, K., Clack, C.: Haploid genetic programming with dominance. Technical Report RN/97/121, (1997)
10. Brandejsky, T.: Small populations in GPA-ES algorithm. In: Matousek, R. (ed.) 19th International Conference on Soft Computing Mendel 2013, pp. 31–36, Brno University of Technology (2013). ISSN 1803-3814. ISBN 978-80-214-4755-4
11. Korns, M.: Extremely accurate symbolic regression for large feature problems. https://doi.org/10.1007/978-3-319-16030-6_7 (2015)
12. Brandejsky, T.: Multi-layered evolutionary system suitable to symbolic model regression. In: Recent Researches in Applied Informatics, Praha, 2011-09-26/2011-09-28, pp. 222–225. WSEAS Press, Athens (2011). ISBN 978-1-61804-034-3

An Interval Valued Triangular Fuzzy Soft Sets and Its Application in Decision-Making Process Using New Aggregation Operator

D. Nagarajan, M. Lathamaheswari, Said Broumi, Florentin Smarandache and J. Kavikumar

Abstract In decision-making process, aggregation plays a vital role as it is a process of combining various numerical scores with respect to all the criteria using aggregation operators. Hence, in this paper, new aggregation operators have been proposed, namely Interval Valued Triangular Fuzzy Soft Weighted Arithmetic (IVTFSWA) operator and Interval Valued Triangular Fuzzy Soft Weighted Geometric (IVTFSWG) operator for Interval Valued Triangular Fuzzy Soft Numbers (IVTFSNs) using Frank triangular norms and applied them in a decision-making problem to identify the patient who has more illness. This application shows the soundness of the presented operators.

Keywords Fuzzy soft set · Aggregation operator · Decision-making process · Type-2 fuzzy model · Triangular norms · Interval valued triangular soft fuzzy number

D. Nagarajan (✉) · M. Lathamaheswari
Department of Mathematics, Hindustan Institute of Technology and Science,
Chennai 603103, India
e-mail: dnrmsu2002@yahoo.com

M. Lathamaheswari
e-mail: lathamax@gmail.com

S. Broumi
Laboratory of Information Processing, Faculty of Science Ben M'Sik, University Hassan II,
B.P 7955 Sidi Othman, Casablanca, Morocco
e-mail: Broumisaid78@gmail.com

F. Smarandache
Department of Mathematics, University of New Mexico, 705 Gurley Avenue,
Gallup, NM 87301, USA
e-mail: fsmarandache@gmail.com

J. Kavikumar
Department of Mathematics and Statistics, Faculty of Applied Science and Technology,
Universiti Tun Hussein Onn, Parit Raja, Malaysia
e-mail: kavi@uthm.edu.my

© Springer Nature Singapore Pte Ltd. 2020   493
S. S. Dash et al. (eds.), *Artificial Intelligence and Evolutionary Computations
in Engineering Systems*, Advances in Intelligent Systems and Computing 1056,
https://doi.org/10.1007/978-981-15-0199-9_42

1 Introduction

Triangular inequalities were protracted by the theoretical concept of triangular norms which are introduced from the scope of prospect metric [1]. Many of the realistic optimization problems are described by flexibility of the constraint which can be used for further decision between correcting the impartial function and fulfilling the constraints [2]. Fuzzy set theory contributes a methodology of representing and handling flexible or soft constraints [3]. SS theory applications normally solve the problems using rough and FSSs [5]. T-norms and t-conorms are the general additional operations in the interval [0, 1] ever present in the theory and applications of Type-2 FS (T1FSs) [7]. Till date, there are many operators which have been introduced to aggregate the information [8]. In DM process, information is collected from a number of experts by preparing survey that involves imprecise words and linguistic terms which have immeasurable domain, and the conclusions of T1 fuzziness are uncertain boundaries of FSs. This can be solved by T2 fuzziness, where the FSs have degrees of membership that themselves fuzzy [9]. In T2FSs, there are two membership functions (MFs) available, namely primary and secondary. For all the value of the primary variable x on the universal set X, the function will be the membership rather than a characteristic value. For secondary MFs, domain is the set of primary membership value. The secondary MF provides three-dimensional T2FS, where the third dimension affords a certain degree of freedom to deal with uncertainties [10]. The soft set theory can be utilized as a common mathematical tool to handle uncertainty [13]. Representation of SSs is nothing but soft matrices which have many advantages like it is easy to collect and apply matrices. Construction of decision making using SSs is more realistic and applied in a fruitful manner to various problems which have uncertainties [18]. The mathematical functions which are applied to integrate the information are called aggregation operators. DM process using fuzzy logic is a sizzling area of the research community [19]. It contains the method of choosing the best alternatives from possible alternatives in consideration of many attributes, where the information of the decision generally with fuzziness is contributed by a number of experts in fuzzy setting. Uncertainty of the real world with its related problems can be dealt well with SS theory. Since there is an increasing difficulty of the real world environment and the inadequate data of the problem, decision-makers may give their preferences about alternatives in the form of Interval Valued Fuzzy Numbers. Matrices are playing a crucial role in the wide area of engineering and science. The conventional theory of matrices could not solve the problems which contain impreciseness [20–22].

2 Review of Literature

The authors of [1] examined a method of DM problem for many attributes using triangular interval type-2 fuzzy numbers (TIT2FNs). Rajarajeswari and Dhanalakshmi [2] defined IVIFS matrix, its types with illustrations, few operators using weights, and their properties. Tripathy and Sooraj [3] conferred few operations of IVIFSS and its application in DM process. Khalil et al. [4] corrected some points on Interval Valued Hesitant FSSs. Sudharsan and Ezhilmaran [5] presented weighted arithmetic averaging (WAA) operator using IVIF values and applied in MCDM in choosing the best investment. Mohd and Abdullah [6] surveyed about the application of aggregation methods in GDM process for the period of 10 years. Shenbagavalli et al. [7] applied FSSs and its matrices in MAGDM process with the partial information of the data. Selvachandran and Sallah [8] proposed IVCFSSs by combining T2 complex FSs and SSs. Hermandez [9] proposed and examined the expansion of some types of FNs, namely T1, T2, and generalized one. Castillo et al. [10] reviewed the applications of T2 fuzzy system in image processing. Dinagar and Rajesh [11] discussed a methodology for a MCDM problem using the AOs of IVFSM. Qin et al. [12] established a new method for decision making which has the capability of solving some complicated problems. Nazra et al. [13] extended hesitant IFSSs to a generalized case and obtained an algebraic structure. Alcantud and Torrecillas [14] combined the aspects of SSs and FSSs and solved a decision-making problem. Selvachandran and Singh [15] handled complex data with the help of the properties of IVCFSs. Dayan and Zulqarnain [16] proposed the notions of generalized IVFSMs and discussed a few of its types, operational laws, and similarity measure. Mahmooda et al. [17] proposed the notion of lattice-ordered IFSS and its operational laws and applied in DM process. Arora and Garg [18] proposed aggregation operators using IFSSs and applied in DM process. Arora and Garg [19] did a comparative analysis of two IF soft numbers and presented weighted averaging and geometric AOs with their different properties. This literature study is the motivation of the present study. Nagarajan et al. [20] analyzed traffic control management using aggregation operators under interval type-2 fuzzy and interval neutrosophic environments. Nagarajan et al. [21] presented edge detection method using triangular norms under type-2 fuzzy environment. Nagarajan et al. [22] introduced image extraction methodology for DICOM image based on the concept of type-2 fuzzy.

2.1 Preliminaries

The following basic result has been given as preliminaries for better understanding of the paper.

2.2 Interval Valued Triangular Fuzzy Weighted Averaging/ Geometric (IVTFWA/IVTFWG) Operator [Qin and Liu 2014]

Let $F_i = \left\langle \left[l_{F_i}^-, l_{F_i}^+ \right], m_{F_i}, \left[r_{F_i}^-, r_{F_i}^+ \right] \right\rangle, i = 1, 2, 3, \ldots, n$ be a set of Interval Valued Triangular Fuzzy Numbers (IVTFNs) and consider TIT2FWA : $\Delta^n \to \Delta$, if

$$\text{IVTFWA}_\rho(F_1, F_2, \ldots, F_n) = \rho_1 \cdot F_1 \oplus \rho_2 \cdot F_2 \oplus \ldots \oplus \rho_n \cdot F_n (\text{IVTFWA}) \quad (1)$$

$$\text{IVTFWG}_\rho(F_1, F_2, \ldots, F_n) = F_1^{\rho_1} \otimes F_2^{\rho_2} \otimes, \ldots, \otimes F_n^{\rho_n} (\text{IVTFWG}) \quad (2)$$

where $\rho = (\rho_1, \rho_2, \ldots, \rho_n)^T$ is the weight vector (WV) F_i, $\rho_i \geq 0$. If $\rho = (1/n, 1/n, \ldots, 1/n)^T$ then IVTFWA/IVTFWG operator will become Interval Valued Triangular Fuzzy Arithmetic/Geometric Averaging (IVTFAA/IVTFGA) operator.

2.3 Frank Triangular Norms [Qin and Liu 2014]

Frank triangular norms include Frank sum and product which are the examples of t-norm and t-conorm, respectively. Here t-norm is Frank product, and t-conorm is Frank sum, and is defined by

$$m \oplus n = 1 - \log_\beta \left(1 + \frac{\left(\beta^{1-m} - 1 \right) \left(\beta^{1-n} - 1 \right)}{\beta - 1} \right),$$

$$m \otimes n = \log_\beta \left(1 + \frac{(\beta^m - 1)(\beta^n - 1)}{\beta - 1} \right) \quad (3)$$

$$\beta > 1, \forall (m, n) \in [0, 1]^2$$

2.4 Ranking Formula (Score Function) for Interval Valued Triangular Fuzzy Numbers [Qin and Liu 2014]

For the type-2 triangular fuzzy number, the ranking formula is defined by

$$S(\chi) = \left(\frac{a_{ij}^- + c_{ij}^+}{2} + 1 \right) \times \frac{a_{ij}^- + a_{ij}^+ + c_{ij}^- + c_{ij}^+ + 4b_{ij}}{8} \quad (4)$$

2.5 Aggregation Operators for Interval Valued Triangular Fuzzy Soft Numbers (IVTFSNs)

In this section, two aggregation operators, IVTFSWA, have been proposed.

2.5.1 Interval Valued Triangular Fuzzy Soft Weighted Average (IVTFSWA) Operator

Let $F_{e_{ij}} = \left\langle \left(a_{ij}^-, a_{ij}^+ \right), b_{ij}, \left(c_{ij}^-, c_{ij}^+ \right) \right\rangle, i = 1, 2, \ldots, n \& j = 1, 2, \ldots, m$ be an IVTFSNs and φ_j, ρ_i be the weight vectors for the parameters e_j's and experts x_i, respectively; then

$$\text{IVTFSWA}(F_{e_{11}}, F_{e_{12}}, \ldots, F_{e_{nm}}) = \overset{m}{\underset{j=1}{\oplus}} \varphi_j \left(\overset{n}{\underset{i=1}{\oplus}} \rho_i F_{e_{ij}} \right) \tag{5}$$

and satisfying the following conditions: $\sum_{j=1}^{m} \varphi_j = 1$ and $\sum_{i=1}^{n} \rho_i = 1$, $\varphi_j > 0, \rho_i > 0$.

2.6 Theorem

Let $F_{e_{ij}} = \left\langle \left(a_{ij}^-, a_{ij}^+ \right), b_{ij}, \left(c_{ij}^-, c_{ij}^+ \right) \right\rangle$ be an IVTFSNs, the aggregated value by IVTFSWA operator is also an IVTFSN and is given by

$$
\begin{aligned}
\text{IVTFSWA}\left(F_{e_{11}}, F_{e_{12}, \ldots}, F_{e_{nm}}\right) = \Big\langle &\Big[\Big(1 - \log_\beta \Big(1 + \wp_j \Big(\mathbb{P}_i \big(\beta^{1-a_{ij}^-} - 1 \big)^{\rho_i} \Big)^{\varphi_j} \Big) \Big), \\
& 1 - \log_\beta \Big(1 + \wp_j \Big(\mathbb{P}_i \big(\beta^{1-a_{ij}^+} - 1 \big)^{\rho_i} \Big)^{\varphi_j} \Big) \Big) \Big], \\
& 1 - \log_\beta \Big(1 + \wp_j \Big(\mathbb{P}_i \big(\beta^{1-b_{ij}} - 1 \big)^{\rho_i} \Big)^{\varphi_j} \Big), \\
& \Big(1 - \log_\beta \Big(1 + \wp_j \Big(\mathbb{P}_i \big(\beta^{1-c_{ij}^-} - 1 \big)^{\rho_i} \Big)^{\varphi_j} \Big) \Big), \\
& 1 - \log_\beta \Big(1 + \wp_j \Big(\mathbb{P}_i \big(\beta^{1-c_{ij}^+} - 1 \big)^{\rho_i} \Big)^{\varphi_j} \Big) \Big) \Big] \Big\rangle
\end{aligned}
\tag{6}
$$

Proof For $n = 1, \rho_1 = 1$:

$$\text{IVTFSWA}\left(F_{e_{11}}, F_{e_{12}}, ..., F_{e_{1m}}\right) = \overset{m}{\underset{j=1}{\oplus}} \varphi_j F_{e_{1j}}$$

$$= \left\langle \left[\left(1 - \log_\beta\left(1 + \wp_j\left(\beta^{1-a_{1j}^-} - 1\right)^{\varphi_j}\right)\right), \right.\right.$$

$$\left. 1 - \log_\beta\left(1 + \wp_j\left(\beta^{1-a_{1j}^+} - 1\right)^{\varphi_j}\right)\right) \right],$$

$$1 - \log_\beta\left(1 + \wp_j\left(\beta^{1-b_{1j}} - 1\right)^{\varphi_j}\right),$$

$$\left(1 - \log_\beta\left(1 + \wp_j\left(\beta^{1-c_{1j}^-} - 1\right)^{\varphi_j}\right)\right),$$

$$\left. 1 - \log_\beta\left(1 + \wp_j\left(\beta^{1-c_{1j}^+} - 1\right)^{\varphi_j}\right)\right) \right] \right\rangle$$

For $m = 1, \varphi_1 = 1$

$$\text{IVTFSWA}\left(F_{e_{11}}, F_{e_{21}}, ..., F_{e_{n1}}\right) = \overset{n}{\underset{i=1}{\oplus}} \rho_i F_{e_{i1}}$$

$$= \left\langle \left[\left(1 - \log_\beta\left(1 + \mathbb{P}_i\left(\beta^{1-a_{i1}^-} - 1\right)^{\rho_i}\right)\right), \right.\right.$$

$$\left. 1 - \log_\beta\left(1 + \mathbb{P}_i\left(\beta^{1-a_{i1}^+} - 1\right)^{\rho_i}\right)\right) \right],$$

$$1 - \log_\beta\left(1 + \mathbb{P}_i\left(\beta^{1-b_{i1}} - 1\right)^{\rho_i}\right),$$

$$\left(1 - \log_\beta\left(1 + \mathbb{P}_i\left(\beta^{1-c_{i1}^-} - 1\right)^{\rho_i}\right)\right),$$

$$\left. 1 - \log_\beta\left(1 + \mathbb{P}_i\left(\beta^{1-c_{i1}^+} - 1\right)^{\rho_i}\right)\right) \right] \right\rangle$$

Hence, the result is hold for $n = 1, m = 1$.
Consider the result is hold $m = k_1 + 1, n = k_2$ and $m = k_1, n = k_2 + 1$

$$\overset{k_1+1}{\underset{j=1}{\oplus}} \varphi_j\left(\overset{k_2}{\underset{i=1}{\oplus}} \rho_i F_{e_{ij}}\right) = \left\langle \left[\left(1 - \log_\beta\left(1 + \wp_j\left(\mathbb{P}_i\left(\beta^{1-a_{ij}^-} - 1\right)^{\rho_i}\right)^{\varphi_j}\right)\right), \right.\right.$$

$$\left. 1 - \log_\beta\left(1 + \wp_j\left(\mathbb{P}_i\left(\beta^{1-a_{ij}^+} - 1\right)^{\rho_i}\right)^{\varphi_j}\right)\right) \right],$$

$$1 - \log_\beta\left(1 + \wp_j\left(\mathbb{P}_i\left(\beta^{1-b_{ij}} - 1\right)^{\rho_i}\right)^{\varphi_j}\right),$$

$$\left(1 - \log_\beta\left(1 + \wp_j\left(\mathbb{P}_i\left(\beta^{1-c_{ij}^-} - 1\right)^{\rho_i}\right)^{\varphi_j}\right)\right),$$

$$\left. 1 - \log_\beta\left(1 + \wp_j\left(\mathbb{P}_i\left(\beta^{1-c_{ij}^+} - 1\right)^{\rho_i}\right)^{\varphi_j}\right)\right) \right] \right\rangle$$

And

$$\overset{k_1}{\underset{j=1}{\oplus}} \varphi_j \left(\overset{k_2+1}{\underset{i=1}{\oplus}} \rho_i F_{e_{ij}} \right) = \left\langle \left[\left(1 - \log_\beta \left(1 + \wp_j \left(\mathbb{P}_i \left(\beta^{1-a_{ij}^-} - 1 \right)^{\rho_i} \right)^{\varphi_j} \right) \right), \right.\right.$$
$$1 - \log_\beta \left(1 + \wp_j \left(\mathbb{P}_i \left(\beta^{1-a_{ij}^+} - 1 \right)^{\rho_i} \right)^{\varphi_j} \right) \right],$$
$$1 - \log_\beta \left(1 + \wp_j \left(\mathbb{P}_i \left(\beta^{1-b_{ij}} - 1 \right)^{\rho_i} \right)^{\varphi_j} \right),$$
$$\left(1 - \log_\beta \left(1 + \wp_j \left(\mathbb{P}_i \left(\beta^{1-c_{ij}^-} - 1 \right)^{\rho_i} \right)^{\varphi_j} \right), \right.$$
$$\left.\left. 1 - \log_\beta \left(1 + \wp_j \left(\mathbb{P}_i \left(\beta^{1-c_{ij}^+} - 1 \right)^{\rho_i} \right)^{\varphi_j} \right) \right) \right] \right\rangle$$

Therefore, it is hold for $m = k_1 + 1, n = k_2 + 1$. Hence, by method of induction, the result is true for all the values of $m, n \geq 1$.

2.7 Proposed Methodology for Decision-Making Process

Step 1. Gather the information associated with all the alternatives under various criterions/parameters in the form of Interval Valued Triangular Fuzzy Soft Matrix (IVTFSM) is SD and defined

by $SD = \left\langle \left(a_{ij}^-, a_{ij}^+ \right), b_{ij}, \left(c_{ij}^-, c_{ij}^+ \right) \right\rangle_{n \times m}$

Step 2. Normalize the matrix SD by converting the grade values into benefit type of the cost parameters by applying the following formula.

$$R_{ij} = \begin{cases} F_{e_{ij}}, & \text{for benefit type parameters} \\ F_{e_{ij}}^c, & \text{for cost type parameters} \end{cases} \tag{7}$$

where $F_{e_{ij}}^c = \left\langle \left(1 - c_{ij}^-, 1 - c_{ij}^+ \right), 1 - b_{ij}, \left(1 - a_{ij}^-, 1 - a_{ij}^+ \right) \right\rangle$ is the complement of $F_{e_{ij}}$.

If all the parameters are of the same type, then normalization is not necessary.

Step 3. Aggregate the IVTFSNs $F_{e_{ij}}$ for all the alternatives $A_k (k = 1, 2, \ldots, u)$ into collective decision matrix ξ_k using proposed IVTFSWA operator.

Step 4. Calculate the score value of all the alternatives $A_k (k = 1, 2, \ldots, u)$.

Step 5. Using score value of the alternatives, select the best one.

Step 6. End.

2.7.1 Decision-Making Problem

By considering a practical example from [20], the above decision-making process has been illustrated for medical diagnosis. The board of four doctors d_1, d_2, d_3, d_4 whose weight vector is $\rho = (0.3, 0.2, 0.1, 0.4)^T$ will award their preference values for four patients A_1, A_2, A_3, A_4 under some criterions/parameters $E = \{$ "Temperature (e_1), Headache (e_2), Stomach Pain (e_3), Chest Pain $(e_4)\}$ with the weight vector $\varphi = (0.35, 0.25, 0.25, 0.15)^T$. The result obtained using the methodology is as follows.

Step 1. The four doctors d_i will measure the illness of the four patients in terms of IVTFSNs. Parameters and their values of rating are summarized in Tables 1, 2, 3 and 4.

Step 2. Here every parameter is of the same type; hence, it is not necessary for normalization.

Step 3. The different opinions of the doctors for every patient $A_k, k = 1, 2, 3, 4$ are aggregated by using the Eq. (6). The aggregated values for the four patients are

$$\xi_1 = \langle (0.4354, 0.5392), 0.6247, (0.6951, 0.8049) \rangle$$
$$\xi_2 = \langle (0.3974, 0.5010), 0.5989, (0.6940, 0.8059) \rangle$$
$$\xi_3 = \langle (0.3439, 0.4477), 0.5400, (0.6415, 0.7534) \rangle$$
$$\xi_4 = \langle (0.4421, 0.5455), 0.5972, (0.6500, 0.7494) \rangle$$

Step 4. The score values are $SV(\xi_1) = 1.0073$, $SV(\xi_2) = 0.9597$, $SV(\xi_3) = 0.8414$, $SV(\xi_4) = 0.9527$ and the ranking is $SV(\xi_1) > SV(\xi_2) > SV(\xi_4) > SV(\xi_3)$.

Step 5. Therefore, the patient, A_1 has more illness.

Table 1 Interval valued triangular FSM for the patient A_1

$\mathbb{U}$	e_1	e_2	e_3	e_4
d_1	$\langle (0.4, 0.5), 0.6, (0.7, 0.8) \rangle$	$\langle (0.5, 0.6), 0.7, (0.8, 0.9) \rangle$	$\langle (0.6, 0.7), 0.8, (0.8, 0.9) \rangle$	$\langle (0.6, 0.7), 0.7, (0.8, 0.9) \rangle$
d_2	$\langle (0.5, 0.6), 0.7, (0.8, 0.9) \rangle$	$\langle (0.5, 0.6), 0.7, (0.8, 0.9) \rangle$	$\langle (0.4, 0.5), 0.6, (0.6, 0.7) \rangle$	$\langle (0.6, 0.7), 0.8, (0.8, 0.9) \rangle$
d_3	$\langle (0.3, 0.4), 0.5, (0.5, 0.6) \rangle$	$\langle (0.4, 0.5), 0.6, (0.7, 0.8) \rangle$	$\langle (0.2, 0.3), 0.4, (0.5, 0.6) \rangle$	$\langle (0.3, 0.4), 0.5, (0.6, 0.7) \rangle$
d_4	$\langle (0.2, 0.3), 0.4, (0.5, 0.6) \rangle$	$\langle (0.6, 0.7), 0.7, (0.7, 0.8) \rangle$	$\langle (0.3, 0.4), 0.5, (0.6, 0.7) \rangle$	$\langle (0.4, 0.5), 0.6, (0.7, 0.8) \rangle$

Table 2 Interval valued triangular FSM for the patient A_2

$\mathbb{U}$	e_1	e_2	e_3	e_4
d_1	$\left\langle \begin{array}{c}(0.5,0.6),0.7,\\(0.8,0.9)\end{array}\right\rangle$	$\left\langle \begin{array}{c}(0.4,0.5),0.6,\\(0.7,0.8)\end{array}\right\rangle$	$\left\langle \begin{array}{c}(0.5,0.6),0.7,\\(0.8,0.9)\end{array}\right\rangle$	$\left\langle \begin{array}{c}(0.4,0.5),0.6,\\(0.7,0.8)\end{array}\right\rangle$
d_2	$\left\langle \begin{array}{c}(0.6,0.7),0.8,\\(0.8,0.9)\end{array}\right\rangle$	$\left\langle \begin{array}{c}(0.6,0.7),0.7,\\(0.8,0.9)\end{array}\right\rangle$	$\left\langle \begin{array}{c}(0.5,0.6),0.7,\\(0.8,0.9)\end{array}\right\rangle$	$\left\langle \begin{array}{c}(0.5,0.6),0.7,\\(0.8,0.9)\end{array}\right\rangle$
d_3	$\left\langle \begin{array}{c}(0.2,0.3),0.4,\\(0.5,0.6)\end{array}\right\rangle$	$\left\langle \begin{array}{c}(0.5,0.6),0.7,\\(0.8,0.9)\end{array}\right\rangle$	$\left\langle \begin{array}{c}(0.1,0.2),0.3,\\(0.4,0.5)\end{array}\right\rangle$	$\left\langle \begin{array}{c}(0.2,0.3),0.4,\\(0.5,0.6)\end{array}\right\rangle$
d_4	$\left\langle \begin{array}{c}(0.3,0.4),0.5,\\(0.6,0.7)\end{array}\right\rangle$	$\left\langle \begin{array}{c}(0.4,0.5),0.6,\\(0.7,0.8)\end{array}\right\rangle$	$\left\langle \begin{array}{c}(0.2,0.3),0.4,\\(0.5,0.6)\end{array}\right\rangle$	$\left\langle \begin{array}{c}(0.1,0.2),0.3,\\(0.4,0.5)\end{array}\right\rangle$

Table 3 Interval valued triangular FSM for the patient A_3

$\mathbb{U}$	e_1	e_2	e_3	e_4
d_1	$\left\langle \begin{array}{c}(0.4,0.5),0.6,\\(0.7,0.8)\end{array}\right\rangle$	$\left\langle \begin{array}{c}(0.5,0.6),0.7,\\(0.8,0.9)\end{array}\right\rangle$	$\left\langle \begin{array}{c}(0.4,0.5),0.6,\\(0.7,0.8)\end{array}\right\rangle$	$\left\langle \begin{array}{c}(0.5,0.6),0.7,\\(0.8,0.9)\end{array}\right\rangle$
d_2	$\left\langle \begin{array}{c}(0.3,0.4),0.5,\\(0.6,0.7)\end{array}\right\rangle$	$\left\langle \begin{array}{c}(0.6,0.7),0.7,\\(0.8,0.9)\end{array}\right\rangle$	$\left\langle \begin{array}{c}(0.4,0.5),0.6,\\(0.7,0.8)\end{array}\right\rangle$	$\left\langle \begin{array}{c}(0.6,0.7),0.7,\\(0.8,0.9)\end{array}\right\rangle$
d_3	$\left\langle \begin{array}{c}(0.1,0.2),0.3,\\(0.4,0.5)\end{array}\right\rangle$	$\left\langle \begin{array}{c}(0.6,0.7),0.8,\\(0.8,0.9)\end{array}\right\rangle$	$\left\langle \begin{array}{c}(0.3,0.4),0.5,\\(0.6,0.7)\end{array}\right\rangle$	$\left\langle \begin{array}{c}(0.4,0.5),0.6,\\(0.7,0.8)\end{array}\right\rangle$
d_4	$\left\langle \begin{array}{c}(0.2,0.3),0.4,\\(0.5,0.6)\end{array}\right\rangle$	$\left\langle \begin{array}{c}(0.2,0.3),0.4,\\(0.5,0.6)\end{array}\right\rangle$	$\left\langle \begin{array}{c}(0.1,0.2),0.3,\\(0.4,0.5)\end{array}\right\rangle$	$\left\langle \begin{array}{c}(0.3,0.4),0.5,\\(0.6,0.7)\end{array}\right\rangle$

Table 4 Interval valued triangular FSM for the patient A_4

$\mathbb{U}$	e_1	e_2	e_3	e_4
d_1	$\left\langle \begin{array}{c}(0.3,0.4),0.5,\\(0.5,0.6)\end{array}\right\rangle$	$\left\langle \begin{array}{c}(0.5,0.6),0.7,\\(0.8,0.9)\end{array}\right\rangle$	$\left\langle \begin{array}{c}(0.3,0.4),0.5,\\(0.6,0.7)\end{array}\right\rangle$	$\left\langle \begin{array}{c}(0.2,0.3),0.4,\\(0.5,0.6)\end{array}\right\rangle$
d_2	$\left\langle \begin{array}{c}(0.5,0.6),0.7,\\(0.8,0.9)\end{array}\right\rangle$	$\left\langle \begin{array}{c}(0.4,0.5),0.6,\\(0.6,0.7)\end{array}\right\rangle$	$\left\langle \begin{array}{c}(0.4,0.5),0.5,\\(0.5,0.6)\end{array}\right\rangle$	$\left\langle \begin{array}{c}(0.3,0.4),0.5,\\(0.6,0.8)\end{array}\right\rangle$
d_3	$\left\langle \begin{array}{c}(0.4,0.5),0.6,\\(0.6,0.7)\end{array}\right\rangle$	$\left\langle \begin{array}{c}(0.7,0.8),0.8,\\(0.8,0.9)\end{array}\right\rangle$	$\left\langle \begin{array}{c}(0.6,0.7),0.7,\\(0.8,0.9)\end{array}\right\rangle$	$\left\langle \begin{array}{c}(0.1,0.2),0.3,\\(0.4,0.5)\end{array}\right\rangle$
d_4	$\left\langle \begin{array}{c}(0.5,0.6),0.6,\\(0.7,0.8)\end{array}\right\rangle$	$\left\langle \begin{array}{c}(0.6,0.7),0.7,\\(0.7,0.8)\end{array}\right\rangle$	$\left\langle \begin{array}{c}(0.5,0.6),0.6,\\(0.7,0.8)\end{array}\right\rangle$	$\left\langle \begin{array}{c}(0.3,0.4),0.5,\\(0.5,0.7)\end{array}\right\rangle$

2.7.2 Comparative Analysis

In [19], decision making is done to find the patient with more illness using intuitionistic fuzzy soft weighted arithmetic operator. As the expert's decision may be in terms of interval values, the existing method is not able to deal with interval data. The methodology proposed in this paper is able to sort out this issue, and hence, the proposed method can be applied and deal more uncertainties in the decision-making process.

2.8 Conclusion

Decision making is an essential task in Science and Engineering. Since most of the real-world problems have uncertainty in nature, making decision is challengeable one for the decision-makers. At this junction, fuzzy sets provide mathematical tool which handles the uncertainty in an efficient way. In dealing with impreciseness/uncertainty, Interval Valued FSs perform well than Type-1 FSs. Due to this reason, IVTFSS has been considered in this paper and proposed aggregation operator IVTFSWA for IVTFSNs using Frank triangular norms. The proposed operator is applied in decision-making process to identify the patient who has more illness as the medical diagnosis which proves the validity of the proposed operator. Also comparative analysis has been done with existing method. In future, more aggregation operators can be obtained under various fuzzy, soft fuzzy environments.

Here FSM is Fuzzy Soft Matrix.

References

1. Qin, J., Liu, X.: Frank aggregation operators for triangular interval type-2 fuzzy set and its application in multiple attribute group decision making. J. Appl. Math, 1–24 (2014)
2. Rajarajeswari, P., Dhanalakshmi, P.: Interval-valued intuitionistic fuzzy soft matrix theory. Int. J. Math. Arch. 5(1), 152–161 (2014)
3. Tripathy, B.K., Sooraj, T.R.: On interval valued fuzzy soft sets and their application in group decision making. In: National Conference on Analysis and Applications, in Project Decision Making and Interval Valued Intuitionistic Fuzzy Soft Set, pp. 1–5 (2015)
4. Khalil, M., Zhang, H., Xiong, L., Ma, W.: Corrigendum to on interval-valued hesitant fuzzy soft sets. Math. Prob. Eng. 1–3 (2016)
5. Sudharsan, S., Ezhilmaran, D.: Weighted arithmetic average operator based on interval-valued intuitionistic fuzzy values and their application to multi criteria decision making for investment. J. Inf. Opt. Sci. 37(2), 247–260 (2016)
6. Mohd, W.R.W., Abdullah, L.: Aggregation methods in group decision making: a decade survey. Informatica 41, 71–86 (2017)
7. Shenbagavalli, R., Balasubramanian, G., Solairaju, A.: Attributes weight determination for fuzzy soft multiple attribute group decision making problems. Int. J. Stat. Syst. 12(3), 517–524 (2017)

8. Selvachandran, G., Sallah, A.R.: Interval-valued complex fuzzy soft sets. In: International Conference on Mathematical Sciences, pp. 3–8 (2017)
9. Hermandez, P., Cubillo, S., Blanc, C.T., Guerrero, J.A.: New order on type-2 fuzzy numbers. Axioms. 6(22), 1–19 (2017)
10. Castillo, O., Sanchez, M.A., Gonzalez, C.I., Martinez, G.E.: Review of recent type-2, fuzzy image processing applications. Information 8(97), 1–18 (2017)
11. Dinagar, D.S., Rajesh, A.: A new approach on aggregation of interval-valued fuzzy soft matrix and its applications in MCDM. Int. J. Math. Arch. 9(6), 16–22 (2018)
12. Qin, Y., Liu, Y., Liu, J.: A novel method for interval-value intuitionistic fuzzy multicriteria decision-making problems with immediate probabilities based on OWA distance operators. Math. Prob. Eng. 1–11 (2018)
13. Nazra, A., Syafruddin, Wicaksono G.C., Syafwan, M.: A study on generalized hesitant intuitionistic fuzzy soft sets. In: International Conference on Mathematics, Science and Education, pp. 1–6 (2017)
14. Alcantud, J.C.R., Torrecillas, M.J.M.: Intertemporal choice of fuzzy soft sets. Symmetry 10(371), 1–18 (2018)
15. Selvachandran, G., Singh, P.K.: Interval-valued complex fuzzy soft set and its application. Int. J. Uncertainty Quant. 8(2), 101–117 (2018)
16. Dayan, F., Zulqarnain, M.: On generalized interval valued fuzzy soft matrices. Am. J. Math. Comput. Model. 3(1), 1–9 (2018)
17. Mahmooda, T., Alib, M.I., Malika, M.A., Ahmeda, W.: On lattice ordered intuitionistic fuzzy soft sets. Int. J. Alg Stat. 7(1–2), 46–61 (2018)
18. Arora, R., Garg, H.: Prioritized averaging/ geometric aggregation operators under the intuitionistic fuzzy soft set environment. Scientia Iranica. 25(1), 466–482 (2018)
19. Arora, R., Garg, H.: Robust aggregation operators for multi-criteria decision-making with intuitionistic fuzzy soft set environment. Scientia Irancia. 25(2), 931–942 (2018)
20. Nagarajan, D., Lathamaheswari, M., Broumi, S., Kavikumar, J.: A new perspective on traffic control management using triangular interval type-2 fuzzy sets and interval neutrosophic sets. Oper. Res. Perspect. (2019). https://doi.org/10.1016/j.orp.2019.100099
21. Nagarajan, D., Lathamaheswari, M., Sujatha, R., Kavikumar, J.: Edge detection on DIOM image using triangular norms in type-2 fuzzy. Int. J. Adv. Comput. Sci. Appl. 9(11), 462–475 (2018)
22. Nagarajan, D., Lathamaheswari, M., Kavikumar, J., Hamzha.: A type-2 fuzzy in image extraction for DICOM image. Int. J. Adv. Comput. Sci Appl. 9(12), 352–362 (2018)

Handwriting Recognition Using Active Contour

Pinaki Saha and Aashi Jaiswal

Abstract Handwriting recognition is the process of recognizing handwritten or printed alphabets in various documents and even old manuscript. This method also helps in preserving the texts and writings of ancient times. This paper aims to represent the work related to recognition of cursive handwriting of different languages. Cursive handwritings are connected to each other, hence segmentation is required. Segmentation is used for extraction: line segmentation to extract sentence, word segmentation to extract words and character segmentation to extract individual letters. The segmentation involves dividing based on contours by setting different kernel size. For classification, we have used classifiers—convolution neural networks. We carried our experiment using datasets collected from E-MNIST, UCI. The experimental accuracy for the E-MNIST dataset is 79.3% and for UCI Devanagari dataset is 93%.

Keywords CNN · Segmentation · Contour · Handwriting recognition · Character detection · Pre-processing

1 Introduction

Handwriting recognition deals with paper documents, photographs and other devices which contain text written in any kind of language. The algorithm used to recognize such text uses the computer's ability to interpret the language written in such documents. The text collected from the sources is passed on to the various steps of the algorithm used in the handwriting recognition. pre-processing → line segmentation → word segmentation → character segmentation → classification

P. Saha (✉) · A. Jaiswal
Computer Science and Engineering, SRM Institute of Science and Technology, Chennai, India
e-mail: psmailpinaki@gmail.com

A. Jaiswal
e-mail: aashijaiswal2109@gmail.com

© Springer Nature Singapore Pte Ltd. 2020
S. S. Dash et al. (eds.), *Artificial Intelligence and Evolutionary Computations in Engineering Systems*, Advances in Intelligent Systems and Computing 1056, https://doi.org/10.1007/978-981-15-0199-9_43

and recognition. In any data analysis, the quality of the data is the most important factor to be considered. Improper data may produce misleading results. Under such circumstances, it is very important to remove such irrelevant and unreliable data. This filtering process of data is termed as pre-processing. Separation of one line from another is termed as Line Segmentation. On applying this step, individual line is passed one by one to the next step. A single line which is passed into the system executes the next step assigned to it. Each word of the line is separated from another word. This word goes to the next step of the process.

Each character of the word is separated from one another. This is the last step of the segmentation process which gives individual characters from the text passed on.

CNN is used for the classification purpose. It is the most efficient algorithm for analysing the image. Multilayer perceptron is used while implementing this algorithm. It is a class of artificial neural network called as feed-forward artificial neural network. This MLP consist of nodes divided into three layers named as input layer, hidden layer and a output layer. As compared to other classification technique, the main advantage of CNN is it uses less pre-processing. The CNN architecture looks like a box with various layers accepting the input and gives classified images as its output. Among the various layers present, each layer is one convolutional layer. There are multiple convolution layers that are put together in a network.

2   Related Works

(a) Handwriting Recognition Databases

The Extended-MNIST (E-MNIST) Dataset [11]

It is a set of handwritten character digits derived from the MNIST Special Dataset and converted to a 28×28 pixel image format. The E-MNIST letters contain 145,600 characters and 26 balanced classes. The E-MNIST digit contains 280,000 characters and ten balanced classes.

UCI Dataset [12]

For handwritten, Devanagari characters UCI dataset is used. There are 46 classes of characters with 2000 examples each. The dataset is split into training set (85%) and testing set (15%). The images are of 32×32 resolutions and are of.png format.

Kaggle [13]

Russian handwritten letters-This contains an image database of the Russian handwritten characters. There are a total of 14,190 instances for a total of 33 classes. The images are of 32×32 resolutions.

(b) Handwriting Recognition and Extraction Techniques

El Bahi et al. [1] make an attempt to identify handwritten characters by extraction and classification processes. It has employed various methods like grey level co-occurrence matrix, Zernike moments, Gabor filters, zoning, projection histogram and distance profile for the extraction purpose. This paper mainly tries to identify characters from image that are clicked from camera phone. The result of this paper has shown multilayer perceptron with Gabor filters and zoning gives the highest accuracy of 97.43%.

Dhande et al. [2] attempt to identify cursive English handwriting. It first attempts a segmentation process as the characters are joined. The extraction process and classification methods are then performed. Horizontal and vertical projections are used for the segmentation process; convex-hull is used for the extraction purpose.

Mukherjee et al. [3] aim to extract the characters from the script by representing a online word sample to an online and offline features. Analysing the offline feature is continued by the extraction of the core region.

Izumi' et al. [4] attempted to obtain a low computational cost by using auto-correction matrix and Fourier expansion together. It uses simple linear regression model to decide the weight-coefficient. Karhunen–Loeve expansion and Linear regression is also used and mentioned in this paper.

Arica et al. [5] attempt to remove maximum errors produced by segmentation and HMM by using efficient graph-search algorithm. For offline character recognition: firstly, parameters which are having the global nature such as height, width are estimated. Secondly, for combining greyscale and binary information, segmentation is used. Thirdly, for shape recognition Hidden Markov Model (HMM) is used. Graph optimization problem is combined with lexicon information and Hidden Markov Model in order to get word-level recognition. This paper has given maximum recognition rates using these algorithms.

Kotak et al. [6] deal with to develop a suitable algorithm to detect strokes and improve the efficiency of character recognition. Many grammatical errors, misplacement of punctuation marks, spelling mistakes occur which are needed to be recognized by non-human intervention. Hence, a real-time-based system for character recognition has been developed. A digital pen carrying accelerometer recognizes the letter with the accuracy of 84%. This efficiency can be increased by using nearest mapping algorithm which makes the efficiency up to 96%.

Zhu et al. [7] attempt to recognize the online handwritten English characters using segmentation-free Markov random field (MRF) and pseudo 2D bi-moment normalization (P2DBMN). It uses two types of features: unary features and binary features. Each character is represented as an MRF and by using tree-lexicon, MRF characters are combined for recognition. A P2DBMN-MQDF recognizer is also combined along with other algorithms to recognize characters especially Chinese and Japanese text. This paper made the segmentation-free recognition possible using MRF model. A high efficiency has been observed during Japanese character recognition.

Lee et al. [8] combine a letter-spotting technique with an island-driven lattice search algorithm for recognition. Markov model is used for spotting the letter of the script. An island-driven lattice search algorithm searches the possible word in the lattice. This method gives 85.4% accuracy in recognition. The Markov model can be used further to improve the efficiency while during further research.

Han and Sethi [9] attempt to use segmentation in different approach. A set of heuristic rules to determine the letters of the image captured has been used. Geometric and topological features of the letter have certain associations determined which helps to set the heuristic rule. A segmentation system with this approach is used for recognition. Pre-processing and normalization steps are used for character extraction from the image. This paper can accurately recognize the letter boundaries which can be used for future references.

Dhande et al. [10] aim to recognize the cursive English handwritten characters. For text-line segmentation horizontal projection method is used. For word segmentation, vertical projection histogram method is used. For classification, SVM is used. For feature extraction convex-hull algorithm is used. This work is inclined more towards the medical purpose for recognizing the text written in the medical prescription. High accuracy of 95% is achieved. Future work includes to get the more accurate result in the noisy environment.

3 Proposed System for Identification and Classification of Characters Using Contour Method

The system includes the process of reading an image which is in from of a paragraph or a manuscript. Since the input image is an entire paragraph, the image is segmented multiple times to extract the characters and then classified. The process contains a pre-processing phase and then segmentation techniques and then finally, followed by classification stage. In the following discussion, we have used mainly three datasets: E-MNIST dataset, the Russian alphabet dataset and the Devanagari dataset. The E-MNIST contains images of digits as well as English alphabets which are classified into 47 classes as all the alphabets are not present in the dataset. The images are of 28×28 dimensions. The Hindi Devanagari UCI dataset contains Hindi handwritten images of numbers as well as the alphabets totalling into 46 classes where each class is having 2000 samples. The images are also of 28×28 dimensions. The Kaggle Russian alphabet dataset contains all the Russian alphabets and is classified into different classes. This dataset is available in csv format and the images of the dataset can be by plotting them using matplotlib module.

Pre-processing Phase

The pre-processing phase mainly includes the reading of the image and then converting it into grayscale. By converting to grayscale, it is easier and more convenient to make further processing. The image can also be smoothened using various filters like the Gaussian algorithm filter. The image is then reshaped before it is used for segmentation.

Line Segmentation

The image contains lines which is first separated and then saved in the disk or in an array. The image is first dilated to join the characters so that it is easy to find the contours. The image can also be eroded, but it gives better results with dilation. A kernel size of (5, 20) is used as lines are big in length and short in height. The lines that are segmented are then sorted along y-axis so that the first line is saved first, then the second and so on. A numpy array of ones is created of the kernel size and then dilated to extract the lines as shown in Fig. 1.

Word Segmentation

The segmented line is then read one by one and similar process to segment word is used. The kernel size to find active contour is set to (5, 5). This kernel size will fragment out average size words. The process of dilation or erosion is also implemented in this stage. The contours are also sorted in along x-axis so that words

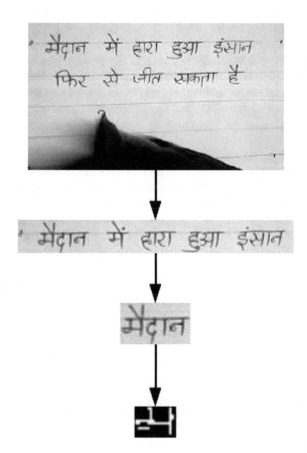

Fig. 1 Devanagari characters segmentation

are saved in the order that appear in the sentence. The contours detected are then used to find a rectangle that shows each of the segmented words. The coordinates of this rectangle are then used to extract the word from the line image (Fig. 1).

The images of words are then stored in the drive in a particular folder or can even be stored in a multi-dimensional array.

Character Segmentation

The segmented word images are then read one by one and the process of segmenting the character is done in similar process by setting the kernel size to (3, 3). The average size of characters is mostly separated using this kernel size. The images are then dilated or eroded and then contours are found which is then sorted along *x*-axis to keep them in order. The contours are then used to find the rectangle similar to word segmentation which will bound each character in the word. The coordinates of the rectangle are then used fragment the character images from the word and then stored in a multi-dimensional array. Figures 1 and 2 shows the segmented characters from the words. The final characters that are segmented are saved in as binary inverted condition, so that each character can be predicted easily. The characters are written in white with lack background that is similar to the training E-MNIST dataset.

This processing of image is for test image. The process uses same algorithm for different segmentation but only uses different kernel size while finding the contours. After the characters are segmented, each of these is then predicted by the convolutional neural network classifier, and then displayed together.

Classification

The classification makes use of convolutional neural network (CNN). The network starts with a convolutional layer with layer size of 64, filter kernel size of (3, 3) and strides of 2. This convolutional layer has relu as an activation function. This is followed by a maxPooling2D layer of pooling size (2, 2) without any padding and this reduces the image array into a smaller-dimensional array. This is again followed by a convolutional network of filter size (3, 3) and a same maxPooling2D

Fig. 2 English character segmentation

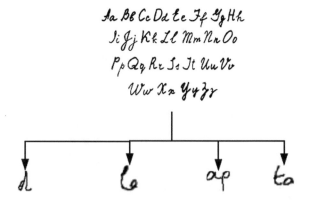

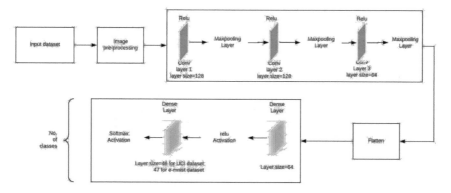

Fig. 3 Model of proposed system

layer having the same relu activation in the convolution layer. The convolutional and maxPooling layer of same kernel size, layer size and same activation function is used one more time. This is then followed by a flatten layer which in fact flattens all the layers before it is fed into a dense layer of layer size 64. This is again followed by a flatten layer which is then finally fed into the dense layer with 47 layer size for classifying into 47 different classes. Figure 3 shows the entire proposed model representing all the layers in the classifier as well.

This overall model is then compiled using a sparse categorical crossentropy loss function and a Adam optimizer. The convolutional neural network layer can be also be used after padding to keep the resultant matrix in the same dimensions. The learning rate is set to 0.01 which is a default learning rate for Adam optimizer.

4 Implementation

The above technique is implemented in Jupyter notebook using Python language with the help of OpenCV. The OpenCV is basically used to implement the findContour function and segment according to the contour. TensorFlow and Keras API are also used to define the classifier model. Numpy is used to create numpy ndarray for handling the images. The images are converted to numpy images. The hardware used in this is i7-5500 intel processor as CPU having 2.40 GHz and having 4 GB RAM. The device used also had Nvidia GeForce Graphics card but graphics card was not used for training the model. Using graphics card increases the speed and the training time will be much less, thus better the hardware better and more efficient will be the results.

On applying the segmentation process on the UCI Devanagari dataset, the lines are segmented properly in most of the cases, but input image should be nearly to 50 × 50 resolution images. If the resolution varies from this, the segmentation process will not be as efficient. In case of splitting Devanagari characters which contain "shirorekha", i.e. a line on every word, this line needs to be removed first in particular so that further segmentation can be done. After the line segmentation and word segmentation the "shirorekha" is removed. For our convenience, all the part above the "shirorekha" is removed. After this the image is saved in binary inverted form, i.e. the writings are in white in a black background.

In case of English characters in E-MNIST dataset, the three segmentation process shows good efficiency but in case of character segmentation, few of the characters are segmented together as the letters are overlapped or are very close to each other as shown in step 2 of figure …

5   Result and Discussion

On classifier discussed in Fig. 3 is trained using both the UCI Devanagri dataset as well as E-MNIST dataset separately. The E-MNIST contains 60,000 images each of 28 × 28 dimensions. These images are then passed to through the different layers of the classifier to train it. While training the dataset, it is split into train dataset and validation dataset. The validation dataset is 0.1% of the training dataset. After training for 5 ephocs, a training accuracy of 81.5% is obtained and validation accuracy of 58% is obtained. The UCI Devanagari dataset contains 118,500 images each of 32 × 32 dimensions. These images are then trained in the same classifier model by splitting the dataset into training and validation dataset. The validation dataset contains 0.1% of the dataset. It gives a training accuracy of 95% and validation accuracy of 96%.

Figure 4 shows a graph to show the relation between accuracy and with no. of epochs for both the UCI dataset as well as E-MNIST dataset. According to the

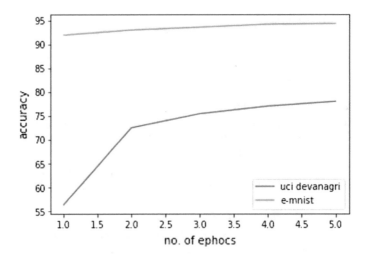

Fig. 4 Graph between no. of epochs versus accuracy

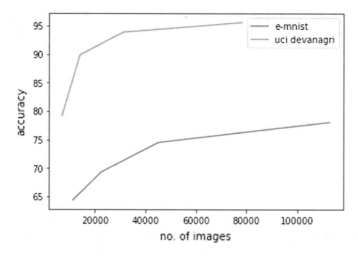

Fig. 5 Graph between no. of images versus accuracy

graph, the accuracy is seen to increase every epochs both for E-MNIST dataset as well as UCI dataset. But the increase in accuracy is more for UCI dataset than for E-MNIST dataset. The various data points are highlighted at shown by an asterisk on the graph line.

Figure 5 shows a graph showing relation between no. of images and accuracy. The model, each trained each time with different no. of images and the accuracy is noted. Two different lines show the relation, one for E-MNIST dataset and second, the UCI dataset. The various data points are shown by asterisk on the lines in the graph. According to the image, the accuracy is more obtained for the UCI Devanagari dataset as compared to the E-MNIST dataset.

6 Conclusion

In this paper, a new classification model with CNN is implemented as well as image segmentation technique for English and Devanagari characters is discussed and implemented. The difference in the training accuracy is because of the difference in quality of the two datasets. So the model proves to be good for UCI Devanagari dataset as it gives a training accuracy of 93% and validation accuracy of 95%. This model gives a training accuracy of 79% and validation accuracy of 81% which is a moderate performance for this dataset. In both the cases, the accuracy is seen to increase with number of epochs and thus the model does not overfit. In case of overfitting, the accuracy would be seen falling down and loss increasing in every ephoc. The number of ephocs can be increased if accuracy still increases with more increase in number of epochs. The segmentation technique works well under few constraints; first, the dimensions of the input image should be close to 150×300

resolution pixels, second in case of Hindi, input image all the letters should be nearly of the same height, the characters should be clear and distinguishable properly. On removing the "shirorekha", we also remove any marks on the top. If there are any lower that is not removed as it does not influence in predicting the character. Predicting the binary inverted image that is obtained on character segmentation gives a better performance. The character segmentation does not work well if the "shirorekha" is not removed; all the characters are then not segmented. On removing this line, the characters get segmented for the same image.

References

1. El Bahi, H., Mahani, Z., Zatni, A.: An offline handwritten character recognition system for image obtained by camera phone. Université Ibn Zohr, ESTA, Laboratoire Matériaux, Systèmes et Technologies de l'information
2. Dhande, P., Kharat, R.: Recognition of cursive english handwritten characters. In: International Conference on Trends in Electronics and Informatics (2017)
3. Mukherjee, P.S., Bhattacharyav, U., Parui, S.: Segmentation free recognition. In: 2018 13th IAPR International Workshop on Document Analysis Systems
4. Izumi', T., Hattorit, T., Kitajimat, H., Yamasak, T.: Characters recognition method based on vector field and simple linear regression model. In: International Symposium on Communications and Information Technologies 2004 (ISCIT 200.1), Sapporo, Japan, 26–29 Oct 2004
5. Arica, N., Yarman-Vural, F.T.: Optical character recognition for cursive handwriting. IEEE Trans. Pattern Anal. Mach. Intell. **24**(6) (2002)
6. Kotak, N.A., Roy, A.K.: An accelerometer based handwriting recognition of English alphabets using basic strokes. In: Proceedings of the 2017 IEEE Region 10 Conference (TENCON), Malaysia, 5–8 Nov 2017
7. Zhu, B., Shivram, A., Setlur, S., Govindaraju, V., Nakagawa, M.: Online handwritten cursive word recognition using segmentation-free MRF in combination with P2DBMN-MQDF. In: 2013 12th International Conference on Document Analysis and Recognition
8. Lee, S.-H., Lee, H., Kim, J.H.: On-line cursive script recognition, using an Island-driven search technique. KAIST Center for Artificial Intelligence Research, The Engineering Research Center of Excellence
9. Han, K., Sethi, I.K.: Off-line cursive handwriting segmentation. Vision and Neural Networks Laboratory. Department of Computer Science, Wayne State University, Detroit, MI 48202
10. Dhande, P.S., Kharat, R.: Character recognition for cursive English handwriting to recognize medicine name from doctor's prescription. In: 2017 Third International Conference on Computing, Communication, Control and Automation (ICCUBEA)
11. EMNIST DATASET-https://www.nist.gov/itl/iad/image-group/emnist-dataset
12. DEVANAGARI DATASET-https://archive.ics.uci.edu/ml/datasets/Devanagari+Handwritten+Character+Dataset
13. KAGGLE-https://www.kaggle.com/olgabelitskaya/classification-of-handwritten-letters

Design and Analysis of Customized Hip Implant Using Finite Element Method

M. Anu Priya, U. Snekhalatha, R. Mahalakshmi, T. Dhivya
and Nilkantha Gupta

Abstract In total hip replacement, the damaged bone and the cartilage are removed and replaced with the prosthetic implant. Though various innovations have come with implant sterilization, designing, fixation and robotic surgery, the long-term issue is to find an optimum patient-specific hip implant which suits the patient's requirement. The aim and objective of the proposed study was to design a highly accurate patient-specific hip implant by standardizing the existing design and to prove that customized design is better than the conventional design. The geometric measurements of hip were performed in hip CT image using MIMICS 20.0 software. The implant design was carried out using Solid-works. The designed implant is meshed and analysed using FEA. From the comparative study of the FEA analysis, it was proved that customized implant was best fit for the patients than the conventional implant.

Keywords Total hip arthroplasty · Patient-specific implant · Conventional implant measurements · Meshing · Finite element analysis

1 Introduction

Hip arthroplasty or hip replacement surgery is an operative procedure wherein the surgeons remove the damaged hip (either total hip or a part of hip) and replace it with a commercially available implant. These implants are globally imported from various countries like North America, Europe, South America, Middle East and Africa. The hip replacement surgery has shown tremendous evolution since 1840.

M. Anu Priya · U. Snekhalatha (✉) · R. Mahalakshmi · T. Dhivya
Department of Biomedical Engineering, Faculty of Engineering and technology,
SRM Institute of Science and Technology, Chennai, India
e-mail: sneha_samuma@yahoo.co.in

N. Gupta
Centre for Environmental Nuclear Research, SRM Institute of Science and Technology,
Chennai, India

© Springer Nature Singapore Pte Ltd. 2020   515
S. S. Dash et al. (eds.), *Artificial Intelligence and Evolutionary Computations in Engineering Systems*, Advances in Intelligent Systems and Computing 1056,
https://doi.org/10.1007/978-981-15-0199-9_44

This kind of complicated surgery could be incorporated in patients during the failure of intake of medications, topical ointments and rehabilitation therapies. The study reveals that about 5.5 times the weight of the body is being barred by pelvis and femur during day-to-day activities [1].

The various diseases which cause the replacement of hip include injuries (tendinitis, inguinal hernia, fracture or dislocation, hip labral tear, bursitis); cancer (bone cancer, leukaemia); arthritis (septic, rheumatoid, osteo, psoriatic); osteoporosis; sciatica; osteonecrosis; osteomyelitis; synovitis; and sacroiliitis. The four different types of fracture patterns are (i) femoral head fracture, involving the femoral head, (ii) femoral neck fracture, (iii) subtrochanteric fracture involving the shaft of the femur immediately below the lesser trochanter and (iv) due to diseases such as osteoporosis [2]. Post-operative surgeries were mainly due to aseptic loosening, dislocation and infections around the implant. The different types of prosthesis used in total hip replacement are cemented, uncemented, hybrid and resurfacing. The uncemented prosthesis is highly used when compared to cemented prosthesis. High survival rate was existing only in the prosthesis with metal-on-metal resurfacing than cemented/uncemented prosthesis.

The major need for the patient-specific implant is to minimize the risk and complications associated with the imported implants from other countries. These conventional implants are imported for Indian patients may vary due to different anatomical size and structure between various population across the world. Hence, patient-specific implants are preferred rather than the conventional implant because of perfect and accurate fitting of implant according to the size and structure of the patient. It was evident from frost et al. that the THR market was growing annually at a compound rate of 26.7% from 2010 to 2017 [3]. Almost 70,000 hip replacement surgeries have been done in India during the year 2011. The Indian society of hip surgeons has a data of about 34,478 patients having the record of hip replacement surgery from the year of 2005 to till date.

One of the major processes in patient-specific hip replacement surgery is pre-operative planning which is useful for eliminating the post-operative problems encountered after the surgery. The pre-operative study includes extracting the various parameters such as the study of acetabular anatomy, implant sizing, implant positioning, pre-operative measures of bone type and length of legs [4]. The successful pre-operative planning minimizes the risk of complications after the surgery. To perform the hip implant surgery, the patient's physiological conditions such as height, weight, age, blood pressure, temperature and oxygenation level were checked to ensure that the patient could undergo hip replacement surgical procedure [5]. The aim and objective of the proposed study was to design a highly accurate patient-specific hip implant by standardizing the existing design and to prove that customized design is better than the conventional design.

2 Literature Review

Jashun Ro et al. developed a hip prosthesis model in a proximal femur region to measure the stress involved in femoral neck. They calculated the stress using finite element method, and the range of motion of neck diameter was obtained using mathematical formulas. They proposed a linear equation to complete the optimum neck diameter with respect to the head diameter of femoral region [6].

Coigny et al. developed software for automated process of designing the patient-specific hip implant. Initially, they load the CT images, comparing the missing component with the normal standardized CT data. Their developed algorithm could able to modify the implant geometry, with respect to the size and position of implant. Finally, they displayed the 3D model of patient-specific implant which could be viewed from all the directions. They developed a web platform which could upload the design file, production of implant, status detection and delivery report [7].

Steen et al. established a system on 3D visualization for pre-operative planning of total hip replacement (THR) surgery. They proposed an application to display the conditions and system parameters that fit the implants by determining the distance between end-to-end parts of the femur bone. The pre-operative planning is essentially done in 2D X-ray images. They also compared the outcomes of developed hip model in both 2D and 3D templates. Their main objective for pre-operative planning in THR is to estimate the best fit patient-specific implant model; they developed an ideal model in association with spectra medical systems AB for 3D visualization of implant. Features like bone contours, implant shading and implant contour in multi-planar ray (MPR) view were executed to find the best fit of hip implant model [8].

Hussam et al. implemented various methods for structural analysis of orthopaedic implants. They compared the stress distributions of hip implant with respect to failure models of hip implant. To overcome the drawbacks of failure model of hip implant, a patient-specific hip implant was designed and executed. They use finite element method to analyse validity and accuracy of hip implant model. The material characterization and selection were also studied in stress comparison. This study exhibits loading in two types: statically and dynamically to analyse the hip implant longevity. In dynamic loading, material selection and prostheses fixation are also studied. The whole process was conducted by 3D femur cemented implant prostheses with loading condition using LS-DYNA3D software [9].

3 Methodology

3.1 Subjects

Computed tomography (CT) scan data of the hip joint is collected from a subject of age 34. The patient's CT data is fractured case wherein the right portion of hip is fractured and the left portion is normal. This CT data has 128 slices of image in Di-com format. From the CT data of the subject, 3D hip model is extracted using MIMICS. The measurements required for the implant design were done using 3D viewer. Simultaneously, the dimensions of conventional implant were acquired. Then, the dimensions of both customized and conventional implants were designed using solid-works. The designed implants are meshed and analysed for FEA using ANSYS. The overall workflow of this proposed study was given below in Fig. 1.

3.2 Image Analysis

In hip fractured CT image, the left portion is normal and right side is fractured, and hence, the normal left portion of the image cropped. The bone part of the image was segmented. The thresholding is done by enhancing the Hounsfield unit between 400 and 1000 HU. This method extracts the bone region from soft tissue and ligaments. The extracted bone image was smoothened in the following tools in a sequential manner. Dynamic region growing is performed after segmentation which is based

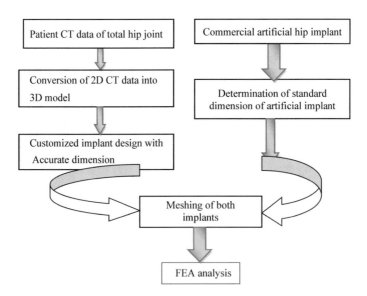

Fig. 1 Flow chart for the proposed work

on grey value connectivity. In the next step, spilt mask was used to separate the femur bone region from the pelvic girdle. The separated femur bone region is superimposed with the new mask image. Then, the smoothening operation was carried out on the new mask image.

3.3   Measurement of Parameters

The extracted 2D femur bone image was converted into 3D using mimics software. The measurements can be taken either directly from the 3D model or it can be taken from 3D viewer. In 3D viewer, the texture of the femur bone was enhanced, and the parameters measured are list as follows:

3.3.1   Femoral Head Diameter

The femoral head region is the highest part of femur bone. Since it is not in a shape of perfect sphere, the dimension may vary at each plane. Hence, diameter is measured across the largest width.

3.3.2   Canal Width

Canal width is the length of the femur bone decreasing from top to bottom. The width is measured from the level of lesser trochanter till the end of the shaft at four different points.

3.3.3   Shaft Angle

Shaft angle is the angle measured between the shaft axis and neck axis. The two standard angle values available in market are 127 and 132°.

3.3.4   Femoral Length

Femoral length is the maximal length of femur bone from top to bottom along the shaft axis. The length varies between 10 and 15 cm from the isthmus. Similarly, the same parameters can be measured in mimics software and can be compared with 3D viewer for accuracy. The parameters which are measured in mimics are shown in Fig. 2; the measurements taken from the materialized software is shown in Fig. 3.

Fig. 2 Parameters
considered for designing an
implant

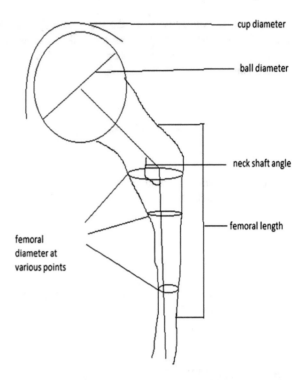

Fig. 3 Dimensions of femur
bone

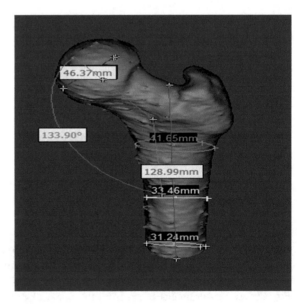

3.4　Implant Design

The design of implant was in done in Solid-works. The obtained measurements from mimics are used to design the hip implant. The design process is subdivided into five parts: acetabular cup, plastic liner, femoral head, femoral shaft and femoral length. The five portions of hip implant are designed separately. Initially, the design is done in a 2D front plane using commands like circle, hemisphere and line. The 2D design in the front plane is converted into 3D design by using the command extrude. The neck-shaft angle is the most important criterion in hip implant design as it varies from each person. When all the five parts of hip implant are designed, they can be combined as a single assembly.

3.5　Material Selection

Stainless steel and titanium alloy are the most commonly used biomaterials for the hip implant prosthesis. Titanium alloy has lighter weight and offers an advantage of better CT/MRI images after implantation, whereas stainless steel causes blurring of images. Modulus of elasticity for titanium is lesser than stainless steel. Lesser modulus of elasticity helps in reducing the possibility of stress shielding effect. It also has a good corrosion resistance property. But it has certain disadvantages like poor wear and low fracture toughness. Considering all the aspects, titanium is the good choice for hip implant prosthesis [10].

3.6　Analysis of Implant

The implant is analysed by meshing and finite element method. Meshing is defined as the process of converting a whole product into a number of elements for uniform load distribution. In meshing, each element has its own stiffness value of loading. Another aspect of meshing is accuracy. The finite element method is the numerical method for solving problem like structural analysis. To resolve the problem, it subdivides a large system into smaller units that are called finite elements. The method uses variational methods to estimate a solution by minimizing a related error function. By analysing the conventional and customized hip implants, we can prove that customized implant is best fit and is patient-specific implant. The deformed part on the right side of the pelvis CT bone is removed. The implant dimension for the right side was taken by mirroring the left side of pelvis CT image. From the acquired measurements, the implant was designed and fixed using materialized software. The finite element analysis was done using ANSYS software. The designed hip implant is meshed. By fixing the bottom of the femoral shaft, a load of 600 N is applied on the femoral head as shown in Fig. 5.

4 Result and Discussion

In this study, the dimensions of the femur bone were extracted from the CT image using materialized software as shown in Table 1.

The customized implant was designed with the parameters such as acetabular cup diameter, femoral ball diameter, neck-shaft angle, and length of the femur, measured from the left side of the femoral bone which is mirrored for right-side fractured bone measurements. The designed right hip implant is shown in Fig. 4. The designed patient-specific implant was fitted on to the right side of the fractured bone as shown in Fig. 4a, b.

The damaged pelvic bone and the prosthetic implant pelvic bone is shown in Fig. 5a, where the right portion of the hip region is fixed with patient-specific hip implant as seen in Fig. 5b.

The designed customized hip implant is analysed using meshing and finite element method. The meshed patient-specific hip implant is shown in Fig. 6. The FEA process was carried out using ANSYS Software by applying 600 N force on the top of the femoral head indicated by red shade. The magnitude acting on implant causes prosthesis stress field changes and instability, thus leading to implant failure [11]. At the bottom side, the femoral shaft is fixed which was indicated in blue colour shade.

The analysis was done based on the stress and displacement values for both fractured and normal bone of the hip. The displacement analysis of implanted and normal bone is shown in Fig. 7a, b. The stress analysis of the implanted and normal bone is shown in Fig. 7c, d.

Table 1 Comparison of customized and conventional implant dimensions

Type of implant	Acetabular cup diameter (mm)	Femoral head diameter (mm)	Neck shaft angle (°)	Femoral length (mm)
Conventional	56	50	127	140
Customized	54	48.71	136.25	131.36

Fig. 4 Designed with the customized right hip implant measurements **a** Enlarged image of the implanted right bone with bone shaded in blue colour **b** Implant fixed on the right side of hip joint with the left side being normal

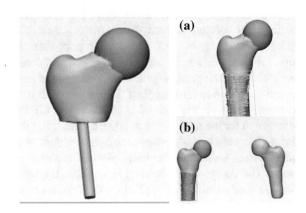

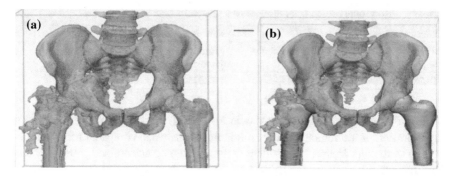

Fig. 5 **a** Image of right hip fractured pelvic bone before implant **b** Image of implanted pelvic bone in the right hip region

Fig. 6 Meshed right-side implanted bone with load applied

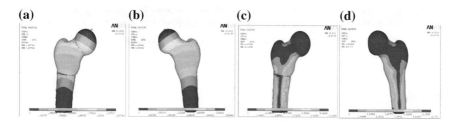

Fig. 7 **a** Displacement analysis of implanted bone **b** Displacement analysis of normal bone **c** Stress analysis of implanted bone **d** Stress analysis of normal bone

Table 2 FEA results of stress and displacement analysis

Stress analysis		Displacement analysis	
Right hip with implant	Left hip without implant	Right hip with implant	Left hip without implant
10.2894	12.775	0.267544	0.252464

From the comparative study as shown in Table 2, it is evident that the implanted right portion of hip region acquires minimum stress when compared to left normal region of the hip. Hence, it is evident that the designed customized patient-specific hip implant is best fit for the subject.

Kumar et al. performed the finite element analysis on cementless hip prosthesis [12]. They computed the von Mises stress and total deformation rate on both head and neck region of hip implant. They applied loads varying from 300 to 1800 N on the top of the femoral head. At the 600 N load, they obtained the stress value as 0.0086 at head region and 2.37 at neck region of the femoral bone. They calculated the total deformation rate at head and neck region as 0.34498 and 0.44366 respectively. In the proposed study, a constant load of 600 N is applied on the femoral head. The stress values obtained at the head and neck region are 0.009 and 2.28, respectively. The deformation rates obtained are 0.59454 and 0.267544 at head and neck region, respectively, which are in agreement with the findings of Kumar et al. The limitation of the proposed study practically needs more detailed validation before it is clinically implemented.

5 Conclusion

As mentioned, the study was carried out for designing a suitable sized hip implant specific for Indian patients. Patient-specific implant leads to lengthy pre-operative planning. Though the surgical duration is reduced, there is a need to trade-off between pre-operative planning and surgical duration. Patient-specific implant has the ability to accurately align implants to the kinematic axis of the patient. It was reported that the technology of patient-specific implant leads to the long-term survival of patient; hence, post-operative surgery was reduced.

References

1. Hodaei, M., Farhang, K., Maani, N.: A contact model for establishment of hip joint implant wear metrics. J. Biomed. Sci. Eng. **7**, 228–242 (2014)
2. Ganapathi, S., Malaikannu, T., Anandan, Kavitha: Musculoskeletal modeling of hip joint and fracture analysis for surgical planning using FEA. EJBI **9**(2), 27–36 (2013)
3. Pachore, J.A., Vaidya, S.V., Thakkar, C.J., Bhalodia, H.P., Wakankar, H.M.: ISHKS joint registry: a preliminary report. Indian J Orthop **47**, 505–509 (2013)

4. Knight , J.L., Atwater , R.D.: Preoperative planning for total hip arthroplasty. J. Arthroplasty **7**(Supplement), 403–409 (1992)
5. Siopack, J.S., Jergesen, H.E.: Total hip athroplasty. Med **162**, 243–249 (1995)
6. Ro, J., Kim, P., Shin, C.S.: Optimizing total hip replacement prosthesis design parameter for mechanical structural safety and mobility. Int. J. Precis. Eng. Manuf. **19**(1), 119–127 (2018)
7. Coigny, F., Todor, A., Rotaru, H., Schumacher, R., Schkommodau, E.: Patient-specific hip prostheses designed by surgeons. Curr. Dir. Biomed. Eng. **2**(1), 565–567 (2016)
8. Steen, A., Widegren, M.: 3D visualization of pre-operative planning for orthopaedic surgery. SIGRAD, pp. 1–8 (2013)
9. El-Shiekh, H.E.D.F.: Finite Element Simulation of Hip Joint Replacement Under Static and Dynamic Loading. School of Mechanical and Manufacturing Engineering, Dublin City University, March 2002
10. Maji, P.K., Banerjee, P.S., Sinha, A.: Designing Suitable Sizes of Hip Implants for Indian Patients. **17**(4), 335–42 (2007)
11. Colica, K., Sedmakb, A., Grbovicb, A., Tatica, U., Sedmaka, S., Djordjevica, B.: Finite element modeling of hip implant static loading. Katarina Procedia Eng. **149**, 257–262 (2016)
12. Bubesh Kumar, D., Muthurajan, K.G.: Finite element analysis of equivalent stress and deformation of cement less hip prosthesis. A Rev. Energy Saving Green Comput. Syst. IJEDR. 93–97 (2012)

Modern Parking Business Using Blockchain and Internet of Things (IoT)

Varun V. Narayanan, M. V. Ranjith Kumar, Kartik Saxena, P. Madhavan and N. Bhalaji

Abstract This paper aims to deliver a sustaining business model by analyzing the market-trends of the bitcoin, develop a blockchain-based transaction system to run the errand and deploy an IoT-based, low-investment platform for ensuring profits in the long run. The existing monetary systems and investment platforms are discussed in detail and various inferences have been drawn from the trend of cryptocurrencies and world economy. An example of running a fully-automated, modern parking facility is used to demonstrate how this business would continue to grow as the demand-supply chain can be partially controlled. To the best of our knowledge from extensive literature survey, it can be a very promising investment for corporates and governments.

Keywords Blockchain · Bitcoin · Financial product management · Fiat money · Internet of things

V. V. Narayanan · M. V. Ranjith Kumar (✉) · K. Saxena · P. Madhavan
Department of CSE, SRM Institute of Science and Technology, Kancheepuram, India
e-mail: ranjithm1@srmist.edu.in

V. V. Narayanan
e-mail: varunv997@outlook.com

K. Saxena
e-mail: kartiksaxena2014@gmail.com

P. Madhavan
e-mail: madhavap@srmist.edu.in

N. Bhalaji
Department of IT, SSN College of Engineering, Kalavakkam, India
e-mail: bhalajin@ssn.edu.in

© Springer Nature Singapore Pte Ltd. 2020
S. S. Dash et al. (eds.), *Artificial Intelligence and Evolutionary Computations in Engineering Systems*, Advances in Intelligent Systems and Computing 1056,
https://doi.org/10.1007/978-981-15-0199-9_45

1   Introduction

The world is sceptical about a new P2P open-source money—the cryptocurrency. Bitcoin was the first cryptocurrency and is still the most popular. It started an entire industry of other digital currencies and has challenged all notions of how money and value work. The underlying technology, blockchain is a public ledger containing a record of all transactions and viewable to everyone. The main reason for this innovation to catch up was that it allowed cryptos to be decentralized. On the invent of such a technology, many banks and governments are worried about the misuse of the technology for illegal activities as it involves no intermediary in transactions [1]. However, another school of thought believes that such a digital currency can replace gold in the current market and set up a new standard of international exchange. But currently, a huge trade happens in cryptocurrencies known or unknown to the public. However, businesses are just at the emerging stage of using cryptocurrencies for value exchange. Keeping in mind the underlying technology—the blockchain which can be termed as a "secure but inefficient database" in layman terms, we have come up with a business model that can suit the current market and allows the business to take advantage of bitcoin's nature [2]. A thorough market study has been carried out and a feasible business model has been developed which can enable the companies to grow faster and raise more capital [3].

2   The Market System

2.1   The Law of Demand

The law of demand states that if all the other factors remain constant, the higher the price of an entity, the less people will buy that entity, or the quantity purchased will be low. The amount of an entity that buyers purchase at a higher price is less because as the price of a good goes up, so does the opportunity cost of buying that good. As a result, the persons will naturally avoid buying a product that will force them to forgo the consumption of something else they value more [4].

The DC is a down-sloping curve (see Fig. 1) which shows the inverse relationship between the quantity demanded and the price of the entity [4].

1. The shift in demand curve: The DC can shift towards the right or left (see Fig. 2) depending upon various factors such as:

a. Tastes/preferences: Suppose if a study says that children who have milk in the morning do better at schools, then the demand for milk would go up and hence would shift the curve to the right.

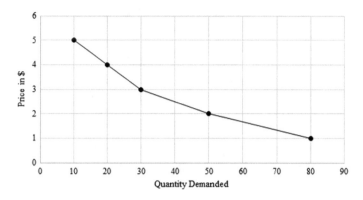

Fig. 1 Milk demand cure

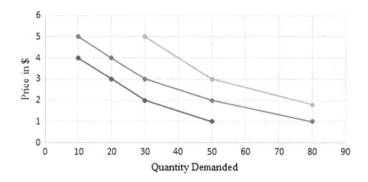

Fig. 2 Shift in milk demand cure

b. Number of consumers: If suddenly, new immigrants come into the town, then the demand for milk would increase and hence would shift the curve to the right. Similarly leaving population would shift it to left.
c. Price of related goods: Almond milk and cow's milk are substitutes for each other. If price of the almond milk goes up, more customers will tend towards buying cow's milk which would increase the demand. The law of supply [5].

The law of supply states that there is a direct relationship between price and quantity supplied, i.e., higher the price, the higher the quantity supplied which allows the producer to generate more revenue. This results in an upward sloping curve (see Fig. 3) between quantity supplied and the price per unit [6].

1. The shift in supply curve: The SC can shift towards the left or right (see Fig. 4) depending upon the various factors such as:

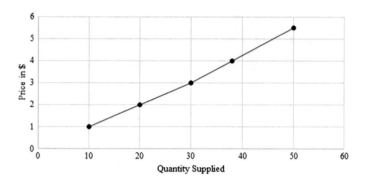

Fig. 3 Milk supply cure

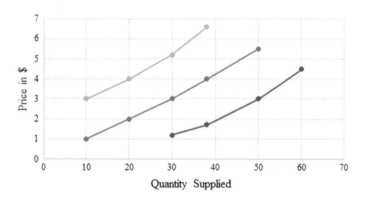

Fig. 4 Shift in milk supply cure

    a. Price of resources: If the price of dairy cows increases, then the number of milk-producing cows will decrease, which results in less supply of milk and will shift the supply SC to the left.

    b. Number of producers: If the number of dairy farmers increases, the supply of milk will also increase and vice versa. The SC will shift towards the right and left, respectively.

    c. Technology: If new technology is used to increase milk production, then that would shift the SC to the right.

2.2 The Modern Parking Lot

Smart cities are being planned and parking is a major issue in Tier-I cities like Bengaluru, Chennai, Mumbai, Delhi, etc. The number of thefts of cars is increasing by the day and personnel cannot be trusted with the safety of the vehicles. It is often seen that these intermediaries rent-out cars at night for extra income, hence compromising the ownership privileges [7]. The parking lots in various places house vehicles parked in random order and this results in congestion. One of the reasons for this is the fact that the visitor cannot know in advance whether space is available or not. These are the people who eventually end up parking their cars in no-parking zones [8].

3 Existing Monetary System

3.1 Supply-Demand Equilibrium

The supply-demand equilibrium is achieved at the point of intersection of the SC and the DC (see Fig. 5). This can be interpreted as the point where the market is self-sustaining, where all the goods produced by the suppliers and bought by the consumers. The market automatically works in such a way that this equilibrium is achieved over time.

3.2 Fiat Money and Gold-Based Economy

Historically, most currencies were based on physical commodities such as gold or silver, but fiat money is based solely on faith and credit of the economy. Most modern paper currencies are fiat currencies; they have no intrinsic value and are used solely as a means of payment [9] (Table 1).

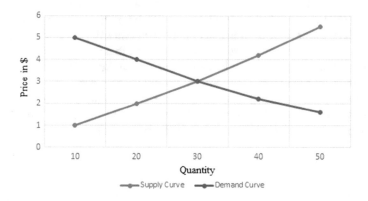

Fig. 5 Supply-demand equilibrium curve

Table 1 Fiat money versus gold-based economy

S. No.	Fiat money system	Gold-based economy
1	The fiat money system can stabilize an economy when there is a time of high demand for currency notes	The gold standard runs out of money as enough currency notes may not be available as there is gold
2	It is regulated by the government of the country and therefore provides an edge over maintaining the economy	It is completely dictated by the conversion of gold existing with the banks or the government
3	Since any amount of money can be printed, it can lead to high inflation	Since the total amount of gold is limited, therefore the amount of inflation is under control

4 Proposed Business Model

The bitcoin has a decreasing inflation curve and being limited in number, it is promised to grow in value in the future. Replacing the bitcoin as the global standard of currency exchange and trade is out of the scope of this paper; however, a promising investment to return only profits over the long run is the goal of this business model [10]. A parking lot is an investment that never runs out of demand and therefore investing in this scenario is a thoroughly calculated risk.

4.1 Methodology

A consumer in big cities would not mind buying a parking ticket when she has to commute to work on a daily basis. The business model proposes to charge the person a constant fee of say, Rs. 20. Now, instead of getting the rupee equivalent, buy bitcoin (or satoshi) equivalent of the amount and accumulate them over time by not releasing them. The fact that bitcoins are limited in number will hence take out bitcoins from the market continuously hence increasing the demand for the same. This in-turn will increase the value of the bitcoin and that is the time to release them into the market, when the supply curve will be tending to move upward. This manipulation of market can, therefore, ensure great profits and nearly no loss, as the value of bitcoin as an entity is sure to go up in the coming years. With proper consensus, the satoshi, which is one millionth of a bitcoin, can be further broken down into denomination. However, that is for future analysis.

1. Demand over time:

$$\frac{dy}{dt} \geq 0$$

2. Supply over time:

$$\frac{dz}{dt} \geq 0.$$

Note that the demand as well as the supply is modelled to increase over time. This comes from the assumption that the cities will continue to grow, and more people will commute in private vehicles also that the company will continue building new parking lots and provide it as a service to keep the business growing.

5 System Design

An IoT-based system has been proposed to be used by the company to provide the service. The parking lot will be hosted on a server with free available slots represented by a hash-map data structure. The customer can view all the available slots on his phone via an app/Web site. She can select a slot and book it for a specified time within which the car is to be moved. If she fails to do so, a fine will be imposed. Based on this data, multiple customers can book their slots. Each of this booking is recorded as a separate transaction on a blockchain hosted on a distributed p2p network [10]. The blockchain can be referred to whenever there is a suspicion of fraud and every transaction made this way is guaranteed to be secure. The hash generated is validated using a Kerberos engine that acts as an intermediate between the user and the slot.

5.1 Architecture

The Raspberry pi has been chosen to create the prototype and RESTful APIs have been written to facilitate the communication between the devices on the network. The data is constantly monitored and stored for future analytics [11]. Servo motors will be installed at each slot connected to an Arduino with which the Raspberry pi communicates serially to open and close the doors [12].

Python has been chosen as the back-end language as it is simple to code in and can facilitate rapid development [13]. More time can be spared for system design instead of going too deep in programming paradigms (see Fig. 6). Moreover, python provides numerous official packages that can be used for our purpose. These packages are well documented and provide an easy interface in terms of easy-to-use classes and methods [14].

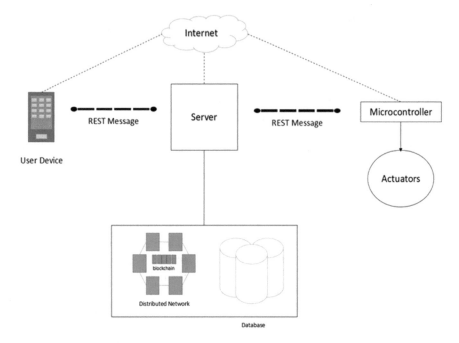

Fig. 6 Architectural diagram

6 Conclusion

The cryptocurrency market has been thoroughly studied and weighed against its advantages and disadvantages. The most fascinating fact about bitcoin is its limited quantity and the ever-increasing value as a result of a negative inflation curve. A highly profitable business model has been proposed for companies and governments to create a subset (a modern parking facility) of a "smart city". This can be innovation with low investment and at the same time provide the service to the consumers only a tap away. Certain assumptions have been made such as taking the upward trend of bitcoin for granted. However, it may not be the case. So, further work has been done on devising a procedure for mitigation in case of a falling market, a better return on investment. Data analytics can be applied to study and predict the market much better and devise the right time for selling and buying coins. This requires the expertise of financial advisors and economists of the country. Security is of utmost importance and data related to the market should stay away from hackers. Highly secure methods have to be implemented to control transactions and store data in decentralized networks. The blockchain is designed to avoid discrepancies in data but not to hide data.

References

1. Miraz, M.H., Ali, M.: Blockchain enabled enhanced IoT ecosystem security. In: Proceedings of the First International Conference on Emerging Technologies in Computing 2018 (iCETiC '18), London, UK, 23 Aug 2018
2. Kshetri, N.: Can blockchain strengthen the internet of things? IT Prof. **19**(4), 68–72 (2017)
3. Miraz, M.H., Ali, M., Excell, P., Rich, P.: A review on internet of things (IoT), internet of everything (IoE) and internet of nano things (IoNT). In: The Proceedings of the Fifth International IEEE Conference on Internet Technologies and Applications (ITA 15), Wrexham, UK, pp. 219–224 (2015)
4. Miraz, M.H.: Blockchain: Technology Fundamentals of the Trust Machine. Machine Lawyering, Chinese University of Hong Kong, 23 Dec 2017
5. Hjálmarsson, F.Þ. Hreiðarsson, G.K., Hamdaqa, M., Hjálmtýsson, G.: Blockchain-based e-voting system. In: Proceedings of 2018 IEEE 11th International Conference on Cloud Computing (CLOUD), San Francisco, CA, pp. 983–986 (2018). https://doi.org/10.1109/CLOUD.2018.00151
6. Nakamoto, S.: Bitcoin: A Peer-to-Peer Electronic Cash System. http://bitcoin.org/bitcoin.pdf (2008)
7. Liang, X. et al.: ProvChain: a blockchain-based data provenance architecture in cloud environment with enhanced privacy and availability. In: Proceedings of the 17th IEEE/ACM International Symposium on Cluster, Cloud and Grid Computing (CCGrid '17), pp. 468–477, Madrid, Spain, 14–17 May 2017
8. Madhavan, P., Thamizharasi, V., Ranjith Kumar, M.V., Suresh Kumar, A., Jabin, M.A, Sampathkumar, A.: Numerical investigation of temperature dependent water infiltrated D-shaped dual core photonic crystal fiber (D-DC-PCF) for sensing applications. Results Phys. **13**. (2019). ISSN 2211-3797. https://doi.org/10.1016/j.rinp.2019.102289
9. Chen, B., Tan, Z., Fang, W.: Blockchain-based implementation for financial product management. In Proceedings of 2018 28th International Telecommunication Networks and Applications Conference (ITNAC) (2018)
10. Kumar, M.R., Bhalaji, N., Singh, S.: An augmented approach for pseudo-free groups in smart cyber-physical system. Cluster Comput. 1–20 (2018)
11. Miller, D.: Blockchain and the internet of things in the industrial sector. In: Proceedings of IBM Internal
12. Arumugam, P., Panchapakesan, M., Balraj, S, Subramanian, R.C.: Reverse search strategy based optimization technique to economic dispatch problems with multiple fuels. J. Electr. Eng. Technol. **14**(2), 595–601 (2019)
13. Fraser, J.G., Bouridane, A.: Have the security flaws surrounding bitcoin effected the currency's value? In: Proceedings of 2017 Seventh International Conference on Emerging Security Technologies (EST) (2017)
14. Florea, B.C.: Blockchain and internet of things data provider for smart applications. In: Proceedings of 2018 7th Mediterranean Conference on Embedded Computing (MECO), Budva, Montenegro, 11–14 June 2018

AWS Infrastructure Automation and Security Prevention Using DevOps

Siva Vignesh and B. Rajesh Kanna

Abstract In day-to-day process lead by organizations while creating new infrastructures, needs to have same software configurations on all the servers to maintain their security and privacy policies. Manually doing this process leads to configuration errors and takes an ample amount of time to achieve the desired configurations. In this paper, we have proposed an automation process of converting Open Virtualization Application/Appliance (OVA) to Amazon machine image (AMI) to provide security (IP-based attacks and port scan attacks) and to maintain privacy by creating user-defined AMIs and to reduce time consumption in setup installation and configuration by creating AMIs from templates of existing infrastructure. The whole process is been orchestrated with the help of some DevOps tools. The tools used in this process are Jenkins, Terraform, Chef, and AWS EC2 instance. By this proposed method, all servers on AWS will have the desired same configuration using the template OVA including the security and privacy prospects.

Keywords Cloud computing · DevOps · Amazon machine image · AWS elastic cloud computing · Open virtualization application/appliance

1 Introduction

Cloud computing provides on-demand services for computing, storage, and networking, typically over the Internet and on a pay-as-you-go basis. Instead of creating one's own infrastructure or data centers, companies can rent data centers from service providers. On using cloud computing services organizations can reduce the cost of having and maintaining their own servers. There are many service providers for computing services such as Amazon Web Services (AWS), Azure, Google Cloud, and so on. Amazon Web Services is the most widely used service provider which provides services for compute, storage, and networking. AWS consists of

S. Vignesh · B. R. Kanna (✉)
School of Computer Science & Engineering, Vellore Institute of Technology, Chennai, India
e-mail: rajeshkanna.b@vit.ac.in

© Springer Nature Singapore Pte Ltd. 2020
S. S. Dash et al. (eds.), *Artificial Intelligence and Evolutionary Computations in Engineering Systems*, Advances in Intelligent Systems and Computing 1056,
https://doi.org/10.1007/978-981-15-0199-9_46

many services like simple storage service, RDS, Elastic Beanstalk, Elastic compute cloud (EC2), etc. EC2 instances are the virtual servers that can be launched by using Amazon Machine Images (AMIs) provided by AWS. Amazon Machine Images are patterning that are configured with an operating system, and some other software packages which regulates the operating environment of the user. AMI can be grouped into different types based on the regions, operating system, system architecture and whether they are backed by Amazon EBS or backed by the instance store. We can create our own AMI with customized configuration and the customized AMI can be used to launch instances. AMI can be copied in the same region and to different regions. The created AMI can either be public or private. Private AMI can be used only by the user, whereas public AMI can be displayed as a part of AMI lists to all users. Registered AMIs can also be deregistered if the AMI is no longer used.

Cloud computing consists of three service models: Software as a Service (SaaS), Platform as a Service (PaaS), and Infrastructure as a Service (IaaS). These models are separated based on the building blocks that can be rented. The building blocks are applications, data, runtime, middleware, O/S, virtualization, servers, storage, and networking. In SaaS all the building blocks are provided by the service provider itself. In PaaS, developers can run their applications provided with all the other building blocks, whereas IaaS is mostly widely popular with most organizations as only the basic building blocks can be rented. This will be useful when an organization wants to work from the very ground up and wants to have control over the whole infrastructure.

Cloud computing services basically reduce the cost of getting and maintaining physical servers. It also reduces lots of manual works which in turn reduces the number of human effort needed. Business agility is one of the most specified benefits of using cloud computing services. The traditional waterfall model works fine but had some challenges for both developers and operational teams. From the developer's perspective, there was an immense time taken for code deployment and work stress on pending and new code was enormous as development and deployment time are high. From operations perspective, maintaining 100% uptime of production environment was difficult, the complexity of monitoring servers increased with time and eventually it was hard to supply feedback and discover product issue. This is where DevOps came into the picture. DevOps is an operational theory that advances better correspondence among development and operations, cross-common sense, and better working relationships. Even though cloud computing provides many advantages, it also gives rise to several concerns about security threats such as data breaches, human error, malicious insiders, account hijacking, and DDoS attacks.

- Data breaches: This is the primary concern of most of the cloud users as most of the sensitive data has been moved to the cloud.
- Human error: A survey describes that most of the failures occur due to the users fault.

- Data loss: Lack of proper backup may lead to permanent data loss due to accidents.
- DDOS attack: Distributed denial-of-service attacks lead to significant risks for both the providers and users which include reputational damage and exposure to customer data, etc.
- Account hijacking: This will happen when attackers gain access to a particular account with stolen credentials which in turn compromising the confidentiality, integrity, and availability of those services.

2 Motivation Background

Amazon Web Services (AWS) is an immense service provider which provides many services for computing, storage, and networking. This whitepaper brings out the benefits of using AWS services, security, and compliance, AWS management console, AWS software development kits (SDKs), application integration, cost management, etc. It also discusses about compute services like Amazon Lightsail, AWS Batch, AWS Elastic Beanstalk, AWS Fargate, AWS Lambda, AWS Serverless Application Repository and database services like Amazon RDS, Amazon DynamoDB, Amazon ElastiCache, and Amazon Timestream [1]. By considering these services and their features, in this paper, most of the computing and storage services are used (e.g., ec2 instances, s3). There are many research works that take place on cloud deployment and development. To understand the different cloud-based technologies, a survey and a comparative study of different cloud development tools based on characteristics such as applicability, and editing capability presented in this paper. They further explain about the available pro-gramming environments, repositories, cloud software modeling, processing and documentation tools, management and orchestration tools. From the comparative study, they concluded that there is no programming environment which provides an internal repository. Instead, they use external repositories like GitHub, nexus, etc. [2]. As most of the enterprises started moving their infrastructure from on-premise to cloud environment, there are several security aspects that are to be included. The comparison made between investing in on-premise and in cloud infrastructure. They also characterize a capacity-based pricing strategy from which they observed that at higher available capacity, the cloud provider tends to charge lower prices and also claim that one of the challenges faced by businesses is the optimal configu-ration of on-demand and on-premise cloud infrastructure [3]. A cloud architecture reference model that includes a wide range of security controls and best practices was presented. In this paper, Cloud-Trust calculates high-level security metrics to measure the standard of confidentiality and integrity provided by a cloud service provider (CSP). They included four cloud building designs based on the reference model with gently more security controls [4]. As cloud services and cloud service providers (CSPs) are expanded, the unreliability and threats for companies using cloud services also elevated [5]. This paper addresses the aspects of cloud service providers and current customer specifications and obstacles of using cloud services from a provider's perspective. And further explains the current and future challenges of cloud service providers.

Cloud environment is the primary target of most attackers as it is the place where a large amount of data can be retrieved. Enterprises that are shifting on-premise to on cloud needs to concentrate more on security aspects of their data. Enterprises use almost all three types of service models. Security aspects of SaaS enterprise cloud services have been focused on this paper [6]. Here, an application framework is been created which allocates resource and apply data exchange. The data and logs generated by public clouds will be collected. Then these data collection will be used for detecting abnormalities which in turn uses support vector machine to instruct multiple classifiers from observed data for multiple system environments [7]. As cloud technology also processes a large amount of sensitive data, it draws attention to the security aspect of it [8, 9]. This paper explains one of the most common attacks called port scanning attack. In this attach, whenever any port is open, the attackers try to send a series of messages through that port. Those messages contain threats which will try to break into the users system/server. All these messages ping simultaneously at all the open ports of that system/server. This attack happens at any stage at any point of time. The author has described the overview and methodologies of this attack. Basically, the attack has been categorized based on the approach of the attack and its strategy. Types of scan namely Vanilla, Strobe, Stealth, FTD Bounce scan, Fragmented Packets, UPD, and Sweep for which one the solution they came up for detection of scanning is flow-based port scan detection. They have also described the most venerable point where most of the attack happens for which we have applied our solution in this paper. One of the other high-security threats is data breaching [10]. In this paper, explains the package called Fail2ban and also explains the usage of it. Fail2ban is an open-source package which is used for break-in prevention from threats. It runs 24×7 in the background and as soon as it finds threats such as bad IPs which are trying to breach, it blocks them and updates the firewall. Once the firewall gets updated, there is a time variable that can be set if user wants to block certain threats temporarily which means those threats will be back to active mode after that time duration. There is one more feature through which the user can manually block and unblock threats and IPs according to their need. The idea behind using Fail2ban package is been applied in this paper to prevent unauthorized IPs from accessing/breaching ec2 instances.

3   Experimental Setup and Architecture

In this chapter, we are going to discuss about the experimental setup and its architecture and as well as the algorithm to operate the automation of infrastructure.

3.1 Experimental Setup

Basically, AWS provides their own Amazon Machine Image (AMI) to use on all ec2 instances. As those are all public images, it impacts on the security prospective. To solve the security prospective, the user creates their own OVA file will all the required configurations and uploads to AWS to use in all ec2 instances as shown in Fig. 1.

For that there are three major components: source, computing, and verification. The OVA file which needs to be converted is all the source component. This OVA file is been uploaded to the S3 bucket in AWS. Once its has been uploaded, user will go the Jenkins pipeline and mention the location of the source file which is the S3 bucket name and the file name which is the OVA file name and BUILD that job. Once the job is triggered, it comes to the computing stage which includes AWSCLI, Terraform, and Chef tools. This inputs will be passed to the AWSCLI which will

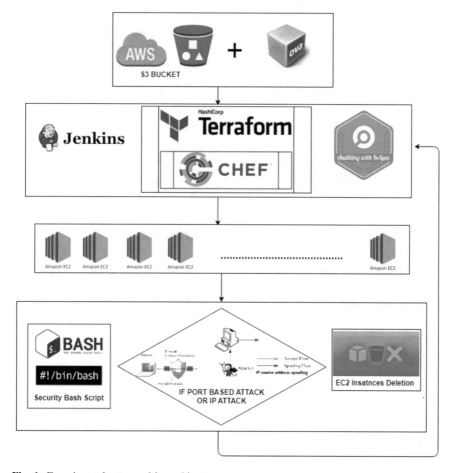

Fig. 1 Experimental setup and its architecture

create the IAM roles and policies to convert the OVA file. In AWS, the user needs privileges and roles to do a certain task. That is called IAM policies. So once the IAM is configured the actually process of conversion starts.

This process of conversion contains pending—converting—updating—updated —preparing to boot—booting—booted—preparing ami—completed. In case if the process fails at any stage of it, the developer team will be notified and the process will terminate. But if the status shows "completed", that means the OVA got successfully converted to AMI and is ready to use. Once AWSCLI phase completes the conversion process, pass AMI-ID as input to the Terraform. This Terraform will trigger ec2 instances according to that particular AMI-ID. Till now the ec2 instances have only been triggered but the security prospective is still pending. For that Chef comes into the picture. To configure security related to port scan and IP-based attacking, a chef cookbook is written which will be called from the userdata section of the Terraform. This cook will install all necessary configurations while booting the ec2 instances. So as soon as these security concepts get deployed, now we need to check if it really got configured. To do that Chef Inspec execute a script which will check if the Chef cookbook has configured it all or not.

Once all process, gets successfully configured, it will be ready for the end users to use. All the ec2 instances will be up and running. Now comes to the crucial part. It is the time to check if the security part is working or not. So there is one script running on the background of the system. If there is any tries to open and unauthorized port or it anyone tries to access the server with and unauthorized IP. The system will automatically detect and trigger the Jenkins back with API call which will trigger the Terraform to destroy all those instances which got trigger. And once destroyed, it will again spin up those ec2 instances with new configurations up and running.

3.2 Algorithm for Automation Infrastructure

Below algorithm is used for automation of infrastructure where two pipelines will be required 'Pipeline 1' for converting the OVA to AMI and generating the AMI-ID and 'Pipeline 2' for creating and destroying EC2s.

pipeline 1:

Node(Master)

step1:declare BucketName

step2:declare OVAName

step3:defining IAM role Policy for accessing resources

step4:defining trust policy based on IAM role policy for connecting to the Amazon management console

step5:declare containers for storing OVA file from S3 Bucket

stage(AWS-CLI Commands)

step6:execute shell script for getting updated os package version

    a. IF stdout is true THEN

        return output

step7:execute shell script for installing awscli

    a. IF stdout is true THEN

        return output

step8:connect to the Aws management console using access key and secret key

step9:creating trust policy

step10:creating role policy from json file

step11:attach trust policy for user

step12:declare region name into which ova file is stored

step13:declare name of the ova file which to be converted into image file

step14:execute shell script for converting image file

    a.IF stdout is true THEN

        return output

step15:checks for the processStatus

step16:repeat untill processStatus not equal complete

    a.converting the consolelog file into json format

    b.assign convert status to the variable

    c.IF status is equal to completed THEN

        i.print the status

    d.ElSE IF status is equal to error THEN

        i.print the error message

        ii.EXIT

    e.ELSE

        i.print process status

        ii.sleep for three minutes

stage(build)

step17: calling the job for building terraform

pipeline 2:

Node(Master)

step1:assign amiid to variable

step2:assign destroyvar value to variable

step3:IF destroyvar is equal to 1 THEN

    a.excute the shell script command for destroying ec2instances

    b.print output

step4:execute shell script command for intializing packages

step5:execute shell script command for checking plan

step6:execute shell script command for building the plan

step7:triggers user data in ec2instances

    a.execute shell script to download chefDK package

    b.execute command to unzip

    c.execute shell script to download portchecker cookbook tar file

    d.execute command to untar file

    e.execute cookbook

        i.triggers command to install firewall package and Fail2ban

        ii.checks for unauthorized ipaddress

        iii.store values in Fail2banlist

        iv.check for port opened and closed by firewall

        v.store values in portlist

        vi.IF portlist is not equal to open or Fail2banlist is not authorizedlist

            a.triggers a jenkins process

4 Experimentation

As we discussed above in architecture there are three major components: source, computing, and verification (Fig. 2). To cover all this components Jenkins are used as the building tool. There are two major pipelines. Before beginning to building the pipelines, user needs to store the OVA file on the same AWS account in which user wants to launch ec2 instances. For that steps are as follows:

1. Create a Bucket or can also use existing Bucket in S3.
2. Upload the OVA file with.OVA extension in that particular bucket in S3.

Soon as OVA is uploaded the user is ready to trigger the Jenkins Pipeline Jobs.

4.1 Pipeline AWSCLI

First pipeline consists of AWSCLI section. Here, user inputs are fed. Two inputs, OVA file and the Bucket name. Once inputs are passed, it will start processing.

Process Flow Diagram:

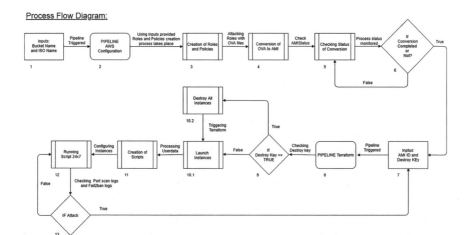

Fig. 2 Process flow diagram from conversion of OVA to AMI till executing Terraform

AWSCLI will declare all the permissions required to convert the OVA to AMI. Permission in AWS, AWS uses a service called identity access management (IAM). In that first step is to create a policy, then to attach that policy to a role and then assign that to a particular user. Same way:

1. Create an IAM role with VIM import assume role. This role will help in accessing the object.
2. Create a role attaching that policy.

Note to remember is these conversions and AMI will be present in one particular region. Once AMI is loaded to AWS on that region, it will not be able to access through other regions of AWS. So it is mandatory to mention that region in the policy. Once the IAM is configured, the command for conversion process is executed which will start converting on one particular region of AWS. This conversion process has many stages in it. Stages are "pending" > "converting" > "updating" > "updated" > "preparing to boot" > "booting" > "booted" > "preparing ami" > "completed".

Soon as the command is triggered, the first stage it enters is pending as it is under process of validating all the dependencies. If all the dependencies are in correct way then it will start converting. Actual conversion process starts now. So once the conversion over it needs to update—prepare booting and once it is done it will start preparing the AMI of it. Last step it covers is the completed stage which the final step. If there is any error or any failures in-between these process, it will not prepare any AMI's. Then we need to trigger it again.

4.2 Pipeline Terraform

So now Pipeline AWSCLI job is completed as the AMI is ready successfully. Now it is the time to trigger the Pipeline Terraform. This pipeline is responsible for all Terraform-related process. From Terraform plan–apply–destroy all. So Pipeline AWSCLI will pass AMI-ID to this job, which it will use to create the ec2 instances. But this pipeline is little different. First thing it checks if this BUILD is for destroying ec2 instances or not. If no then it will accept the request and pass the AMI-ID to the Terraform code and execute the code.

So as soon as Terraform is triggered for the first time, it will initialize the environment using Terraform in it. Then it will validate using Terraform Plan. Once both of this gets successful then only if will start applying. Now Terraform apply will trigger with that particular AMI-ID and launch all ec2 instances.

While the ec2 instances are still under booting phase, the userdata of Terraform will trigger the Chef cookbook which will configure the security of the ec2 instances. Cookbook will start configuring the instances:

1. Update the OS.
2. Install Fail2ban.
3. Configure Fail2ban.
4. Allow desired ports.
5. Block unwanted ports.
6. Create script under profile to monitor port and Fail2ban logs.

So once cookbook configured all the above steps, the script will run on background 24 × 7. As the end user is normally using the system, the attackers will manage to find the vulnerable port numbers of the ec2 instance and there is chance that as AWS console authorization is been compromised and someone has open port from console. For all those possibilities, we have manually blocked unwanted ports. Even though any unauthorized port is been scanned, this script will call the Jenkins API which will call the Terraform destroy from the Pipeline Terraform which will destroy all the ec2 instances and launch again all those ec2 instances. Same goes with IP attacks, now some ports have to be open and people will try to attack from there so that time Fail2ban will use its feature of recording those IP address and blocking it. So till the time is blocked, again the script will call Jenkins API for Pipeline Terraform which will again call Terraform destroy to destroy all those ec2 instances and simultaneously launch more. By this way, data will be secure and developers can have records which are the vulnerable ports and places of server which can be enhanced.

5 Validation and Performance

For validation, we are using Terraform plan to check and validate if all the information about the instances which are going to be launch is correct or not. Terraform plan will give exactly what will happen once we say Terraform apply to create all the instances.

Other than Terraform plan, we are also using Chef Inspec Fig. 3 which is specially used for validation and testing purpose. It check whether the cookbooks have configured properly or not.

For performance, if this process is done manually it will take a lot of time (approx. 1 h as conversion of OVA to AMI itself takes 15–20 min). But here, as all process is automated so all process completes within 14–15 min (Fig. 4 pipeline AWS takes 12 min and 14 s and Fig. 5 pipeline Terraform takes 1 min).

```
ubuntu@ip-172-31-31-190:~/chefrepo/cookbooks/sanfinalthesis/test/integration/default$ inspec exec default_test.r
b

Profile: tests from default_test.rb (tests from default_test.rb)
Version: (not specified)
Target:  local://

  User root
     □
  Port 80
     □
  Port 81
     ✓  should not be listening
  Port 54
     ✓  should not be listening
  Port 124
     ✓  should not be listening
  Port 444
     ✓  should not be listening
  Port 8080
     ✓  should be listening
  Port 22
     ✓  should be listening

Test Summary: 6 successful, 0 failures, 2 skipped
ubuntu@ip-172-31-31-190:~/chefrepo/cookbooks/sanfinalthesis/test/integration/default$ ▌
```

Fig. 3 Executing Chef Inspec

Fig. 4 Time consumed by Terraform pipeline

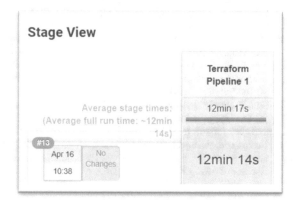

Fig. 5 Time consumed for deleting/launching instances

6 Conclusion and Future Enhancements

The future enhancements will be how more secure we can make it. Right now it is ideal for cases for reducing deployment time, preventing the error which takes place while doing it manually and two types of attacks in real world. But as there are a lot of other types of security threats which are as equally harmful like these, we will continue this research and try to focus on how to include more security prospective into this.

References

1. Overview of Amazon Web Services.: https://d1.awsstatic.com/whitepapers/aws-overview.pdf (2018)
2. Jain, T., Hazra, J.: On-demand pricing and capacity management in cloud computing. J. Revenue Pricing Manag. (2018). https://doi.org/10.1057/s41272-018-0146-0
3. Fylaktopoulos, G., et al.: An overview of platforms for cloud based development. SpringerPlus (SpringerOpen) (2016). https://doi.org/10.1186/s40064-016-1688-5
4. Gonzales, D., Kaplan, J., Saltzman, E., Winkelman, Z., Woods, D.: Cloud-trust—a security assessment model for infrastructure as a service (IaaS) clouds. IEEE Trans. Cloud Comput. (2015). https://doi.org/10.1109/tcc.2015.2415794
5. Hentschel R, et al.: Current cloud challenges in Germany: the perspective of cloud service providers. J Cloud Comput. 7(1) (2018). https://doi.org/10.1186/s13677-018-0107-6
6. Niu, D.-D., Liu, L., Zhang, X., Lu, S., Li, Z.: Security analysis model, system architecture and relational model of enterprise cloud services. Int. J. Autom. Comput. (2017). https://doi.org/10.1007/s11633-016-1014-2
7. Sun, D., et al.: Non-intrusive anomaly detection with streaming performance metrics and logs for DevOps in public clouds: a case study in AWS. IEEE Trans. Emerg. Top. Comput 4(2), 278–289 (2016). https://doi.org/10.1109/tetc.2016.2520883
8. Deshpande, P., Aggarwal, A., Sharma, S.C., Kumar, P.S., Abraham, A.: Distributed port-scan attack in cloud environment. In: 2013 Fifth International Conference on Computational Aspects of Social Networks, pp. 27–3. IEEE (2013)

9. Ring, M., Landes, D., Hotho, A.: Detection of slow port scans in flow-based network traffic. PLoS One **13**(9), e0204507 (2018)
10. Ford, M., Mallery, C., Palmasani, F., Rabb, M., Turner, R., Soles, L., Snider, D.: A process to transfer Fail2ban data to an adaptive enterprise intrusion detection and prevention system. In: Southeast Conference 2016, pp. 1–4. IEEE (2016)

Recommendation System for Big Data Software Using Popularity Model and Collaborative Filtering

Shreayan Chaudhary and C. G. Anupama

Abstract A recommender system is a model which has the ability to predict the list of items according to the user's preferences or ratings. Recommender systems have huge areas of application ranging from music, books, movies, search queries, and social sites to news. A recommender system filters information into various categories which may be personalized or non-personalized. Recommendation systems are very useful for the user in discovering those items which they may not have found else ways. These above-mentioned systems are quite similar to search queries except that search queries need to be explicitly provided by the user but recommender system may do it implicitly without the user having prior knowledge about its working. There are many algorithms used in recommendation systems such as restricted Boltzmann machines (RBM), slope one, singular value decomposition, alternating least squares. This paper presents an in-depth study and analysis of the following two algorithms: Review-based popularity model and collaborative filtering model. Popularity model is a straightforward model based on the popularity and rating of an item given to it by the user. Collaborative filtering model is a very efficient and popular model and is used by many popular organizations like Netflix, Amazon, and Facebook.

Keywords Recommender system · Machine learning algorithms · Big data system · Data scientist · Popularity model · Information filtering · Collaborative filtering

S. Chaudhary (✉) · C. G. Anupama
Department of Software Engineering, SRM Institute of Science and Technology,
Chennai, India
e-mail: shreayan98c@gmail.com

C. G. Anupama
e-mail: anupama.g@ktr.srmuniv.ac.in

© Springer Nature Singapore Pte Ltd. 2020
S. S. Dash et al. (eds.), *Artificial Intelligence and Evolutionary Computations
in Engineering Systems*, Advances in Intelligent Systems and Computing 1056,
https://doi.org/10.1007/978-981-15-0199-9_47

1 Introduction

With the rapid industrialization taking place in the twenty-first century, technology all around the world has been increasing at an exponential rate. For example, more than a billion YouTube users viewed about 300 h of videos daily and uploaded about 300 h of video daily in 2015 [1]. As a result, this increases the focus on data generated in all the information systems which are related to software systems. The data being generated constantly needs to be collected, cleaned, stored, processed, visualized, managed, and modified. The data may occur as structured data which is machine readable or as unstructured data which is human readable. A data can be classified as big data if the data has 5 V's, i.e., volume, velocity, variety, veracity, and value [2]. Many organizations and expert data scientists can enhance the quality of big data by making accurate predictions to maximize the profit of the organization [3]. A data scientist can use various tools to clean, visualize, manage, and carefully study the data. He/she can also use the various tools and algorithms of machine learning to classify or predict data based on previous experience on which the machine is trained on. After carefully studying and inspection the data of a big data system, a data scientist can find out the correlation among the data using which he can predict outcomes. Interpreting the user's unique preference and ratings, a recommender system is able to predict the items. In this paper, we have implemented and compared the various algorithms of recommender systems like collaborative filtering and content-based filtering.

2 About Recommender Systems

A recommender systems principal goal is to provide the user personal recommendations based on the previous items or choices or likes of the user. A recommender system falls under the category of information filtering [4]. There is a wide range of applications of the recommender system ranging from music, books, movies, search queries, and social sites to news. There are eight major areas in which recommender systems are used which are listed as follows: e-government, e-business, e-commerce/e-shopping, e-library, e-learning, e-tourism, e-resource services, and e-group activities [5]. These are very useful since they help the user to discover those items which they may have not found otherwise. Recommendation systems are similar to search queries except that search queries have to be explicitly provided by the user but a recommender systems may do it implicitly without the user having the knowledge about its working. According to a study conducted in "Recommender System Application Developments: A Survey" by Lu, Jie and Wu, Dianshuang and Mao, Mingsong and Wang, Wei and Zhang, Guangquan, a survey proved that collaborative filtering and matrix factorization are two of the most widely used important algorithms used in a recommender system [5]. According to "A personalized search algorithm by using content-based filtering" by C. Zeng,

C. X. Xing, L. Z. Zhou, collaborative filtering is used to accurately express the user's interests and improve the accuracy of personalized search [6]. Hence, we have chosen collaborative filtering since its efficiency has been proven. In the paper, "Collaborative filtering algorithm based on representation learning of knowledge graph." Xiyu Wu, Qimai Chen, Hai Liu, and Chaobo, they have mentioned that one of the main disadvantage of collaborative filtering is the infamous cold start problem. Our model will not be able to provide item suggestions to new users since the algorithm is based on the user's search history [7]. To deal with this cold start problem, we have paired the simple popularity model along with the collaborative filtering algorithm.

3 Applications of Recommender Systems

Recommender systems have a wide range of application ranging from recommendations for music, books, movies, search queries, and social sites to news. The applications include [8]:

1. To discover some popular items
2. To discover all highly rated items
3. Recommend a sequence
4. Recommending a bundle
5. Just browsing
6. Influencing others.

4 How Does a Recommender System Work?

A recommender system is made by a collection of one or more machine learning algorithms. There are many algorithms used in recommendation systems such as restricted Boltzmann machines (RBM) [9], slope one [10], singular value decomposition [11], content filtering [6], alternating least squares [12], and matrix factorization [3, 13]. These algorithms are divided into two major categories: personalized algorithms and non-personalized algorithms. The non-personalized algorithms, like popularity-based model, only depend on the highest-rated items or the highest selling items. They do not take user personalization into account. This paper presents an in-depth study and analysis of the following two algorithms: Review-based popularity model [14] and collaborative filtering [7] model. Popularity model is an elementary model based on the popularity and rating of an item given to it by the user. Collaborative filtering [7] model is a very efficient and popular model and is used by many popular organizations like Netflix, Amazon, and Facebook.

5 Popularity Model

Popularity model is a fundamental model based on the popularity and rating of an item given to it by the user. It is a non-personalized algorithm. Popularity model recommends items to the user according to the current trends. It may suggest the user those items which are the most common or the most selling items. The main drawback to this model is the lack of personalization. Even though we know the behavior of the user, we cannot recommend him items accordingly. This model will make the same item recommendations to all the users regardless of their history. Simple popularity model works best for new users or those users using a new account. Since their item history will be null, this model will suggest them those items that are popular in general. The user has a higher chance of liking the item since it has a high rating and is liked in general. Each user will be recommended with the items according to the current trend.

6 Collaborative Filtering

Collaborative filtering is a popular recommender system algorithms that matches a customer preferences to other customers and then makes recommendations accordingly. It is a personalized algorithm. It can produce high-quality recommendations. It is a process of finding out a common pattern and to filter information using various techniques that involves collaboration among the various viewpoints [15]. One of the main drawbacks of this algorithm is that with the increase in the number of products and customers, its performance will deteriorate. Collaborative filtering can be branched into memory-based and model-based.

1. Memory-based collaborative filtering: This type of filtering method uses the item ratings data to compute the correlation or similarities between two users or items. This is also called neighborhood-based collaborative filtering. Memory-based collaborative filtering has its application in the nearest neighbor algorithm. Its main advantage is that it scales very well compared to other models.
2. Model-based collaborative filtering: In this type of filtering method, we build models using various algorithms from machine learning and data mining to predict what rating a user will give to those items which he has not used or rated yet. Model-based collaborative filtering includes algorithms like singular value decomposition, Markov decision process, clustering models, Bayesian networks, Dirichlet allocation, etc. These algorithms require dimensionality reduction to reduce the number of dimensions of its features to improve the accuracy and efficiency of the model. It is works well for sparse matrix hence in some ways it is better than memory-based collaborative filtering.

In our case study, we have used the memory-based collaborative filtering. There are various methods for calculating the similarity between the items based on user ratings or even similarity between the users such as Pearson correlation, cosine similarity, Euclidean similarity, and adjusted cosine similarity. We have used cosine similarity because it works best for a sparse matrix. In our case, we have two million ratings distributed over 27,000 movies by 138,000 users, which means that we will have a sparse matrix [16].

$$\text{sim}(u_1, u_2) = \cos \theta = \frac{r_u \cdot r_{u'}}{\| r_u \| \| r_{u'} \|}$$

From the above relation, we can calculate the rating given to a movie i by user u by taking the weighted sum of other users u' who have given ratings to that movie. This is a direct approach which requires very little to no training or optimization hence it is an efficient algorithm.

7 Case Study

In this paper, we have considered the MovieLens dataset containing 20 million movie ratings given to 27,000 movies by 138,000 users [16]. We have used collaborative filtering and simple popularity method to create our movie recommendation system. The simple popularity model will recommend the users the most popular and the highest-rated movies and TV shows and the collaborative filtering model will provide suggestions to the user based on the history of shows and movies he has already watched. We have performed an in-depth analysis of the above algorithms and analyzed it carefully. We have also compared the above algorithms in terms of accuracy, precision, recall, and the outputs of both the algorithms (Fig. 1).

8 Results

The confusion matrix of a classifier is given as (Fig. 2):

TP—Correctly predicted positive results	FP—Incorrectly predicted positive results
FN—Incorrectly predicted negative results	TN—Correctly predicted negative results

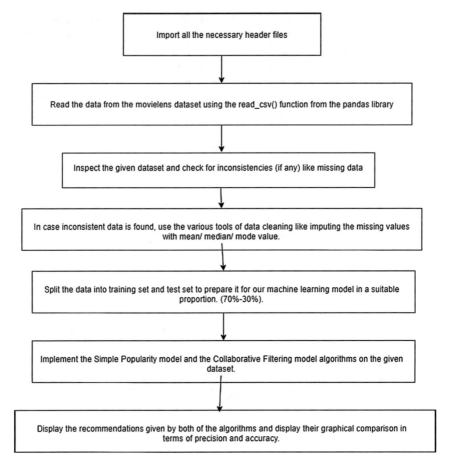

Fig. 1 Flow chart of the algorithms used

We can define the following terms to evaluate or results:

Precision of a machine learning model is the fraction of correct predictions made by the model upon the total number of predictions marked as positive by the model.

Accuracy of a machine learning model is fraction of correctly marked predictions upon the total number of predictions made by the model.

Recall of a machine learning model is the fraction of the relevant results which was correctly predicted by or model (Fig. 3).

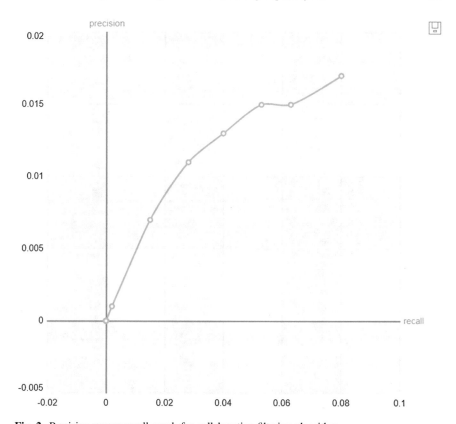

Fig. 2 Precision versus recall graph for collaborative filtering algorithm

9 Conclusion

The results of the research paper "Performance of recommender algorithms on top-n recommendation tasks" by Paolo Cremonesi, Yehuda Koren, Roberto Turrin show that improvements in root-mean-squared error (RMSE) often do not translate into improvements in accuracy [14].

We can observe from the above graphs that the precision and recall values of our model extremely small in the order of 10^{-7}.

$$\text{Precision} = 7.22271978736 \times 10^{-7}$$

$$\text{Recall} = 4.94706834751 \times 10^{-7}$$

We can also observe from the above graph that the precision of simple popularity model is higher since the recommendations made by that model will be the same for everyone.

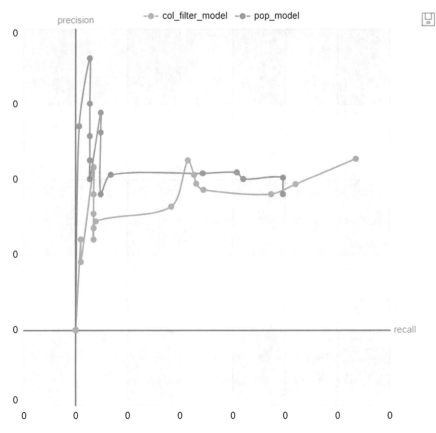

Fig. 3 Comparison of precision versus recall graph for popularity model and collaborative filtering

We can notice that the mean recall for collaborative filtering model is higher since the model will make more relevant predictions to the users as it is a personalized algorithm.

For our dataset, the performance of collaborative filtering model is better than simple popularity model since its recall is higher and precision is almost equal to the latter model.

The research paper "A Survey on RSS and its Applications" by P. N. Vijaya Kumar, Dr. V. Raghunatha Reddy proves that using hybrid approach for recommender systems can potentially overcome the limitations of the other approaches [15].

Hence, we can conclude that each algorithm has its own pros and cons and we should use these algorithms together as a hybrid algorithm to improve the performance of our model.

Acknowledgements We would like to express our sincerest gratefulness to everyone who have helped us with this research directly, or indirectly.

A very special thanks to my brother who has helped me review this paper and fix all the errors in it. This work would not have been possible without the constant support from my parents who helped us a lot for this research. I would also like to thank my fellow batch mates and faculties of the software engineering department for their continuous guidance and assistance.

References

1. Gorton, I., Bener, A.B., Mockus, A.: Software engineering for big data systems, pp. 32–35. IEEE Software (2016)
2. Bagriyanik, S., Karahoca, A.: Big data in software engineering: a systematic literature review. Glob. J. Inf. Technol. **6**(1), 107–116 (2016)
3. Altarturi, H.H., Ng, K., Ninggal, M.I.H., Nazri, A.S.A., Ghani, A.A.A.: A requirement engineering model for big data software. In: 2017 IEEE Conference on Big Data and Analytics (ICBDA), pp. 111–117. Kuching (2017). https://doi.org/10.1109/icbdaa.2017.8284116
4. Ricci, F., Rokach, L., Shapira, B.: Introduction to recommender systems handbook. In: Recommender Systems Handbook, pp. 1–35. Springer (2011)
5. Lu, J., Wu, D., Mao, M., Wang, W., Zhang, G.: Recommender system application developments: a survey. Decis. Support Sys. **74** (2015). https://doi.org/10.1016/j.dss.2015.03.008
6. Zeng, C., Xing, C.X., Zhou, L.Z.: A personalized search algorithm by using content-based filtering. J. Soft. **14**(5), 999–1004 (2003)
7. Wu, X., Chen, Q., Liu, H., He, C.: Collaborative filtering recommendation algorithm based on representation learning of knowledge graph. Comput. Eng. **44**(2), 226–232, 263 (2018)
8. Herlocker, J.L., Konstan, J.A., Terveen, L.G., Riedl, J.T.: Evaluating collaborative filtering recommender systems. ACM Trans. Inf. Sys. **22**(1), 5–53 (2004)
9. Salakhutdinov, R., Mnih, A., Hinton, G.: Restricted Boltzmann machines for collaborative filtering. In: Proceedings of the 24th International Conference on Machine Learning, pp. 791–798 (2007)
10. Lemire, D., Maclachlan, A.: Slope one predictors for online rating-based collaborative filtering. In: Proceedings of the 2005 SIAM International Conference on Data Mining, pp. 471–475 (2005)
11. Cherry, S.: Some comments on singular value decomposition analysis. J. Clim. **10**, 1759–1761 (1997). https://doi.org/10.1175/1520-0442(1997)010%3c1759:SCOSVD%3e2.0.CO;2
12. Felsenstein, J.: An alternating least squares approach to inferring phylogenies from pairwise distances. Syst Biol. **46**(1), 101–111 (1997). https://doi.org/10.1093/sysbio/46.1.101
13. Koren, Y., Bell, R., Volinsky, C.: Matrix factorization techniques for recommender systems. Computer **42**(8) (2009)
14. Cremonesi, P., Koren, Y., Turrin, R.: Performance of recommender algorithms on top-n recommendation tasks. In: Proceedings of the Fourth ACM Conference on Recommender systems. Barcelona, Spain, 26–30 Sept 2010. https://doi.org/10.1145/1864708.1864721
15. Kumar, P.N.V, Reddy, V.R.: A survey on recommender systems (RSS) and its applications. Int. J Innov. Res. Comput. Commun. Eng. **2**(8)
16. Harper, F.M., Konstan, J.A.: The movieLens datasets: history and context. ACM Trans. Interact. Intell. Sys. (TiiS) **5**(4), 19, Article 19 (2015). http://dx.doi.org/10.1145/2827872

Co-simulation of Fuzzy Logic Control for a DC–DC Buck Converter in Cascade System

Mohammed Kh. Al-Nussairi and Ramazan Bayindir

Abstract The multi-converter system has increasingly used in the aerospace ships, sea ships, electric vehicles, and microgrids. The multi-converter is a cascade system of power electronic converter. The second stage is represented as a load to the first stage. At some, the power electronic loads behave as a constant power load. The constant power load makes system under negative damping and unstable situation. The instability effects of constant power loads are caused by incremental negative resistance. This paper introduces the fuzzy logic control to stabilize the DC–DC buck converter, which is the first stage of the cascade system. The second stage has a mixed load (buck converter, constant power load, and resistive load). The function control of fuzzy logic control is regulated the voltage terminals of the DC–DC buck converter. Firstly, the fuzzy logic controller is simulated with the suggested system by using the MATLAB/Simulink environment. Moreover, then the implementing and simulation method of fuzzy logic control by using FPGA-in-the loop (FIL)-based. The simulation environment is MATLAB/Simulink, which is used for building and implementing the behavioral model. The behavioral model has used the Verilog HDL language. The results of this work compared between the simulations of the fuzzy logic toolbox and FIL-based fuzzy logic controller and have the same results have obtained.

Keywords Buck converter · Mixed load · Fuzzy logic control · Verilog HDL · FPGA-in-the loop

M. Kh. Al-Nussairi
Department of Electrical Engineering, College of Engineering,
University of Misan, 62001 Al Amarah, Misan, Iraq
e-mail: muhammed.kh@uomisan.edu.iq

R. Bayindir (✉)
Department of Electrical and Electronic Engineering, Faculty of Engineering,
Gazi University, 06500 Besevler, Ankara, Turkey
e-mail: bayindir@gazi.edu.tr

© Springer Nature Singapore Pte Ltd. 2020   561
S. S. Dash et al. (eds.), *Artificial Intelligence and Evolutionary Computations
in Engineering Systems*, Advances in Intelligent Systems and Computing 1056,
https://doi.org/10.1007/978-981-15-0199-9_48

1 Introduction

At the recent, the power system has a multi-converter system in its structure. The multi-converter has a multi-stage level of power converter systems, electronic loads, and passive elements [1]. Some electronic loads at some time behave as constant power load (CPL) or need to consume constant power. These types of load may be DC–DC converter with resistive load or DC–AC inverter like as motor driver as mention in [2–4]. The CPLs make the system under negative damping, and system may be going to failure.

In this work, the modeling and co-simulation of the fuzzy logic control for DC–DC buck converter for stabilizing and regulation of voltage control under destabilizing effects caused by constant power load. Verilog HDL model designs the behavioral model.

The modeling and co-simulation of fuzzy logic control by using FPGA-in-the loop for DC–DC boost converter loaded constant power load (CPL) introduced in [5]. It discussed the regulated voltage of boost converter, which is feeding constant power load. The 3-level space vector modulation NPC inverter for a permanent magnet synchronous machine (PMSM) is presented in [6]. The co-simulation environment implemented the proposed control by using an FPGA-in-the loop. The speed control is presented by the improvement of adaptive of the fuzzy logic control for DC motor-based FPGA based on hardware implementation in [7]. The optimization of fuzzy logic used as MPPT of a stand-alone photovoltaic system using FPGA presented in [8].

At present, the field-programmable gate arrays (FPGAs) have widespread applications in power system because they have faster processing [9]. Also, the internal connections of FPGAs can change as required functions due to the interconnections between logic blocks of these devices are electrically programmable. For these reasons, the FPGAs differ from another integrated circuit [10]. FPGAs present the solutions to overcome the limitations of hardware and software-based solutions to implement the fuzzy logic control. The hardware-based solution has a fast processing of inference achievement, but it is not flexible, and the opposite is true for a software-based solution [10–14].

However, the buck converter steps down the source voltage to the desired voltage. The terminals voltage of the buck converter is connected to mixed loads, as shown in Fig. 1. One of these loads is a constant power load (CPL) that will be unstable if any load variations happen at CPL. At nature, CPL is stable, but any changing of the load makes voltage increases/decreases as current decreases/ increases.

The output voltage is a function of duty cycle which is generated from fuzzy logic, as shown in Fig. 1. The error that is the result of reference and measurement values and change of error inputted to fuzzy logic control. Fuzzy logic control generated the proper duty cycle to switch on/off MOSFET switch for the required value of voltage. FPGA-in-the-loop (FIL) is used to implement and model fuzzy logic via the MATLAB/Simulink environment.

Fig. 1 DC–DC buck converter with a mixed load using fuzzy logic controller

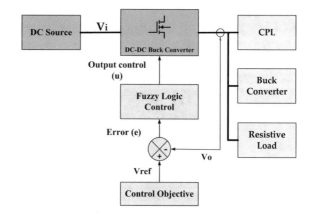

The concept of CPL and its destabilizing effect on the buck converter in section II. FLC structure described in section III. The implementation and modeling of the fuzzy logic controller using the FPGA-in-the loop (FIL) presented in section IV. The simulation results and its discussed in section V. The conclusion is in section VI.

2   Constant Power Load

Some power electronic loads behave as constant power loads under tight control. The incremental negative resistance (INR) is a phenomenon caused by the constant power load. The INR means the change of voltage to change of current is a negative value, as shown in Fig. 2 [2]. The INR impacts the stability of the system and makes under oscillations, and the system may go to failure [15].

From Fig. 3, the equations that described as the behavior of a DC–DC buck converter are feeding a CPL give below:

Fig. 2 The relationship between voltage and current in the constant power loads. The voltage decreases with increasing current to keep the power constant

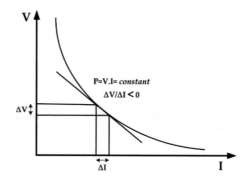

Fig. 3 Buck converter with CPL

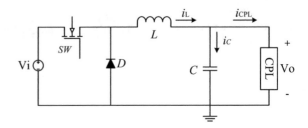

$$\frac{di_L}{dt} = \frac{1}{L}[uV_i - v_o] \qquad (1)$$

$$\frac{dv_o}{dt} = \frac{1}{C}[i_L - i_{CPL}] \qquad (2)$$

Where i_L, V_i, v_o, L, C, i_{CPL}, and u are inductor current, the control function within values [0, 1], the input DC voltage, the capacitor and output voltage, the inductance and capacitance values and the current of CPL which equal to (P_{CPL}/v_o), respectively.

In the [16] that has proved the buck converter will be unstable when are sourcing the CPL and the linear control does not guarantee the stability of source converter with constant power if any step change happens in the load because of the non-linearity of system and INR destabilizing effects [17].

3 Fuzzy Logic Controller

The fuzzy logic is an intelligent and nonlinear controller, which its concept was introduced by Lotfi Zadeh, a professor at the University of California at Berkley [18]. FLC is not needed an exact mathematical model like other control methodologies [18], it requires understanding of the system, which is to controlled [19, 20]. According to [19], the fuzzy control has the best performance among other controllers in the DC–DC converter.

The FLC is proper for complicated and large power system because it has advantages as: is not need accurate mathematical models. Therefore, for a large system, it is useful when the mathematical model becomes difficult for modeling. FLC is depended on deduction so able to incorporate human intuition and experience.

Figure 4 shows the complete structure and principal elements of a typical fuzzy logic controller. These elements are described in [18] as fuzzification module (fuzzifier), knowledgebase, rule base, inference engine, and defuzzification module (defuzzifier).

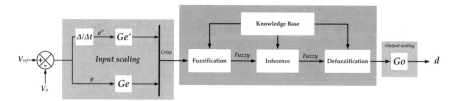

Fig. 4 Structure of fuzzy logic control

- Fuzzification: in this step, classification and transformation of crisp input data into proper linguistic values or sets are done.
- Knowledgebase: The knowledgebase has a database of the system. It has all the necessary definitions for fuzzification processing and membership functions.
- Inference engine: the decision processing is executed.
- Defuzzification: the desired linguistic values are converted into a crisp value.

The base rules of fuzzy logic write in the form of IF … THEN statements. The rules are depended on the fuzzy inference understanding, and the conditions and output are associated with linguistic variables. For example:

IF e is negative big	Rule antecedent
THEN u is positive big	Rule consequent

The input (e) and the output (u) are fuzzy variables (linguistic variables) while Negative Big and Positive Big are the linguistic values.

In this work, the input of fuzzy logic is two inputs error and changing of error. These can calculate as follows:

$$e(n) = V_{\text{ref}} - V_o(n) \tag{3}$$

$$e' = e(n) - e(n-1) \tag{4}$$

The five memberships for each input and divided into *Negative Big (NB)*, *Negative Small (NS)*, *Zero (ZZ)*, *Positive Small(PS)*, and *Positive Big (PB)*. The rules can be written, as shown in Table 1. The error and change of error are two

Table 1 Base rule of fuzzy logic controller

e	e'				
	NB	NS	ZZ	PS	PB
NB	NB	NB	NB	NS	ZZ
NS	NB	NB	NS	ZZ	PS
ZZ	NB	NS	ZZ	PS	PB
PS	NS	ZZ	PS	PB	PB
PB	ZZ	PS	PB	PB	PB

inputs to the fuzzy logic block, and one output (duty cycle) forms the fuzzy logic block to switch the buck converter. The Ge, Ge, and Go are normalization parameters of error, the variation of error and output, respectively as shown in Fig. 4.

4 FPGA-in-the Loop of FLC

First, the FLC is designed in MATLAB/Simulink environment by FLC toolbox. The Verilog HDL module will be generated by using HDL coder tool in Simulink. The Verilog HDL supports different types of data: integer, real and Boolean and std_logic, but it does not support floating-point data type. For this reason, the implementation of Verilog HDL should be done in the fixed-point algorithm.

Singed 20-bit fixed-point data of outputs of rule base to map [−1, 1]. The first bit is a sign (1 for negative and 0 for positive); however, the maximum and minimum expected values for the variables can be calculated as below:

Maximum value = 2^{N-1-1} ($N = 20$, it equals to 1,048,574)
Minimum value = -2^{N-1} ($N = 20$, it equals to −1,048,575)

In FIL, every step of time, the system model will be simulated in the MATLAB/Simulink. Thus, FPGA receives data from the MATLAB/Simulink; it implements a Verilog HDL module for each sample interval. However, the FPGA sends the required control function during the same step to the system model in MATLAB/Simulink. Now, one cycle of the sample time of the FPGA-in-the loop is done.

5 Result and Analysis

The implementation of suggested work uses DE1-SoC Altera FPGA board and MATLAB/Simulink to achieve the Verilog HDL implementation of the fuzzy logic control for DC–DC buck converters. The parameters of the simulation of the system are Vin = 314 V, L = 1.6 mH, C = 150 μF, PCPL = 200 W, RL = 25 Ω, and Switching Frequency = 10 kHz.

After the Verilog HDL module generated, the FIL Wizard will open the command window. The generated block can be replaced with FLC in MATLAB/Simulink. The Simulink model modified by replacing FLC by a generated block from HDL coder.

The step voltage command is designed by changing from 160–180–200–180–160 V with 0.1 s period time, and the results for reference voltage and voltage measurement, which is controlled by Verilog HDL module is shown in Fig. 5. The result presents a proper voltage following response without overshoot occurred and matched the reference value of voltage.

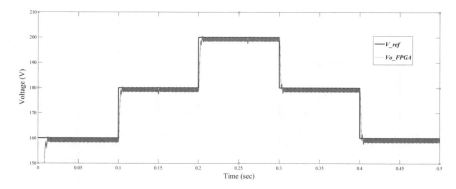

Fig. 5 Step voltage respond

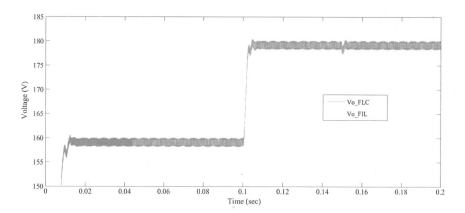

Fig. 6 Comparison of results

Also, the simulation results are included in a comparison between the FLC toolbox and Verilog HDL in Fig. 6. The two results are very similar and have the same rising time, steady-state, and response time values.

6 Conclusion

This work has been successfully introduced a voltage control and stabilization by using the fuzzy logic control for DC–DC buck converter drive and demonstrated its performance from co-simulation to FPGA realization. The fuzzy logic control has been programmed in Verilog HDL. After this, fuzzy logic control is implemented in the DE1-SoC Altera FPGA board. During the modeling, the co-simulation between FLC toolbox and Verilog HDL model is done.

References

1. Emadi, A., Ehsani, M.: Multi-converter power electronic systems: definition and applications. In: IEEE 32nd Annual, Power Electronics Specialists Conference, 2001. PESC. 2001, pp. 1230–1236 (2001)
2. AL-Nussairi, M.K., Bayindir, R., Padmanaban, S., Mihet-Popa, L., Siano, P.: Constant power loads (cpl) with microgrids: problem definition, stability analysis and compensation techniques. Energies 10(10), 1656 (2017)
3. Emadi, A., Khaligh, A., Rivetta, C.H., Williamson, G.A.: Constant power loads and negative impedance instability in automotive systems: definition, modeling, stability, and control of power electronic converters and motor drives. IEEE Trans. Veh. Technol. 55(4), 1112–1125 (2006)
4. Rivetta, C., Williamson, G. A., Emadi, A.: Constant power loads and negative impedance instability in sea and undersea vehicles: statement of the problem and comprehensive large-signal solution. In: Electric Ship Technologies Symposium, 2005 IEEE, pp. 313–20 (2005)
5. Al-Nussairi, M. K., Bayindir, R.: DC-dc boost converter stability with constant power load. In: 2018 IEEE 18th International Power Electronics and Motion Control Conference (PEMC), pp. 1061–66 (2018)
6. Jiang, S., Liang, J., Liu, Y., Yamazaki, K., Fujishima, M.: Modeling and cosimulation of FPGA-based SVPWM control for PMSM. In: 31st Annual Conference of IEEE Industrial Electronics Society, 2005. IECON 2005, pp. 6 (2005)
7. Ramadan, E., El-Bardini, M., Fkirin, M.: Design and FPGA-implementation of an improved adaptive fuzzy logic controller for DC motor speed control. Ain Shams Eng. J. 5(3), 803–16 (2014)
8. Messai, A., Mellit, A., Guessoum, A., Kalogirou, S.: Maximum power point tracking using a GA optimized fuzzy logic controller and its FPGA implementation. Solar Energy 85(2), 265–77 (2011)
9. Akbatı, O., Üzgün, H.D., Akkaya, S.: Hardware-in-the-loop simulation and implementation of a fuzzy logic controller with FPGA: case study of a magnetic levitation system. Trans. Inst. Meas. Control. https://doi.org/10.1177/0142331218813425 (2018)
10. Sulaiman, N., Obaid, Z.A., Marhaban, M., Hamidon, M.: FPGA-based fuzzy logic: design and applications-a review. Int. J. Eng. Technol. 1(5), 491 (2009)
11. Oliveira, D.N., de Souza Braga, A.P., da Mota Almeida, O.: Fuzzy logic controller implementation on a FPGA using VHDL. In: 2010 Annual Meeting of the North American, Fuzzy Information Processing Society (NAFIPS), pp. 1–6 (2010)
12. Hamed, B., Almobaied, M.: Fuzzy PID controllers using FPGA technique for real time DC motor speed control. Intell. Control Autom. 2(3), 233 (2011)
13. Anand, M.S., Tyagi, B.: Design and implementation of fuzzy controller on fpga. Int. J. Intell. Syst. Appl. 4(10), 35 (2012)
14. Cabrera, A., Sánchez-Solano, S., Brox, P., Barriga, A., Senhadji, R.: Hardware/software codesign of configurable fuzzy control systems. Appl. Soft Comput. 4(3), 271–85 (2004)
15. Ghisla, U., Kondratiev, I., Dougal, R.: Protection of medium voltage DC power systems against ground faults and negative incremental impedances. In: Proceedings of the IEEE SoutheastCon 2010 (SoutheastCon), pp. 259–63 (2010)
16. Grigore, V., Hatonen, J., Kyyra, J., Suntio T.: Dynamics of a buck converter with a constant power load. In: 29th Annual IEEE, Power Electronics Specialists Conference, 1998. PESC 98 Record, pp. 72–78 (1998)
17. Emadi, A., Ehsani, M.: Negative impedance stabilizing controls for PWM DC-DC converters using feedback linearization techniques. In: Energy Conversion Engineering Conference and Exhibit, 2000 (IECEC) 35th Intersociety, pp. 613–20 (2000)
18. Cirstea, M., Dinu, A., McCormick, M., Khor, J.G.: Neural and fuzzy logic control of drives and power systems. Journal (2002)

19. Verma, S., Singh, S., Rao, A.: Overview of control techniques for DC-DC converters. Res. J. Eng. Sci. **2278**, 9472 (2013)
20. Patil, B.U., Jagtap, S.R.: Adaptive fuzzy logic controller for buck converter. In: 2015 International Conference on Computation of Power, Energy Information and Communication (ICCPEIC), pp. 0078–82 (2015)

Intelligent Web-History Based on a Hybrid Clustering Algorithm for Future-Internet Systems

S. Selvakumara Samy, V. Sivakumar, Tarini Sood and Yashaditya Singh Negi

Abstract Proposition systems can abuse semantic reasoning abilities to crush ordinary obstacles of current structures and improve the recommendations' quality. In this paper, we present an altered proposition structure, a system that makes usage of depictions of things and customer profiles subject to ontologies in order to outfit semantic applications with redid organizations. The recommender uses zone ontologies to improve the personalization: from one point of view, customer's interests are shown in an inexorably amazing and exact way by applying a space-based inducing system; on the other hand, the stemmer estimation used by our substance-based filtering approach, which gives an extent of the prejudice between a thing and a customer, is updated by applying a semantic likeness procedure Web Usage Mining accepting a basic occupation in recommender structures and web personalization. In this paper, we propose a feasible recommender structure subject to logic and Web Usage Mining. The underlying advance of the technique is isolating features from web files and building imperative thoughts. By then gather logic for the site use the thoughts and basic terms removed from reports and used for analysis. As demonstrated by the semantic closeness of web reports to amass them into different semantic points, the assorted subjects propose particular tendencies. The proposed technique consolidates semantic learning into Web Usage Mining and personalization shapes.

Keywords Semantic learning · Web Usage Mining · Recommender structures

S. Selvakumara Samy (✉) · V. Sivakumar · T. Sood · Y. S. Negi
Department of Software Engineering, SRM Institute of Science and Technology,
Chennai, India
e-mail: selvakus1@srmist.edu.in

V. Sivakumar
e-mail: sivakumv1@srmist.edu.in

T. Sood
e-mail: tarinisood09@gmail.com

© Springer Nature Singapore Pte Ltd. 2020   571
S. S. Dash et al. (eds.), *Artificial Intelligence and Evolutionary Computations*
in Engineering Systems, Advances in Intelligent Systems and Computing 1056,
https://doi.org/10.1007/978-981-15-0199-9_49

1 Introduction

Proposition structures can abuse semantic reasoning abilities to overcome ordinary obstructions of current systems and improve the recommendations' quality. In this paper, we present a redid proposition system, a structure that makes usage of depictions of things and customer profiles reliant on ontologies in order to outfit semantic applications with altered organizations. The recommender uses region ontologies to improve the personalization: from one point of view, customer's interests are shown in an undeniably incredible and exact way by applying a space-based deriving methodology; of course, the stemmer figuring used by our substance-based filtering approach, which gives an extent of the inclination between a thing and a customer, is overhauled by applying a semantic likeness system Web Usage Mining accepting a basic occupation in recommender structures and web personalization. In this paper, we propose a suitable recommender system reliant on theory and Web Usage Mining. The underlying advance of the procedure is isolating features from web chronicles and building critical thoughts. By then collect theory for the site use the thoughts and basic terms removed from reports for the analysis of the web usage mining. As shown by the semantic closeness of web reports to gather them into different semantic points, the assorted subjects propose unmistakable tendencies. The proposed philosophy joins semantic learning into Web Usage Mining and personalization shapes.

2 Related Work

C. Biancalana, A. Micarelli, "Social naming in request advancement: another course for altered web look", Proc. Int. Conf. Comput. Sci. Eng., pp. 1060–1065, 2009. Relational associations and helpful naming structures are rapidly getting a reputation as fundamental techniques for organizing and sharing data: customers mark their bookmarks in order to streamline information dispersing and later question. Social bookmarking organizations are useful in two basic respects: first, they can empower an individual to review the visited URLs, and second, names can be made by the system to oversee customers towards gainful substance. In this paper, we revolve around the last use: we present a novel strategy for tweaked web look for using request expansion [1].

M. Drinking binge et al., "Abusing social relations for inquiry expansion and result situating", Proc. IEEE 24th, Int. Conf. Data Eng. The workshop, pp. 501–506, 2008. Online social order has starting late transformed into a pervasive mechanical assembly to circulate and looking substance, similarly concerning finding and interfacing with various customers that share ordinary interests. The substance is usually customer made and fuses, for example, singular web diaries, bookmarks and propelled photos. A particularly spellbinding sort of substance is customer-delivered remarks (marks) for substance things, as these compact string portrayals consider clarifications about the interests of the customer who made the substance, yet furthermore about the customer who made the remarks. This paper shows a structure to cast the unmistakable substances of such frameworks into a bound together outline exhibit addressing the regular associations of customers, substance and marks [2].

A. Zubiaga, A. P. Garcia-Plaza, V. Fresno, R. Martinez, "Content-based gathering for name cloud portrayal", Proc. Int. Conf. Advances Soc. Netw. Butt-driven. Mining, pp. 316–319, 2009. Social naming systems are transforming into a captivating technique to recoup web information from as of late remarked on data. These areas present a name cloud made up by the most standard marks, where neither name gathering nor their relating content is considered. We present a framework to secure and imagine a dimness of related marks reliant on the usage of self-dealing with maps, and where the relations among names are set up thinking about the abstract substance of named reports [3].

S. Aliakbary, H. Abolhassani, H. Rahmani, B. Nobakht, "Site page Classification Using Social Tags", Proc. IEEE Int'l Conf. Computational Science and Eng., vol. 4, pp. 588–593, 2009. Social naming is a system in which various customers add metadata to a topical substance. Through the past couple of years, the noticeable quality of social marking has created on the web. In this paper, we analysed the usage of social names for webpage portrayal, adding new site pages to a present web list [4].

M. Vojnovic, J. Voyage, D. Gunawardena, P. Marbach, "Situating and Suggesting Popular Items", IEEE Trans. Data and Data Eng., vol. 21, no. 8, pp. 1133–1146, Aug. 2009. We consider the issue of situating the unmistakable quality of things and suggesting surely understood things reliant on customer analysis. Customer input is gotten by iteratively showing a lot of proposed things and customers picking things reliant on their own tendencies either from this suggestion set or from the course of action of each possible thing. The goal is to quickly get acquainted with the certified commonness situating of things (fair-minded by the made proposals) and propose authentic understood things. The inconvenience is that creating proposition to customers can reinforce the universality of specific things and wind the ensuing thing situating. The portrayed issue of situating and proposing things rises in grouped applications, including look request proposition and name suggestions for social marking systems [5].

P. Symeonidis, A. Nanopoulos, Y. Manolopoulos, "A Unified Framework for Providing Recommendations in Social Tagging Systems Based on Ternary Semantic Analysis", IEEE Trans. Learning and Data Eng., vol. 22, no. 2, pp. 179–192, Feb. 2010. Social naming is the methodology by which various customers incorporate metadata as catchphrases, to clarify and arrange things (songs, pictures, web joins, things, etc.). Social naming structures (STSs) can give three particular sorts of proposition: They can endorse (1) marks to customers, in the light of what names distinctive customers have used for comparable things; (2) things to customers, in perspective on names they share for all plans and reason with other relative customers; and (3) customers with ordinary social interest, in perspective on customary names on near things [6].

J. Xia, K. Wen, R. Li, X. Gu, "Improving Academic Conference Classification Using Social Tags", Proc. IEEE Int'l Conf. Computational Science and Eng., pp. 289–294, 2010. Automatically portraying insightful social event into semantic subject ensures improved educational sweep and scrutinizing for customers. Social naming is a certainly pervasive technique for portraying the purpose of academic gathering [7].

3 Proposed System

- In proposed structure, we present a modified recommendation system, a system that makes usage of depictions of things and customer profiles reliant on ontologies in order to outfit semantic applications with tweaked organizations (Fig. 1).

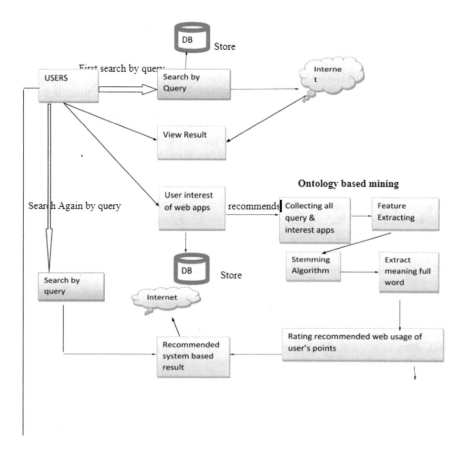

Fig. 1 Ontology based mining

- The semantics system is achieved by using two particular techniques. A region-based system makes acceptances about customer's interests, and a logical classification-based likeness strategy is used to refine the thing customer planning computation, improving as a rule result. The recommender proposed is without region and is completed as a web organization, and usages both expressed and comprehended analysis aggregation strategies to get information on customer's interests.

- Proposed recommender structure is reliant on rationality and Web Usage Mining. The underlying advance of the procedure is expelling features from web chronicles and creating huge thoughts. By then develop logic for the site use the thoughts and basic terms removed from reports and used for further web Usage Mining. According to the semantic closeness of web chronicles to cluster them into different semantic subjects, the various points propose particular tendencies.

3.1 Advantages

- Integrating zone learning with web use data improves the execution of recommender systems using cosmology-based web mining strategies.

- The improvement in this model is semi-modernized with the objective that the headway tries from architects can be diminished.

- The customer profile learning figuring, responsible for expanding and keeping up excellent the whole deal customer's interests, uses a zone-based determination system in blend with other congruity analysis techniques to populate even more quickly the customer profile and thusly reduce the normal cold-start issue.

- The filtering count, that seeks after a stemming approach, makes usage of a semantic comparability procedure reliant on the different levelled structure of the power to refine the thing customer organizing score estimation.

3.1.1 Creating Search History

Any up close and personal records, for instance, examining history and messages on a customer's PC could be the data hotspot for customer profiles. This revolves around persistent terms compelling the dimensionality of the chronicle set, which further gives an undeniable depiction of customers' leverage. This module empowers the web crawler to all the almost certain fathom a customer's session and conceivably tailor that customer's interest involvement according to his or her needs. At the point when request packs have been recognized, web records can have a good depiction of the chase setting behind the present inquiry using request and snaps in the looking at inquiry gathering.

3.1.2 Query Gathering

Customer's request can be portrayed in different request gatherings. Thought-based customer profiles are used in the grouping technique to achieve personalization sway. The most relative pair of thought is centre points, and from that point onwards, mix the most equivalent pair of request centre points and so forth. Each individual request set up together by each customer is treated as an individual centre point and each inquiry with a customer identifier. We play out the social affair in a near ground-breaking plan, whereby we in front of every other person present request and snaps into an inquiry gathering.

3.1.3 Query Reformulation

To ensure that every request accumulated contains immovably related and relevant request and snaps, it is fundamental to have a fitting congruity between the present inquiry social affairs. We acknowledge that customers all around the issue in a general sense are equivalent to request and snaps inside a brief time span. The interesting history of innumerable contains motions with respect to request congruity, for instance, which request will when all is said in done be issued solidly together. This gets the association between inquiries a great part of the time inciting taps on relative URLs. Question reformulation outline and the request click graph from chase logs, and how to use them to choose congruity between the request or question clusters inside a customer's history.

3.1.4 History Get-Together

Question bundles are to at first treat every request in a customer's history as a request gathering and a short time later merge these inquiry groups in an iterative style (in a k-infers). Regardless, this is unfeasible in our circumstance for two reasons. First, it may have the undesirable effect of changing a customer's present inquiry social events, conceivably fixing the customer's very own manual undertakings in dealing with his or her history. Second, it incorporates a high computational cost, since we would need to go over a broad number of request collect closeness estimations for each new inquiry.

3.2 Pseudo-document

In this paper, we have to plot session to pseudo-records User Search objectives. The structure of a pseudo-record joins two stages. One is tending to the URLs in the investigation session. URL in an investigation session is tended to by a little substance fragment that contains its title and piece. At that point, some printed procedures are executed to those substance territories, for example, changing the

Fig. 2 Search sequence

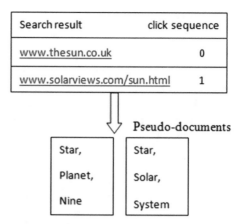

majority of the letters to lowercase, stemming and evacuating stop words. Another is forming pseudo-report dependent on URL delineations. So as to get the part delineation of an investigation session, we propose an improved technique to join both clicked and unclicked URLs in the data session (Fig. 2).

3.3 User Search Goals

We assemble pseudo-archives by K-recommends grouping, which is principal and persuading. Since we don't have the foggiest idea with respect to the reasonable number of client look focuses for every request, we set K to be five uncommon qualities and perform gathering dependent on these five attributes, only. Coming about to pressing all the pseudo-records, each social affair can be considered as one client look objective. The middle inspiration driving a group is set up as the standard of vectors of all the pseudo-narratives on the social occasion.

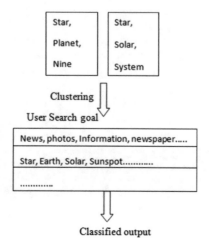

4 Experiments

We have created a login portal for the user to enter his or her details after registration. Later after logging in successfully, the user will enter his desired search query. The first search engine result will depend upon the ranking algorithm, which is used by the systems these days. After submitting the click-through link, the user will get the different outputs as results based on the hybrid clustering algorithms (Figs. 3 and 4).

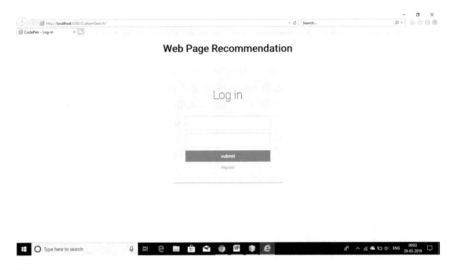

Fig. 3 Login page

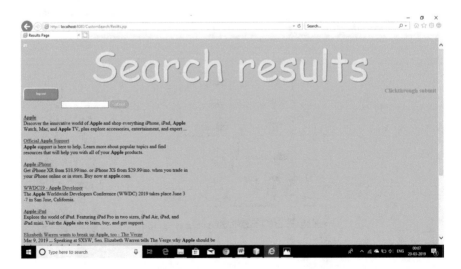

Fig. 4 Search queries

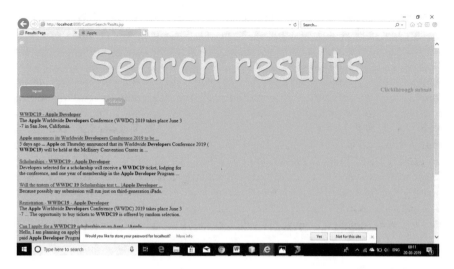

Fig. 5 Search results

For Example: Once the user successfully registers online, he will get personalized outputs for their search queries (Fig. 5).

4.1 Mathematical Models

First, we interpret the majority of our characters to numbers, '*a*' = 0, '*b*' = 1, '*c*' = 2, …, '*z*' = 25. We would now be able to speak to the Caesar figure encryption work, $e(x)$, where x is the character we are encoding, as:

$$e(x) = (x + k)(\mathrm{mod}\, 26) \tag{4.1}$$

where k is the key (the move) connected to each letter. In the wake of applying this capacity, an outcome is a number which should then be made an interpretation of once more into a letter. The decoding capacity is:

$$e(x) = (x - k)(\mathrm{mod}\, 26)$$

5 Result Analysis

On the basis of our project development and research, we have made a comparative study between the present system and the proposed system.

In the proposed system, we present a personalized-recommendation system, a system that makes use of representations of items and user profiles based on ontologies in order to provide semantic applications with personalized services. The semantics method is achieved by using two different methods. A domain-based method makes inferences about user's interests, and a taxonomy-based similarity method is used to refine the item-user matching algorithm, improving overall results. The recommender proposed is domain-independent, is implemented as a web service and uses both explicit and implicit feedback-collection methods to obtain information on the user's interests. Proposed recommender system is based on ontology and Web Usage Mining. The first step of the approach is extracting features from web documents and constructing relevant concepts. Then build ontology for the web site use the concepts and significant terms extracted from documents. According to the semantic similarity of web documents to cluster them into different semantic themes, the different themes imply different preferences.

ADVANTAGES

- Integrating domain knowledge with web usage knowledge enhances the performance of recommender systems using ontology-based web mining techniques.
- The construction of this model is semi-automated so that the development efforts from developers can be reduced.
- The user profile learning algorithm, responsible for expanding and maintaining up to date the long-term user's interests, employs a domain-based inference method in combination with other relevance feedback methods to populate more quickly the user profile and therefore reduce the typical cold-start problem.
- The filtering algorithm, which follows a stemming approach, makes use of a semantic similarity method based on the hierarchical structure of the ontology to refine the item-user matching score calculation.

6 Conclusion

In this paper, a novel methodology has been proposed to reason customer look destinations for a solicitation by clustering its examination sessions tended to by pseudo-records. In any case, we familiarize examination sessions with the neediness is stricken down to infer customer look focuses rather than location things or clicked URLs. Both the clicked URLs and the unclicked ones going before the last snap are considered as customer evident wellsprings of data and considered to make examination sessions. Along these lines, input sessions can reflect customer information needs more productively. Second, we map examination sessions to pseudo-records to questionable target messages in customer minds. The pseudo-reports can actuate the URLs with additional scholarly substance including the titles and bits. In setting on these pseudo-records, customer look for targets would then be able to found and outlined with express catchphrases. Finally, another establishment stemming is proposed to survey the execution of customer scan for target inciting. Test results on customer examine logs from a business web searcher demonstrate the abundance of our proposed structures.

References

1. Biancalana, C., Micarelli, A.: Social naming in request advancement: another course for altered web look. In: Proceedings of International Conference Computing in Science & Engineering, pp. 1060–1065 (2009)
2. Drinking Binge, M., et al.: Abusing social relations for inquiry expansion and result situating. In: Proceedings of IEEE 24th International Conference on Data Engineering Workshop, pp. 501–506 (2008)
3. Zubiaga, A., Garcia-Plaza, A.P., Fresno, V., Martinez, R.: Content-based gathering for name cloud portrayal. In: Proceedings of International Conference on Advances Social Network Butt-driven. Mining, pp. 316–319 (2009)
4. Aliakbary, S., Abolhassani, H., Rahmani, H., Nobakht, B.: Site page classification using social tags. In: Proceedings IEEE International Conference on Computational Science and Engineering, vol. 4, pp. 588–593 (2009)
5. Vojnovic, M., Voyage, J., Gunawardena, D., Marbach, P.: Situating and suggesting popular items. IEEE Trans. Data Data Eng. **21**(8), 1133–1146 (2009)
6. Symeonidis, P., Nanopoulos, A., Manolopoulos, Y.: A Unified framework for providing recommendations in social tagging systems based on ternary semantic analysis. IEEE Trans. Learn. Data Eng. **22**(2), 179–192 (2010)
7. Xia, J, Wen, K., Li, R., Gu, X.: Improving academic conference classification using social tags. In: Proceedings of IEEE International Conference Computational Science and Engineering, pp. 289–294 (2010)

Cluster Algorithm Integrated with Modification of Gaussian Elimination to Solve a System of Linear Equations

Tatyana A. Smaglichenko, Alexander V. Smaglichenko, Wolfgang R. Jacoby and Maria K. Sayankina

Abstract The data accumulation and their inhomogeneous distribution lead to the issue of large and sparse systems solving in various fields: industrials, emergency management, etc. Complex structure in the data error creates additional risk to obtain an adequate solution. To facilitate problem-solving, we describe the technique that is based on intellectual division of data with following application of cluster algorithm and the modification of Gaussian elimination to different portions of data. In this paper, we present results of developed technique that was applied to samples of synthetic and real data. We compare them with outcomes of other algorithms (intelligence and classical) by using of numerical estimates and graphical format.

Keywords Large and sparse system · Intelligence algorithms · Gaussian elimination

T. A. Smaglichenko (✉) · M. K. Sayankina
Russian Academy of Sciences, Research Oil and Gas Institute, IPNG RAS Moscow, Moscow, Russia
e-mail: t.a.smaglichenko@gmail.com

M. K. Sayankina
e-mail: msayankina@gmail.com

A. V. Smaglichenko
Russian Academy of Sciences, V.A. Trapeznikov Institute of Control Sciences, IPU RAS, Moscow, Russia
e-mail: losaeylin@gmail.com

Institute of Seismology and Geodynamics, V.I. Vernadsky Crimean Federal University, Simferopol, Republic of Crimea

W. R. Jacoby
Institute of Earth Sciences, Johannes Gutenberg University Mainz, Mainz, Germany
e-mail: jacoby@uni-mainz.de

© Springer Nature Singapore Pte Ltd. 2020
S. S. Dash et al. (eds.), *Artificial Intelligence and Evolutionary Computations in Engineering Systems*, Advances in Intelligent Systems and Computing 1056,
https://doi.org/10.1007/978-981-15-0199-9_50

1 Introduction

The system unknown parameter to be determined requires the data from relevant measurements. A generally accepted opinion is that the more measurements are used, the better the system will be resolved. Thus, the system is over-determined because the number of unknowns is much less than the number of equations. At the same time, the main feature of modern data is their clustering, homogeneity, and consequently the information has such a property as repeatability. And hence for some cases the system can be transformed from over-determined to under-determined [1]. In computing algebra, a solving systems of linear equations are performed by applying classical methods that include direct techniques based on Moore-Penrose inverse [2–4], iterative [5, 6] and projection techniques [7, 8] and also by using intelligence methods [9], which involve genetic [10], swarm [11] and memetic [12, 13] algorithms.

If the system is consistent and well-conditioned, then any method is able to construct a single and stable solution. However in practice, which deals with measurements under an influence of various factors the large consistent systems are absent. The first goal of the proposed approach is to overcome the causes that may lead to inconsistency of the system. It is obvious that the main reason is a sufficiently large number of zeros in the original matrix, its sparseness. Therefore the choice of data must be made in such a way as to solve filled subsystems. Idea of nonzero elements processing has been utilized solving systems with sparse and band matrix long time ago. The scheme was based on reorder of initial matrix, on storage and operating with nonzero diagonal elements [14]. Next idea of a block scheme was applied for solving a cluster problem for systems with symmetric matrix [15]. Note in both cases unknowns are obtained in substitution classical process without any transitional estimates. Similar example for two blocks of arbitrary matrix one can find in [1].

When working with large and sparse arbitrary matrix, the filled subsystems may be found by selecting the cluster data that can be homogeneous data. Such data exist for instance in nature as a distribution of microseismicity. Namely, the earthquake hypocenters having a small magnitude are often clustery located in small swarms. They are registering by seismic stations on the surface of the Earth. Considering the same station, we are dealing with homogeneous, repetitive linear equations of seismic rays. Other examples can be taken from the biological world. Bats may be considered as simultaneous acoustic sources and receivers. They move in space to search of prey, release an echo signal and receive it without changing of the information until they meet the prey (heterogeneity) or some obstacle. Dolphins, whales, some birds and even mice use echolocation also.

In the next section, the developed formulas are given in order to process homogeneous data. Then we apply these formulas to the sample of artificial data published by other researchers and analyze a difference of solutions obtained by the genetic, memetic and proposed algorithm. In Sect. 4, we provide a sample of a real microseismic data and illustrate the result of their processing.

2 Cluster Algorithm

2.1 Description of Basic Formulas

The cluster algorithm is originated from the relaxation method that uses the iterative expansion of columns of an initial matrix by adding known column vector with following application of the principle of relaxation and selection of the maximal vector [16]. The relaxation method has been farther developed and applied to seismic data in [17] demonstrating its efficiency in comparison with the Singular Value Decomposition (SVD) direct technique if individual inhomogeneous structure is reconstructing. At each iteration step, a candidate solution is the separated component that provides a local minimum of the functional in a sense of least squares. However, if we are dealing with cluster homogeneous data, then for subsystem $Bu = v$ having repeated rows all components pretend to this role. Therefore, the first approximation, which is calculated by the following formula, makes it possible to estimate the value of each homogeneous component of the solution:

$$u_k = \frac{(b_k, \tilde{v}_k)}{(b_k, b_k)}, \tag{1}$$

where b_k is kth column of matrix B, $\tilde{v}_k$ is equal to component wise multiplication of vectors v and $\hat{b}_k$, $\hat{b}_k$ is equal to $b_k / \sum_{m=1}^{M} b_{mk}$, here b_{mk} is element of kth row of matrix B, M be a number of elements in kth row, $(\,,\,)$ denotes a scalar product.

A check of the solution is performed through calculating resolution parameters, which show how one component or another ensures that the difference between the left and right parts of the system is close to zero:

$$u_k = \frac{(b_k, \tilde{v}_k)^2}{(b_k, b_k)(\tilde{v}_k, \tilde{v}_k)}, \tag{2}$$

The cluster algorithm has been applied to the synthetic data of a hydrocarbon simulation model, and its effectiveness with respect to the classical direct SVD method was again confirmed [18]. Homogeneous part of the model that imitated clay (shale) was reconstructed due to homogeneous components defined from formula (1), while the SVD solution determined many heterogeneous components. However, it should be noted that the SVD resolution parameters showed bad resolution for these components. This confirms the importance of estimating the resolution parameters when solving a problem by any method.

Table 1 Outcomes of classical and intelligence algorithms [12] and the cluster algorithm result

Equations of the system	Gauss-elimination	Gauss-Jordan	Genetic algorithm	Memetic algorithm	Proposed cluster algorithm
$2x + 4y + z = 5$ $4x + 4y + 3z = 8$ $4x + 8y + z = 9$	0.5, 0.75, 1.0	0.50, 0.74, 0.99	0, 1, 1	0.55, 0.72, 0.97	0.66, 0.66, 1

Table 2 Deviations from the accurate solution

Equations of the system	Genetic algorithm	Memetic algorithm	Proposed cluster algorithm
$2x + 4y + z = 5$ $4x + 4y + 3z = 8$ $4x + 8y + z = 9$	0, 1, 0	0.05, 0.01, 0.07	0, 0.3, 0

2.2 Simple Numerical Experiment to Compare the Cluster Algorithm with Others

In this paper, we conduct a new numerical experiment on the example of the simplest system of equations in order to compare solutions obtained by the cluster algorithm with outcomes of other intellectual algorithms: genetic and memetic. To do this, we used the study [12] performed by the authors L.O. Mafteiu-Scai and E. J. Mafteiu-Scai. From a set of small systems that were analyzed in [12], we selected the one that is characterized by a presence of two dependent rows that are repeating rows under $z = 1$. The use of Eq. (1) gave the result, which is shown in Table 1. One can see that the cluster algorithm provides an opportunity for an additional solution.

Table 2. illustrates deviations of solutions from the accurate one for each intelligence algorithm. The memetic algorithm provides the most minimal deviation, and, perhaps, it is a good advantage over others. However, the solution proposed by this algorithm is not homogeneous. Components of the solution that correspond to homogeneous Eqs. (1) and (3) are different. This means that despite the fact that from a formal point of view the memetic solution is more accurate it does not characterize the cluster nature of the data distribution, their similarity.

3 Integrated Algorithm

Clearly, large and sparse systems contain inhomogeneous parts besides of homogeneous subsystems. In the introduction section, we have considered various algorithms including classical and intelligence. Some of them use regularization operators to stabilize the inversion solution. They try to overcome the data error and ill-conditioning in the system. General assumption is that the data have some small level of noise and solution must be stable under a small deviation of the noise in the data. However, in practice there are many sources of error in the data. In a case of microseismic data, the error structure is complex and it consists of not only from noise caused by measurements but also from mistake in readings of seismograms, from the parametrization error, which depends on the configuration of the experiment. In order to decrease an influence of unpredictable error on the inversion outcome, the classical Gaussian elimination was revised by the first author [19]. Normally, Gauss scheme transforms the initial matrix to a triangular form to eliminate step by step the unknown. Applying Modification of Gaussian Elimination (MGE), we divide the ill-conditioned sparse matrix into a set of sub-matrixes and solve a set of subsystems. Selecting solutions of consistent subsystems for the same unknown, one can estimate the probability of a stable solution for this unknown.

Intellectual division of input data into homogeneous and inhomogeneous portions gives a possibility to integrate various inversion outcomes for the same unknown component. In general, the structure of the proposed approach can be presented as following steps.

1. Input nonzero elements of initial matrix and known column of the system;
2. Select filled subsystems having defined dimension;
3. Choose homogeneous subsystems and apply the cluster algorithm;
4. Choose other subsystems, find their consistent parts and apply the Gauss-Jordan method;
5. Using the result of items 3 and 4, choose solutions for the same unknown component;
6. Determine a probability of near equal values for the unknown component and set the stable value;
7. Eliminate the found components from equations of the initial system;

Return to item 2 and repeat items 3–7 till the filled subsystems will be determined.

4 Application to Real Microseismic Data

The MGE approach integrated with the cluster algorithm has been applied to solve the system with the 4000×380 matrix. Each row of the matrix is in correspondence with the seismic ray that goes from the microearthquake hypocenter to seismic station at the Earth surface. Seismic events are mainly happened from 1986 to 1989 y. along Grimsey Lineament (GL), the Northeast Iceland. Each column of the matrix conforms the block of a geological medium, through which a seismic ray penetrates. Each element of the matrix is a length of seismic ray in a block. Components of unknown vector are P-wave seismic velocity perturbations in

Table 3 Solution of subsystem. Example of real data

Velocity perturbations, values of vector u	Resolution parameter
−0.0072, −0.0076, −0.0075	0.9995, 0.9996, 0.9994

blocks, owing to which values of seismic velocity V_P are calculated. Components of known vector are travel time residuals of seismic rays, which can be calculated taking data of the visual analysis of seismograms.

Let us provide an example of subsystem $Bu = v$ that was selected by the integrated algorithm.

$$
B = \begin{vmatrix} 7.0743 & 0.8728 & 6.9276 \\ 6.9566 & 0.9001 & 6.1843 \\ 7.0806 & 1.2535 & 5.9371 \\ 6.9701 & 1.1311 & 6.2822 \end{vmatrix} \quad v = \begin{vmatrix} -0.1109 \\ -0.1041 \\ -0.1078 \\ -0.1064 \end{vmatrix} \tag{3}
$$

Values of unknown vector were obtained owing to Eq. (1), while values of resolution parameter were calculated using Eq. (2) (see Table 3).

5 Discussion

Figure 1. illustrates the inversion result in the depth range 5–10 km. Line AB corresponds to the profile, along which we study the character of the velocity distribution. In upper right corner of Fig. 1, one can see two neighboring velocity anomalies denoted black and white colors.

Next, we investigate the result of classical LSQR algorithm [7], which was applied for the same area to a big data set, which involved microearthquakes recorded between 1994 and 2002 [20]. The number of rows of the matrix was equal to 44,794. Thus, the data set was 11 times larger than the data set that was utilized by the integrated algorithm. Figure 2 displays the result obtained along the line AB by authors C. Riedel et al. in [20]. Lower picture clearly shows that the distribution of seismic velocity in the depth range 5–10 km is very similar to those found by the integrated algorithm along the same line AB.

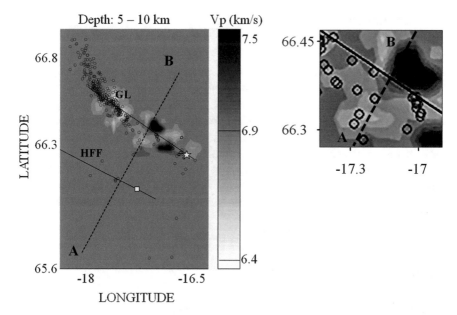

Fig. 1 Distribution of *P*-wave seismic velocity, V_P found as a solution of the system by the integrated algorithm. Hypocenters of microseismic events are noted by open circles

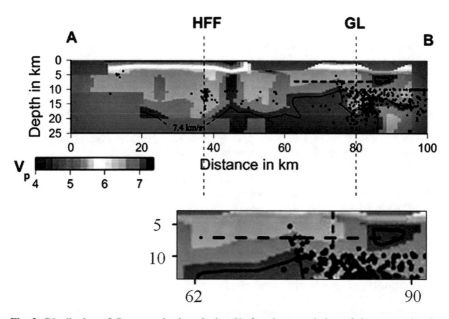

Fig. 2 Distribution of *P*-wave seismic velocity, V_P found as a solution of the system by the classical LSQR algorithm [20]. Hypocenters of microseismic events are noted by dotes

6 Conclusion

Comparison of the cluster algorithm result and outcomes of other intelligence algorithms shows that the important feature of the proposed algorithm is the possibility of constructing an additional solution under the existing problem of non-uniqueness. Comparison of the application results for the developed integrated technique and classical LSQR algorithm reveals the advantage of the integrated strategy that is in a possibility to construct the adequate solution by using a smaller data set than in the case of LSQR algorithm.

We thank DAAD foundation for support, due to which the part of results presented in this paper. The research was carried out within the framework of the state projects № AAAA-A19-119013190038-2, № AAAA-A19-119101690016-9, № 10.331-17, № 5.6370.2017/BCh.

References

1. Lanczos, C.: Applied Analysis. Prentice-Hall. New Jersey: Englewood Cliffs, p. 524, (1956)
2. Moore, E.H.: On the reciprocal of the general algebraic matrix. Bull. Am. Math. Soc. **26**, 394–395 (1920)
3. Penrose, R.A.: Generalized inverse for matrices. Proc. Cambridge Philos. Soc. **51**, 406–413 (1955)
4. Matlab copyright. Version 7.13.0.564 (R2011b), Ram disc (DVD), (2011)
5. Cormack, A.: Representation of a function by its line integrals with some radiological application. J. Appl. Phys. **34**(9), 2722–2727 (1963)
6. Golub, G.H., Overton, M.L.: The convergence of inexact Chebyshev and Richardson iterative methods for solving linear systems. Numer. Math. **53**(5), 571–593 (1988)
7. Paige, C.C., Saunders, M.A.: LSQR: an algorithm for sparse linear equations and sparse least squares. ACM Trans. Math. Softw. **8**, 43–71 (1982)
8. Larsen, R.M.: Lanczos bidiagonalization with partial reorthogonalization. Department of Computer Science, Aarhus University, Technical Report, DAIMI PB-357, (1998). The PROPACK package realizing this algorithm is available at http://sun.stanford.edu/~rmunk/PROPACK/lanbpro.txt. Website last updated May 16, 2012
9. Terfaloaga, I.M.: Solving systems of equations with techniques from artificial intelligence. Analele universității "Eftimie Murgu" Reşiţa anul **XXII**(1), 307–404 (2015)
10. Abiodun, I.M., Olawale, L.N., Adebowale, A.P.: The effectiveness of genetic algorithm in solving simultaneous equations. Int. J. Comput. Appl. **14**(8), 38–41 (2011)
11. Zhou, Y., Huang, H., Zhang, J.: Hybrid artificial fish swarm algorithm for Solving Ill-conditioned linear systems of equations. In: ICICIS 2011 Proceedings, Part 1, pp. 656–662.Springer (2011)
12. Mafteiu-Scai, L.O., Mafteiu-Scai, E.J.: Solving linear systems of equations using a memetic algorithm. Int. J. Comput. Appl. **58**(13), 16–22 (2012)
13. Mafteiu-Scai, L.O.: A new approach for solving equations systems inspired from brainstorming. Int. J. New Comput. Architectures Appl. (2015). (The Society of Digital Information and Wireless Communications)
14. Ogbuobiri, F.C., Tinney, M.F., Walker, J.W.: Sparsity directed decomposition for Gausson Elimination on matrixes. IEEE Trans. **PAS-89**(1), 141–150 (1970)

15. Gomez, A., Franqueio, L.G.: Multi-processor architectures for solving sparse linear systems. Appl. Load Flow Problem IFAC Proc. **19**(13), 315–320 (1986)
16. Householder, A.S.: Principles of Numerical Analysis. McGraw-Hill, New York (1953)
17. Smaglichenko, T.A., Nikolaev, A.V., Horiuchi, S., Hasegawa, A.: The method for consecutive subtraction of selected anomalies: the estimated crustal velocity structure in the 1996 Onikobe (M = 5.9) earthquake area, northeastern Japan. Geophys. J. Int. **153**, 627–644 (2003)
18. Smaglichenko, T.A., Sayankina, M.K., Smaglichenko, A.V.: Computational tactic to retrieve a complex seismic structure of hydrocarbon model. Emergence, Complexity, Computations. Book Series, vol. 5, pp. 215–237. Springer Publisher, Heidelberg, New York, Dordrecht, London (2013)
19. Smaglichenko, T.A.: Modification of Gaussian elimination to the complex seismic data. In: AIP Conference on Proceeding of Numerical Analysis and Applied Mathematics, ICNAAM 2011, vol. 1389, pp. 1003–1006 (2011)
20. Riedel, C., Tryggvason, A., Dahm, T., Stefanson, R., Bodvarson, R., Gudmundsson, G.B.: The seismic velocity structure north of Iceland from joint inversion of local earthquake data. J. Seismolog. **9**, 383–404 (2005)

Simulation of Quantum Channel and Analysis of Its State Under Network Disruption

S. Praveen Kumar, Aishwarya Balaje, Ananya Banerjee, Induvalli, T. Jaya and Sonali Sharma

Abstract Presently the work done in the field of quantum cryptography is primarily theoretical, the purpose of this paper is to showcase its implementation in an observable manner. Thus, this paper establishes a more secure network using quantum key distribution (QKD) for data transfer between sender and receiver and also enables the quick identification of an eavesdropper in the said network. An analysis of the quantum channel traffic at the ideal state and also during network disruption (i.e. when the quantum state collapses) has been carried out. Due to the complex nature of quantum networks, a physical implementation of the same is not feasible. Hence, a simulation has been implemented via the use of NS-3 (Network Simulator Version 3).

Keywords Quantum key distribution · Quantum cryptography · Simulation · NS-3 · Quantum channel

S. Praveen Kumar (✉) · T. Jaya
Department of ECE, VELS Institute of Science, Technology and Advanced Studies, Chennai, India
e-mail: praveenkumar.se@ktr.srmuniv.ac.in

T. Jaya
e-mail: jaya.se@velsuniv.ac.in

A. Balaje · A. Banerjee (✉) · Induvalli · S. Sharma
Department of ECE, SRM Institute of Science and Technology, Chennai, India
e-mail: nishupiki@gmail.com

A. Balaje
e-mail: aishwaryabalaje@gmail.com

Induvalli
e-mail: induvalli9@gmail.com

S. Sharma
e-mail: sonali3006sharma@gmail.com

© Springer Nature Singapore Pte Ltd. 2020   593
S. S. Dash et al. (eds.), *Artificial Intelligence and Evolutionary Computations in Engineering Systems*, Advances in Intelligent Systems and Computing 1056,
https://doi.org/10.1007/978-981-15-0199-9_51

1 Introduction

In the current age, data security has become one of the foremost priorities of existing organizations; this is majorly due to the value that is attached to information. Hence, the existence of nefarious individuals whose sole purpose is to obtain such data is inevitable. This led researchers to look towards other means of encrypting data, thus quantum particles due to their fragile nature were considered appropriate for this purpose.

Quantum communication takes advantage of the laws of quantum physics to protect data. These laws allow particles to transmit data along optical cables. These particles are known as quantum bits or qubits. Their super-fragile quantum state "collapses" to either 1 or 0, if a hacker tries to observe them in transit. This enables the quick identification of an eavesdropper in the network.

The purpose of this paper is to implement quantum key distribution in an observable manner via the means of a simulation. This is done in order to understand the nuances in the establishment and the working of a quantum as well as a public channel. This is done using NS-3 and QKDNetSim environment that had been built into it.

This paper's contents have been divided into the following sections: Sect. 2 describes the process of intruder identification via QKD. Section 3 describes the system model utilized for simulation. In Sect. 4, the obtained results have been discussed. Section 5, the discussion, addresses the issues faced when simulating the network in the quantum channel versus the public channel. Section 6 deals with the existing developments that can shape future research. We have summarized our contribution in Sect. 7.

2 Intruder Identification via QKD

Quantum key distribution refers to a method in which a private key is shared between two parties using the quantum channel; this is authenticated via the public channel. The key is used to encode and decode messages exchanged by both parties over the public/classical channel. Due to the fragile nature of quantum particles, the presence of an eavesdropper can be accurately identified if it intercepts the quantum channel. In this case, the generation of the key is terminated by the QKD protocol. QKD is one of the most widely known methods of quantum cryptography; it provides information-theoretic security (ITS) solution [1] to the key exchange problem.

As mentioned earlier, QKD process depends on the laws of quantum physics, which have been discussed in the following section:

- Decoherence: This refers to the property that causes qubits to decay and ultimately disappear while interacting with the environment hence making only point to point communication between two nodes feasible.

Fig. 1 Orthogonal state
representation

$$|\varphi\rangle = \alpha|0\rangle \pm \beta|1\rangle,$$

$$|\emptyset\rangle = \alpha|0\rangle \pm \beta|1\rangle.$$

- Superposition: Qubits can represent multiple combinations of 1 and 0 simultaneously.
- Uncertainty principle: Measurement of properties in quantum physics cannot be done in the same way as classical physics. In the quantum scale, some of the physical properties of certain pair of particles are complementary. This statement defines Heisenberg's uncertainty Principle. The primary property utilized QKD is photon polarization.
- Entanglement: This describes a state in which two or more quantum particle's physical properties are strongly correlated. This property can facilitate the research in long-distance quantum key distribution [2] (Fig. 1).

The uncertainty principle-based QKD protocol BB84 developed by Charles H. Bennett and Gilles Brassard is applied in this paper in order to visualize the process of data transfer through a quantum channel. We choose this protocol due to its prompt identification of the presence of an intruder in the quantum channel via this, and because its implementation in a simulation environment when compared to other protocols is easy [3, 4]. It has two bases of measurements (orthogonal states) and four photon polarization states. It begins with the transmission of photons which have four random quantum states, relating to two mutually conjugate bases, rectilinear and diagonal [5]. The rectilinear basis has two polarizations namely 0° represented horizontally and 90° represented vertically. The diagonal basis has 45° and 135°. The measurement of these polarizations cannot be done simultaneously, as if done they randomize each other. Hence, if an intruder attempts to access information from the quantum channel this will change the polarization of the intercepted photon. Thus, alerting the users of the presence of an eavesdropper in the network.

3 System Model

3.1 Theoretical Process

The quantum key distribution process consists of primarily three steps [6]. Although the QKD protocols only define the first two stages:

- Key Exchange: The raw key is generated and is exchanged in the form of Qubits between the two parties.
- Key Sifting: Certain cases from the raw key are selected and checked if they are in perfect correlation between the sender and the receiver. After the sifting step,

both parties share the sequence of correlated bits, called the sifted key. The information revealed ensures that an intruder does not get any access to the secret key.

- Key Distillation: The shifted key is jointly processed by the sender and the receiver jointly to extract the secure sequence of bits called secret key. It consists of three steps: error correction, privacy amplification and authentication.

3.2   Simulation Methodology

The QKDNetSim module that has been built into NS-3 enabled us to simulate and analyse characteristics of the quantum channel. This contains the following features [7]:

- QKD Key: This describes the key being used for encryption.
- QKD Buffer: The keys are stored in the buffer. The analysis of the buffer concentration describes the traffic in the channel. More about this is discussed in a later section.
- QKD Crypto: This class of the module used to perform encryption, decryption, authentication and reassembly of previously fragmented packets.
- QKD Virtual Network Device: It facilitates the operation of the overlay routing protocol.
- QKD Post-processing Application: It deals with the extraction of the secret key from the raw key transmitted over the quantum channel.
- QKD Graph: This class of the module enables the easy extraction of the graphs related to the QKD buffer states.

The element of this module that this paper is primarily focussed on is the QKD Buffer. This has many variables that further facilitate the process of analysing the quantum channel. Endpoints of links which contain the buffer are gradually filled with the new key material and subsequently used for the encryption/decryption of data flow [8]. The key consumption rate depends on the encryption algorithm used and the network traffic, while the key rate of the link determines the key charging rate. If there is not enough key material in the storage, encryption of data flow cannot be performed [9] and QKD link can be characterized as "currently unavailable". Key material storage has a limited capacity and QKD devices constantly generate keys at their maximum rate until key storages are filled. Hence any disruption in the link can also compromise the key rates, these changes can be observed in the QKD buffer graphs. The variables used to define said graphs are, M_{cur}—current buffer capacity, M_{min}—minimum pre-shared key material, M_{max}—maximum storage depth, M_{cur}—current key concentration in the buffer, M_{thr}—threshold value.

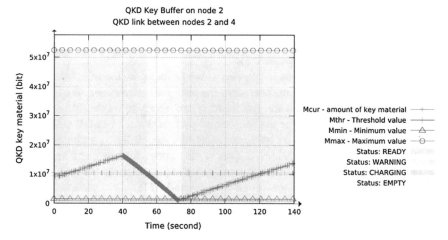

Fig. 2 Graphical representation of QKD buffer

The QKD buffer can be in one of the following states:

- Ready: $M_{cur}(t) \geq M_{thr}$,
- Warning: $M_{thr} > M_{cur}(t) > M_{min}$, the previous state was ready,
- Charging: $M_{thr} > M_{cur}(t)$, the previous state was empty,
- Emtpy: $M_{min} \geq M_{cur}(t)$, the previous state was warning or charging (Fig. 2).

The process used for implementing the simulation of the networks in public as well as quantum channel in NS-3 follows the steps shown in the flowchart. These steps are common for implementation of any network on this platform (Fig. 3).

- Topology Definition: since it is possible to implement these networks in public as well as the quantum channel, it is also possible to integrate the nodes of the network to form various arrangements (like bus, ring, star, mesh) with the help of links.
- Model Development: this speaks of the various protocols (TCP, UDP) that can be implemented in the network in the NS-3 by programming it along with the nodal orientation.

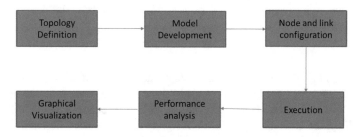

Fig. 3 Network model implementation flowchart

- Node and Link Configuration: these are additional node and link specifications (Baud rate, Bit rate, public channel congestion control, etc.).
- Execution: when the implemented network is simulated in the NS-3, the results are twofold (ASCII and graphical).
- Performance Analysis: this is the ASCII result that tells the client-server relation, the total number of packets sent, the total data transmitted, the client and server address along with the transmitting and receiving ports. In the case of public channel, it also tells the number of packets dropped and at what time.
- Graphical Visualization: by using two software two different graphical visualization of the network is achieved. One is a network animator that simulates the data flow for the total time duration and the other, is a plotting aid that is used to determine the total data transmitted (in bits) and BER.

4 Results

The implementation of this paper has been done in two parts: firstly, we have tried to implement a normal public channel at ideal conditions as well as a public channel network with congestion control. Secondly, we have tried to run a simple quantum channel network at ideal conditions, after which we have tried to implement a more complicated overlay network of the same to identify the changes in the channel traffic.

The ideal values of M_{min}, M_{max}, M_{cur} can be set in the program used to design the network. The M_{thr} depends on the network topology. It can be calculated using specific formulae [10]. The threshold value M_{thr} is proposed to increase the stability of QKD links, where it holds that $M_{thr} \leq M_{max}$.

- Each node a calculates value L_a summarizing the M_{cur} values of links to its neighbours j and dividing it with the number of its neighbours N_a, that is:

$$L_a = \frac{\sum_j^N M_{cur,a,j}}{N_a}, \quad \forall j \in N_a \tag{1}$$

- Then, each node exchanges calculated value L_a with its neighbours. The minimum value is accepted as the threshold value of the link, that is:

$$M_{thr,a,b} = \min\{L_a, L_b\} \tag{2}$$

By using M_{thr}, the node gains information about the statuses of network links. The higher the value, the better the state of links that are more than one hop away. Since the protocol used by the networks in both channels is Open Shortest Path First (OSPF), the node of higher threshold value can be chosen. Depending on this, paths can be rerouted or terminated. Hence, the QKD buffer capacity graph enables

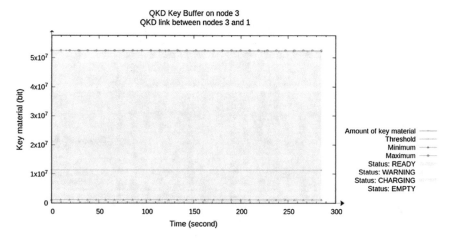

Fig. 4 Ideal quantum channel buffer graph

us to analyse the traffic in the current path, and choose a more efficient path for data exchange. A dip in the QKD buffer capacity indicates that parts of the key have been dropped. This can be due to congestion or the presence on an intruder in the path. Path congestion in the quantum channel is unlikely since the connection between the nodes is point to point. This is because the key material is primarily in the form of qubits, and it follows quantum properties. The addition of constraints due to the environment need not be implemented in the simulation, as data transfer via the quantum channel is done using fibre optic cables. The images shown below are the results that we have obtained via simulating data transfer through a six-node mesh network in both channels (Figs. 4, 5 and 6).

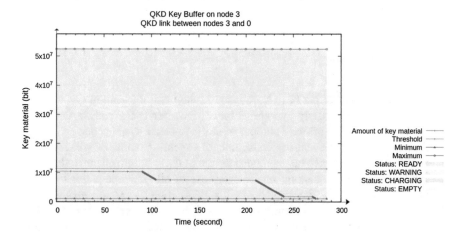

Fig. 5 Buffer graph due to disruption in quantum channel

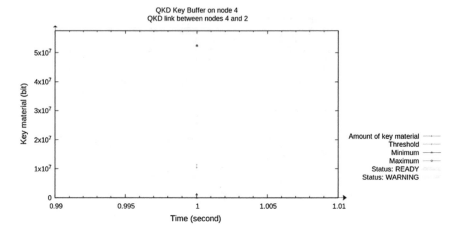

Fig. 6 Buffer graph when connection has been terminated

5 Discussion

In the process of implementing the simulation of the QKD network, we had encountered certain hurdles on the basis of which we made the following observations:

- Even though there is no limit to the number of nodes that can be added to an executable network in the public channel, in the quantum channel of our network this number was limited to 6.
- Also, unlike the public channel networks, the quantum channel ones are not so complicated as they do not exhibit any congestion control features. Hence increasing its ability to detect intruders more efficiently in simpler networks.

6 Future Research

Multiple advancements that have been made in the field of quantum cryptography even though they are purely hypothesis based, they depict the path in which this research is headed. The primary purpose of this is to create systems that can closely emulate and counter the real-life constraints of execution; some of which are, the effect of fibre birefringence on data transmission in the quantum channel, the use of the implementation of quantum repeaters by use of all-graphene solid-state components [11, 12]. Also, the implementation of satellite to ground quantum key distribution on the basis of entanglement, as well as the implementation of QKD in IoT to increase the security of wireless sensor-based networks as well as cellular networks is being researched [13, 14].

7 Conclusion

In this paper, we have presented a practical realization of QKD networks. We have also analysed the factors that affect traffic in the quantum channel. Hence, a relation between the variation of the QKD Buffer with the presence of network disruptions in the form of an intruder has been established. This simulation has been done by utilizing the BB84 protocol due to its ease of implementation, although other protocols can also be implemented within the developed simulation environment. The simulations carried out in this paper have been done using NS-3 with QKDNetSim built into it. Hence, the main purpose of this project was to prove theoretical concepts in the form of a simulation which closely emulates the real-life constraints of data transmission. Thus, proving that QKD is the most secure means of data encryption.

Acknowledgements We would like to express our gratitude to our University SRM Institute of Science and Technology, for giving us access to the resources required to undertake this project. We would also like to thank our guide Mr. S. Praveen Kumar, who is a faculty in this institution, for his valuable guidance, constant encouragement, timely help and for providing us an opportunity to do research in this field.

References

1. Mehic, M., Fazio, P., Voznak, M., Chromy, E.: Toward designing a quantum key distribution network. Adv. Electr Electron. Eng. **14**(4), 413–420 (2016)
2. Houshmand, M., Hosseini-Khayat, S.: An entanglement-based quantum key distribution protocol." In: IEEE 8th International ISC Conference on Information Security and Cryptology, pp. 45–48. IEEE (2011)
3. Abdulbast, A., Elleithy, K.: QKDP's comparison based upon quantum cryptography rules. In: IEEE Long Island Systems, Applications and Technology Conference (LISAT), pp. 1–5. IEEE (2016)
4. Trizna, A., Ozols, A.: An overview of quantum key distribution protocols. Inf. Technol. Manage. Sci. **21** (2018)
5. Padamvathi, V., Vishnu Vardhan, B., Krishna, A.V.N..: Quantum cryptography and quantum key distribution protocols: a survey. In: IEEE 6th International Conference on Advanced Computing (IACC), pp. 556–562 (2016)
6. Jasim, O.K., Abbas, S., El-Horbaty, E.-S.M., Salem, A.-B.M.: Quantum key distribution: simulation and characterizations. Procedia Comput. Sci. **65**, 701–710 (2015)
7. Mehic, M., Maurhart, O., Rass, S., Voznak, M.: Implementation of quantum key distribution network simulation module in the network simulator NS-3. Quantum Inf. Process. **16**(10), 253 (2017)
8. Kollmitzer, C., Pivk, M. (eds.): Applied quantum cryptography, vol. 797. Springer (2010)
9. Elliott, C.: Building the quantum network. New J. Phys. **4**(1), 46 (2002)
10. Mehic, M., Niemiec, M., Voznak, M.: Calculation of the key length for quantum key distribution. Elektron. Elektrotechnika **21**(6), 81–85 (2015)
11. Lobino, M., Zhang, P., Martín-López, E., Nock, R.W., Bonneau, D., Li, H.W., Niskanen. A. O.: Quantum key distribution with integrated optics. In: IEEE 19th Asia and South Pacific Design Automation Conference (ASP-DAC), pp. 795–799 (2014)

12. Wu, G.Y., Lue, N.-Y.: Graphene-based qubits in quantum communications. Phys. Rev. B **86** (4) (2012)
13. Yin, J., Cao, Y., Li, Y.-H., Ren, J.-G., Liao, S.-K., Zhang, L., Cai, W.-Q., et al.: Satellite-to-ground entanglement-based quantum key distribution. Phys. Rev. Lett. **119**(20) (2017)
14. Routray, S.K., Jha, M.K., Sharma, L., Nyamangoudar, R., Javali, A., Sarkar, S.: Quantum cryptography for IoT: A perspective. In: IEEE International Conference on IoT and Application (ICIOT), pp. 1–4 (2017)

Estimation of Air Pollution Concentration in Sub-urban of Chennai City and Validation of the Same Using Air Quality Models

J. S. Sudarsan, R. Saravana Kumar, K. Prasanna, S. Karthick and G. S. Suprajha

Abstract The rapid growth in population and industrial urbanization caused a massive rise in the number of automobiles which are the prime source of air pollution. It is estimated these automobiles contribute around 60% of air pollution in urban areas due to road traffic. A series of case studies is required to predict and foresee the CO levels along the highways, so as to draft a set of traffic administration measures in order to maintain the air quality levels within the standard tolerable limits. Thus, there is an immediate and alarming need to develop the checking of emission inventory competences in these urban cities which are essential for the formulation of a wide range of air pollution management strategies. There is also a critical need to develop modelling capabilities based on Indian standards. The National Highway, NH-45 between Tambaram and Chengalpet, of the Kanchipuram District, Tamil Nadu, Chennai, India Was selected for the study. Carbon monoxide (CO) concentration has been monitored at six interfaces. The CO concentration values were compared with NAAQS, and the same CO values were predicted using California LINE Source Dispersion Model, version 4 (CALINE4), American Meteorological Society/Environmental Protection Agency Regulatory Model (AERMOD). This specific study has concluded that the two models, CALINE4 and AERMOD, are capable of producing accurate predictions to a great extent.

J. S. Sudarsan (✉)
National Institute of Construction Management and Research (NICMAR),
Balewadi, Pune 411045, India
e-mail: ssudarsan@nicmar.ac.in

R. Saravana Kumar
Chennai Metro Rail Limited, Chennai, India

K. Prasanna · G. S. Suprajha
Department of Civil and Environmental Engineering, SRM Institute of Science
and Technology, Chennai, India

S. Karthick
Department of Software Engineering, SRM Institute of Science and Technology (SRMIST),
SRM Nagar, Kattankulathur, Kancheepuram District, Tamilnadu 603203, India

© Springer Nature Singapore Pte Ltd. 2020   603
S. S. Dash et al. (eds.), *Artificial Intelligence and Evolutionary Computations
in Engineering Systems*, Advances in Intelligent Systems and Computing 1056,
https://doi.org/10.1007/978-981-15-0199-9_52

Keywords Air pollution · Carbon monoxide · Urbanization · Air quality ·
Automobiles

1 Introduction

Air pollution due to transportation is one of the main concerns of urban planners
and environmental administrators in developing countries for decades and causes
human health effects [1]. Similarly, with other developing cities, National Highway
between Tambaram and Chengalpet attract huge traffic especially during the peak
hours which relatively has a large volume of traffic. Carbon monoxide (CO) is an
odourless, colourless and poisonous gas which is formed by the burning of fossil
oils such as petrol, diesel and gasoline and is produced primarily from cars and
trucks [2]. Air dispersion models have been widely used to identify the carbon
monoxide in the air due to its release from vehicles [3, 4]. The main purpose of this
research was to identify CO hotspots for the period of typical peak hours and to
estimate the acquiescence of expected levels by comparing with the recommended
National Ambient Quality Standards (NAAQS).

Software Modelling

- CALINE4 the standard modelling tool is used by Caltrans to assess the amount
  of carbon monoxide which impacts the transportation facilities, based on the
  Gaussian diffusion equation to characterize pollutant dispersion. It predicts air
  concentrations of carbon monoxide (CO), nitrogen dioxide (NO_2) and sus-
  pended particles near the site [5, 6]. Modelling is done near the intersections,
  elevated parking lots, and depressed freeways.
- The AERMOD is a type of dispersion model having an integrated system that
  includes three modules [7]. It is a dispersion model designed for small-range
  dispersion of air contaminant releases from static manufacturing sources.
  A meteorological data pre-processor (AERMET) that receives meteorological
  data, higher air soundings, and data from on-site instrument towers. It also
  includes an algorithm for modelling the effects of downwash called Plume Rise
  Model Enhancements (PRME) [8, 9].

2 Materials and Methods

The major municipalities in the south of Chennai City namely Tambaram and
Chengalpet are the developing towns near Chennai city in the state of South India. It is
located on the Coromandel Coast of Bay of Bengal. Due to the restraints of time, data
collection was limited to six links on National Highway 45 between Tambaram and
Chengalpet. The Traffic Survey is done manually and the CO measurement was

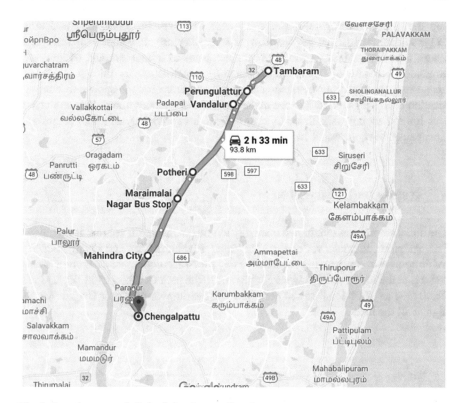

Fig. 1 Location map of all the links. Source: Google maps

determined using CALINE4 and AERMOD. The detailed sampling location and research methodology for the study are discussed in Figs. 1 and 2.

Description of the Study Area

The area covering Tambaram to Chengalpet is one of the most developing regions located near the Chennai city, of the northeast of the state of *Tamil Nadu* in India,

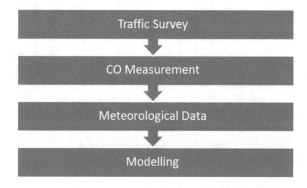

Fig. 2 Methodology (Compiled by authors). Source: Authors

and the study area location was very close to Mahindra world city and other technological parks, has witnessed a massive increase in vehicles in the recent years. This extraordinary growth of vehicles leads to a huge amount of vehicular emissions. Because of this situation, existing infrastructure facilities was not sufficient and the situation of existing road infrastructure for transport and communication has ultimately lead to the deterioration of the environment. Data collection was limited to only six links in National Highway 45 between Tambaram and Chengalpet as represented in Fig. 1.

3 Results and Discussion

The traffic survey was conducted manually to find out the traffic volume at six links—Perungalathur, Vandaloor, Potheri, MM Nagar, Mahindra City, and Chengalpet. The traffic count was done for the consecutive day for 24 h continuously. Figure 3 clearly states that the traffic volume is very high during the periods 08:00 to 11:00 A.M and 03:00 P.M to 07:00 P.M. It was observed that the number of Motor Cycles (MC) is high at all the links.

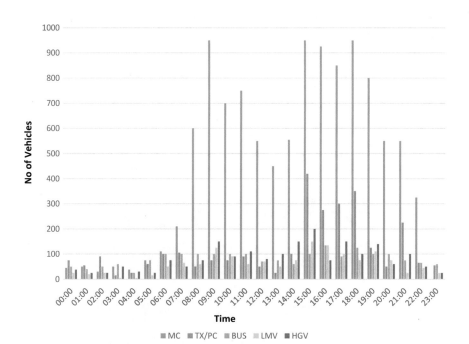

MC- Motor Cycle, TX/PC- Taxi/Cars, LMV- Light Motor Vehicle, HGV- Heavy Motor Vehicle

Fig. 3 Average hourly traffic volume at all links

Table 1 Observed and predicted CO value for AERMOD and CALINE4 model

Model	AERMOD		CALINE4	
Links	Observed (ppm)	Predicted (ppm)	Observed (ppm)	Predicted (ppm)
Perungalathur	0.8	0.7	0.8	0.6
Vandaloor	0.5	0.6	0.5	0.5
Potheri	0.5	0.7	0.5	0.8
MM Nagar	0.5	0.7	0.5	0.6
Mahindra City	0.7	0.7	0.7	0.9
Chengalpet	0.8	0.8	0.8	0.7

The overall results show that there is maximum traffic volume at Perungalathur since it is very near to Chennai city boundary. Also, the traffic volume is very low at MM Nagar and Mahindra City links. Two-wheelers are most predominantly more in number compared to other types of vehicles. It was observed that the CO pollutant is very high compared to other vehicles and the emission factor is very low for HGV vehicles.

Table 1 shows observed and predicted values for CO using AERMOD and CALINE4 model. Based on the model trails, CO concentration found seems to be maximum at Chengalpet using AERMOD model and with the help of the CALINE4 model; it was maximum in Mahindra city. It was observed that the observed and the predicted values at Mahindra City were the same for AERMOD whereas there is a difference of 0.2 ppm between the observed and predicted values by CALINE4. CALINE4 model does not consider the vegetation present along the roadway since trees and soil are the major sinks of CO pollutant. It is necessary to upgrade the software with some more extended features.

The number of taxies and private cars is found to be maximum in Perungalathur link as represented in Fig. 4 and similarly a number of motorcycles in the same link

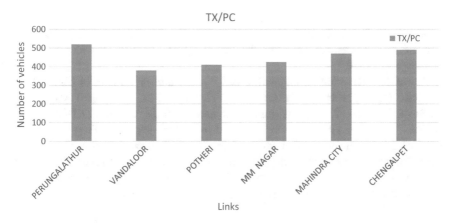

Fig. 4 Number of taxi/private cars during peak hours

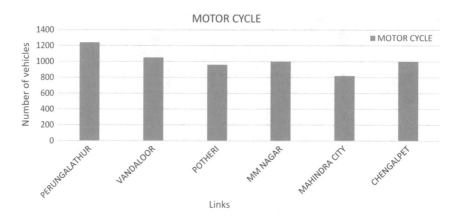

Fig. 5 Number of motor cycles during peak hours

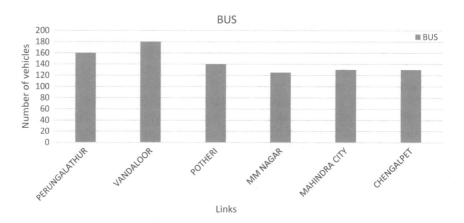

Fig. 6 Number of buses during peak hours

as shown in Fig. 5. Coming to a number of buses, it is more in Vandaloor link because there is connectivity state highway which connects OMR with GST and it is shown in Fig. 6.

It was observed that Perungalathur link has the highest number of Light Motor Vehicles (LMV) after Vandaloor link as shown in Fig. 8. Figure 7 shows that Perungalathur has the highest number of heavy goods vehicles (HGV) in comparison with other all links, whereas Vandaloor is the second highest in terms of number. Based on the research study in the six links, it was clear that all the CO concentration is maximum during peak hours and sometimes it was also exceeding the threshold limits. It can be controlled by increasing the vegetative cover and by regulating the traffic volume during peak hours.

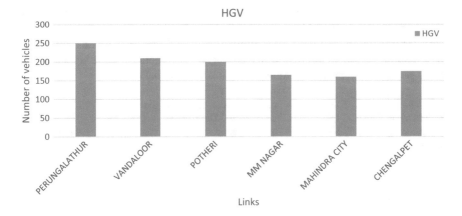

Fig. 7 Number of HGV during peak hours

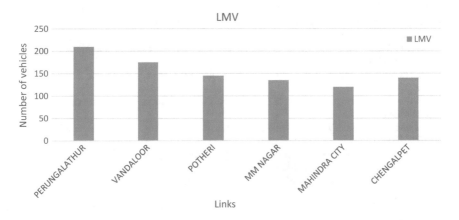

Fig. 8 Number of LMV during peak hours

4 Conclusion

Based on the detailed analysis, it was concluded that Perungalathur and Vandaloor have the highest Carbon monoxide (CO) concentration among all the six links that were studied. It may be due to the bifurcation of national highway and also due to the railway overbridge construction activities at Perungalathur and also due to increase with the number of four-wheelers, three-wheelers, and two-wheelers mainly due to industrial activities and urbanisation. It was estimated that CO pollutant is higher for two-wheeler than the other types of vehicles. The concentration of CO increases with a decrease in the speed of the vehicle due to more time in the traffic stream. The same data trail was carried out with CALINE4 and AERMOD. The output results were observed the predicted value was slightly higher than the observed value and if the models are calibrated with respect the location-specific better results can arrive. It was found that AERMOD model is more efficient than CALINE4 in the estimation of CO. Similar study performed throughout the city for an extended period of time would facilitate in developing a data inventory that could propose mitigative measures for controlling air pollution.

Acknowledgements The author's thanks to the Director general of NICMAR and the Management, Director (Engg. and Tech), Head, Department of Civil Engineering, SRMIST, Kattankualthur, Tamil Nadu, India for their support in executing this joint multidisciplinary research work.

References

1. Namdoo, A., Mitchell, G., Dixon, R.: TEMMS: an integrated package for modelling and mapping urban traffic emissions and air quality. J. Environ. Model. Softw. **17**, 179–190 (2002)
2. Ghose, Mrinal K., Paul, R., Banerjee, S.K.: Assessments of impacts of vehicular emissions on urban air quality and its management in Indian context. J. Environ. Sci. Policy **7**, 345–351 (2004)
3. Kho, F.W.L., Law, S.H.: Carbon monoxide levels along roadway. Int. J. Environ. Sci.Technol. **4**, 27–34 (2007)
4. Venkatram, A., Horst, T.W.: Approximating dispersion from a finite line source. Atmos. Environ. **40**, 2401–2408 (2005)
5. Chock, D.P.: A simple line source model for dispersion near roadways. Atmos. Environ. **12**, 823–829 (1998)
6. Nagendra, S.M.S., Khare, M.: Line source emission modelling. Atmos. Environ. **36**, 2083–2098 (2002)
7. Calori, G., Finardi, S., et al.: Air quality integrated modelling in Turin urban area. J. Environ. Model. Softw. **21**, 468–476 (2006)
8. Samson, Si P., et al.: A study of pollutant dispersion near highway. Atmos. Environ. **13**, 669–685 (1979)
9. Guan, H., Bergstorm, R.W.: Modeling the effect of plume-rise on the transport of carbon monoxide over Africa with NCAR CAM. J. Atmos. Chem. Phys. **8**, 6801–6812 (2008)

An Investigation on Performance of Mobile RPL in Linear Topology

M. Shabana Parveen and P. T. V. Bhuvaneswari

Abstract The traditional RPL supports static environment. In this chapter, the impact of mobility on RPL is investigated in linear topology with varied node density. Mobility model is developed with feature support randomness, and it is manually configured. The performance of mobility enabled RPL is simulated using COOJA and compared with traditional static RPL in terms of packet delivery ratio (PDR), control overhead and latency. From the simulation results, it is inferred that PDR is less and control overhead and latency are higher in mobility enabled RPL.

Keywords RPL · Mobility · Packet delivery ratio · Control overhead · Latency

1 Introduction

Evolution of Internet of things (IoT) leads to several opportunities in different applications including smart home, smart city, healthcare, smart agriculture, sports monitoring, animal tracking and many other fields. They involve networking of smart objects that are small, low powered and inexpensive. In order to enable the communication between the objects, Internet Engineering Task Force (IETF), Routing Over Low-power and Lossy networks (ROLL), Working Group has proposed IPv6 Routing Protocol for Low-Power and Lossy Networks (RPL) [1, 2]. It is a standard routing protocol for IPv6 routing in low-power and lossy networks. Using RPL, IPv6 enabled objects can communicate at the network level and thus be easily connected to the Internet. Hence, RPL can be considered as the routing protocol for IoT applications.

M. S. Parveen (✉)
Electronics and Communication Engineering, Sri Sairam Engineering College, Chennai, India
e-mail: Shabanaafsar08@gmail.com

P. T. V. Bhuvaneswari
Department of Electronics Engineering, MIT Campus, Anna University, Chennai, India
e-mail: ptvbmit@annauniv.edu

© Springer Nature Singapore Pte Ltd. 2020
S. S. Dash et al. (eds.), *Artificial Intelligence and Evolutionary Computations in Engineering Systems*, Advances in Intelligent Systems and Computing 1056, https://doi.org/10.1007/978-981-15-0199-9_53

RPL is a routing protocol that was designed for static networks. After the connections are established, it assumes that the network is in a steady state and does not consider the mobility of nodes. There are many applications where the nodes are mobile, namely healthcare monitoring, industrial automation, body sensor network, and home automation. The traditional RPL working mechanism needs to be made adaptive to the dynamic changing topology in order to reduce packet loss and delays.

Thus, the main objective of the proposed work is to evaluate the performance of RPL in static and dynamic environment with varied network densities in terms of packet delivery ratio (PDR), control overhead and end-to-end delay. The rest of the paper is organized as follows: Sect. 2 elaborates the literature survey related to the proposed work. Section 3 explains the proposed methodology. Section 4 discusses the inference of the obtained results. Finally, Sect. 5 concludes the paper with the future scope.

2 Literature Survey

This section discusses the various methods and analyzes presented in the literature related to performance evaluation of RPL. In [3], the authors have investigated the performance of RPL network construction process. Simulation is done in COOJA. The author has analyzed the impact of number of nodes, number of network roots, the objective function and the packet loss incurred during the network construction process. It is found that the power consumption in the nodes is reduced when the network is sparse. Increasing the number of directed acyclic graph (DAG) root improves the network performance. Two objective functions (OF), namely OF0 with hop count metric and OF1 with expected number of transmission (ETX), as metric are formulated. OF0 has lower hop distance and lower power consumption when compared to OF1. Sparse network has high packet loss due to the non-availability of alternative parents in the case of node failure. Designing of the repair mechanism in RPL is considered as the future work.

In [4], the performance of RPL is evaluated in comparison with Ad Hoc On-Demand Distance Vector (AODV) and Dynamic MANET On-demand (DYMO) protocols in terms of average end-to-end delay and routing overhead. From the results, it is inferred that RPL outperforms when compared to AODV and DYMO routing protocols.

In [5], the authors have studied the working of RPL with two different OFs, namely objective function zero (OF0) and Minimum Rank with Hysteresis Objective Function (MRHOF). Evaluation is done in both random and grid topologies by varying the reception ratio in terms of power consumption and PDR. From the results, it is inferred that the performance of RPL is appreciable in both OFs when the reception ratio is 60–100% and node density is maintained between 30 and 40. The present work can be extended to medium and high density of nodes.

In [6], the authors have analyzed the performance of RPL in a multisink scenario. They have considered two OF formulated with hop count and ETX as metrics. The results show that an increase in sink node leads to increase in PDR and decrease in energy consumption.

From the above literatures, it is inferred that the performance evaluation of RPL has been made in varied scenarios by modifying the factors, namely OF, increasing the number of sink node and measuring the QoS parameters, namely PDR, energy consumption and control overhead.

In the proposed work, the performance of RPL is done in both static and dynamic environments and is evaluated in terms of PDR, latency and control overhead. Major contribution claimed is that random mobility model is developed where the randomness is induced manually. The behavior of traditional static RPL configured with the developed mobility plug-in is investigated for linear topology with varied node density.

3 Proposed Methodology

The aim of the proposed work is to examine the impact of mobility on RPL performance. Simulation is done in COOJA.

3.1 System Model

Consider 'N' number of IP enabled 6LoWPAN nodes, out of N nodes 1 node is configured as server and $(N - 1)$ node as clients. In COOJA simulator, a node is configured as server using '*udp-server.c*' library file, while all client using udp-client.c library file. These nodes are deployed in linear topology as shown in Fig. 1 in $n \times m$ terrain. Let R be the transmission range of each node and I be interference range experienced by each node. Contiki Test Editor Plugin is used to set the simulation time. This plug-in creates a log file which contains the outputs of all nodes for specified simulation time. The output so obtained is further analyzed using PythonScript.

3.2 Initialization

When the simulation is started, the first step is all the nodes get its node id and IP6 address. Figure 2 illustrates the address configuration procedure involved in ith client. Initially, the MAC address of the client node is fetched and node ID is assigned for the corresponding client. Then, CSMA Contiki MAC protocol is invoked, and the channel check rate, radio channel number and status of the channel

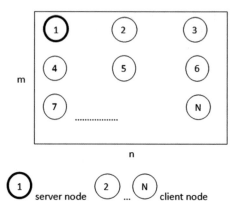

Fig. 1 Deployment of nodes in linear topology

```
00:00.622  ID:4  Rime started with address 0.18.116.4.0.4.4.4
00:00.630  ID:4  MAC 00:12:74:04:00:04:04:04 Contiki-2.6-900-ga6227e1 started. Node id is set to 4.
00:00.639  ID:4  CSMA ContikiMAC, channel check rate 8 Hz, radio channel 26, CCA threshold -45
00:00.643  ID:4  RPL started
00:00.651  ID:4  Tentative link-local IPv6 address fe80:0000:0000:0000:0212:7404:0004:0404
00:00.653  ID:4  Starting 'UDP client process'
00:00.655  ID:1  Rime started with address 0.18.116.1.0.1.1.1
00:00.655  ID:4  UDP client process started
00:00.660  ID:4  Client IPv6 addresses: aaaa::212:7404:4:404
00:00.662  ID:4  fe80::212:7404:4:404
00:00.664  ID:1  MAC 00:12:74:01:00:01:01:01 Contiki-2.6-900-ga6227e1 started. Node id is set to 1.
00:00.668  ID:4  Created a connection with the server :: local/remote port 8765/5678
00:00.673  ID:1  CSMA ContikiMAC, channel check rate 8 Hz, radio channel 26, CCA threshold -45
00:00.676  ID:1  RPL started
00:00.684  ID:1  Tentative link-local IPv6 address fe80:0000:0000:0000:0212:7401:0001:0101
00:00.686  ID:1  Starting 'UDP server process'
00:00.688  ID:1  UDP server started
00:00.691  ID:1  created a new RPL dag
00:00.696  ID:1  Server IPv6 addresses: aaaa::212:7401:1:101
00:00.698  ID:1  aaaa::ff:fe00:1
00:00.701  ID:1  fe80::212:7401:1:101
00:00.707  ID:1  Created a server connection with remote address :: local/remote port 5678/8765
```

Fig. 2 Initialization of motes

busy or ideal information are obtained. Then, RPL procedure is initialized, and IPV6 address is assigned to the ith client. Through UDP client process, a connection is established between client and server. This procedure is repeated for all $(N - 1)$ clients.

3.3 Destination-Oriented Directed Acyclic Graph (DODAG) Construction and Maintenance

RPL constructs a Destination-Oriented Directed Acyclic Graph (DODAG). Each DODAG has a route from all the nodes to the sink node. Each node in the DODAG is designated with a rank [1]. Rank of node tells about its individual

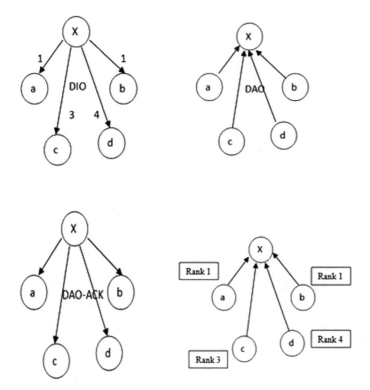

Fig. 3 Construction of DODAG with respect to server

position with respect to other nodes. The rank increases in downstream direction of the Directed Acyclic Graph (DAG). Nodes on top have smaller rank. The computation of rank depends on DAG's objective function (OF). An OF defines the routing metrics.

Construction of DODAG with respect to server is illustrated in Fig. 3. Initially, the server node multicast DIO message which contains the rank information of the server to the entire client node which is within the coverage.

Upon reception of DIO message, the client node computes their rank with respect to server using OF where ETX is taken as metric. The client node acknowledges the message through Destination Advertisement Object (DAO) message. The server node acknowledges the message through DAO-ACK message. Thus, DODAG is constructed with respect to the server. This procedure is repeated downwards to the rest of the topology where the client in the downstream computes their rank with respect to the parent client.

Consider 4 client nodes a, b, c, d and 1 server or root node X. X multicast DIO which consist of rank information and all other node receives DIO and in turn will know the distance which is 1, 1, 3, 4, respectively. Regardless of the distance, the

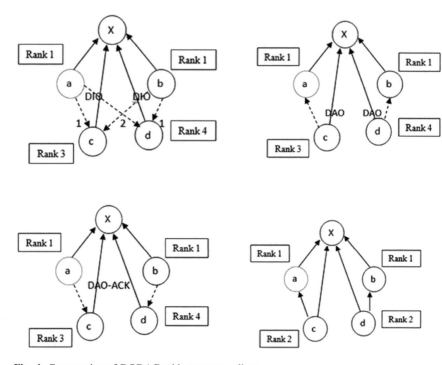

Fig. 4 Construction of DODAG with respect to client

node will try to join. The client sends DAO to the server, and the server node sends DAO-ACK to the client node.

The next set of nodes nearest to the server which has rank 1 a and b will start sending DIO as shown in Fig. 4. c receives the message and figure out that its distance from a and b is 1 and 2, respectively. d receives the message and figure out the distance is 2 and 1, respectively. c sends DAO to a, and d sends DAO to b. DAO-ACK is sent by a to c and b to d.

If a new client wants to join an already defined network the client sends a DODAG information solicitation (DIS) message. The DIS message requests DIO message from its neighbor. When DIS is received, the neighbor node sends DIO immediately, and from the DIO message, it can select the preferred parent using the rank information present in DIO.

To maintain the DODAG, both server and client send a DIO message periodically to their neighbors. The interval between each DIO is determined based on Trickle timer [7]. Trickle timer is a timer which will be minimum or maximum and is used to schedule the transmission of DIOs. Maximum timer will be applied when the network is static and the timer increases when the network becomes unstable. Periodic Trickle timer t is bounded by the interval $[I_{min}, I_{max}]$.

Where I_{min} is minimum interval defined in ms which is $I_{min} = 2^{12} = 4096$ ms. $I_{max} = I_{min} * 2^{I_{doubling}}$ assuming $I_{doubling} = 4$.

3.4 Data Transfer

After the construction of DODAG, data is transferred from the client to server node by using the routing table constructed in each client. Figure 5 shows the data transfer message between the client and server.

3.5 Mobility Management

COOJA simulator does not support node mobility by default. However, it can be made possible by enabling the available mobility plug-in [8]. By this, the position information of each node during the simulation time can be obtained as shown in Fig. 6.

The first field gives the node ID, second field gives the time and third and fourth fields provide the x- and y-axis position. The position of $(N - 1)$ mobile client can be obtained in the similar manner.

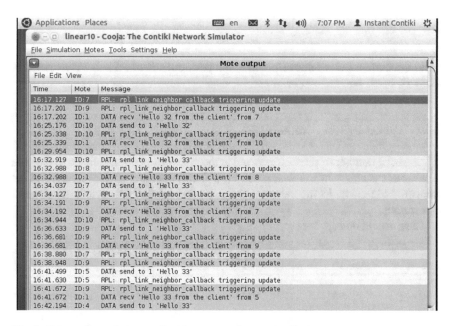

Fig. 5 Data exchange between client to server and server to client

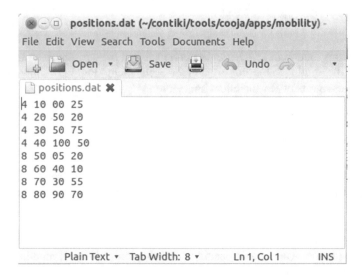

Fig. 6 Position information of mobile client

3.6 Analysis of RPL

After the completion of above process, the performance of RPL is investigated in terms of PDR, control overhead and latency.

3.6.1 Packet Delivery Ratio (PDR)

It is the total number of packets received at the server from all the clients to the total number of packets sent to the server. Let 'p' be the number of packets transmitted from $(N-1)$ client to the server. Let 'q' be the number of packets received at server node.

Then, the PDR is

$$\varepsilon = (q/p) * 100 \tag{1}$$

3.6.2 Control Overhead

The various control messages used are DIO, DIS and DAO. Let 'α' be the number of DIO messages sent from parent to client where the parent can be either server node or client node. Let 'β' be the number of DIS messages sent from client to the server and γ be the number of DIO messages sent from the client to server. The total number of control messages sent is given by

$$\eta = \sum_{i=1}^{N} \alpha_i + \sum_{i=2}^{N} \beta_i + \sum_{i=2}^{N} \gamma_i \tag{2}$$

3.6.3 Latency

Latency is the amount of time taken by a packet from client to reach the server. The total latency is the sum of latency experienced by all the packets in the network.

Let '$\tau(k)$' be the transmitting time of Kth data packet from client node to the server and '$\mu(k)$' be the receiving time at the server. Then, the network latency is computed using

$$\text{Totallatency} = \sum_{k=1}^{m} (\mu(k) - \tau(k)) \tag{3}$$

where 'm' is the total number of packets received successfully.

4 Results and Discussion

4.1 Simulation Parameters and Network Setup

The simulation is done under COOJA [9]. It is an extensively used sensor network simulator under Contiki [10] operating system.

The parameter for simulation and its environment are shown in Table 1 (Fig. 7).

Table 1 Simulation parameters

Parameters	Values
Area of deployment ($n * m$) m^2	100 * 100
Radio environment	Unit disk graph medium: distance loss
Server node	1
Number of client nodes	5, 10, 20, 30
Mote type	Sky mote
TX range (R) m	50
INT range (I) m	100
RX ratio	100%
Mote start-up delay (ms)	1000
Scenarios considered	Linear
Simulation time (minutes)	20

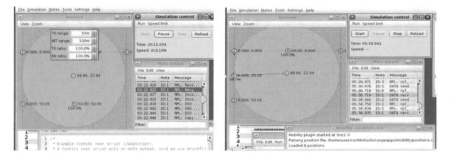

Fig. 7 Screen shot of static 5 clients with 1 server and screen shot of 5 clients node with 1 client mobile and 1 server

4.2 Simulation Metrics

The performance of RPL in both mobile and static environments is simulated in COOJA and performance is compared. The node density is varied, and their influence on PDR, control overhead and latency were analyzed.

Simulation result in Fig. 8 illustrates the impact of PDR on client density. Simulation is carried for static and dynamic environment. The PDR is calculated using Eq. (1).

From the result obtained, it is observed that when the number of client increases, PDR increases. This is because data is sent through more reliable path created by neighbors. But reduction in PDR is experienced in mobile environment due to the fact that when the nodes are mobile, its neighbor node and parent node changes stochastically which may lead to loss of data.

The simulation result in Fig. 9 is plotted between control overhead and client density in static and dynamic environments. The control overhead is calculated on the basis of Eq. (2).

From the results obtained, it is observed that when number of client increases, control overhead increases. This is because when number of node increases each node will transmit control overhead messages, namely DIO, DIS to maintain connectivity. RPL exhibits increasing control overhead with the mobile environment when compared to static.

Fig. 8 PDR in different client density for static and mobile networks

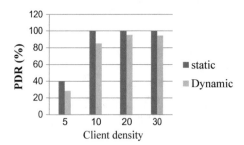

Fig. 9 Total control overhead in different client density for static and mobile networks

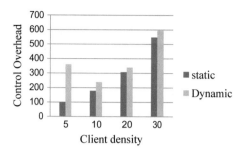

Fig. 10 Latency in different client density for static and mobile networks

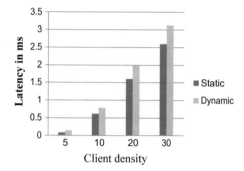

It is due to flat fading environment experienced in the channel. The medium characteristic exhibits staticness which will limit the periodicity of updates.

Simulation result in Fig. 10 is plotted between latency and client density in static and dynamic environments. The latency is calculated according to Eq. (3). From the results obtained, it is inferred that latency increase in latency is witnessed when the client intensity increases. The time taken for each data sent by the client to the server is more when the number client increases. This is because of more collision when there is heavy traffic. So the packets need to be buffered and sent. From the comparison, it is inferred that the latency is higher for mobile environment than static environment.

5 Conclusion

From the proposed research, the performance of the RPL with mobility is analyzed in both static and mobile environments in terms of PDR, control overhead and latency. The analysis performed in COOJA simulator. The client and server nodes are configured with sky mote specification. The clients are arranged in linear topology. They are varied from 5 to 20 in steps of 5. The clients are made mobile by configuring the mobility plug-in in the nodes. This introduces randomness in their position. From the simulation result, it is inferred that reduction in PDR with increased control overhead and latency is experienced by mobile clients when they are in the dynamic environment.

References

1. Winter, T., Thubert, Brandt, A., Hui, J., Kelsey, R.: RPL: IPv6 Routing Protocol for Low Power and Lossy Networks. IETF6550
2. Gaddour, O., Koubáa, A.: RPL inanutshell: asurvey. Comput. Netw. **56** (14) pp. 3163–3178 (2012)
3. Gaddour, O., Koubâa, A., Chaudhry, S., Tezeghdanti, M., Chaari, R., Abid, M.: Simulation and performance evaluation of DAG construction with RPL. In: Third International Conference on Communications and Networking. Hammamet, pp. 1–8 (2012)
4. Xie, H., Zhang, G., Su, D., Wang, P., Zeng, F.: Performance evaluation of RPL routing protocol in 6lowpan. In: 2014 IEEE 5th International Conference on Software Engineering and Service Science, Beijing, pp. 625–628 (2014)
5. Qasem, M., Altawssi, H., Yassien, M.B., Al-Dubai, A.: Performance Evaluation of RPL Objective Functions. In: 2015 IEEE International Conference on Computer and Information Technology; Ubiquitous Computing and Communications; Dependable, Autonomic and Secure Computing; Pervasive Intelligence and Computing, Liverpool, pp. 1606–1613 (2015)
6. Zaatouri, I., Alyaoui, N., Guiloufi, A.B., Kachouri, A.: Performance evaluation of RPL objective functions for multi-sink. In: 2017 18th International Conference on Sciences and Techniques of Automatic Control and Computer Engineering (STA), Monastir, pp. 661–665 (2017)
7. Levis, P., Clausen, T., Hui, J., Gnawali, O., Ko, J.: The Trickle algorithm. RFC 6206 (2011)
8. https://anrg.usc.edu/contiki/index.php/Mobility_of_Nodes_in_Cooja
9. Contiki operating system. http://www.contiki-os.org/
10. https://anrg.usc.edu/contiki/index.php/Cooja_Simulator

Plant Detection and Classification Using Fast Region-Based Convolution Neural Networks

Raja Naga Lochan, Anoop Singh Tomar and R. Srinivasan

Abstract The identification of plants is an important task for preserving plants, which are very crucial for the existence of humans on earth. Many of the species are at the stage of extinction and are present in very discreet location. Their identification and protection are one of the major concerns. Many of these plants have great medicinal value, so saving them becomes very important. Plants can be identified using their leaves, bark, seed, fruit, flower, etc. The methodology described in this paper considers the identification of plants using the features of its leaves. The plants considered are the medicinal plants which can be presented in discreet locations like the Himalayas or can be presented in the kitchen garden. In this paper, we have used regional convolution neural network (RCNN) for the identification of plants. The system uses fast RCNN (fast RCNN) model using convolution networks to extract features and for classification support vector machine (SVM) issued.

Keywords Medicinal plants · RCNN · Fast RCNN · Regional convolution neural network · Relu · Softmax · ROI (Regions of interest) · Support vector machine (SVM) · Kernel · Regularization parameter · Margin · Gamma · Hyperplane · TensorFlow

1 Introduction

Plants are one of the very important life forms present on the earth's ecosystem. Plants provide food, medicine, fuel and all other valuables which are vital for the existence of all the life forms on the earth. India has always been rich in plants, i.e., medicinal plants, and our ancestors used these herbs for treatment of all their ailments, as a result of which India is recognized as the origin of Ayurveda.

R. N. Lochan (✉) · A. S. Tomar · R. Srinivasan
Department of Computer Science and Engineering, SRM Institute of Science
and Technology, Kattankulathur 603203, India
e-mail: rn.lochan@gmail.com

© Springer Nature Singapore Pte Ltd. 2020
S. S. Dash et al. (eds.), *Artificial Intelligence and Evolutionary Computations
in Engineering Systems*, Advances in Intelligent Systems and Computing 1056,
https://doi.org/10.1007/978-981-15-0199-9_54

Ayurveda is considered as an alternative for allopathic medicine and has proved its worth since a very long time. In modern days, the knowledge of medicinal herbs needs to be preserved. Hence, the task to identify and classify the plants becomes very important. Plants can be classified based on the textures, sizes, colors and structures of the leaves, bark, flowers, seed, fruits and morph. But if the classification of plants is based on the two-dimensional structures, it will be very difficult to study the shape and size of the fruits and flowers. Hence, this increases the complexity of the classification. We take plant leaves for classification because of their two-dimensional structure, which holds the important properties such as textures, shape, size, etc. through which the plant can be identified. Thus, classification and identification are based on the features of leaves [1–3].

Wilson [4] shows that biodiversity is in a huge crisis. Many important medicinal plants are on the verge of extinction. To improve the drug industry and household treatment of diseases, a good knowledge of plant is needed [5]. In this paper, we used Tensor Flow (tf) object detection along with RCNN to classify the leaf accurately. This method gives better predictions even when the test image is not very sharp or multiple images are present in the frame. Thus, this methodology gives better results when the background is noisy.

In the following section 'Literature Survey,' various other works on the leaf identification have been summarized. These other methodologies give alternate options to identify and classify the plant leaf images. The section 'Image Detection' contains the methodology used by us to identify the plant. The section 'Data collection' describes the sources of our data and how did we extracted the images, the section 'Image Pre-processing' describes the methodology used for pre-processing of the leaf dataset and the bounding boxes across the required area of the images decreases the computation. Section RCNN tells about the working of RCNN and architecture and later describes the working of fast RCNN in Sect. 3.4. Section 4 contains the implementation of the above approaches and answers to how we have used the above algorithms mentioned in previous sections to obtain better results.

2 Literature Survey

A lot of researches have been carried out during the last century resulting in development of computer vision and artificial intelligence. These systems were used for investigating various classification tasks and pattern recognition tasks [6].

Artificial neural network (ANN) has gained a lot of popularity in pattern recognition and has been used a lot in various applications because of its ease of implementation, usage and flexibility. The development in the processor technology has also encouraged the use of neural networks in many applications [7–9]. Neural networks imitate the structure of human brain and use mathematical calculations to recognize the patterns and apply them based on the previous experience to differentiate whether they have a similar pattern or not.

A lot of work has been done in recognizing and identification of different plants. Wu et al. [10] used 32 different types of leaves. They used the morphological features of the leaves for recognition and got the accuracy of 90.3%. Prasad et al. [11] got 95% accuracy in recognizing 23 different types of leaves by using support vector machine. Gray-level co-occurrence matrix was used by Ehsanirad and Kumar [12] to recognize 13 different types of plant leaves and got 78.46% recognition rate. Cope et al. [13] used the image dataset of 32 plant leaves to get the recognition rate of 85.16% using Gabor co-occurrences. Kumar et al. [14] used plant leaf images to identify the plant species on the curvature-based shape features by using integral measures to compute functions of the curvature at boundary. Nearest neighbor is used for identification of these plant leaves. Some of the others used moment invariants, multi-scale distance matrix [15, 16]. Compared the different classification techniques of different models classifying on the bases of various geometrical features, vein features, texture features, morphological features, geometrical features, shape, color, volume, etc. [17], Boran et al. used dataset of 27 different types of leaves using back-propagation neural networks and get the recognition rate of 97.2%.

3 Data Collection

Dataset was collected in two forms.

1. Data was scrapped from the internet from different sources using python crawlers as shown in Fig. 1.

Fig. 1 Data collected (*source* Internet)

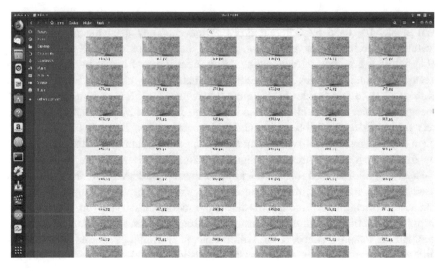

Fig. 2 Data collected (*source* Digital Camera)

2. Data was also made by taking videos using mobile camera of the plant leaf and extracting each frame from the video as shown in Fig. 2.

Bounding boxes around the leaf of the plant in the image were obtained by using contour detection and color filtering using OpenCV. Contours are curves which join the continuous points all together (along the boundary), which have the same color or having same intensity. These contours are very useful tools when it comes to applications like shape analysis, object recognition and object detection, etc. For color detection, in Range() function is used. The in Range() function checks whether array A1 elements are lying in between the elements of two other different array elements.

As a result of which the function checks the range, method of which is mentioned in the following line:

For the given each element of input array belonging to the single channel:

$$\text{dst}(I) = \text{lower}b(I)0 \leq \text{src}(I)0 \leq \text{upper}b(I)0$$

For two-channel arrays:

$$\text{dst}(I) = \text{lower}b(I)0 \leq \text{src}(I)0 \leq \text{upper}b(I)0 \leq \text{lower}b(I)1 \leq \text{src}(I)1 \leq \text{upper}b(I)1$$

Once the data was collected, image location, bounding values of the object within the image and labels were stored in CSV files. The CSV files are used as an input for converting the images into tensor objects, as tensor objects are in binary vector form and occupy less space. The tensor objects which are in the binary form are provided as the input for fast RCNN model.

4   Image Pre-processing

The images used for training and testing are pre-processed. The pre-processing decreases the number of computations and increases the speed of processing, to make the classification more real time. All images are first converted into grayscale as shown in Fig. 3.

The images are then filtered using morphological functions:

1. Erosion and dilation, i.e., opening morphological function;
2. Dilation and erosion, i.e., closing morphological function.

The erosion of the binary image A by the structuring element B is defined by:

$$A \oplus B = \{z \in E | Bz \subseteq A\}$$

where Bz is the translation of B by the vector z, i.e.,

$$Bz = \{b + z | b \in B\}, \forall z \in EE$$

The dilation operation makes use of a structuring element for probing tasks and for expanding the shapes contained in the input image.

The dilation of A by B is defined by

$$A \oplus B = \cup b \in BAb, Ab$$

where Ab is the translation of A by b, i.e., these filters help us in removing tiny patches and unwanted holes in the image.

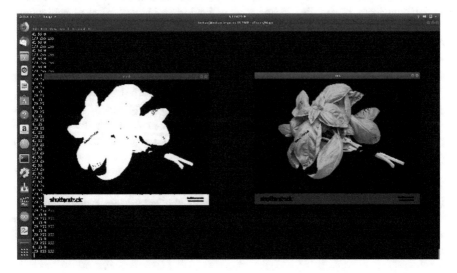

Fig. 3 Grayscale output image

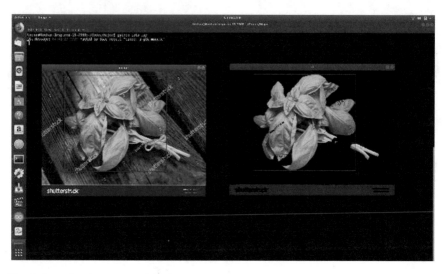

Fig. 4 Bounding boxes around the image

The images are then converted into one common size and pixel density, i.e.,
128 × 128 (Fig. 4).

5   Convolution Neural Network

Convolution neural networks (CNNs) are a class of machine learning networks
which are commonly applied to image visualization problems such as classification.
The CNNs got influenced by the connections of the neurons and synapses in the
human brain. These networks are made by the connection of a serial arrangement of
convolutional layers, pooling layers and fully connected layers. The convolutional
layer, as its name indicates, makes use of the filters as it applies a number of
convolution filters to these input images which in turn acquire the learning
parameters for the neural network. A number of pooling layers are placed in
between these convolution layers, which are useful in reducing the number of
parameters which will be useful for learning, and thus reduce the computation
required.

Finally, fully connected layers are full connections to the previous layer, rather
than the small window the convolution layers are connected to in the input. Any
input given to this neural network goes through the hidden layer where an activation function named Relu is applied to the images, which gives us feature vectors
or feature maps as an output. Each feature map then goes through the phase of
pooling where the image size is made sure that it is not compromised (Fig. 5).

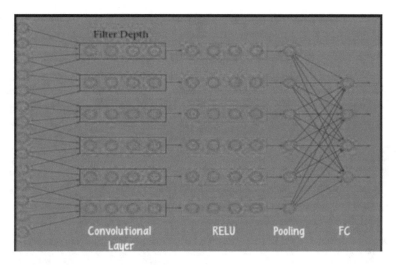

Fig. 5 Convolution neural network

6 Fast Regional Convolution Neural Network

Region-based neural networks. In region-based CNN, the initial prediction made depends on the regions of interest. For every image, regions of interests are obtained and passed onto the hidden layer where the Relu activation function is applied and the same process as CNN is followed. In fast RCNN, the model performs better and more quickly as the regions of interest are found using a selective search method and all the regions of interests are found at once for an image, unlike CNN which finds ROI (regions of interest) and applies Relu on each ROI separately which is more time-consuming and slower. Fast RCNN of the initial prediction made depends on the regions of interest of the image. The regions of interests are found using selective search. Selective search is based on computing hierarchical grouping of similar regions based on color, texture and size and shape compatibility. It starts by over-segmenting the image based on intensity of the pixels using a graph-based segmentation method by Felzenszwalb and Huttenlocher.

Selective search merges more and more regions based on the color, texture and other features which give us regions of interests. These ROIs are passed through a CNN upon which activation function is applied on each layer. At each layer, a better feature vector is extracted. Convolution feature kernels are extracted as output once Relu activation function is applied to the ROI on different layers of CNN. The kernels are then classified by the SVM classifier. After classification, a corrective measure is added and again ROIs are obtained. The bounding boxes and the size of ROI are calculated and passed onto the fully connected layers. The fully connected layers now search for features with similar size and features and then determine the softmax. The softmax is used to determine the class of the object in

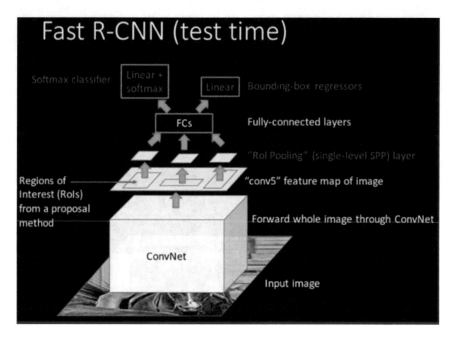

Fig. 6 Fast RCNN architecture

the image. The fully connected layer also provides a bounding loss. The loss obtained is then used as a corrective feature for improving the classification and detection (Fig. 6).

7 Softmax Classifier

Softmax regression algorithm is a generalization of logistic regression that is used by the machine learning experts in various applications where multi-class classification is needed. Assuming that classes are mutually exclusive. Hence, we use logistic regression model classifying two classes separately, i.e., binary classification image below gives us a more detailed idea about the working of logistic regression how it works so as to get a better understanding of softmax algorithm and how it is different than logistic regression algorithm. In logistic regression, we take the set of training examples and try to make a regression curve which fits the training data well. Now, the important task of the algorithm is to reduce the cost. Cost of the algorithm is defined as the root mean square of the example and the predicting curve. This is done by making a curve equation in which the weights and bias as corrected again and again to reduce the task. Now, these optimized weights and biases give the equation of the most optimized curve according to the data. In this activation, functions are also used in logistic regression sigmoid function is also used, which is an activation function after which the error is sent back to the previous step which changes the weights and biases.

In linear regression, we use sigmoid function whereas in softmax regression this is replaced by the sigmoid function which is mentioned below.

8 Implementation

The model was trained successfully using the TensorFlow API. A fast RCNN model was used to train it with a total of 10,000 images and 50,000 training steps. The model was then tested on real-time data and sample data, and it was able to detect and classify the images successfully with hit percentage averaging to 96%. Data was collected and converted into tensor records. The records are used as input for the model. Regions of interests were then found based on selective search. Each region of interest was then passed through the SVM classifier to classify the ROI based on the features extracted. The loss is then compared between the actual value and the expected value. Accordingly, to loss activation function value is changed. Once the loss is below 1, the training is paused and the model is extracted. After extracting the model, it is then tested on real-time data (Table 1).

9 Results

Image in Fig. 7a Coriander leaf—97%
Image in Fig. 7b Mint leaf—96%
Image in Fig. 7c Aloe vera—98%
Image in Fig. 7d Neem—92% (Table 2).

Table 1 Implementation details

Training dataset	8,500
Test dataset	1,500
Total number of classes for training model	10
Classes of images	Ajwain, aloe vera, coriander, bryophyllum, mint, neem, spinach, basil, methi, thyme
Image size	600 × 1024 pixels
Average images per class for training	850
Average images per class for testing	150

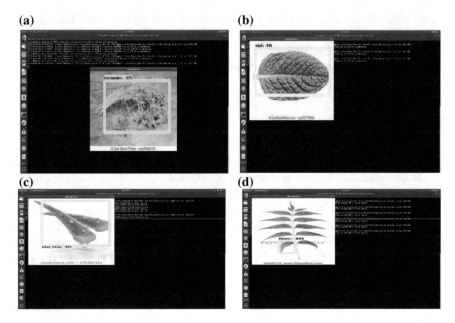

Fig. 7 **a** Coriander. **b** Mint. **c** Aloe vera. **d** Neem

Table 2 Comparative results

	Number of images	Number of classes
Training dataset	8500	10
Test dataset	1500	10
Images per class	1000	1
Images classified correctly	1443	10
Images classified incorrectly	57	10
Avg accuracy per class	95.7	1
Accuracy obtained for all classes	96.2	10

10 Conclusion

Fast RCNN of the initial prediction made depends on the regions of interest of the image. The regions of interests are found using selective search. Selective search is based on computing hierarchical grouping of similar regions based on color, texture and size and shape compatibility. It starts by over-segmenting the image based on intensity of the pixels using a graph-based segmentation method. This improves the efficiency of training process and classification by 200 times and also provides an accuracy of (96–97)% with a dataset of 10,000 images and 10 classes.

In the future, we want to try to incorporate the image classifier and detector I to a mobile or a web app which will give information about the medicinal values of the plant and also give the methodology to use it as a medicine. This will make our coming generations less dependent on allopathic medicines. This will also motivate the upcoming generation for plant preservation as they will understand its value. Also, the knowledge for which our country has always been known for will be preserved and successfully transferred to the next generation. Such a mobile app can also help the ayurvedic medicine manufacturing companies in reducing the human error possibility as the plant to be processed for medicine can be identified more accurately.

References

1. Heywood, V.H., Watson: Global Biodiversity Assessment. Cambridge University Press, Cambridge, UK (1995)
2. Loreau, M., et al.: Diversity without representation. Nature **442**, 245–246 (2006)
3. Pimm, S.L., Russell, G.J., Gittleman, J.L., Brooks, T.M.: The future of biodiversity. Science **269**, 347–350 (1995)
4. Wilson, E.O.: The current state of biological diversity. In: Wilson, E.O. (ed.) Biodiversity, pp. 3–18. National Academy Press, Washington, DC (1988)
5. Cope, J.S., Corney, D., Clark, J.Y., Remagnino, P., Wilkin, P.: Plant species identification using digital morphometrics: a review. Expert Syst. Appl. **39**(8), 7562–7573 (2012)
6. Setzer, V.W.: Artificial intelligence or automated imbecility, Can machines think and feel (2003)
7. Jyotismita, C., Ranjan, P.: Plant leaf recognition using shape based features and neural network classifiers. Int. J. Adv. Comput. Sci. Appl. **2**, 10 (2011)
8. Ekshinge, S., Sambhaji, D.B., Andore, M.: Leaf recognition algorithm using neural network based image processing. Asian J. Eng. Technol. Innov. **10**, 16 (2014)
9. Wu, Q., Changle, Z., Chaonan, W.: Feature extraction and automatic recognition of plant leaf using artificial neural network. AvancesenCiencias de la Computación, 5–12 (2006)
10. Wu, S.G., Forrest, S.B., Xu, E.Y., Wang, Y.X., Chang, Y.F., Xiang, Q.L.: Leaf recognition algorithm for plant classification using probabilistic neural network. In: IEEE International Symposium on Signal Processing and Information Technology (2007)
11. Prasad, S., Kudiri, K.M., Tripathi, R.C.: Relative sub-image based features for leaf recognition using support vector machine. In: Proceedings of the International Conference on Communication (2011)
12. Ehsanirad, A., Kumar, S.: Leaf recognition for plant classification using GLCM and PCA methods. Oriental J. Comput. Sci. Technol. **3**, 1 (2010)
13. Cope, J.S., Remagnino, P., Barman, S., Wilkin, P.: Plant texture classification using gabor cooccurrences. In: Advances in Visual Computing (2010)
14. Kumar, N., Belhumeur, P.N., Biswas, A., Jacobs, D.W., Kress, W.J., Lopez, I.C., Soares, J.V. B.: Leafsnap: A computer vision system for automatic plant species identification. In: ECCV, pp. 502–516. Springer, 2012
15. Kalyoncu, C., Toygar, Ö.: Geometric leaf classification. Comput. Vis. Image Underst. http://dx.doi.org/10.1016/j.cviu.2014.11.001

16. Radha, N., Rehana Banu, H.: A survey on approaches used in classification of leaf Images. https://www.ijcaonline.org/archives/volume182/number19/radha-2018-ijca-917954.pdf
17. Şekeroğlu, B., İnan, Y.: Leaves recognition system using a neural network. Procedia Comput. Sci. **102**, 578–582 (2016). https://doi.org/10.1016/j.procs.2016.09.445
18. Savan Patel article on Medium—https://medium.com/machine-learning-101/chapter-2svm-support-vector-machine-theory-f0812effc72

Building an Ensemble Learning Based Algorithm for Improving Intrusion Detection System

M. S. Abirami, Umaretiya Yash and Sonal Singh

Abstract Intrusion detection system (IDS) alerts the network administrators against intrusive attempts. The anomalies are detected using machine learning techniques such as supervised and unsupervised learning algorithms. Repetitive and irrelevant features in data have posed a long due to speed bump in efficient network traffic classification. This issue could be resolved by reducing the dimensionality of feature space using feature selection method wherein it identifies the important features and eliminates irrelevant ones. An intrusion detection system named Least Square Support Vector Machine (LSSVM-IDS) is built using this feature selection algorithm. It is tested on intrusion detection data set like KDD Cup 99, NSK-KDD and Kyoto 2006+ data set. This LSSVM machine has accuracy of 95%, i.e. it has predicted correct output with 95% of time. To avoid the intrusion detection system from getting obsolete, to adapt it with newer attack resistance feature and also to make it less expensive, we applied ensemble learning algorithm on UNSW-NB15 data set, using a stacking classifier method. We have combined random forest, support vector machine and Naive Bayes methods using logistic regression as meta-classifier and have achieved 95% accuracy.

Keywords Intrusion detection · Feature selection · Hybrid method · Lasso · Ensemble learning · Support vector machine

M. S. Abirami (✉) · U. Yash · S. Singh
Department of Software Engineering, SRMIST, Chennai, India
e-mail: abirami.srm@gmail.com

U. Yash
e-mail: umaretiyayash_ji@srmuniv.edu.in

S. Singh
e-mail: sonalsingh_sujitkumar@srmuniv.edu.in

S. S. Dash et al. (eds.), *Artificial Intelligence and Evolutionary Computations in Engineering Systems*, Advances in Intelligent Systems and Computing 1056, https://doi.org/10.1007/978-981-15-0199-9_55

1 Introduction

There is an ever-increasing demand of a robust security to be implied upon the evolving technology. As the need for a robust security increases the conventional solutions will not be capable of providing security to advanced threats like denial-of-service (DoS) attacks, bugs and malwares. For this, making efficient and flexible security methods are more important. The first line of security defence methods such as user authentication, firewall and data encryption are not enough to cover the entire system's security needs and also covering challenges from ever-lasting intrusion methods. Therefore, we need better version of security such as intrusion detection system.

Intrusion detection system is a software or a device which audits data of network or system for any suspicious activity or action which is malevolent. All the details about suspicious activities will be given to Security Information and Event Management (SIEM) system. The filtering techniques alert user about suspicious activity and separate true alarm from false one.

Intrusion detection system is useful for single computers as well as for large networks. We can divide IDS in two types: which are known as network intrusion detection system (NIDS) and host intrusion detection system (HIDS). HIDS is a system which organizes important operating system files; while NIDS is useful for evaluating traffic of incoming network. IDScan also be categorized as signature-based detection, i.e. to detect bad pattern, for example, malware, is mostly used while anomaly-based detection [1] is used for detection of good traffic. Generally, machine learning algorithms are used for anomaly detection.

Intrusion detection systems [2] are situated at unique strategic point or sometime at some point inside network in order to analyse all traffic going in and coming out of all devices on the network. This system conducts an analysis of traffic on the entire subnet and matches it with the traffic that is passed on the subnets to the library of known attacks. After the attack is identified, or suspicious behaviour is analysed, an alert is generated and can be sent to the admin. Firewalls are located on NIDS to check whether any attacker is seeking to penetrate into the system. NIDS can also be merged with other technologies so that improving the rate of detection and prediction of different types of attacks.

IDS based on neural network and machine learning [3–6] can able to monitor huge amount of data in such a way that it self-organizes itself and detects intrusion patterns in a much better way. On the other hand, host-based IDS provides security to individual host or device. It manages the packet going to and coming from device and gives an alarm to the user for any malicious activity. A snapshot of current system file is doubled with the previous snapshot. If any of this snapshot or files are modified or removed, it will cause an alert which will be sent to admin for further investigation. Mission critical machines, which are not supposed to make any changes in their properties or configure, are examples of HIDS.

Wrapper methods are very computationally expensive because it will train new model for all the subset, but it also gives best performing features as a result for specific type of model. On the other hand, filter methods score features based on some measures. These measures are selected to be better to compute, while it also considers advantages of feature set.

Feature set generated as a result of filter methods is not specific to any model. As the feature sets are different for each model, predictive performance of filter methods is bad compared to wrapper method. Also, the set of features does not have any assumption of predictive model, so it will be used for getting idea about relationship between features rather than its usefulness. A feature ranking is preferred by many filter methods in place of exact list of best features, and measure of ranking can be selected using cross-validation methods. For pre-processing, filter methods are used before wrapper methods, which are efficient to work with larger problems. Recursive Feature Elimination algorithm is another popular approach which is combined with SVM to build a model where no features are having low weights.

There are mainly two detection methods, namely signature based and anomaly based. We address intrusions detection using anomaly-based feature selection method. The main objectives of this proposed hybrid feature selection intrusion detection system are:

- To build a feature-based IDS in which the relevant features are narrowed down from the wide features using machine learning algorithms and also to optimize the process
- All the data are trained using different classification algorithms for more accuracy and later tested with test data
- For better accuracy, the algorithms are combined using ensemble learning technique. This process is done using stacking classifier.

2 Background and Related Work

In feature selection, the features which are not important and repeated are removed. This means they are either irrelevant or redundant. After that best features are selected in order to create better characterization of pattern of different classes. There are mainly two ways in which these methods are classified such as filter and wrapper methods. Filter algorithms use autonomous measures as criterion to get estimate of relation of features, and wrapper methods use specific learning algorithms to assess value of each features. The wrapper methods are expensive compared to filter methods especially when dealing with high-dimensional data. In this work, filter methods are used for intrusion detection system.

Ambusaidi et al. [7] presented mutual information-based feature selection algorithm they achieved 92% accuracy in LSSVM-based algorithm trained and tested on KDD Cup 99 data set. Su et al. [8] proposed an algorithm named learning automata-based feature selection for NIDS. Learning automaton (LA) [9, 10] method is a decision maker with adaptive nature. It was designed to learn the behaviour of biological tissues. It constantly interacts with external environment in order to learn stochastic behaviour and maximize benefits. They have used learning automata-based algorithms and achieved accuracy of 90%.

Amriti et al. [11] suggested a method known as forward feature selection algorithm using mutual information-based method. In this method, features were selected using LSSVM classifier. Horng et al. [12] proposed IDS, which combined two methods known as SVM and clustering. Hierarchical clustering is used to generate highly classified data and improving detection time. Experimental method on KDD data set was giving 94.32% accuracy with false positive rate of 0.7. So many IDS has been proposed using ensemble learning algorithm [13–15]. Shrivas and Dewangan [16] developed an ensemble model which is based on the two classifiers artificial neural network (ANN) [17] and Bayesian net, and they combined them using Gain Ratio with Feature Selection Technique. This system was tested on KDD99 and NSL-KDD data set. This method gave better results than any single classifier. They have extracted important features from NSL-KDD data set and then make separate data set from it and for better results. It has 98% true positive rate. They have used Naive Bayes, Bayesian net and J48 combined using ensemble method [18].

Chaurasia and Jain [19] used two base classifier and bagging method of ensemble learning: k-nearest neighbours and neural networks. This paper shows better result in terms of accuracy then single classifier. Govindarajan [20] presented on topic of an ensemble model using Arcing (Adaptive Resampling and Combining) technique with two algorithms for base classifiers such as radial basis function neural network (RBF) and support vector machine (SVM). The proposed Arcing technique uses a sequential construction for base classifiers, which means the next classifier construction would depend on the performance of all previous classifiers. The classification accuracy on NSL-KDD data set showed the outperformance of the ensemble model relative to single classifiers (RBF and SVM).

There are so many hybrid techniques [21–23] which were produced to make IDS better. Ensemble learning combines models using ensemble techniques like bagging, boosting, stacking, while this hybrid model can use different model for both feature selection and building classifier. Several novel approaches were applied to build a hybrid IDS, such as principal component analysis (PCA), stochastic variant of Primal estimated sub-gradient solver in SVM (SPegasos), Ensembles of Balanced Nested Dichotomies (END) and Grading. The experimental results showed that the combination of random forest and END provided the highest detection rate with a low false alarm rate.

Association rule mining (ARM) can also be applied for IDS by many researchers. This method is used for replicating UNSW-NB15 data set on KDD Cup 99 data set to find effective data set. Moustafa and Slay [24] have used Naive Bayes and EM clustering methods in order to judge them based on accuracy and efficiency.

Most of the papers related to NIDS and supportive algorithms are tested on KDD Cup 99 data set. In this paper, UNSW-NB15 [25] data set have been used throughout research because UNSW-NB15 has newer and updated results which help to improve FAR and increase efficiency.

3 Proposed Methodology

Different types of classification algorithms are applied on data set. The proposed system architecture for intrusion detection system is shown in Fig. 1. Data are extracted from open data set. Then, principal component analysis and random forest methods are applied to find the important features of the data set. Also to get better accuracy, some of the features which are dependent on other features of data set or unrelated features for this detection system have been removed from the data set.

Data acquisition model means extracting data which is required for intrusion detection. In this scenario, this extraction has been done by Cyber Range Lab of the Australian Centre for Cyber Security (ACCS). ACCS has generated UNSW-NB15 data set and made it public. From that data, the best features are selected. Principal Component Analysis is a statistical process that uses an orthogonal alteration to change an observation set of probably correlated variables (entities each of which takes on various arithmetic values) to value set of linearly uncorrelated variables which are called as principal components. This method is applied for getting the list of best features.

Once the best features are selected, different classification methods can be applied. The four algorithms random forest, linear support vector machine (linear SVM), Naive Bayes and logistic regression are applied on the pre-processed data set. All these methods have been tested in analyser state. All unknown types of attacks are known as anomaly detections. For any type of known attacks, the required action has

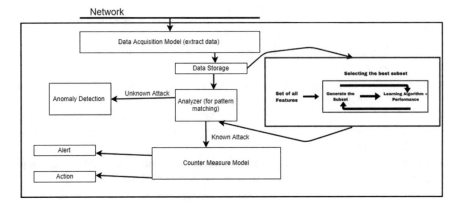

Fig. 1 System architecture for intrusion detection system

to be taken. All the algorithms are trained on trained data set, and later, it can be predicted that whether attack has happened or not using test data set.

Then, stacking classifier algorithm [14, 26] is used for combining all classification methods. One of the methods of ensemble learning is known as stacking, and it takes more than one classification or regression model and then combines the result of each model using meta-classifier or meta-regressor. First classification or regression model are trained on full training data set by classifier of that type and later result of all these base-level model are considered as feature of test data and be tested using meta-classifier or meta-regressor.

4 Experimental Set-Up

4.1 Data Set Description

This data set is named as UNSW-NB15 is generated by a tool known as IXIA PerfectStrom tool in order to get idea about new types of attacks. There is a tool which named as tcsdump, was operated to get hundred GB of raw network traffic which is also known as pcap files. Each file has 1000 MB to make an easy analysis of packets. Two techniques named as Bro-IDS and Argus and twelve methods were used in parallel in order to get 49 features with labels. This data set is supervised (labelled) so that there are two features named as attack_cat to define type of attack and one binary column to define whether or not attack happened. This data set has 2,540,044 records which are saved in four different CSV files. It has two files divided as test and trained data set which are cleaned and can be used for analysis [25]. There are nine types of attacks which are listed as follows:

- Fuzzers—In this, attacker try to find loopholes of secured system, OS or network by providing it a large set of data which is random and try to crash it.
- Analysis—In this, different groups of instruction try to get into web application using emails or port, sometimes web scripts.
- Backdoor—This type of attack normally bypasses authentication and gets unauthorized access of system remotely.
- DoS—This intrusion rattles the system resources using memory aimed to stop authorized request by user to access system.
- Exploit—In this, string of instruction get support of bug or vulnerability in order to create unsuspected behaviour on network of host.
- Generic—In this type of attack, it creates against each block cipher with the help of hash function in order to make collision without considering the configuration of block cipher.
- Reconnaissance—It is known as probe and in this it finds out network data and uses it to penetrate in computer network of secured system.
- Shellcode—It is a type of attack where attacker gets into system to find some part of code from shell to make system vulnerable and get the control of it.

- Worm—In this, people who want to attack the network depict their selves to increase reach over all network. Later, increment of this attack's effect is based on security of host system.

The features of the UNSW-NB15 data set are categorized into six different types as described below and list of these features are shown in Table 1.

- Flow features—In this, all the features which identify the connection are in this group like client to client.
- Basic features—For representation of protocol connection, this category will be used.
- Content features—This type is used to group features of TCP/IP, and also, it contains details about http services.
- Time features—This type has details about features related to time, for example, source to destination time, jitter time, etc.
- Additional generated features—This type is divided further into two categories: (1) general purpose features (36–40) here each feature has unique scope to secure feature services; (2) connection features (41–47) were framed using last time feature which was gathered using hundred connection records.
- Labelled features—As our data set is supervised, these two are output (label) remaining 47 features.

Table 1 shows the list of features of UNSW-NB15 data set. Then, few classification algorithms and ensemble algorithm are applied on this data set, which is described in next section. There are 1,93,254 rows for training the model and 64,419 for testing the model are used.

4.2   Stacking Algorithm

Three different methods are used for classification named as random forest, Naive Bayes and linear SVM. Also, logistic regression is used. These algorithms are explained as shown below. In Spyder tool, python is used as development language and the classifier function, i.e. inbuilt function for each method, which is going to ensemble for the stacking algorithm.

4.2.1   Random Forest

Step 1: Divide data in test and train 80:20 ratio.
Step 2: Scale all the variable using standard Scaler.
Step 3: Import RandomForestClassifier and apply.
"from sklearn.ensemble import RandomForestClassifier as rf"
Step 4: Use fit and score function of classifier.

Table 1 Features of dataset

#	Name	Description
1. Flow features		
1	scrip	Source IP address
2	sport	Source port number
3	dstip	Destinations IP address
4	dsport	Destination port number
5	proto	Protocol type, such as TCP, UDP
2. Basic features		
6	state	The states and its dependent protocol e.g., CON
7	dur	Row total duration
8	sbytes	Source to destination bytes
9	dbytes	Destination to source bytes
10	sttl	Source to destination time to live
11	dttl	Destination to source tune to live
12	sloss	Source packets retransmitted or dropped
13	dloss	Destination packets retransmitted or dropped
14	service	Such as http, ftp, smtp, ssh, dns and ftp-data
15	sload	Source bits per second
16	dload	Destination bits per second
17	spkts	Source to destination packet count
18	dpkts	Destination to source packet count
3. Content features		
19	swin	Source TCP window advertisement value
20	dwin	Destination TCP window advertisement value
21	Stcpb	Source TCP base sequence number
22	dtcpb	Destination TCP base sequence number
23	smeansz	Mean of the packet size transmitted by the srcip
24	dmeansz	Mean of the packet size transmitted by the dstip
25	trans_depth	The connection of http request/response transaction
26	res bdy len	The content size of the data transferred from http
4. Time features		
27	sjit	Source jitter
28	djit	Destination jitter
29	stime	Row start time
30	ltime	Row last time
31	sintpkt	Source inter-packet arrival time
32	dintpkt	Destination inter-packet arrival tune
33	tcprtt	Setup round-trip tune, the sum of 'synack' and 'ackdat'
34	synack	The time between the SYN and the SYN_ACK packets
35	ackdat	The time between the SYN_ACK and the ACK packets
36	is_sm_ips_ports	If srcip (1) = dstip (3) and sport (2) = dsport (4), assign 1 else 0

(continued)

Table 1 (continued)

#	Name	Description
5. Additional generated features		
37	ct_state_ttl	No. of each state (6) according to values of sttl (10) and dttl
38	ct_flw_http_mthd	No. of methods such as Get and Post m http service
39	is_ftp_login	If the ftp session is accessed by user and password then 1 else 0
40	ct_ftp_cmd	No of flows that has a command in ftp session
41	ct_srv_src	No. of rows of the same service (14) and srcip (1) in 100 rows
42	ct_srv_dst	No. of rows of the same service (14) and dstip in 100 rows
43	ct_dst_ltm	No. of rows of the same dstip (3) m 100 rows
44	ct_src_ltm	No. of rows of the srcip (1) in 100 rows
45	ct_src_dport_ltm	No of rows of the same srcip and the dsport in 100 rows
46	ct_dst_sport_ltm	No of rows of the same dstip (3) and the sport (2) m 100 rows
47	ct_dst_src_ltm	No of rows of the same srcip (1) and the dstip (3) in 100 records
6. Labelled features		
48	Attack_cat	The name of each attack category
49	Label	0 for normal and 1 for attack records

4.2.2 Linear SVM

Step 1: Divide data in test and train 80:20 ratio.
Step 2: Scale all the variable using standard Scaler.
Step 3: Import LinearSVC and apply.
   "from sklearn.svm import LinearSVC as svm"
Step 4: Use fit and score function of classifier.

4.2.3 Naive Bayes

Step 1: Divide data in test and train 80:20 ratio.
Step 2: Scale all the variable using standard Scaler.
Step 3: Import GausianNB and apply.
   "from sklearn.naive_bayes import GaussianNB"
Step 4: Use fit and score function of classifier.

4.2.4 Logistic Regression

Step1: Divide data in test and train 80:20 ratio.
Step 2: Scale all the variable using standard Scaler.
Step 3: Import LogisticRegression and apply.
          "from sklearn.linear_model import LogisticRegression"
Step 4: Use fit and score function of classifier.

4.2.5 Stacking Algorithm

Input: UNSW-NB15 data set
Output: Fitted Classifier
Step 1: Break data set into 80:20: train: test data
Step 2: Create classifier for each algorithm
Step 2.1: Use all three classifier with for random forest, naïve bays and liner svm.
Step 2.2: Define logistic regression classifier which will be used as meta-regressor
Step 3: Train data set based on base classifier
Step 4: Construct new data based on prediction.
Step 5: Learn a meta-classifier
Step 6: Fit data using train x and train y.

In stacking algorithm, logistic regression is used as meta-regressor. Three different models are combined, i.e. random forest, SVM and Naive Bayes. Also logistic regression is used as meta-classifier. If this method contains M models which has a rate of error less than half and also if that base models' error is not dependent on any other factor, then we can conclude that probability of model M making error is equivalent to probability of half of model (M/2) is misclassified. Logic behind stacking algorithm is that if pair of input and output represented as (a, b) is not part of training set g_i, the output b can be used to find out error of model. Also we can say since a and b is not part of g_i, $g_i(a)$ will have different result than b. So based on this we can train new classifier which can be represented by b—$g_i(a)$. In this way, a second classifier can be trained to learn error from other classifier. Using both these methods, we can get better classification decision. This algorithm generates stacking classifier score, and later, we can check results using accuracy score.

5 Results and Discussion

For all the three random forest, SVM and Naïve Bayes algorithms, the confusion matrix is generated. The ROC curve and the confusion matrix for these algorithms are shown in Figs. 2, 3 and 4.

Fig. 2 ROC curve for random forest

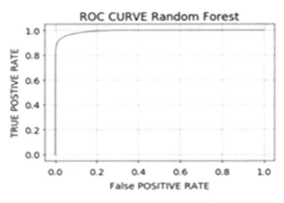

TP = 17434   FP = 1150   TN = 1316   FN = 31635

Fig. 3 ROC curve for SVM

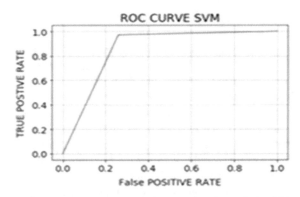

TP = 13727   FP = 4802   TN = 931   FN = 320

Fig. 4 ROC curve for Naive Bayes

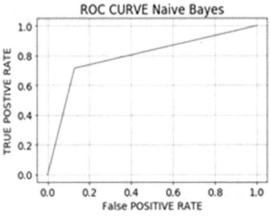

TP = 16103   FP = 2377   TN = 9427   FN = 31635

Table 2 Accuracy measure of different classifiers

Classifier	Accuracy
Random forest	0.93 (±0.02)
Linear SVM	0.85 (±0.05)
Naive Bayes	0.79 (±0.03)
Logistic regression	0.80 (±0.05)

- True positive (TP): includes the number of attack successfully detected by IDS.
- False positive (FP): includes the normal behaviour which was detected as attack/abnormal.
- True negative (TN): included normal or non-intrusive type of behaviours, which are also identified as same.
- False negative (FN): includes attacks which are classified as normal behaviour.

The accuracy of these algorithms can be measured using F1 score as shown in Table 3. For logistic regression, accuracy can be measured using R2 score (Table 2).

Stacking normally has two steps. In first step, all classifiers which are at base level participate in training using CV (cross-validation) in which one vector will be the final result. It takes the form of $<(x'0...x'm), xk>$, where k is total fold of cross-validation and $x'm$ is the output which is predicted for mth classifier and xk is the output which is expected for the same. In the next step, this result will be transferred using meta-learning algorithm which solves the bugs or errors in a way that it is optimized the whole process of model. Also the process will be repeated again using k-fold CV to get the last optimized model (Fig. 5).

All the separate results can be combined using stacking classifier. The meta-level classifier using logistic regression is applied for prediction. The result of stacking classifier is shown in Table 3.

The accuracy of stacking classifier is 95%. The comparison of accuracy between ensemble and other algorithms are shown in Fig. 6. When comparing with other algorithms, ensemble algorithm is achieving highest accuracy of 95%.

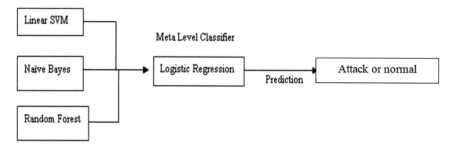

Fig. 5 Ensemble methodology

		Precision	Recall	F1 Score
Table 3 Classification result of stacking classifier	0	0.91	0.93	0.92
	1	0.96	0.95	0.95
	Micro-average	0.94	0.94	0.94
	Macro-average	0.94	0.94	0.94
	Weighted average	0.94	0.94	0.94

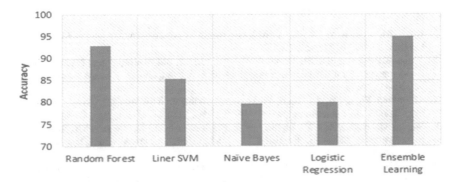

Fig. 6 Comparison of accuracy of ensemble algorithm with other classifiers

6　Conclusion and Future Work

The stacking classifier or ensemble learning algorithm is a better option among many classification, decision tree or regression algorithms. Hybrid method of feature selection also comprises L1 regularization-based Lasso regression [27]. LASSO is actually abbreviation for Least Absolute Shrinkage and Selection Operator. It can be used for regression using L1 regularization, i.e. that can be done by adding a factor of sum of absolute value of coefficients in the optimization objective.

Lasso algorithm randomly chooses one feature among the highly correlated ones and makes coefficients of others to zero. Also, these variable changes arbitrarily when model parameters change. But this is not better compared to ridge regression.

Even though UNSW-NB15 data set has labelled (yes/no) data column, Lasso algorithm is applied using SVM. The accuracy of Lasso on scaled data set can be measured using R2 score evaluation. R2 score for Lasso is 59% on test data set.

Achieving embedded methodology for feature selection using L1-norm-based Lasso classification algorithm and maintaining same accuracy as ensemble learning algorithm are considered as future work.

References

1. Yolacan Esra, N., Kaeli David, R.: A framework for studying new approaches to anomaly detection. Int. J. Inf. Secur. Sci. (2016)
2. Singh, P.L., Akhilesh, T.: Survey on intrusion detection using data mining methods. Int. J. Sci. Adv. Res. Technol. (2016)
3. Kabir, E., Hu, J., Wang, H., Zhuo, G.: A novel statistical technique for intrusion detection systems. Future Gener. Comput. Syst. **79**, 303–318 (2018)
4. Shah, S.A.R., Issac, B.: Performance comparison of intrusion detection systems and application of machine learning to snort system. Future Gener. Comput. Syst. **80**, 157–170 (2018)
5. Umer, M.F., Sher, M., Bi, Y.: Flow-based intrusion detection: techniques and challenges. Comput. Secur. **70**, 238–254 (2017)
6. Vieira, T.P.B., Tenrio, F.D., Costa, J.P.C., et al.: Model order selection and Eigen similarity based framework for detection and identification of network attacks. J. Netw. Comput. Appl. **90**, 26–41 (2017)
7. Ambusaidi, M.A., He, X., Nanda, P., Tan, Z.: Building an intrusion detection system using a filter-based feature selection algorithm. IEEE Trans. Comput. **65**(10), 2986–2998 (2016)
8. Su, Y., Qi, K., Di, Ma, Y., Li, S.: Learning automata based feature selection for network traffic intrusion detection. In: IEEE Third International Conference on Data science in Cyberspace (2018)
9. Zhang, Y., Di, C., Han, Z., Li, Y., Li, S.: An adaptive Honeypot deployment algorithm based on learning automata. In: IEEE Second International Conference on Data Science in Cyberspace, pp. 521–527. IEEE Computer Society, China (2017)
10. Narendra, K.S., Mandayam, A.L.: Thathachar: learning automata: an introduction. Dover Books on Electrical Engineering, Dover (2012)
11. Amiri, F., Yousefi, M.R., Lucas, C., Shakery, A., Yazdani, N.: Mutual information-based feature selection for intrusion detection systems. J. Netw. Comput. Appl. **34**(4), 1184–1199 (2011)
12. Horng, S.J., Su, M.Y., Chen, Y.H., Kao, T.W., Chen, R.J., Lai, J.L., Perkasa, C.D.: A novel intrusion detection system based on hierarchical clustering and support vector machines. Expert Syst. Appl. **38**(1), 306–313 (2011)
13. Pham, N.T., Foo, E., Suriadi, S., Jeffrey, H., Lahza, H.F.M: Improving performance of intrusion detection system using ensemble methods and feature selection. In: ACSW 2018: Australasian Computer Science Week (ACSW), January 29–February 2, 2018, Brisbane, QLD, Australia (2018)
14. Zhou, Z.-H.: Ensemble Methods: Foundations and Algorithms. CRC Press (2012)
15. Panda, M., Patra, M.R.: Ensemble of classifiers for detecting network intrusion. In: Proceedings of the International Conference on Advances in Computing, Communication and Control, pp. 510–515. ACM (2009)
16. Shrivas, A.K., Dewangan, A.M.: An ensemble model for classification of attacks with feature selection based on KDD99 and NSL-KDD data set. Int. J. Comput. Appl. **15** (2014)
17. Akashdeep, Manzoor, I., Kumar, N.: A feature reduced intrusion detection system using ANN classifier. Expert Syst. Appl. **88**, 249–257 (2017)
18. NSL-KDD datasey, http://www.unb.ca/cic/research/datasets/nsl.html (2017)
19. Chaurasia, Shalinee, Jain, Anurag: Ensemble neural network and KNN classifiers for intrusion detection. Int. J. Comput. Sci. Inf. Technol. **5**, 2481–2485 (2014)
20. Govindarajan, M.: Hybrid intrusion detection using ensemble of classification methods. Int. J. Comput. Netw. Inf. Secur. **6**(2) (2014)
21. Haq, N.F., Onik, A.R., Shah, F.M.: An ensemble framework of anomaly detection using hybridized feature selection approach (HFSA). In: SAI Intelligent Systems Conference, pp. 989–995. IEEE (2015)

22. Sumalatha, P., Namita, P.: A review on hybrid intrusion detection system using TAN & SVMIPASJ. Int. J. Comput. Sci. (IIJCS) (2015)
23. Li, L., Yu, Y., Bai, S., Cheng, J., Chen, X.: Towards effective network intrusion detection—a hybrid model integrating Gini Index and GBDT with PSO. J. Sens. **18** (2018)
24. Moustafa, N., Slay, J.: The significant features of the UNSW-NB15 and the KDD99 data sets for network intrusion detection systems. In: International Workshop on Building Analysis Datasets and Gathering Experience Returns for Security. IEEE (2015)
25. Moustafa, N., Slay, J.: UNSW-NB15: a comprehensive dataset for network intrusion detection systems. In: Military Communications and Information Systems Conference (MilCIS). IEEE, Australia (2015)
26. Kuncheva, L.: Combining Pattern Classifiers: Methods and Algorithms. Wiley (2014)
27. Fonti, V.: Feature selection using LASSO in research paper of business analytics, VU Amsterdam, March (2017)

Damped Sinusoidal Exploration Decay Schedule to Improve Deep Q-Networks-Based Agent Performance

Hriday Nilesh Sanghvi and G. Chamundeeswari

Abstract Environments in which the rewards are sparse or occur rarely prove to be some of the most challenging for reinforcement learning (RL) agents and can take a long time to train. The decision between attempting to gain new experience, or using the knowledge already acquired to perform what is thought to be the optimal action in a given state, is computed by the exploration factor. This problem has been debated as the exploration–exploitation trade-off, which simply means choosing between what the agent already knows, and new things the agent can potentially discover. In this paper, an attempt is made to construct a new sinusoidal ε-decay function, which steadily decreases the exploration factor and increases performance. This approach is inspired from the sinusoidal curve which causes an "oscillation" of values. It is expected to stabilize and speed up a DQN-based agent's learning process, preventing it from diverging even after long periods. The new exploration decay equation can be adopted for other RL algorithms as well.

Keywords Reinforcement learning · Exploration decay · Exploration versus exploitation · Deep Q-learning · DQN · DDQN

1 Introduction

The way an RL agent decides what action to take is by first deciding if it must explore or exploit. Here, exploration would mean to perform actions to gain new experiences or see observations/states that it has never encountered before. Exploitation would mean that the agent makes use of what it already knows and chooses the most optimal action for the given state according to its neural network.

H. Nilesh Sanghvi (✉) · G. Chamundeeswari
Department of Computer Science and Engineering, SRM Institute of Science
and Technology, Chennai, India
e-mail: hriday_nilesh@srmuniv.edu.in

G. Chamundeeswari
e-mail: chamundeeswari.g@vdp.srmuniv.ac.in

© Springer Nature Singapore Pte Ltd. 2020
S. S. Dash et al. (eds.), *Artificial Intelligence and Evolutionary Computations in Engineering Systems*, Advances in Intelligent Systems and Computing 1056,
https://doi.org/10.1007/978-981-15-0199-9_56

This decision of exploration versus exploitation is commonly known as the exploration–exploitation trade-off. In Q-learning, the most common approach to make this decision is the ε-greedy strategy. If the number generated from a uniform distribution over the interval $[0, 1)$ is smaller than ε which is initially set to 1, then the agent decides to explore, but otherwise, the agent decides to exploit. For example, if ε is brought down to, say, 0.4, there is a 40% chance that the agent will choose to explore, instead of exploiting its existing knowledge. A vital component of using the ε-greedy strategy is the decay of the exploration factor ε. As the training progresses and the agent becomes increasingly more mature, it has to steadily move from pure exploration to pure exploitation. This happens through a process called ε-decay. ε can be decayed using a constant factor or as a function of another variable. The problem with using a constant factor is that it does not take into account the agent's stage during the training process, and as such, is not preferable. As a function, the ε-decay could work with timesteps, average rewards, epochs, and other indicative metrics of the agent's training progress. The effectiveness of the proposed sinusoidal decay equation is examined by conducting experiments on three distinct environments provided by the OpenAI Gym research framework that is often used for testing RL algorithms. The environments in order of increasing complexity of the environment and its inputs, are Cart Pole, Mountain Car and Pong. The novel sinusoidal damping equation is introduced as the exploration decay function and is expected to bring down training time of the agent.

2 Background

2.1 RL Systems

As can be seen from Fig. 1, the agent performs an action A_t based on an observation S_t it makes of the environment. The effect of the action leads to the environment

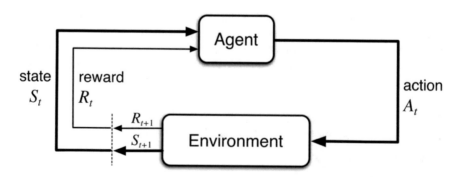

Fig. 1 Generic RL system

reaching a new state S_{t+1}, passing it and the reward R_{t+1} associated with reaching that state back to the agent.

2.2 Training Environments

In Cart Pole, the goal or task of the agent is to balance the pole for as long as it can, without it tipping too far away from its center axis, or going out of the screen. The longer it can do this, using two possible actions, move left and move right, the more reward it collects. In Mountain Car, the task becomes a little more complex. To win, the agent simply has to reach the flag at the top of the mountain/hill. But the engine of the car is not strong enough to climb over the slope of a hill; it will need to build some momentum before it can do so. The faster it can do this, using two possible actions, move left and move right, the lesser the penalty it evokes. In Pong, the player controls a paddle and tries to hit the ball to ensure that his opponent cannot volley it back. If the player misses the ball, he loses a point. And if his opponent is unable to volley the ball, he gains a point. The first one to reach a score of 21 wins the game. The agent is pitted against a hard-coded AI which simply follows the position of the ball at a constant speed, trying to ensure that the central axis of the paddle aligns with that of the ball.

2.3 Concepts and Equations for Q-Learning

The role of the Q-function is to predict a "Q-value" gained from taking a particular action in a given state. In other words, the Q-function predicts the Q-values of all possible actions that can be taken in a given state, as a 1-D vector as follows:

$$Q(s, a) = E\left[R_t + \gamma R_{t+1} + \gamma^2 R_{t+2} + \cdots | (s, a)\right] \tag{1}$$

where

$Q(s, a)$ Q-value for action a taken at state s,
R_t Immediate reward at timestep t,
γ Discount factor.

Based on the Bellman Eq. (1), recursive form of the equation is:

$$Q(s, a) = R + \gamma \max_{a'} Q(s', a') \tag{2}$$

where

$Q(s, a)$	Q-value for action a taken at state s,
R	Immediate reward,
γ	Discount factor,
$\max\limits_{a'} Q(s', a')$	Maximum future reward assuming optimal action a taken in next state s'.

In deep Q-networks (DQN) [1], the core update equation which predicts the true Q-value label for the optimization algorithm to learn from, resembles the traditional gradient descent update rule and is based on Eq. (2):

$$Q'(s, a) = Q(s, a) + \alpha \left[R' + \gamma \max_{a'} Q(s', a) - Q(s, a) \right] \tag{3}$$

where

$Q'(s, a)$	Q-value for action a taken at state s,
$Q(s, a)$	Current Q-value,
α	Learning rate,
R'	Immediate reward for the new state,
γ	Discount factor,
$\max\limits_{a} Q(s', a)$	Maximum future reward assuming optimal action a' taken in next state s'.

The double DQN (DDQN) [2], on the other hand, incorporates two networks, the local and the target network. The local network is used for the actual prediction of Q-values for actions while the target network is used to estimate the optimal Q-values during training of the local network. This decoupling of prediction and estimation substantially reduces the overestimation of values that tends to occur when using a lone DQN.

An experience replay [3] is a powerful concept used in offline reinforcement learning. This technique utilizes a "replay buffer" memory from which random subsamples are drawn in minibatches during training. This tends to break strong correlations between samples and acts in accordance with the I.I.D. [4] assumption. Another vital process in Q-learning is the ε-decay in ε-greedy exploration, where the ε-value starts at 1 and slowly decays until it reaches a minimum value that is defined:

$$\text{action} = \begin{cases} \text{random action,} & \text{if } x < \varepsilon \\ \text{optimal action from network,} & \text{otherwise} \end{cases} \tag{4}$$

where

ε	Exploration factor $\in [\varepsilon_{min}, 1]$,
x	Generated random value $\in [0, 1)$,
ε_{min}	Minimum value of ε.

3   Related Work

With advancements in computer vision technology like convolutional neural networks (CNNs), it provided a way to extract features from the stack of image frames close to how humans would see the game to some extent. Deep Q-learning with this CNN-based architecture came to be known as DQN [1]. An interesting idea was introduced recently, where recurrent neural networks (RNNs), known for extracting temporal relations between sequential data, was used. The fully connected layer being replaced by a long short-term memory (LSTM) unit to extract accurate temporal information saw an increase in DQN performance. They called it deep recurrent Q-networks (DRQN) [5]. A direct improvement of the DQN was the DDQN [2], which used two networks in order to reduce over optimistic Q-value estimation when only one network is used for prediction. The Google DeepMind team extended the DDQN architecture to include an additional "Advantage" stream that assisted in computing even more accurate Q-values. This was called the dueling DQN [6], to express the "dueling" nature of the value and advantage streams. Apart from the architecture, some work also focused on a component called the experience replay [3, 7]. An improved method of sampling experiences from the replay buffer according to a "priority" value that was computed based on a proxy of the TD-error gave an indication of the extent of learning possible from each experience. This was called prioritized experience replay (PER) [8]. Recent work by the OpenAI team [9] shows that a generic way of deriving rewards for sparse-reward environments, using an "exploration bonus", which is based on the degree of newness of an observation, outperforms previous agents in exploration of the "Montezuma's Revenge" environment. There are numerous works surveying various exploration strategies [10, 11], but none of them talk specifically about the improvements in exploration decay schedules required by the ε-greedy exploration approach.

4   Components and Methods

Preprocessing. The amazing thing about RL is that it does not use predefined or labeled data as seen in supervised learning. Nor does it use existing unlabeled data to figure out interesting patterns in the data, commonly seen in unsupervised learning. Instead, it collects its own data during the course of its interaction with an unfamiliar environment as seen in Fig. 1. The "classic control" problems—Cart Pole and Mountain Car—are relatively simple [12, 13] and required little to no preprocessing of the low-dimensional observation data obtained directly from the internal **state** of the simulator. The preprocessing steps for the Pong Atari environment like frame-skipping and element-wise maximum of frames, stacking of frames, grayscale and downsampling, the initial no-op actions and "episodic" life tweak were implemented based on a variety of sources [1, 7, 14].

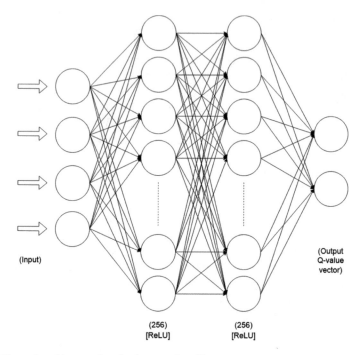

Fig. 2 Network architecture for classic control problems

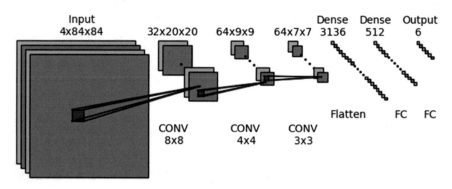

Fig. 3 Network architecture for Atari Pong

Training. The same multilayer perceptron (MLP) architecture, in Fig. 2, was used for both classic control problems (Fig. 3 and Tables 1, 2, 3).

Exploration decay. The simple linear decaying schedule [1] currently used in the ε-greedy strategy is shown below:

Table 1 Network configuration for classic control problems

Classic control		
Layer	Neurons	Activation
Input	#observations	Linear
FC1	256	ReLU
FC2	256	ReLU
Output	#Actions	Linear

Table 2 Network configuration for Atari Pong

Atari environment					
Layer	Shape	#Filters × Kernel	Stride	Padding	Activation
Input	4 × 84 × 84	–	–	–	–
CONV1	32 × 20 × 20	32 × 8 × 8	4	Valid	ReLU
CONV2	64 × 9 × 9	64 × 4 × 4	2	Valid	ReLU
CONV3	64 × 7 × 7	64 × 3 × 3	1	Valid	ReLU
FC1	3136	–	–	–	ReLU
FC2	512	–	–	–	ReLU
Output	#Actions	–	–	–	–

Table 3 Comparison of training hyperparameters

ts = timesteps, eps = episodes	Environments		
Hyperparameters	Cart Pole	Mountain Car	Pong
Total episodes (T)	1000	2000	500
Initial epsilon (ε)	1.0	1.0	1.0
Batch size	20	64	32
Replay buffer size	100,000	400,000	10,000
Learning rate (α)	0.001	0.0001	0.0001
Discount Factor (γ)	0.95	099	0.99
Target network update frequency	10 eps	1 eps	1000 ts
Optimization algorithm	Adam	Adam	Adam

$$\varepsilon_t = \varepsilon \cdot d_f \qquad (5)$$

where d_f is the constant factor by which ε decays at timestep/episode t.

The proposed equation instead relies on a sinusoidal manner of damping/decaying:

Table 4 Experimental values used for hyperparameters/constants in exploration decay

Environment	Hyperparameter/constant			
	d_f	k_1	k_2	ε_{min}
Cart Pole	0.995	0.15 0.20	0.010 0.025	0.001
Mountain Car	0.995	0.20	0.025	0.001
Pong	0.995	0.020	0.025	0.001

$$\varepsilon_t = 0.5e^{(-t/k_1 T)}\left(1 + \cos\left(\frac{t}{k_2 T}\right)\right) \tag{6}$$

where k_1 and k_2 are hyperparameters in the proposed equation. After the decay step, the larger of ε_t and ε_{min} is taken to ensure that exploration factor never drops below a certain threshold (Table 4).

5   Results

The dotted red line and straight blue line indicate proposed equation and linear equation performances, respectively. Here, k_1 is set to 0.20 and k_2 is set to 0.025 because preliminary results on Cart Pole were better for these values.

From Figs. 4, 5, 6, and Table 5, we can see that the increase in training performance is more prominent in sparse-reward environments. Each cell contains the values for the proposed equation in bold (Fig. 7).

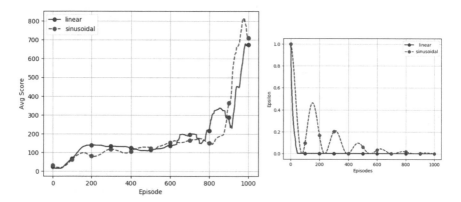

Fig. 4 Cart Pole's mean performance (left) and exploration decay curve (right)

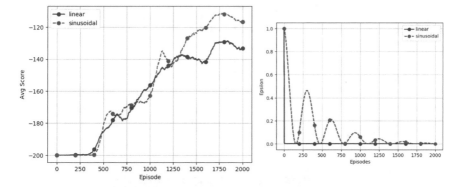

Fig. 5 Mountain Car's mean performance (left) and exploration decay curve (right)

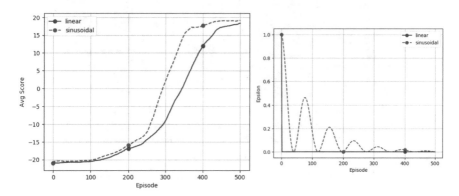

Fig. 6 Pong's mean performance (left) and exploration decay curve (right)

Table 5 Results

| Metric | Environments | | |
	Cart Pole	Mountain Car	Pong
Best mean score	**820**	**−108**	**19.1**
	680	−130	18.0
Final mean score	**720**	**−117**	**19.1**
	680	−131	18.0
Best episode (linear equation)	**945**	**1350**	**400**
	1000	1800	500
% Training speed	**105.5%**	**125.0%**	**120.0%**
	100.0%	100.0%	100.0%

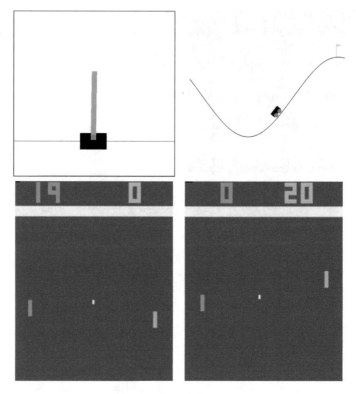

Fig. 7 Screenshots of agent playing in simulation environments for Cart Pole (top left), Mountain Car (top right), Pong before training (bottom left), and Pong after training (bottom right)

6 Discussion

There is a greater improvement in more complex environments like Pong and Mountain Car, when compared to a simple environment like Cart Pole, especially in the later stages of training, where the choice between exploration and exploitation becomes increasingly important. The % training speed is calculated by using the best episode value for the linear equation, and estimating the number of episodes before which the proposed equation had already attained that score. For Cart Pole, the training speed increased by a meager **5.5%**. For Mountain Car, the % increase in training speed was a striking **25.0%**. For Pong, the training speed % saw an increase of **20.0%**, which is quite impressive. The high computing requirements for training the agent on Mountain Car and Pong were overcome by efficient usage of the Google Colab tool.

7 Conclusion

The proposed sinusoidal equation, as hypothesized, improves training performance for DQN-based agents, being more prominent in sparse-reward environments like Mountain Car and Pong. To produce better results, future work can focus on extending training times, incorporation into the more recent DQN-based algorithms and techniques, usage of other exploration strategies that are based on "curiosity" [15] of an agent, and generalization of the proposed equation by elimination of the hyperparameters k_1 and k_2.

References

1. Mnih, V., Kavukcuoglu, K., Silver, D., Graves, A., Antonoglou, I., Wierstra, D., and Riedmiller, M. A.: Playing Atari with deep reinforcement learning. arXiv preprint (arXiv:1312.5602) (2013)
2. van Hasselt, H., Guez, A., Silver, D.: Deep reinforcement learning with double Q-learning. In: AAAI, pp. 2094–2100 (2016)
3. Lin, L.-J.: Reinforcement learning for robots using neural networks. Technical report, DTIC (1993)
4. Epstein, L.G., Schneider, M.: IID: independently and indistinguishably distributed. J. Econ. Theory, 32–50 (2003)
5. Hausknecht, M., Stone, P.: Deep recurrent Q-learning for partially observable MDPs. In: AAAI Fall Symposium Series
6. Wang, Z., Schaul, T., Hessel, M., van Hasselt, H., Lanctot, M., de Freitas, N.: Dueling network architectures for deep reinforcement learning. In: ICML'16 Proceedings of the 33rd International Conference on Machine Learning, vol. 48, pp. 1995–2003 (2016)
7. Mnih, V., Kavukcuoglu, K., Silver, D., Rusu, A.A., Veness, J., Bellemare, G.M., Graves, A., Ried-miller, M., Fidjeland, A.K., Ostrovski, G., Petersen, S., Beattie, C., Sadik, A., Antonoglou, I., King, H., Kumaran, D., Wierstra, D., Legg, S., Hassabis, D.: Human-level control through deep reinforcement learning. Nature **518**(7540), 529–533 (2015)
8. Schaul, T., Quan, J., Antonoglou, I., Silver, D.: Prioritized experience replay. In: Proceedings of the International Conference of Learning Representations (ICLR) (2016)
9. Burda, Y., Edwards, H., Storkey, A., Klimov, O.: Exploration by random network distillation. arXiv preprint (arXiv:1810.12894) (2018)
10. McFarlane, R.: A survey of exploration strategies in reinforcement learning. Unpublished (2018)
11. Bent, O., Rashid, T., Whiteson, S.: Improving exploration in deep reinforcement learning. Unpublished (2017)
12. OpenAI.: OpenAI Gym Cart Pole Wiki. Available on https://github.com/openai/gym/wiki/CartPole-v0
13. OpenAI.: OpenAI Gym Mountain Car Wiki. Available on https://github.com/openai/gym/wiki/MountainCar-v0
14. Seita, D.: Frame skipping and pre-processing for deep Q-networks on Atari 2600 games on GitHub.io blog. Available on https://danieltakeshi.github.io/2016/11/25/frame-skipping-and-preprocessing-for-deep-q-networks-on-atari-2600-games/ (2016)
15. Burda, Y., Edwards, H., Pathak, D., Storkey, A., Darrell, T., Afros, A.A.: Large-scale study of curiosity-driven learning. arXiv preprint (arXiv:1808.04355) (2018)

An Intelligent System for Appraisal of Heart Infirmity to Rescue Future Generation

S. Kaliraj, D. Vivek, M. Uma and P. Balasubramanie

Abstract Prognosis means prediction from past data. Many of the people are not aware of hereditary diseases; due to their unawareness, their life span get reduced. In this paper, we have analyzed the reason for the prognosis of heart problems to rescue future generation. We have done a research on the people based on their age groups and working departments and analyzed how their individual level and emotional level differ from the normal people. We have collected different types of symptoms for various heart-related diseases from affected people and it is placed in database. We also collected the different types of symptoms from experts and categorized the symptoms in the database based on our gathered information that this system is useful to rectify the generic problems and to increase their life time. Hence, the heart disease is predicted with different levels of data analysis with various combinations of attributes and models.

Index Terms CVS · Heart rate · ECG

1 Introduction

One of the important causes of death in the globe is coronary heart disease [1]. The methodology showed that the health system in the globe focuses on various different treatments to cure the diseases. The electrocardiographic constraints are pointers for the prediction of various diseases, which act as the main form and track

S. Kaliraj · D. Vivek (✉) · M. Uma
Department of Software Engineering, SRM Institute of Science and Technology,
Kattankulathur 603203, India
e-mail: vivekprasanth87@gmail.com

S. Kaliraj
e-mail: kaliraj.se@gmail.com

P. Balasubramanie
Department of Computer Science and Engineering, Kongu Engineering College, Erode, India
e-mail: pbalu_20032001@yahoo.co.in

© Springer Nature Singapore Pte Ltd. 2020 663
S. S. Dash et al. (eds.), *Artificial Intelligence and Evolutionary Computations in Engineering Systems*, Advances in Intelligent Systems and Computing 1056,
https://doi.org/10.1007/978-981-15-0199-9_57

of heart problem. The constraints are studied periodically during the observation of the patients, and they can describe the activity of the patient and monitor the patient's level in the heart problems. By predicting the heart diseases earlier, the more effective treatment method is used to improve the patient's life into better class the medical physicians and the doctors, researchers work on the anticipation of heart problems.

The person who learns to change the lifestyle and who helps in recovering and caring the people meets heart problem is called primary anticipation. The patients must learn to recover their lifestyle. A person has previously suffered a heart attack, and the historical medicinal process of that person in a cardiac reintegration program is imported for the analyses of the heart diseases. Herewith, the steady appraisals to detect any early signs of additional attack and the continues lifestyle monitoring in done in the secondary anticipation. To spread this kind of cardiac deterrence, the rest ECG is a low-cost cardiac test that offers respected data to reach the performance level.

The objective of this work is by using the classification of cardihoger, we can able to predict the person who will suffer heart problem in future based on the historical data of his ECG features This kind of education is absorbed on the approximation of cardiac problems.

Figure 1 explains the normal functioning of heart and the affected coronary artery disease.

Fig. 1 Heart image with normal functioning

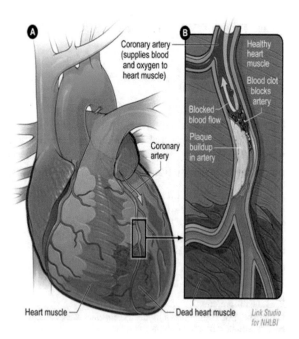

2 Related Works

Coronary artery disease is one of the most common diseases of atherosclerotic occlusion of the coronary arteries. Atherosclerosis is a process that will involve many of the body's blood vessels with a variety of presentations. When the coronary involves these arteries, it results in the coronary artery disease, the cerebral arteries and the cerebrovascular disease the aortic aneurysms [2].

Half of all deaths in the world was made by the cardiovascular disease which is comprised of hypertension and the diseases caused. In this, the agreement in the slope between the subject and population is based on profiles for the healthy group and indicates the population-based profile and it is representative of the individual subjects. This agreement was not found only for the AMI populations and one could speculate that this discrepancy holds diagnostic information [3].

The sphygmogram data sets and support vector machines classifier to the diagnose coronary heart disease. The hemodynamic is parameters it was derived from the sphygmogram. It reflects the status of the human cardiovascular system [4]. Based on homodynamic constraints, the aspect bargain methods and modified support vector machines are classified to prognosis sensitivity and specificity.

The ECG system has 2 main: a portable device for ECG acquisition, henceforth Recorder, and the Windows-compatible software are named as the Analyser it is used to storage and study the signals acquired by the recorders. These are the signals that are used to display the heart rate on the screen.

The storage of data in the SQL with the features of ECG is recorded which helps in producing the amplitude and duration of all recorded ECG of the patients. Hence, the request of the machinist produces the relevant electromagnetic waves which relates with heart diseases. The extracted data stored in the database and patent of the ECG were calculated. To predicate the heart rate, the various QT models are used [5]. The ventricular system of heart and the progression of R-wave amplitude in chest leads are used by calculating the R-value to refer the patients who have suffered a heart attack.

3 Proposed System Architecture Diagram

This prognosis system defines about the symptoms and the group-level problems. This system will provide the database about the disease details and the hereditary details. Each interface system will provide each database about the problems. The details will be collected from the hospital interface system and the group problems based on the heart attack are collected from the survey.

3.1   User Interface System

The user interface system contains the medical research-related data and the genetic disease symptoms collected from the patients. In this, the user will give their data to the hospital and they will give their symptoms to the doctors; the doctors will collect all data and store it in the database.

3.2   Information Collector

This organizes all the information collected from the user interface. The information collector acts as an input to the database analyzer. This information collector will store all the user data.

3.3   USI—User Interface System

The knowledge data analyzer collects the information from the expert, beneficiaries, and user interface and organized those details in the systematic way, and then, it serves as the output for prognosis report generation.

3.4   Prognosis Deposit

The different set of diseases with the influence of age problem is monitored by the prognosis analysis. The cardiac heart disease treatment and processing of these diseases depend on two levels; one is in which signs and symptoms of the warning of the attack got to the patient in the minimum age factor. Second, the person who is greater than 20 age should analyze the previous signs and symptoms of the body, the position of the warning signs. These levels calculate the emergency rate of the persons. In addition, with these levels, the underaged people also tested with the emergency rate based on sex and education level, which overlying 95% of confidence intervals were used to identify the difference in significance of heart attack (Fig. 2).

3.5   Prognosis System

The software tools used here are the front end as asp.net and back end as sql database. All details about the symptoms that occur due to hereditary heart diseases

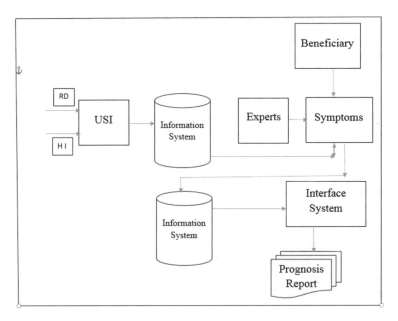

Fig. 2 Sytem architecture for the proposed system

are stored in this database based on age group working areas and eq level. This system will define the solution to rectify that problem. Hence, the process is identified by the data given by the doctor by checking the ECG features.

The various defendant and alertness of major warning in heart disease are provided in Table 1.

The death rate will be reduced if the patient identified the symptoms and received the medical care more priory. The emergency therapies are improved and

Table 1 Alertness of heart disease

Influence	Defendant/alertness	Percentage (%)
Aching discomfort in the jaw, neck, and back	Defendant	58
Sensation weak, dizzy, or faded		62
Chest pain		82
Pain in arm and shoulder		95
Shortness of breath		95
Non-Hispanic whites	Alertness	30.5
College Education		38.4
Women		35.9
Non-Hispanic black		16.6
Men		22.5
High school education		15.7

overall time set for the treatment is approximately decreased, since only one-third of the population knows the exact signs of the high attack and the symptoms of the heart attack indications.

The death rate will be reduced if the patient identified the symptoms and received the medical care more priory. The emergency therapies are improved and overall time set for the treatment is approximately decreased, since only one-third of the population knows the exact signs of the high attack and the symptoms of the heart attack indications.

4   Implementation

This project is done in asp.net as a front end and back end as sql database. This database contains all details about the symptoms and the hereditary details. It also contains the generation problems and hoe to rectify the problems. This system is used to reduce the generation problems based on heart.

This diagram explains the interaction between the user and the actors.

4.1   Design of Prognosis System

See Fig. 3.

4.2   Class Diagram for Overall System

The class diagram explains about the parameter of the hospital management system (Fig. 4).

Here, we have used the front end as asp.net and back end as sql database. This database contains all details about the symptoms that occur due to hereditary heart diseases based on age group working areas and eq level. This system will define the solution to rectify that problem.

In this login page, the user or the generation people will give their details about their working area, age, and gender. This page will contain the name and address details.

This symptom pages show the type disease based on heart attack and what are the symptoms based on that disease will be shown. These details will be stored in the database.

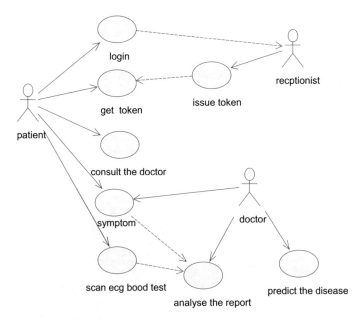

Fig. 3 Hospital interface

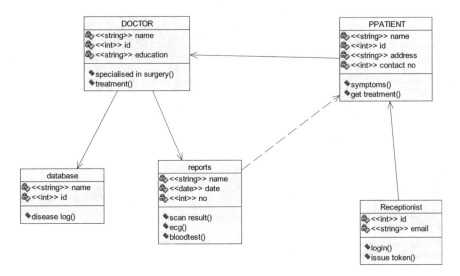

Fig. 4 Classes diagram for hospital interface

5 Result and Discussion

The process is made by analyzing the heart disease data by using Python. The data set collected related with heart disease is processed. In Fig. 5, heat map is generated using the smoke and heart disease prediction. Thus, this paper is used to find the symptoms and how to reduce the heart attack problems. It is useful to live their long life and they can reduce their family health problems also by having good food and by reducing their stress.

The analysis justifies the cigarette smokers and tells the symptoms they attain when they are in the specific age (Table 2).

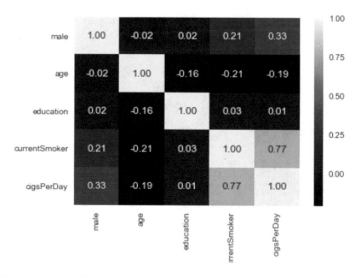

Fig. 5 Comparison of coefficient analysis in heart disease prediction

Table 2 Problem reduction

Generation	Problems (%)	Symptoms	Reduce problems
1st (1930–1960)	23	Back pain	Healthy food
2nd (1960–1990)	50	Heart pain	Less stress level, less work level
3rd (1990–2013)	75	Cardiac arrest, back pain, heart rate increases	To change way of food habits, do meditation

6 Conclusion

Hence, the reason for the prognosis of heart problems to rescue future generation is predicted for the people based on their age groups and working departments and analyzed how their individual level and emotional level differ from the normal people. The different types of symptoms for various heart-related diseases are from affected people and it is placed in database. The different types of symptoms from experts categorized the symptoms in the database based on our gathered information, and this system is useful to rectify the generic problems and to increase their lifetime. In future, this prediction algorithm can be used for the different disease prediction mechanisms with the help of categorized information system.

References

1. Kulakov, A., Davcev, D., Trajkovski, G.: Implementing artificial neural-networks in wireless sensor networks (2005)
2. Stamenkovic, Z., Randjic, S., Santamaria, I., Pesovic, U., Panic, G., Tanaskovic, S.: Advanced wireless sensor nodes and networks for agricultural applications (2016)
3. Kassim, M.R.M., Mat, I., Harun, A.N.: Wireless sensor network in precision agriculture application (2014)
4. Serpen, G., Li, J.: Parallel and distributed computations of maximum independent set by a Hopfield neural net embedded into a wireless sensor network (2011)
5. Adams, M.L., Cook, S., Corner, R.: Managing uncertainty in site-specific management: what is the best model? Precis. Agric. 2, 39–54 (2000)
6. Wang, N., Zhang, N., Wang, M.: Wireless sensors in agriculture and food industry recent development and future perspective, review. Comput. Electron. Agric. 50, 1–14 (2006)
7. Alippi, C., Camplani, R., Galperti, C., Roveri, M.: A robust, adaptive, solar-powered WSN framework for aquatic environmental monitoring. IEEE Sens. J. 11(1), 45–55 (2011)
8. Carpenter, G.A., Grossberg, S.: A massively parallel architecture for a self-organizing neural pattern recognition machine. Computer Vision, Graphics, and Image Processing, vol. 37, pp. 54–115 (1987)
9. Carpenter, G.A., Grossberg, S., Rosen, D.B.: Fuzzy ART: fast stable learning and categorization of analog patterns by an adaptive resonance system. Neural Netw. 4, 759–771 (1991)
10. Carpenter, G.A., Grossberg, S., Markuzon, N., Reynolds, J.H., Rosen, D.B.: Fuzzy ARTMAP: a neural network architecture for incremental supervised learning of analog multidimensional maps. IEEE Trans. Neural Netw. 3, 698–713 (1992)
11. Stankovic, J.A., Wood, A.D., He, T.: Realistic applications for wireless sensor networks. In: Theoretical Aspects of Distributed Computing in Sensor Networks, pp. 835–863 (2011)
12. Ergut, S., Rao, R.R., Dural, O., Sahinoglu, Z.: Localization via TDOA in a UWB sensor network using neural networks. In: IEEE International Conference on Communications (ICC), pp. 2398–2403 (2008). ISBN: 978-1-4244-2075-9
13. Guan, P., Li, X.: Minimizing distribution cost of distributed neural networks in wireless sensor networks. GLOBECOM, pp. 790–794 (2007)
14. Wang, N., Zhang, N., Wang, M.: Wireless sensors in agriculture and food industry-recent developments and future perspective. Comput. Electron. Agric. 50, 1–14 (2006)
15. Pursche, T., Krajewski, J., Moeller, R.: Video-based heart rate measurement from human faces. In: 2012 IEEE International Conference on Consumer Electronics (ICCE)

A Deep Neural Network Strategy to Distinguish and Avoid Cyber-Attacks

Siddhant Agarwal, Abhay Tyagi and G. Usha

Abstract In this paper, we will discuss the incorporation of machine learning techniques with traditional intrusion detection systems so that we can deal with different types of cyber-attacks using a single IDS. By applying machine learning to the IDS, we not only increase the system's sensitivity to malicious packets, but we can also remain secure from highly complex or previously unknown attacks. Initially, this paper introduces the current existing intrusion detection systems and their drawbacks. Secondly, this paper discusses a new system, secured attack-avoidance technique (SAAT), with the incorporation of machine learning and its architecture. Over the course of the paper, it was found that SAAT produced results of high accuracy and less false feedbacks.

Keywords Intrusion · Detection · Cyber · Security · IDS · IPS · Deep learning · Neural network

1 Introduction

Many people rely on the Internet for their day-to-day personal, social, and professional activities. This has led to the presence of a large amount of data on the global network and some people attempt to damage our Internet-connected computers, violate our privacy, and render many Internet services inoperable. Hence, we need security measures and controls in place; otherwise, our data might be subjected to an attack. There are mainly two types of attacks. While some are

S. Agarwal · A. Tyagi · G. Usha (✉)
SRM Institute of Science and Technology, Kattankulathur, India
e-mail: usha.g@ktr.srmuniv.ac.in

S. Agarwal
e-mail: sagarwa97@gmail.com

A. Tyagi
e-mail: abhaytyagi_vi@srmuniv.edu.in

© Springer Nature Singapore Pte Ltd. 2020   673
S. S. Dash et al. (eds.), *Artificial Intelligence and Evolutionary Computations
in Engineering Systems*, Advances in Intelligent Systems and Computing 1056,
https://doi.org/10.1007/978-981-15-0199-9_58

passive, meaning information is monitored; other attacks are active, which means the information is altered with intent to manipulate or damage the data or the network itself. Some of the most prevalent types of attacks today are denial of service (DoS)/distributed denial of services (DDoS), malicious software (malware), viruses, botnets, worms, Trojan, packet sniffing, data manipulation, DNS spoofing, etc. Being subject to any of these attacks can lead to a loss of personal data, irreparable damage to social and professional reputation and monetary loss.

A distributed denial of service (DDoS) attack occurs when the network or resources of the victim are flooded resulting in the crashing of the network. They prevent legitimate users from accessing the network and its resources. According to McAfee Laboratory, nearly one-third of all network attacks in the world can be credited to DoS or DDoS attacks [1]. DDoS defense systems are divided into the following categories: destination-side defenses and source-side defenses. Detection and responses to DDoS attacks are done on the side of the victim in destination side defense systems. While observing received packets, once an attack is detected, these systems can cut off the connection [2–4]. Some of them also use carefully designed protocols or router information to track the attacker, for example, IP traceback using router information [5], management information based traceback [6], or packet marking based tracing [7]. Packet dropping is an immediate way to mitigate these DDoS attacks based on the level of congestion [8].

An intrusion detection system (IDS) is a software that monitors a network for corrupt or malicious packets which are aimed at corrupting, censoring, or destroying either the computers on the network or the network as a whole [9]. Most techniques used in today's IDS are not able to deal with the dynamic and complex nature of cyber-attacks on computer networks. Hence, using efficient adaptive methods like different techniques of machine learning in multiple layers can result in higher detection rates, lower false alarms, and reasonable computation and communication costs.

The system introduced in this paper, SAAT, is based on a layered architecture which will employ different algorithms in each layer. There are three layers in the system. The first layer is a K-nearest neighbors (KNN) [10] classifier, which classifies the incoming packets into categories of attacks. The second layer is a convolutional neural network (CNN) with a layer of long short-term memory (LSTM) [11], which works with the first layer to again classify the incoming packets. The results of the first two layers are then compared, and in case of conflict, the conflicting packet is sent to the third layer which is a random forest classifier. The third layer renders the final decision on the nature of the packet, and the packet is passed through or dropped based on the classification.

The paper is organized as follows. Section 2 discusses the existing work on the topic. We discussed our proposed architecture and implementation in Sect. 3. The results are discussed in Sect. 4. Finally, Sect. 5 concludes our work. This paper, however, does not discuss the general techniques of intrusion detection.

2 Related Works

An intrusion detection system (IDS) is a device or application that analyzes whole packets, both header and payload, looking for known events. When a known event is detected, a log message is generated detailing the event, whereas an intrusion prevention system (IPS) not only detects an intrusion and generates a log message, but also rejects the malicious packet.

There has been a lot of research on implementing machine learning and deep learning with network security. The whole concept was studied by Syam Akhil Rapelle and Venkata Ratnam Kolluru in their survey titled "**Intrusion Detection Systems Using AI and Machine Learning Algorithms**." In this survey, they studied the various possible methods that can be applied to implement into intrusion detection systems. They found that machine learning and deep learning do show great promise in the future of network security and packet filtering [9]. It is difficult, however, to narrow down to a single solution due to the vast number of attacks and the pros and cons of each method implemented. Each model has some downside which affects the system negatively in the long run. The trick is to identify the shortcomings that can be tolerated in the system as a whole.

In specific systems, the DeepDefense system developed by Xiaoyong Yuan, Chuanhuang Li, and Xiaolin Li and illustrated in their paper "**DeepDefense: Identifying DDoS Attack via Deep Learning**" was a good indication of the type of architecture an IDS implementing learning models [11]. They made a learning model using bidirectional recurrent neural networks. They achieved a very high accuracy. The one disadvantage of this system is the possibility of the generation of false positives. Since the system has only one layer, the false positives are allowed to pass through without further hindrance. Our model aims to reduce this by using multiple layers of filters.

Another approach to an IDS was by using Blockchain as illustrated by Guy Zyskind, Oz Nathan, and Alex "Sandy" Pentland titled "**Decentralizing Privacy: Using Blockchain to Protect Personal Data**" [12]. The concept was unique but it had the same drawbacks as any Blockchain, the issue of user trust.

A different approach was described by Wei-Chao Lin, Shih-Wen Ke, and Chih-Fong Tsai in their paper "**CANN: An Intrusion Detection System based on Combining Cluster Centers and Nearest Neighbors**." They used, among many other algorithms, K-nearest neighbors (KNN) for classifying packets [13]. This is used in our model as one of the filter layers. The drawback of using only KNN is that a KNN is incapable of detecting DoS attacks, because of a lack of memory. We improve on this by using another layer with a convolutional neural network (CNN) with a long short-term memory (LSTM) layer in between.

Lastly, a paper written by Zecheng He, Tianwei Zhang, and Ruby B. Lee titled "**Machine Learning Based DDoS Attack Detection From Source Side in Cloud**" helped us learn about the nature of DDoS attacks and their different types of detection techniques which are in use today. They also proposed a source-side machine learning-based detection technique [14].

Based on the works done by people before us, we have developed a new system, SAAT, which eliminates the drawbacks detected by us in the above-mentioned works.

3 Proposed System

Our system, SAAT, uses a layered approach for its filtering. This means that if an intrusion is missed by one layer, another layer might be able to catch it. Figure 1 shows the proposed architecture for the system. Each data packet is made to pass through two layers where each layer ranks the packets as either clean or malicious based on different machine learning algorithms. If both the layers rank the packet as clean, it is allowed to pass through, whereas if both the layers rank it as malicious, the packet is dropped. In case of a mismatch, the packet is then passed through a third layer which casts the deciding vote. The proposed system contains the following modules:

- Multi-layered packet filter
- Data processor and extractor
- Machine learning engine.

Multi-layered Packet Filters

This layer contains the network packet filters in a three-layered structure. Each layer filters the incoming packets based on the different machine algorithms they are

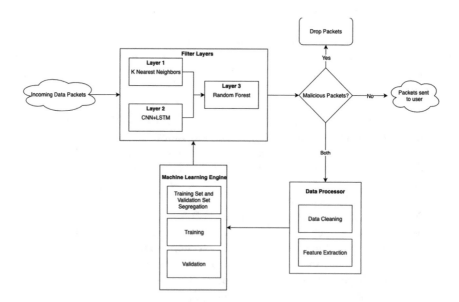

Fig. 1 Proposed system architecture for secured attack-avoidance technique (SAAT)

trained with. Each packet is made to pass through the first two filters. The initial layer (KNN) is used for general categorization. It classifies the input into one of 33 categories defined in the KDD cup 1999 dataset [9, 11]. The output is stored separately. The second layer (CNN + LSTM) is more centered toward detecting DoS attacks. The LSTM layer allows the model to analyze a sequence of data with the added benefit of memory checking. This allows the filter to check the data based on previous packets received. The output of this layer is also stored separately. The output of the two previous filters is then compared and if a conflict is detected, then the conflicting input is sent to the third layer. The third layer is a random forest classifier. The random forest classifier classifies the final class of the conflicting input. The result of the third layer is used as the final output of the IDS.

Machine Learning Engine

For this IDS, a variety of algorithms were tried to make the system as accurate as possible. The first layer was made using a K-nearest neighbors algorithm. This was partially inspired by Lin et al. [13]. The model was trained with the parameter $k = 5$. This was found to be the ideal value in terms of accuracy. The main issue with this approach was the lack of "memory" that is used to evaluate DDoS attacks. To address this issue, the second layer was introduced. The second layer which works concurrently with the first layer is a convolutional neural network (CNN) with a layer of long short-term memory (LSTM). This enables the system to analyze the packets with a consideration of the packets that came before. The CNN had the following layers:

- Four convolution layers
- Two max-pooling layers
- One LSTM layer
- One dropout layer
- One dense layer.

The third layer is useful only in times when the results of the first two layers are in conflict. Thus, the third layer is a random forest classifier. The final output of the conflicting packet states is given by the random forest. The reason we used a random forest classifier is due to its linear nature. The model is very fast to train and deploy. This is due to it been based on simple probability and non-complex calculations.

The CNN + LSTM is also trained simultaneously with incoming data, which should, theoretically, increase efficiency with more use.

Data Processor

This module makes sure that the data entering the machine learning engine are clean and meeting the standards of the system being used. The data processor converts the string features into numeric features so that it is compatible with the machine learning engine.

The above-proposed system was tested using the KDD99 dataset and produced results of high accuracy and less false feedbacks.

4 Results

The first layer of our model has a 97.6 F1 score. The logistic regression model had a 97.7 F1 score, but it took more than thrice the time to train. Thus, the KNN model was chosen as the final model for the first layer. The CNN has a classification accuracy of 97.4% and a logarithmic loss of 0.0467. This accuracy and loss were obtained by training the CNN for 9 iterations. The accuracy rises at a very slow rate, but it can be improved by running more iterations.

The KNN layer produced an accuracy of 97.68% and was highly successful in detecting DoS and Probe attacks (Fig. 2).

The CNN and LSTM layer performed exactly as they were expected to perform with perfect detection rates of DoS attacks along with giving an accuracy of 96.67% in overall detection. The LSTM layer gave the benefit of having a memory which allows us to track unnatural flooding of the network (Fig. 3).

The random forest classifier layer has high detection rates of R2L attack packets and they also help with the classification of DoS packets. This layer had an accuracy of 96.38% (Fig. 4).

```
Accuracy = 0.9768350319904833
Confusion Matrix =
 [[11880    17    12   293     1]
 [    11 44889     6     1     0]
 [    18    27   767    12     0]
 [  1020     5     6  3225     0]
 [    12     0     0     0     4]]
Classification Report =
              precision    recall  f1-score   support

           0       0.92      0.97      0.94     12203
           1       1.00      1.00      1.00     44907
           2       0.97      0.93      0.95       824
           3       0.91      0.76      0.83      4256
           4       0.80      0.25      0.38        16

   micro avg       0.98      0.98      0.98     62206
   macro avg       0.92      0.78      0.82     62206
weighted avg       0.98      0.98      0.98     62206
```

Fig. 2 Results obtained by the first layer (KNN)

```
Accuracy = 0.9667877696685208
Confusion matrix =
[[11396    46     38    723      0]
 [   31  44714    44    118      0]
 [   43     83   685     13      0]
 [  886      0    25   3345      0]
 [   14      0     0      2      0]]
Classification Report =
              precision    recall  f1-score   support

           0       0.92      0.93      0.93     12203
           1       1.00      1.00      1.00     44907
           2       0.86      0.83      0.85       824
           3       0.80      0.79      0.79      4256
           4       0.00      0.00      0.00        16

   micro avg       0.97      0.97      0.97     62206
   macro avg       0.72      0.71      0.71     62206
weighted avg       0.97      0.97      0.97     62206
```

Fig. 3 Results obtained by the second layer (CNN + LSTM)

```
Accuracy = 0.9638137800212199
Confusion Matrix =
[[12056    50     29     68      0]
 [   92  44809     5      1      0]
 [   82     66   643     33      0]
 [ 1785      1    23   2447      0]
 [   13      0     0      3      0]]
Classification Report =
              precision    recall  f1-score   support

           0       0.86      0.99      0.92     12203
           1       1.00      1.00      1.00     44907
           2       0.92      0.78      0.84       824
           3       0.96      0.57      0.72      4256
           4       0.00      0.00      0.00        16

   micro avg       0.96      0.96      0.96     62206
   macro avg       0.75      0.67      0.70     62206
weighted avg       0.97      0.96      0.96     62206
```

Fig. 4 Results obtained by the third layer (random forest classifier)

Overall, all the layers combined provide a high detection rate of almost every kind of attack with an accuracy rate of 97.83%. It is worth noting that DoS and U2R attacks have an almost perfect detection rate (Fig. 5).

```
Accuracy = 0.9783300646239913
Confusion Matrix =
[[11718    14     6   465     0]
 [    7 44896     4     0     0]
 [   19    25   777     3     0]
 [  787     2     2  3465     0]
 [   13     0     0     1     2]]
Classification Report =
                    precision    recall  f1-score   support

                 0       0.93      0.96      0.95     12203
                 1       1.00      1.00      1.00     44907
                 2       0.98      0.94      0.96       824
                 3       0.88      0.81      0.85      4256
                 4       1.00      0.12      0.22        16

      micro avg       0.98      0.98      0.98     62206
      macro avg       0.96      0.77      0.80     62206
   weighted avg       0.98      0.98      0.98     62206
```

Fig. 5 Overall results obtained by the system

The key for the output class numbers is as follows:

- 0 represents normal (clean) packets
- 1 represents packets part of a DoS attack
- 2 represents packets part of a Probe attack
- 3 represents packets part of a R2L attack
- 4 represents packets part of a U2R attack.

5 Conclusion

In our system, the first two layers have high accuracy of 97.68 and 96.67%, respectively; therefore, we decided to pass the incoming network data through both the layers simultaneously. The classification outputs of both the layers are compared and only the conflicts are passed through the random forest classifier which has an accuracy of 96.38%. This classifier gives the final output for the conflicting data. The combined accuracy of all three models working in union is 97.83% with an almost perfect DoS detection rate.

In future updates, an intrusion prevention mechanism will also be incorporated into the system. The prevention module would take the decision of an intrusion based on the number of malicious packets. If the number of malicious packets detected over a period of time is less, the packets are simply discarded (no attack detected), whereas if the number of malicious packets is high, the system not only discards the packets but also closes the connection with the network to prevent the further threat.

References

1. M. L. T. Report, McAfee Lab: [Online]. Available: http://www.mcafee.com/us/resources/reports/rp-quarterly-threat-q1-2015.pdf (2015)
2. Noh, S., Lee, C., Choi, K., Jung, G.: Detecting distributed denial of service (DDoS) attacks through inductive training. In: International Conference on Intelligent Data Engineering and Automated Learning. Springer (2003)
3. Shon, T., Moon, J.: A hybrid machine learning approach to network anomaly detection. Inf. Sci. **177**(18) (2007)
4. Amor, N.B., Benferhat, S., Elouedi, Z.: Naive Bayes vs decision trees in intrusion detection systems. In: Proceedings of the 2004 ACM Symposium on Applied Computing (2004)
5. Zargar, S.T., Joshi, J., Tipper, D.: A survey of defence mechanisms against distributed denial of service (DDoS) flooding attacks. IEEE Commun. Surv. Tutorials **14**(4) (2013)
6. Cabrera, J.B., Lewis, L., Qin, X., Lee, X., Prasanth, R.K., Ravichandran, B., Mehra, R.K.: Proactive detection of distributed denial of service attacks using MIB traffic variables—a feasibility study. In: Integrated Network Management Proceedings. IEEE/IFIP (2001)
7. Peng, T., Leckie, C., Ramamohanarao, K.: Protection from distributed denial of service attacks using history-based IP filtering. In: IEEE International Conference, vol. 1 (2003)
8. Yaar, A., Perrig, A., Song, D.: Pi: a path identification mechanism to defend against DDoS attacks. In: Security and Privacy, IEEE Symposium (2013)
9. Repalle, S.A., Kolluru, V.R.: Intrusion detection system using AI and machine learning algorithm. Int. Res. J. Eng. Technol. **4**(12) (2017)
10. Haq, N.F., Onik, A.R., Hridoy, M.A.K., Rafni, M., Shah, F.M., Farid, D.M.: Application of machine learning approaches in intrusion detection system: a survey. Int. J. Adv. Res. Artif. Intell. **4**(3) (2015)
11. Yuan, X., Li, C., Li, X.: DeepDefense: identifying DDoS attack via deep learning. In: IEEE International Conference on Smart Computing (SMARTCOMP) (2017)
12. Zyskind, G., Nathan, O. Pentland, A.: Decentralising privacy: using blockchain to protect personal data. In: IEEE Security and Privacy Workshops (2015)
13. Lin, W.-C., Ke, S.-W., Tsai, C.-F.: CANN: an intrusion detection system based on combining cluster centers and nearest neighbors. Knowl. Based Syst. (2015)
14. He, Z., Zhang, T., Lee, R.B.: Machine learning based DDoS attack detection from source side in cloud. In: IEEE 4th International Conference on Cyber Security and Cloud Computing (2017)

Secured Congestion Control in VANET Using Greedy Perimeter Stateless Routing (GPSR)

S. Arvind Narayan, R. Rajashekar Reddy and J. S. Femilda Josephin

Abstract Vehicular Adhoc networks (VANET), a subset of mobile adhoc networks (MANET), is revolutionizing the intelligent transport system. With this advancement, the environment around a car can be more effective with transmitting information for emergency messages and also periodical messages like distance, speed, etc. In today's world, due to the increasing vehicles on road, there is a need for traffic congestion control due to instances like road construction and management, accidents and weather conditions. This can be achieved by intimating the driver on board about the traffic ahead. In a dense traffic road, information comes from all the vehicles on the segment, hence this information needs to be processed and shared between all the vehicles in the segment. Therefore, there is need for a protocol that all the adhoc vehicles must adhere and detect if the traffic is greater than the threshold, consequently to avoid growing into severe congestion. Many protocols have been proposed for VANET such as AODV, DSDV and AOMDV. The aim of the research is to analyze different existing protocols over different conditions so as to find the optimal protocol for VANET by validating the metrics of network effectiveness such as throughput, delay, etc., and implement traffic congestion detection and rerouting using GPSR protocol in a secured way using RSA algorithm. The VANET system can be simulated over the software such as SUMO and NS2.

Keywords VANET · NS2 · SUMO · AODV · AOMDV · DSDV · GPSR · RSA

S. Arvind Narayan (✉) · R. Rajashekar Reddy · J. S. Femilda Josephin
Department of Software Engineering, SRM Institute of Science and Technology,
Chennai, India
e-mail: varvind7@gmail.com

R. Rajashekar Reddy
e-mail: rsrreddyr@gmail.com

J. S. Femilda Josephin
e-mail: femildajosephin.j@ktr.srmuniv.ac.in

© Springer Nature Singapore Pte Ltd. 2020
S. S. Dash et al. (eds.), *Artificial Intelligence and Evolutionary Computations*
in Engineering Systems, Advances in Intelligent Systems and Computing 1056,
https://doi.org/10.1007/978-981-15-0199-9_59

1 Introduction

VANET is an intelligent transport system used for wireless communication networks. It uses the Dedicated Short Range Communication Service (DSRC) which uses the frequency range of 5.85–5.92 GHz. VANET is a subset of MANET which aims to increase road safety and ease of transport for road users. The two unique features of VANET are dynamic connectivity and network topology. In MANET, the nodes move freely while in VANET the nodes tend to move in an organized mobility such as road lanes and junctions. VANET has two type of communication— vehicle-to-vehicle (V2V) and vehicle-to-infrastructure (V2I) [1].

Major application of VANET is congestion control. The main aim of congestion control is to allocate resources optimally which would reduce the delay created by traffic and also avoid deadlock. Once a congestion area is detected, the vehicle neighboring congested area is informed. By this information, the vehicles may change their path, thus reducing further congestion. Traffic congestion design can be implemented either in distributed manner where the computation is done on vehicle or a centralized system where most of the information is shared and computed.

To enhance the congestion control, the communication needs to be secured. As VANET is based on wireless communication and the security of VANET greatly depends upon communication medium, there is need for preserving each vehicle's information. One major security issue is delay in transfer of packets, so attackers get more time to analyze the system. Some of the VANET security requirements are authentication, accessibility, verification and privacy. The addition of security leads to communication overhead and computational overhead.

In this paper, we use NS 2.35 and SUMO to study the efficiency of VANET under different traffic conditions and four routing protocols: adhoc on demand distance vector (AODV), destination-sourced distance vector (DSDV), adhoc on demand multipath distance vector (AOMDV), DumbAgent.

For simulating traffic congestion detection and rerouting using greedy perimeter stateless routing (GPSR) protocol, a location-based protocol that communicates only with its closest neighbors that are nearer to the destination are proposed and tested. Security is achieved using RSA algorithm which is a public key infrastructure where messages are encrypted using public key and are decrypted using private key to achieve authentication.

The remainder of the paper is organized as follows. Section 2 shows the study related to VANET protocols, congestion control and security. Section 3 describes the analyzing of suitable protocols. Section 3.2 describes congestion control using GPSR. Section 3.3 Security. Section 4 Implementation details. Section 5-Results and discussions and Sect. 6 concludes the paper.

2 Related Work

In [1], the author discussed various VANET characteristics like topology and connectivity, tested over different protocols—simulated over 20, 40 and 80 vehicles and concluded that the hierarchical architectural protocol is more efficient than well non-hierarchical protocols (AODV, DSDV, and DSR). In [2], performance analysis of AODV and early detection congestion and control routing protocol (EDAODV) in a uni-directional highway with their performance metrics like throughput, packet loss and transmission delay.

In [3], author categorized the routing protocols into five categories and discussed about their merits and demerits and surveyed protocols that fall under these five categories. In [4], the author took three greedy routing algorithms—gateway node-based greedy routing (GNGR), edge node-based greedy routing algorithm (EBGR), and predictive directional greedy routing algorithm (PDGR), and evaluated the network parameters. The simulation results show that GNGR has high packet delivery ratio (PDR).

In [5], the author compared on-demand and table-driven routing protocols and simulated it over NCTUns 4.0 with varying number of nodes. The author concludes that with increasing number of nodes, DSDV would perform better in PDR but it may have considerable routing overhead. In [6, 7], the author simulates MANET routing strategies like DSDV, AODV, Cluster-Based Routing Protocol (CBRP) and extended the simulation to multi-hop wireless networks with physical data link and MAC layer. The author concludes that CBRP improves the correctness of the routing protocol but performs least in end-to-end delay.

In [8, 9], the author considered various scenarios to analyze performance of AODV and GPSR routing protocols. The author performs the simulation using NS2 and VanetMobiSim. The author compares the protocol with respect to average delay and PDR. Author found that with respect PDR, AODV performs better and average delay GPSR performs better. In [10], the author's main aim is to increase data transmission reliability by a method called DSR protocol with the help of intersection node at GPSR protocol.

In [11], the author enhances the above work using MOVE, a mobility model to generate realistic traffic. The author concludes by providing solutions for congestion control through AODV and provides alternate paths to the vehicle incoming. In [12], the author analyzes different congestion control algorithms suitable for VANET and proposes a coordinator-based and segment-based algorithm. Each segment has local coordinator who allocates the time slots to the vehicles in that segment.

In [13], the author surveys different security requirements, types of attackers and attacks in VANET. In [14], the author proposes a security solution based on public key infrastructure (PKI) and group-based trust evaluation. The author takes scenarios for specific attacks like Sybil and black hole attack over the hybrid model. In [15], the author implements RSA-based encryption in VANET and simulated using MATLAB. The public key is shared by the RSU and each vehicle has a private key. The author compares RSA and advanced encryption standard (AES) and concludes that the RSA algorithm is found to be efficient than AES.

3 Proposed Methodology

The proposed methodology entails establishment of VANET environment, congestion detection and control and security using RSA algorithm. OpenStreetMap, SUMO, and NS2 are used to extract road data, simulate traffic and network simulation respectively. For traffic congestion detection and security algorithm, we simulated traffic with an accident scenario on a junction road. Congestion is implemented using the GPSR protocol for communication among vehicles and this message is encrypted and decrypted using RSA algorithm. The graph generated compares the network parameter with security and without security. Figure 1 shows the system architecture for the proposed system.

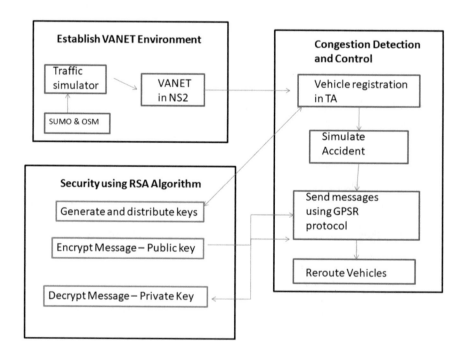

Fig. 1 System architecture

3.1 Analyzing Different Protocols

To evaluate the efficiency of protocols—AODV, AOMDV, DSDV, and DumbAgent—the following parameter metrics are compared. These parameters are calculated using awk script which uses the network simulator-generated trace file and produces the result.

3.1.1 Packet Delivery Ratio (PDR)

The ratio between the number of packets transmitted (P_r) by source to number of packets received (P_s) by a destination node. Loss rate of a protocol is found by this parameter which specifies the correctness of routing protocols. A good packet delivery ratio refers to the lesser loss rate [1, 3].

$$PDR = \frac{\sum P_r}{\sum P_s} \qquad (1)$$

3.1.2 Average Delay (End-to-End)

This calculates the average time taken by the packet to transmit from source to the destination node. It is measured by subtracting the sent time (st) from the arrival time (at). Delays in networks come from sources such as propagation delay, transmission delay, queuing delay, and processing delay. Higher delays result in lesser efficient protocols [1, 2]. Nc-Number of connections

$$D = \frac{\sum pr(at - st)}{\sum Nc} \qquad (2)$$

3.1.3 Average Throughput

Average throughput is the ratio of the total data packets generated (Nb) in the entire session by the total time $(T(i))$. In general, average throughput measures the maximum processing rate of a network protocol. Throughput is usually measured in bits per seconds. Higher throughput refers to the higher bandwidth consumption [1, 2].

$$Thr = \frac{\sum_{i=1}^{n} (Nb(i))}{T(i)} \qquad (3)$$

$$\text{Average(throughput)} = \sum_{i=1}^{n} (\text{Thr}(i)) \tag{4}$$

3.1.4 Overhead

Routing requires routing packets to be delivered for communication ($P_{\text{delivered}}$) and the number of routing packets required for communication (P_{total}) is known as routing overhead. As nodes change their location so often, some no longer new routes are generated and lead to unnecessary routing overhead [1, 2].

$$\text{Overhead} = \frac{P(\text{delivered})}{P(\text{total})} \tag{5}$$

3.2 Congestion Control-GPSR

For Congestion detection, we propose a use of a location-based algorithm called GPSR [9, 10]—greedy perimeter stateless routing. GPSR combines greedy forwarding and perimeter forwarding for routing the destination. The main aim of GPSR is to increase the mobility rate under increasing number of nodes in a network. For the geographic routing to be stateless, it requires the topology information for only a single hop. In GPSR, the position of the node destination and the position of the node next hop are sufficient enough to make routing decisions.

3.2.1 Greedy Forwarding

In GPSR, the position of the destination node and the current node's neighbors are known, a greedy choice is made by choosing the node's neighbor which is nearest to the destination.

The greedy forwarding has one drawback in which the next nearest neighbor is sometimes farther from the current node, so the greedy approach fails in that case.

3.2.2 Perimeter Forwarding

When greedy forwarding fails to choose the nearest neighbor, perimeter forwarding is used. This implements two mechanisms:

(i) Right-hand rule

- When the neighbor nodes are farther from the destination, the right-hand rule comes into play. The next node is selected sequentially counter clockwise to the current node.

(ii) Planarized Graphs

- A planar graph is one where no two edges cross each other. It uses RNG and GG methods to remove crossing edges.

We have simulated a two-lane road with a situation where one node is marked as accident. With the help of GPSR and DSDV, the remaining nodes in that lane recognize a traffic congestion zone and thereby broadcast the emergency messages which help other nodes to take alternate paths.

3.3 Security in Vanet

As VANET technology enables communication among the vehicles and road side units, it is necessary to secure this communication. As the number of vehicles increases, the need for security requirement also increases [15].

3.3.1 Encryption and Decryption Using RSA Algorithm

In this paper, RSA algorithm is used for security in VANET communication. RSA is widely used algorithm for secure transmission of data as it one of the early used public key cryptography. Figure 2 illustrates how RSA algorithm can be used in VANET. RSA algorithm consists of two keys—public key, private key. RSA is done three steps:

(a) **key generation**:

Take two random numbers p and q and compute $n = p * q$. Euler's totient function $\varphi(n)$ is computed as $\varphi(n) = \varphi(p) * \varphi(q) = (p - 1) * (q - 1) = n - (p + q - 1)$. Select integer e such that $1 < e < \varphi(n)$ and $\text{GCD}(e, \varphi(n)) = 1$ which makes e and $\varphi(n)$ coprimes. 'e' is announced as the public key d is considered as the multiplicative inverse of $e(\text{modulo } \varphi(n))$. Public key consists of n and exponent e. Private key consists of n an exponent d.

(b) **Encryption**:

RSA transmits public key (n, e) to all vehicles and keeps the private key (d) as a secret. First the message M is made into an integer m, such that $n > m > = 0$. Then, the cipher text is computed as

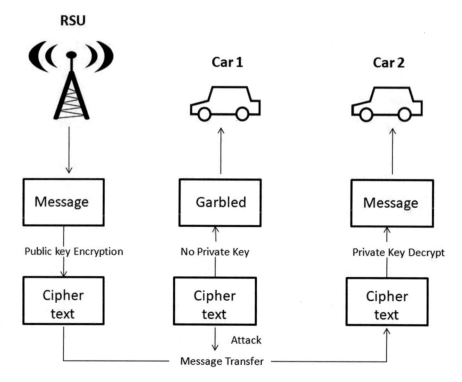

Fig. 2 RSA in VANET

$$c \approx m^e \,(\text{mod}\, n)$$

(c) **Decryption**:

Using the private key d,

$$m \approx c^d \,(\text{mod}\, n)$$

4 Implementation Details

For protocol analysis, we use Open Street Map to generate road layout and convert the OSM file to sumo configurable file to simulate traffic. This configurable file is then used by NS2 to generate NAM and trace files.

Table 1 describes about the simulation parameters of different simulator with the versions:

Table 1 Simulation parameters

Parameter	Simulation value
Network simulator	NS 2.35
Traffic simulator	SUMO 0.26
Map model	OSM
Routing protocol	AODV, DSDV, AOMDV, DumbAgent
Transport protocol	TCP
Number of nodes	50, 100, 150, 200, 250
Simulation time	180 s
Minimum speed	1 kmph
Maximum speed	60 kmph
Traffic type	CBR

In Fig. 3, Data flow diagram (DFD) is shown with the help of Open Street Map which generates manually selected location with coordinates of road, lane in the OSM file which is of xml data. This data is then used by Simulation of Urban Mobility (SUMO), an open-source tool to generate traffic over the map generated by the open street map. SUMO simulates the traffic parameters like speed, lane change, and path. This SUMO file is then used by the Network Simulator (NS2) to simulate the network for the generated traffic by the .sumo file. NS2 can also be used to create a manually generated map, traffic, and simulate traffic. NS2 is mainly run on C++ and oTcl languages .NS2 generates two types of files—nam file and trace file. Network Animator (NAM) visualizes the output of the network simulation and trace file contains the entire data of the network simulation.

5　Results and Analysis

The main goal of this paper is to deduce the optimal protocol, which is simulated and analyzed with different network parameters over multiple topologies with nodes ranging from 50 to 250 nodes for 300 s. The analyzed results from the AWK script is transformed into a line graph with the help of gnuplot.

Figure 4 portrays SUMO configurable file simulated via SUMO-GUI with 50 nodes with random trips and paths.

In Fig. 5, the SUMO file is converted into NS 2 executable file (TCL) and the nodes are visualized without the road layout.

Figure 6, depicts that the packet delivery ratio is preferable in AODV than other protocols because it is reactive to change.

Figure 7, illustrates that AODV has the least delay because it uses only one shortest route for data delivery from source to destination.

Figure 8, represents that AODV has the higher throughput since AODV avoids loops and freshness of routes.

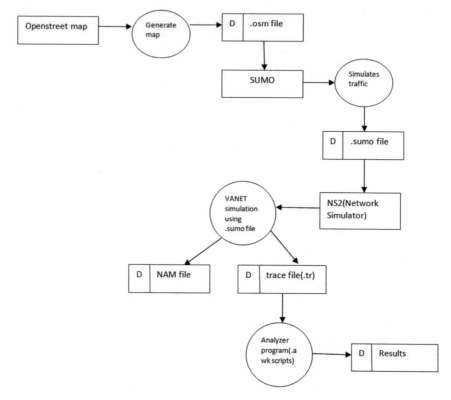

Fig. 3 Dataflow diagram

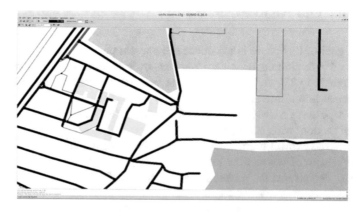

Fig. 4 SUMO visualization

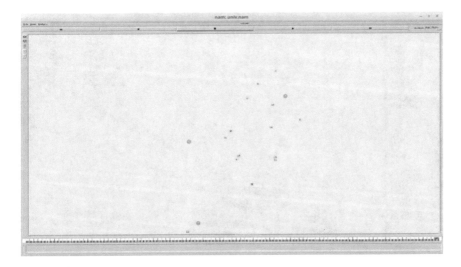

Fig. 5 NS2 visualization

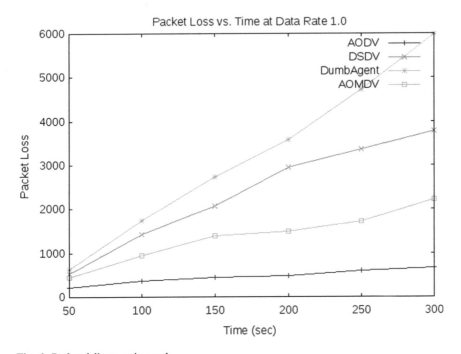

Fig. 6 Packet delivery ratio graph

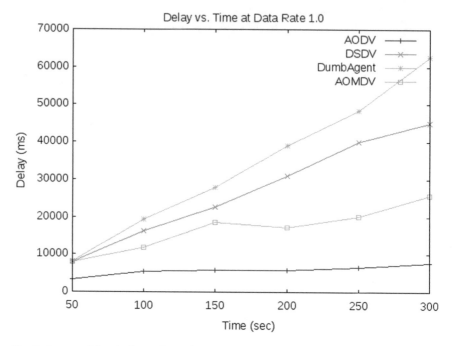

Fig. 7 Average delay (end-to-end) graph

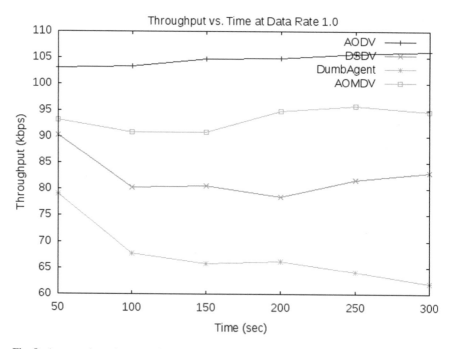

Fig. 8 Average throughput graph

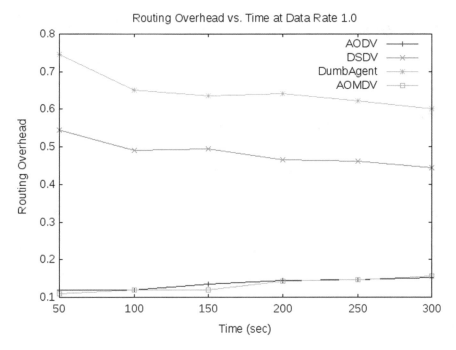

Fig. 9 Routing overhead graph

From Fig. 9, we can interpret that AOMDV has least overhead because the data packet header has the least size.

From the generated graphs, it is evident that AODV has been proven efficient in almost all the networking parameters and DumbAgent fixating to be the least. Also, AOMDV has had better efficiency next to AODV protocol.

For the congestion detection as shown in Fig. 10, we have used GPSR protocol to simulate a two-lane road where an accident occurs and the subsequent vehicles in that lane are intimated with the emergency message and those vehicles stop forwarding the lane.

After implementing congestion detection, we have implemented security in the message communication between V2V and V2I using RSA public key encryption.

A junction road is created with set of vehicles, Road Side Units, Intrusion Detection System for analyzing any outsider attack and trusted authority (TA) that acts as the registry center of RSUs and vehicles, is trusted by all entities in the VANETs, and responsible for distributing key materials to all entities.

From Fig. 11, CAR-7 is depicted as accident vehicle, which transmits secured messages (encrypted using public key) to nearby vehicles—CAR 3 and CAR 11 and also the RSU in the range. The vehicles then decrypt the message using their private key and start to change their next possible path to proceed further. The TA establishes key materials with CAR-4.

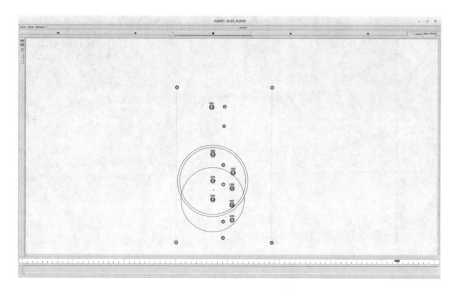

Fig. 10 Congestion detection

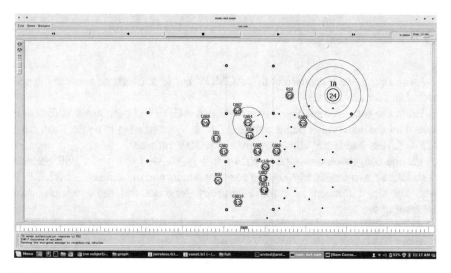

Fig. 11 Occurrence of accident

With the generated trace files of the simulation files with and without security, different network parameters are plotted and compared.

From Fig. 12, the packet delivery ratio is almost similar in both cases.

From Fig. 13, due to computational overhead the delay is significantly more for simulation with security.

Fig. 12 Packet delivery ratio

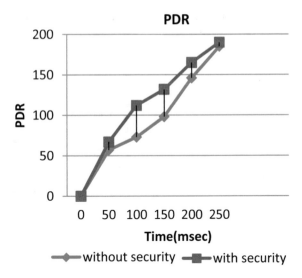

Fig. 13 Delay

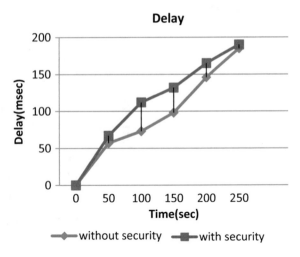

In Fig. 14, the energy consumption for simulation with security is high because it involves a junction road and the distance travelled is more compared to simulation without security. Figure 15 shows that the throughput is comparatively high in simulation with security.

Fig. 14 Energy consumption

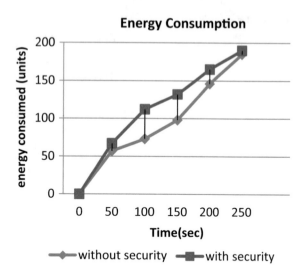

Fig. 15 Throughput

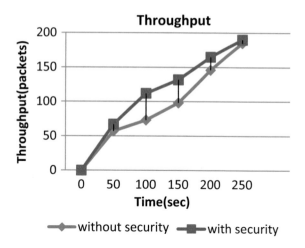

6 Conclusion

This paper aimed to analyze the optimal protocol for VANET over different topologies for various networking parameters. Considering the major limitations in VANET such as high mobility and the road infrastructure, arriving at the best protocol will not be optimal at all conditions. In this study we have focused on a set of protocols over different road maps. From the results of our study, we conclude that AODV is preferable for packet delivery ratio (500 for 300 s simulation time), average end-to-end delay (5000 ms for 200 s simulation time) and throughput (103 kbps for 100 s simulation time) and AOMDV is suitable for least overhead (0.1 for 50 s simulation time) but DSDV is preferable for congestion control using

GPSR because GPSR requires destination to be known for routing and DSDV requires full network topology beforehand for routing.

Congestion detection and control is achieved through GPSR protocol for message communication and RSA algorithm for security. From the plotted graphs, we conclude that the network parameters excluding delay (for 150 s simulation time—43 ms added delay for congestion with security) and energy consumption (for 150 s simulation time—39 units additional energy required for congestion with security), is high for simulation with security.

For the future works, this study can be further extrapolated to higher number of nodes and also can be tested over hybrid protocols for more effective results. Also various PKA security mechanisms can be compared and an optimal security algorithm can be found for a particular scenario.

References

1. Khoza, E., Tu, C., Owolawi, P.A.: comparative study on routing protocols for vehicular ad-hoc networks (VANETs). In: International Conference on Advances in Big Data, Computing and Data Communication Systems (2018)
2. Sailaja, P., Ravi, B., Jaisingh, T.: Performance analysis of AODV and EDAODV routing protocol under congestion control in VANETs. In: Second International Conference on Inventive Communication and Computational Technologies (ICICCT) (2018)
3. Li, L., Wan, F., Wang, Y.: Routing protocols in vehicular ad hoc networks (VANET): a survey. IEEE Veh. Technol. Mag. **2**(2) 12–22 (2014)
4. Rama, K., Lakshmi, K., ManjuPriya, S., Thilagam, K., Jeevarathnam, A.: Comparison of three routing algorithms based on greedy for effective packet forwarding in VANET. Int. J. Comput. Technol. Appl. **3**(1), 146–151 (2013)
5. Narendra Singh, Y., Yadav, R.P.: Performance comparison and analysis of table-driven and on-demand routing protocols for mobile ad-hoc networks. Int. J. Inf. Commun. Eng (2008)
6. Tuteja, A., Gujral, A., Thalia, A.: Comparative performance analysis of DSDV, AODV and DSR routing protocols in MANET using NS2. IEEE Comput. Soc. **333**, 330–333 (2010)
7. Boukhalkhal, A., Yagoubi, M.B., Djoudi, D., Ouinten, Y., Benmohammed, M.: Simulation of Mobile Adhoc Routing Strategies, pp. 128–132. IEEE (2008)
8. Abbas, J., Tarik, T., Ehssan, S., Kazuo, H., Nei, K., Yoshiaki, N.: A stable routing protocol to support ITS services in VANET networks. IEEE Trans. Veh. Technol. **56**(6), 3337–3347 (2014)
9. Bala, R., Krishna, C.R.: Scenario based performance analysis of AODV and GPSR routing protocols in a VANET. In: IEEE International Conference on Computational Intelligence & Communication Technology (2015)
10. Anggoro, R., Husni, M., Bastian, R.: Source route implementation using intersection node on GPSR protocol to increase VANET's data transmission reliability. In: International Conference on Advanced Mechatronics, Intelligent Manufacture, and Industrial Automation (ICAMIMIA) (2017)
11. Bhargavi, G., Saleh, A: Novel approach to improvise congestion control over vehicular ad hoc networks (VANET). In: 3rd International Conference on Computing for Sustainable Global Development (INDIACom) (2016)
12. Kolte, S.R., Madankar, M.S.: Adaptive congestion control for transmission of safety messages in VANET. In: International Conference for Convergence for Technology (2014)

13. Kaur, R., Singh, T.P., Khajuria, V.: Security issues in vehicular ad-hoc network (VANET). In: 2nd International Conference on Trends in Electronics and Informatics (ICOEI) (2018)
14. Hasrouny, H., Samhat, A.E., Bassil, C., Laouiti, A.A.: Security solution for V2V communication within VANETs. In: Wireless Days (WD) (2018)
15. Nema, M., Stalin, S., Tiwari, R.: RSA algorithm based encryption on secure intelligent traffic system for VANET using Wi-Fi IEEE 802.11p. In: International Conference on Computer, Communication and Control (IC4) (2015)

Steganographic Approach to Enhance Secure Data Communication Using Nonograms

K. R. Jansi and Sakthi Harshita Muthusamy

Abstract In the present scenario the security of data is of utmost importance. While cryptographic techniques render the data unintelligible, steganography conceals the presence of the data itself. There are many pre-existing methods of steganography that involve hiding encrypted data in various mediums such as images, videos, audio, etc. In this project, we are going to focus on data hiding in images. There are numerous ways to embed data into images out of which the most widely used is the least-significant-bit substitution, in this method the least-significant bits of each pixel are substituted with the data instead. Nonograms are Japanese grid-based puzzles which when solved produce a picture composed of square blocks. We are using the puzzle to divide the cover image into blocks and then we use k-means clustering to select the pixels for LSB substitution thereby going a step further and hiding the hidden data in RGB images.

Keywords Image steganography · Nonograms · Data hiding · K-means

1 Introduction

These days' data is transmitted through the internet in various forms such as image, audio, video and text and communicating sensitive information over the internet is unsafe and not secure. Therefore, to ensure the confidentiality and security of such data we use cryptographic and steganographic methods. In Steganography, a range of data can be hidden such as images, text, audio and video. The word steganography is derived from the Greek terms 'steganos' and 'graphein' meaning hidden

K. R. Jansi (✉) · S. H. Muthusamy
Department of Computer Science Engineering, SRMIST, Kattankulathur,
Chennai 603203, India
e-mail: jansik@srmist.edu.in

S. H. Muthusamy
e-mail: harshu41195@gmail.com

© Springer Nature Singapore Pte Ltd. 2020
S. S. Dash et al. (eds.), *Artificial Intelligence and Evolutionary Computations in Engineering Systems*, Advances in Intelligent Systems and Computing 1056,
https://doi.org/10.1007/978-981-15-0199-9_60

and writing, respectively. In image steganography, the data is hidden within the image.

The main characteristics of data hiding methods are capacity, robustness and confidentiality. Robustness is the ability or the resistance of the algorithm to withstand attacks to break it. Capacity is the amount of data that can be embedded and confidentiality is the aspect of steganography that hides the presence of the data itself thereby giving it a layer or security.

In today's high-tech world image steganography has many applications. The main concern at present is being privacy and anonymity. As it can be used to communicate secretly and covertly. People can use this method to spread information anonymously in situations where it is not possible to communicate directly; copyright protection on digital files is another possibility. Communication of highly sensitive information between two government entities, to send sensitive medical data or reports online or even to just store them in a more secure manner, this method is extremely useful. Although it can be used for legitimate reasons such a hiding sensitive data by legal bodies, it is also used for nefarious purposes. Steganography when used along with Encryption can be a powerful combination. It can be used by cyber criminals to send malware to compromise machines, and also by terrorists and other criminal organizations that need a safe and secure method to communicate with one another.

There are many existing methods to embed data. Such as LSB Replacement, LSB Matching, Matrix Embedding, DWT, DCT, DFT. The LSB technique involves embedding the data in the LSB of the pixels in the image.

The frequency-domain techniques such as DCT, DWT and DFT embed the data into the different coefficient off the image. The Masking and Filtering Method involves marking the image to hide data. This is generally done by watermarking the image and hiding data in specific areas.

An RGB image has three color planes namely red, blue and green which provides three different planes for embedding the data as compared with a gray-scale image.

Nonograms or Picross or Griddlers are Japanese number puzzles. On solving the puzzle, an image is formed from the colored or blacked-out grids. The two most common algorithms used for solving nonograms are genetic algorithm and depth-first search (Fig. 1).

Components used in image steganography include:

- The data or information that is to be embedded in the cover image is known as the message.
- A cover image is the image that is used to hide the data in.
- The image after the embedding process called as the stego image.

K-means algorithm is a clustering method; it partitions the present observations into x clusters, each cluster thus formed belongs to its nearest mean. RGB images

Fig. 1 Solved nonogram

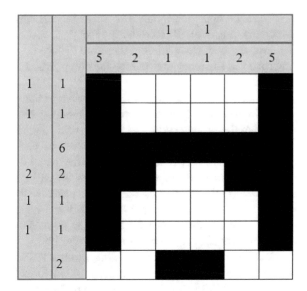

have three planes in which data can be embedded, which increases the capacity. In this approach, we use nonograms to generate blocks and the selected block then undergoes contouring then k-means clustering, and finally, the data is embedded.

2 Literature Survey

G. Manikandan et al. [1] propose a lossless image steganography approach based on contours and clusters. Contouring of the gray-scale cover image is first performed. Followed by clustering of the contoured pixels. The data is then embedded into the appropriate cluster's LSB. The resulting entropy values of the stego image show that information leakage in the embedding process is negligible and the scheme is effective against entropy attacks. The size of the contour randomizes the region for embedding data along with the number of clusters both of which add to the complexity and robustness.

An enhanced content confidentiality using a crypto-stego scheme for RGB images is proposed by Shifa et al. [2]. AES algorithm is used to encrypt the cover image and the key is then embedded in the image by using k-nearest neighbor clustering and least-significant-bit substitution method. Since the deciphering key is also inserted in the image itself no additional exchange of keys is required. The PSNR, MSE, SSIM and histogram values show that this method provides good confidentiality against statistical attacks and stego transparency. Implementation of the given scheme in parallel processing improves efficiency considerably.

A unique approach is taken to solve the nonograms, which are NP-Complete problems. Jing et al. [3] consider the fact that most of Japanese puzzle are compact by structure and are adjoining, therefore deduce constraints that can be applied to improve the efficiency and reduce the computation time when using DFS to solve nonograms.

The first phase uses logical rules to solve as many blocks as possible while the second phase uses DFS to solve the remaining blocks. The experimental results show that the proposed method provides the solution faster.

The different mediums for steganography and various steganographic techniques such as LSB, OPAP, IP LSB, PVD in the spatial domain and DCT, DWT in the frequency domain for image stego are reviewed by Kuramri and Singh [4].

Bobko et al. [5] discuss the use of genetic algorithms to solve nonograms. Nonograms of sizes such as 4×4 and 5×5 coped better with optimization in shorter time as compared with the larger grids. The complexity of NP-complete problems was shown using brute-force algorithms.

A RGB image is embedded within another RGB image without transformation into a gray-scale image. Discrete Cosine Transformation is used in this method by Ghosh et al. [6]. More data is stored due to the presence of three planes Red, Green and Blue data is stored in the lsb of each of these planes in the image.

G. Manikandan et al. [7] use k-means clustering in the gray-scale cover image and the data is then embedded using least-significant-bit substitution. A noise value is then inserted into the pixels. The added noise value provides better security for the embedded pixels. Efficiency depends on number of clusters and chosen pixels.

The results obtained are compared with existing stego techniques, which shows the proposed method achieves not only the same data capacity but also enhances the peak signal to noise ratio of the stego image.

A comparison study between using genetic algorithm and depth-first search for solving a nonogram is performed by Wiggers [8]. The genetic algorithm outperforms the depth-first search by at least three times for all large nonograms; however, the depth-first search algorithm works better for smaller nonograms.

Manikandan et al. [9] propose a scheme that involves dividing the cover image into 64 segments and applying discrete cosine transform for every single grid or segment, a set of discrete cosine coefficient values for each of them is obtained. The values are scrambled using a mapper. The initial position of the Queen is given to 8-Queens Algorithm which generates the solution. The polynomial equation created by the solving the problem is used for embedding information.

A lossless image steganography using spiral scan is put forth by Kiran et al. [10]. Three gray-scale images are embedded into a single RGB cover image. This approach works with both raw and png images. The embedding ratios of the proposed method are much higher than that of the existing one.

3　Proposed Method

The current methods mainly focus on a single segment of the image to embed the data in. They choose a single block or portion to embed the data in. Therefore, if during steganalysis, the hidden data is discovered, it will be easier to extract as they are all in one segment. Nonograms are solved mainly using genetic algorithm and depth-first search. They are normally used in games and puzzle-based applications.

The proposed method uses nonograms to dynamically choose the segments based on the required capacity. Therefore, even if the hidden data is discovered during steganalysis it will be hard to find the order and all the blocks in which the data was embedded in. This improves complexity and robustness of the system. A nonogram of required size is chosen and generated. The RGB cover image is then segmented into the same number of grids as the selected nonogram. Only the grids that fall under the solved segments of the nonogram is selected. The values of the data to be embedded are shifted n spaces, so that the message does not make sense to the third party. Then, the first grid in the solution is taken and k-means is performed. Then data is embedded in the chosen pixel cluster. If there is more data to be stored, then the next grid from the selection is taken and the same operations are performed on it. This process of selecting a grid and performing clustering to embed the data is done until there is no more data to be embedded in the image. The nonogram parameters are shared with the recipient who then performs the same operations; only instead of embedding the data they extract it from the chosen pixels.

Steps Involved:

- Select RGB cover image.
- Use nonogram to generate blocks.
- Select Block.
- Perform k-means clustering.
- Choose a particular cluster to embed in.
- From the result, embed data into the pixels in the required color planes.
- If more data is to be embedded, then repeat from Step 3 (Fig. 2).

The cover image is divided into a $n \times n$ matrix depending on the input parameters. The input is taken as a matrix of the column values followed by the row values. A depth-first search algorithm is used to implement the nonogram solver. The solution to the nonogram is obtained using a depth-first search algorithm. This algorithm is effective for small- and medium-sized nonogram puzzles. The output of the nonogram solver is a $n \times n$ matrix consisting of zeroes and ones. The ones represent the solution grids that we will use in our steganographic method (Figs. 3 and 4).

The solution to the nonogram is obtained using a depth-first search algorithm. This algorithm is effective for small- and medium-sized nonogram puzzles. The

Fig. 2 Flowchart of
proposed nonogram-based
method

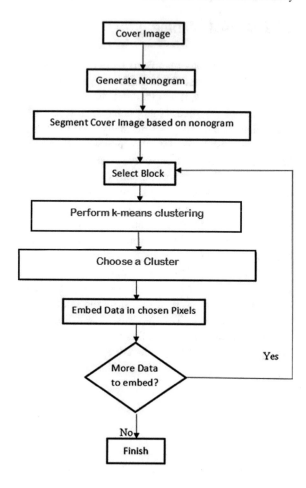

Fig. 3 Nonogram input

$$2\ 0\ 2\ 1\ \#\ 1\ 11\ 110$$

Fig. 4 Nonogram solution

$$0\ 1\ 1\ 0$$
$$0\ 0\ 0\ 0$$
$$1\ 1\ 0\ 0$$
$$0\ 0\ 1\ 0$$

Fig. 5 Division of image for Nonogram

Fig. 6 Grids selected based on Nonogram

output of the nonogram solver is a nxn matrix consisting of zeroes and ones. The ones represent the solution grids that we will use in our steganographic method (Figs. 5 and 6).

The individual letters in the message are shifted based on an agreed value. After performing k-means clustering, a cluster group is chosen at random. The data is embedded in the solution grid's chosen cluster group. If there is more data to be embedded and the next grid in the solution is taken and the same processes are performed again.

4 Experimental Results

MATLAB was used to implement the proposed method. To evaluate the quality of the stego image, PSNR and MSE of the steganographic image were calculated. Sail Boat on Lake (http://sipi.usc.edu), Tree (http://sipi.usc.edu), Lena (http://eeweb. poly.edu) and Marbles (http://www.fileformat.info) were the images used. PSNR is used to measure the quality of the stego image, the higher the value the closer the stego image to the original image. The message size and the PSNR value are inversely related, as the message size increases the PSNR value decreases. We have embedded 400 bits of data in each of the image sizes (Fig. 7).

The lower the value is for MSE, lesser the error in the stego image. The shifting of the letters in the message data converts the message into unintelligible characters (Fig. 8).

The values obtained through the computations in Table 1 are both satisfactory and show higher PSNR values and lesser MSE values compared to existing methods. The shifting of the letters in the message data converts it into unintelligible characters; this along with the usage of nonograms and selection of only one particular cluster to embed the data in increases the robustness and security of the image. The capacity of this method is lesser compared to other RGB steganographic techniques but similar compared to gray-scale techniques. This method is more suited to communicate moderate amount of data. To measure the efficiency of the proposed method, we compare it with methods given in [11–13, 15]. The results of the comparison in Table 2 show that the PSNR and MSE values of the proposed method are more secure (Figs. 9 and 10).

The entropy of the Cover and Stego Images were calculated. Information Entropy is used to calculate the information content of an image 4 [14], as embedding using LSB technique causes an increase in the entropy of the stego image. The larger the difference between the entropy of the stego image to that of the cover image, the more vulnerable it is to entropy-based attacks. From Table 3 the calculated results it can be seen that there is no change in the entropy values between the cover and stego image for three of the images and only an extremely slight change in the Marbles image. This shows that the proposed method is robust and secure against entropy-based attacks.

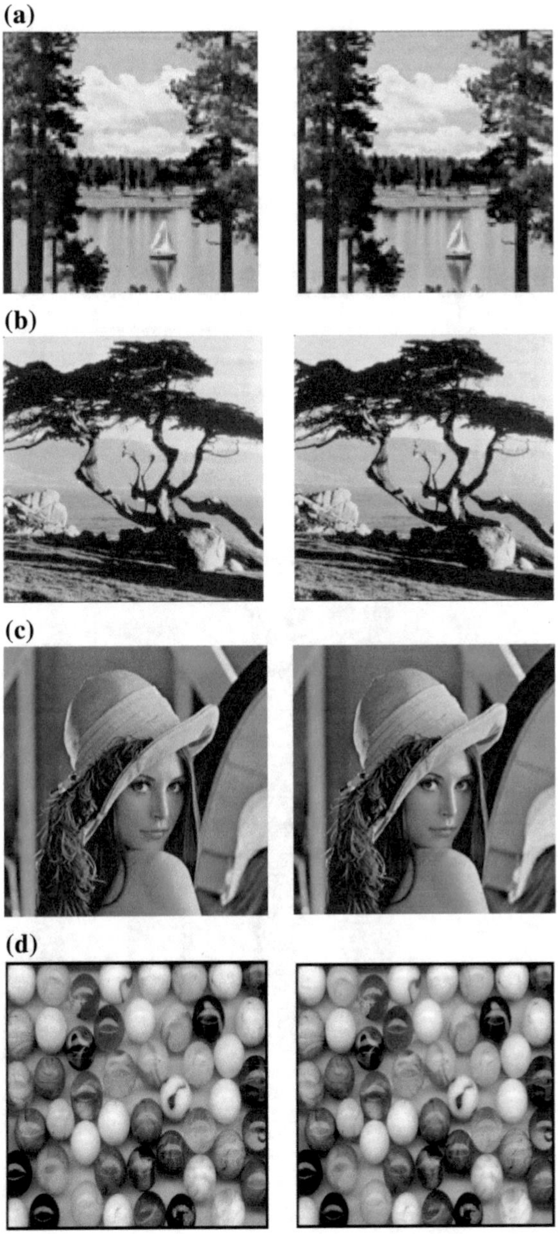

Fig. 7 Original and stego image of **a** Sail Boat **b** Tree **c** Lena **d** Marbles

Image Steganography Technique]□u{y4g□y{u□□{□u□|□4hyw|□}□□y

Fig. 8 Message data before and after shifting

Table 1 PSNR and MSE values

Input image	Image size	PSNR	MSE
Sail Boat	128 × 128	72.3278	0.0041
Tree	256 × 256	77.8862	0.0011
Lena	512 × 512	83.9068	0.00026449
Marbles	1024 × 1024	90.2977	0.000060717

Table 2 Comparison of PSNR and MSE values for Lena (256 × 256)

Capacity	Method	PSNR	MSE
128 (bytes)	LSB [11, 15]	60.3614	0.0625
128 (bytes)	LSB [12, 15]	68.3805	0.0094
128 (bytes)	LSB [13, 15]	70.97	0.0052
128 (bytes)	Proposed nonogram-based LSB method	73.9403	0.0026

Fig. 9 PSNR value comparison

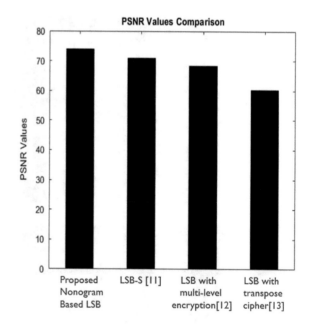

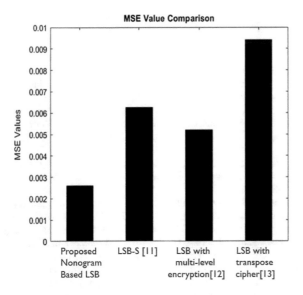

Fig. 10 MSE value comparison

Table 3 Entropy calculation

Input image	Image size	Entropy of cover image	Entropy of stego image
Sail boat	128 × 128	7.7402	7.7402
Tree	256 × 256	7.5371	7.5371
Lena	512 × 512	7.7502	7.7502
Marbles	1024 × 1024	7.6572	7.6573

5 Conclusion

Applications that rely on steganography mainly use it for hiding their data, therefore with imperceptibility as our main objective and by reviewing the existing methods, we have come up with a method that can be used in an effective manner and preserves the likeliness of the stego image to the original image to a greater extent. The use of nonograms and clustering in this method increases the randomness, robustness and the security of the image while still preserving the quality. Future work can focus on improving the embedding capacity, while still maintaining the imperceptibility and robustness of this method. Thus, this system is an effective solution to communicate data in a secure manner.

References

1. Manikandan, G., Bala Krishnan, R., Rajesh Kumar, N., et al.: Multimed Tools Appl. **77**, 32257 (2008). https://doi.org/10.1007/s11042-018-6237-5
2. Shifa, A., et al.: Joint crypto-stego scheme for enhanced image protection with nearest-centroid clustering. IEEE Access **6**, 16189–16206 (2018)
3. Jing, M.Q., Yu, C.H., Lee, H.L., Chen, L.H.: Solving Japanese puzzles with logical rules and depth first search algorithm. Int. Conf. Mach. Learn. Cybernetics **2009**, 12–15 (2009)
4. Kumari, T., Singh K.: A review on information hiding methods. Int. J. Eng. Sci. Comput. (2018, May)
5. Bobko, A., Grzywacz, T.: Solving nonograms using genetic algorithms. In: 2016 17th International Conference Computational Problems of Electrical Engineering (CPEE), pp. 1–4. Sandomierz (2016)
6. Ghosh, E., Debnath, D., Gupta Banik, B.: Blind RGB image steganography using discrete cosine transformation. In: Abraham, A., Dutta, P., Mandal, J., Bhattacharya, A., Dutta S. (eds.) Emerging Technologies in Data Mining and Information Security. Advances in Intelligent Systems and Computing, vol. 814. Springer, Singapore (2018)
7. SaiKrishna, A., Parimi, S., Manikandan, G., Sairam, N.: A clustering based steganographic approach for secure data communication. In: 2015 International Conference on Circuits, Power and Computing Technologies [ICCPCT-2015], pp. 1–5. Nagercoil (2015)
8. Wiggers, W.A.: A comparison of a genetic algorithm and a depth first search algorithm applied to Japanese nonograms. In Twente Student Conference on IT (2004)
9. Manikandan, G., Ramakrishnan, S., BaburamSathiyaNijanthan, P., Harikrishnhaa, R.: A N-queen based polynomial approach for image steganography. Int. J. Eng. Technol. **5**(3), 2828–2831 (2013)
10. Prajapati, H.A., Chitaliya, N.G.: Secured and robust dual image steganography: a survey. Int. J. Innovative Res. Comput. Commun. Eng. **3**, 30–37 (2015)
11. Joshi, K., Yadav, R.: A new LSB-S image steganography method blend with cryptography for secret communication. In: 2015 Third International Conference on Image Information Processing (ICIIP), pp. 86–90. IEEE (2015)
12. Charan, G.S., Nithin Kumar S.S.V., Karthikeyan, B., Vaithiyanathan, V., Divya Lakshmi, K.": A novel LSB based image steganography with multi-level encryption. In: 2015 International Conference on Innovations in Information, Embedded and Communication Systems (ICIIECS), pp. 1–5. IEEE (2015)
13. Jain, M., Lenka, S.K.: Secret data transmission using vital image steganography over transposition cipher. In: 2015 International Conference on Green Computing and Internet of Things (ICGCIoT), pp. 1026–1029. IEEE (2015)
14. Qiao, X., Ji, G., Zheng, H.: A new method of steganalysis based on image entropy. **2**, 810–815 (2007). https://doi.org/10.1007/978-3-540-74282-1_90
15. Al-Omari, Z.Y., Al-Taani, A.T.: Secure LSB steganography for colored images using character-color mapping. In: 2017 8th International Conference on Information and Communication Systems (ICICS), Irbid, 2017, pp. 104–110. https://doi.org/10.1109/iacs.2017.7921954

Machine-Learning-Based Taxonomical Approach to Predict Circular RNA

S. Saranya, G. Usha and Satish Ramalingam

Abstract Machine-learning approach is a good alternative for big-data analytics with high accuracy. Circular RNA plays a major role in development of many diseases such as diabetes, heart failure, osteoarthritis, Alzheimer, cancer, etc. Several circular RNA detection techniques have been developed. In this paper, we provide machine-learning-based taxonomical technique to predict circular RNA sequences from RNA Sequences. Here, we compared various types of circular RNA detection techniques with their performance regarding the metrics such as specificity, sensitivity, accuracy, MCC and precision. Among all of these comparison methods, it is concluded that none of the metrics were dominated.

Keywords Circular RNA · Machine learning · Specificity · Precision · Transcriptome

1 Introduction

Ribonucleic acid plays several important roles such as coding, decoding and also in regulation of gene expression. Coding RNAs (messenger RNA) are conveying genetic information [1] from DNA to ribosome resulting in gene expression. Non-coding RNAs (tRNA, rRNA) are functional RNAs, they are transcribed from DNA but not translated into proteins (Fig. 1).

Recent studies have revealed the presence of various other non-coding RNAs (miRNAs and CircRNAs) that are involved in critical cellular functions; CircRNAs are formed by applying back-splicing technique which is depicted in Fig. 2.

CircRNA [2–4] is closed continuous loop without any free gaps. It is more stable than other non-coding RNA. Hence, identifying circRNA is very important in

S. Saranya (✉) · G. Usha
Software Engineering, SRMIST, Kattankulathur, Chennai, India
e-mail: saranya.it78@gmail.com

S. Ramalingam
Genetic Engineering, SRMIST, Kattankulathur, Chennai, India

© Springer Nature Singapore Pte Ltd. 2020
S. S. Dash et al. (eds.), *Artificial Intelligence and Evolutionary Computations in Engineering Systems*, Advances in Intelligent Systems and Computing 1056,
https://doi.org/10.1007/978-981-15-0199-9_61

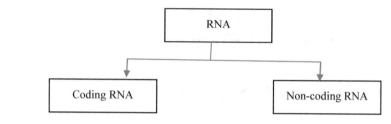

Fig. 1 Types of RNA

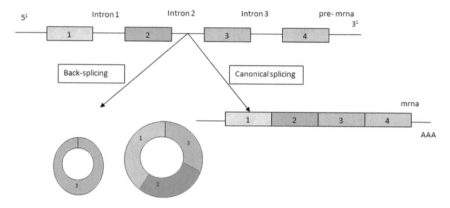

Fig. 2 Back-splicing RNA to form CircRNA

healthcare system to predict the disease [5]. In 2010, tools were developed to detect circRNAs from sequencing data. Various types of techniques are used to detect cicRNA such as polymerase chain reaction, northern blot, 2D electrophorese and H degradation assay. Various types of circular RNAs are existing [6, 7]. Circular RNA is 100 bp to 4 kb in size, which are non-functional by-products of RNA splicing. Many types of circular RNA are found using recent bio informatics technique. Circular RNAs [3] consists of covalently closed loops. It does not contain any tails in the 5′–3′ port. Figure 3 explains basic common types of circular RNA [8].

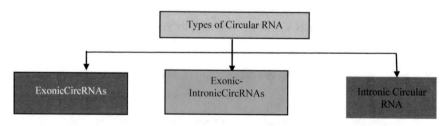

Fig. 3 Types of Circl RNA

Machine-learning algorithms [9] involve in predictions on data, which is an emerging applications for bioinformatics. There are two ways to detect circular RNA. One way is, the structure of circular RNA can be identified from gene expressions [10]. Another way of identifying circular RNA is where machine-learning algorithms can be applied to evaluate pattern recognition, prediction and classification. Machine-learning techniques are used in various types of biological problems such as genetics and genomics and identifying the location of gene transcriptions. Various computation techniques have been proposed in literature to identify circRNA [11, 12]. The paper is formatted as follows: Sect. 2 describes various computational techniques to detect circRNA. Section 3 discusses result analysis and proposes a comparative study. Finally, Sect. 4 of this paper concludes with future perspectives.

2   Machine-Learning Taxonomy to Detect Circular RNA

Here, we concentrate on machine-learning algorithms that are used in the detection of circular RNA. Figure 4 depicts the taxonomical classification techniques to detect circular RNA using random forest, SVM, MKL, HELM and naïve-based techniques. In RNA preprocessing, 5′ and 3′ terminals of one or more exons are combined to form CircRNA [13]. Now, we discuss each of the machine-learning techniques in detail with its proposed solution.

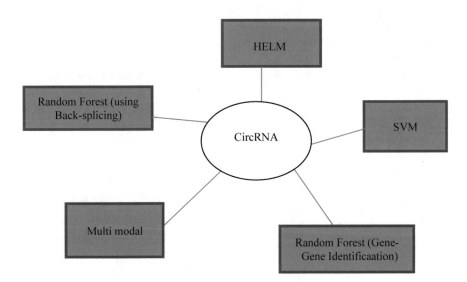

Fig. 4 Taxonomical approach for circular RNA detection

2.1 Random Forest Classification for Gene-Gene Identification (GGI)

In this technique, the human brain tissues are used to detect circular RNA. The entire approach mainly focuses on the formulation of gene-gene interaction identification with particular disease (Alzheimer's disease). Two expression profiles were used to increase the sample size, they are GSE33000 and GSE44770. The dataset uses various statistical measurements and the gene expression in disease and normal conditions. Additionally, they used two dataset; first dataset is human protein interaction. Systematic approach screening technique is used to retrieve the human protein interaction dataset [14]. This first dataset was referred as biophysical protein-protein interaction (bPPI). Second dataset constructed for co-expressed/co-occurring gene pair. To get accurate and meaningful interactions, bPPI/integrated bPPI are combined.

Table 1 explains the feature of gene profile. By using statistical measurement process, the various features were identified as follow. E_{A_L0} denotes expression value of gene A of sample label 0. And E_{B_L0} denotes expression value of gene B of sample label 0. By using these two different gene, the following feature extracted using basic statistical methods.

- Mean
- Standard deviation
- Find the maximum and minimum elements—are calculated to get better reflect in heterogeneous expression of gene profile.
- Welch's t-test—used to discriminate between two gene group. It was more reliable when two samples have unequal variance and unequal sample size.
- Correlation coefficient—It is used to compute correlation between two different classes but same gene.
- Mutual Information (MI)—used to calculate mutual information for gene classes E_{A_L0} and E_{A_L1}. For example, E_{A_L0} contains [2, 3, 15–17] elements and E_{A_L1} contains [4, 11, 12] the mentioned elements. To calculate MI for these two elements, under sampling was performed by randomly pulled elements to form decision tree in order to make two elements in equal size. Hence finally, E_{A_L0} and E_{A_L1} are equal in size, $E_{A\_L0} = [2, 3, 15]$. The same process will be repeated to create many decision tree.

Table 1 Notation of gene expression with class and gene pair

Gene pair	Class label 0 (normal)	Class label 1 (AD)
Gene A	E_{A_L0}	E_{A_L1}
Gene B	E_{B_L0}	E_{B_L1}

Table 2 Dataset values used to model random forest and SVM using back-splicing method (ref Jun Wang)

Dataset	Positive instances	Negative instances
Training	5811	5137
Validation	1208	1055
Test	8260	1117

2.2 Random Forest (RF) Classification for Back Splicing

This technique uses two public datasets: circRNADband circBase. These datasets consist of unique circRNAID, genomic position, gene expression, biosample and references of identification. In this technique, in order to use random forest algorithm, R package is used. RF construct n_{tree} independent decision tree using n_{tree} bootstrap samples. For each step in the sample which are not selected to construct n_{tree} decision tree from sample is appealed Out Of Bag (OOB). The sample data which is present in boostrap sample taken considered as n_{tree} independent decision tree.

Not all data is applied to the decision tree, only randomly selected variable m_{try} feature is used (The total values taken to construct the model is given in Table 2). M_{try} variable is randomly selected. In this decision tree, nodes are separated based on best feature variable among m_{try}. This technique uses only two parameter to predict CircRNA which are: n_{tree}, m_{try}. So these two parameters are fixed for decision tree construction.

In this technique, performance is evaluated based on the following metrics called accuracy, sensitivity, specificity and MCC. The accuracy value is not only sufficient for high performance model selection. Also, sensitivity is based entirely on True Positive Rate, which is a proportion of positive values. Specificity is based on the True Negative value. Mattews coefficient of correlation (MCC) is the correlation between the predicted and the labeled binary classification where the value of the MCC is always −1 to 1.

2.3 Support Vector Machine

In this approach, the model was constructed with data from positive and negative instances for back-splicing prediction. Total training instances was 10,948 and for validation 1208 positive data instances, 1055 negative instances. And for testing 1241 positive instances and 1117 negative instances were used to get high performance. Using the SVM algorithm, the optimal hyperplane is determined [18]. The selection space of the high dimensional feature is separated by the positive and negative instances. Maximizing the margin between positive and negative classes has made the separation.

The implementation of the proposed technique [16] is done using R package "e1071." Radial basis function (RBF) kernal is used to get good performance. Additionally, two parameters are used to predict circRNA that is C and γ. C is capacity constant (which controls trade between decision rule complexity and error frequency. If C value is very large, the result of SVM is non-separable points. So the result is over fitting. If C value is small means, the result of the classifier is under fitting.

γ is a training parameter that defines the classification decision boundary. If γ is too large, it will include only the support vector. Second, if the γ value is too low average, the entire training set will be included in the influence area of the selected support vector.

2.4 Multiple Kernal Learning (MKL)

In this approach, the human circular RNA dataset is used from Circbase database. This dataset, having nearly 90,000 CircRNA transcripts, that was collected experimentally along with genome coordinates. The transcripts that are shorter than 200 nucleotides (nt) was removed from the gene dataset. This process was repeated to remove all short gene expression, then finally 14,084 circular RNA as positive data including the coding and non-coding RNA. In this paper, graph feature plays very important role to represent sequence and structure of RNA.

MKL is first applied in SVM which is used to make kernel matrix to encode the similarities between samples [13]. This technique combines different kernel as it linearly weighted kernels. Multiple features are combined to represent same data to get better feature representation. Hence, this technique uses machine-learning algorithms. Given the value M, kernel k_m is called reproducing kernel from different resources.

For this, with different M, is different reproducing kernel k_m from different resources. In order to solve optimization problem of MKL, the formula is as follows:

$$\min_{\theta} \max_{\alpha} \{ 1^T - \tfrac{1}{2} \sum \alpha_i \alpha_j y_i y_j \sum_m \theta_m k_m$$
$$\text{Subject to } \|\theta\| p \le 1, \theta \ge 0, Y^T \propto = 0, \propto \ge 0, \propto \le c$$

where I and j is used to train data index, p is used in normal vector controlling kernel weight regularization. In this technique, in order to evaluate the performance one more metrics is added which is known as precision. In order to improve predicted performance, fivefold cross validation is used. Compared to other approach, MKL is high in accuracy.

2.5 HELM-Hierarchial Extream Learning Machine Algorithm

In this approach, circbase database and GENCODE datacase are used as datasets [19, 20]. For convenience, the positive dataset considered as CircRNA and negative dataset considered as lncRNA. HELM is feed-forward neural network with multiple hidden layers [21]. For feature selection, it follows the common data mining techniques (Relief, MRMD, Monte Carlo feature selection). HELM process takes two steps to predict CircRNA. First step is unsupervised feature representation; input is given to sparse encoder in order to produce high-level representation output. This high-level representation output is further converted into matrix format and then this matrix is input to the second step [7]. In the second step, initial process is supervised-feature classification. At this stage, it takes only one hidden layer to produce expected output called CircRNA prediction.

Once predicted, the CircRNA [22], it undergoes validation process using tenfold cross validation. At this stage, the entire dataset is divided into ten parts and each part is used as test sample in turn wise. Initially, the first part is considered for sample data, the remaining nine parts are considered for training dataset. The above-mentioned test is repeated ten times to find mean prediction performance.

In this approach, two types of data are available one is positive sample which is considered as CircRNA and negative samples which is considered as lncRNA. Hence, it follows binary classification process to predict CircRNA. The next section compares the performance of various techniques.

3 Comparison of Performance Evaluation

The above-mentioned computation technique uses five metrics, which are used to predict circRNA [23] from RNA sequences. The values of the metrics are explained in Table 3.

Sensitivity [24] is proportion of positive value which is predicted by each classifier. Formula (1) explains how to compute sensitivity

$$\text{Sensitivity} = \frac{\text{TP}}{\text{TP} + \text{FN}} \tag{1}$$

Table 3 Performance metrics values using different computational approach

Dataset	Accuracy	Sensitivity	Specificity	MCC
RF	0.793	0.837	0.747	0.587
SVM	0.778	0.863	0.6934	0.565
MKL	0.801	0.732	0.851	0.589
HELM	0.789	0.703	0.850	0.527

Specificity [25] is proportion of negative value which is predicted as negative by each classifier. Formula (2) explains how to compute specificity

$$\text{Specificity} = \frac{\text{TN}}{\text{TN} + \text{FP}} \tag{2}$$

Accuracy [26] is fraction value of exactly predicted value from all the classifier. Formula (3) explains how to compute Accuracy

$$\text{Accuracy} = \frac{\text{TP} + \text{TN}}{\text{TP} + \text{TN} + \text{FP} + \text{FN}} \tag{3}$$

MCC is measured as correlation between predicted and labeled input data.

$$\text{MCC} = \frac{\text{TP} \times \text{TN} - \text{FP} \times \text{FN}}{\sqrt{(\text{TP} + \text{FP})(\text{TP} + \text{FN})(\text{TN} + \text{FP})(\text{TN} + \text{FN})}} \tag{4}$$

Precision is used to compute the proportion of positive values are correct are not which is explained using the formula (5)

$$\text{Precision} = \frac{\text{TP}}{\text{TP} + \text{FP}} \tag{5}$$

From the above formula, TP is known as True Positive value number, TN is known as True Negative number, FP is known as False Positive value number and FN is known as False Negative value number.

4 Conclusion and Future Perspectives

In this paper, we provide techniques for taxonomic machine learning to detect circular RNA from RNA sequences. We provide an in-depth knowledge in understanding circular RNA from back-splicing process. We focused on machine-learning-based circular RNA detection techniques. We have compared four important metrics such as specificity, sensitivity, accuracy, MCC and precision. We did not find the evidence that the five metrics are neither independent nor dependant. Hence, this study presents the various computational techniques to detect circular RNA.

However, many unknown questions are there to be addressed yet. First, there is a need to form a baseline work to detect the correct circular RNA sequences from RNA sequence data. Second, the data set can be collected from DNA sequences to collect large number of gene features might be useful. Lastly, we have provided a better bioinformatics based machine-learning approach such as deep-learning methods to identify the circular RNA formation.

References

1. Telonis, A.G., Magee, R., Loher, P.: Knowledge about the presence or absence of miRNA isoforms (isomiRs) can successfully discriminate amongst 32 TCGA cancer types. Nucleic Acids Res. **45**, 2973–2985 (2017)
2. Hsu, Y.-H., Si D.: Cancer Type Prediction and Classification Based on RNA-sequencing Data (2013)
3. Pana, X., Kai, X.: PredcircRNA: computational classification of circular RNA fromother long non-coding RNA using hybrid features (2010)
4. Mikolov, T., Chen, K., Corrado, G., Dean, J.: Efficient Estimation of Word Representations in Vector Space. arXiv preprint arXiv:1301.3781 (2013)
5. Roberts, A., Pimentel, H., Trapnell, C., Pachter, L.: Identification of novel transcripts in annotated genomes using RNA-Seq. Bioinformatics **27**, 2325–2329 (2011)
6. Jeck, W.R., Sorrentino, J.A., Wang, K., Slevin, M.K., Burd, C.E., Liu, J., Marzluff, W.F., Sharpless, N.E.: Circular RNAs are abundant, conserved, and associated with ALU repeats. RNA **19**, 141–157 (2013)
7. Xiong, H.Y., Alipanahi, B., Lee, L.J., Bretschneider, H., Merico, D., Yuen, R.K., Hua, Y., Gueroussov, S., Najafabadi, H.S., Hughes, T.R., et al.: RNA splicing. The human splicing code reveals new insights into the genetic determinants of disease. Science **347**, 6218 (2015)
8. Kloft, M.: Ph.D. thesis (2011)
9. Kourou, K., Exarchos, T.P., Exarchos, K.P.: Machine learning applications in cancer prognosis and prediction. Computat. Struct. Biotechnol. J. **13**, 8–17 (2015)
10. Ryvkin, P., Leung, Y.Y., Ungar, L.H., Gregory B.D., Wang, L.S.: 28–35 (2013)
11. Harrow, J., Frankish, A., Gonzalez, J.M., Tapanari, E., Diekhans, M., Kokocinski, F., Aken, B.L., Barrell, D., Zadissa, A., Searle, S., et al.: "Gencode: the reference human genome annotation for the encode project. Genome Research **22**(9), 1760–1774 (2012)
12. Cooper, T.A., Wan, L., Dreyfuss, G.: RNA and disease. Cell **136**(4), 777–793 (2009)
13. Li, P., Chen, S., Chen H., Mo, X., Li, T., Shao, Y., Xiao, B., Guo, J.: Using circular RNA as a novel type of biomarker in the screening of gastric cancer. Clin. Chim. Acta. **444**, 132–136 (2015)
14. Jeck, W.R., Sharpless, N.E.: Detecting and characterizing circular RNAs. Nat. Biotechnol. **32**, 453–461 (2014)
15. Wang, J., Wang, L.: Prediction of Back-Splicing Sites Reveals Sequence Compositional Features of Human Circular RNAs. IEEE (2017)
16. Chaabane, M.: End-To-End Learning Framework for Circular Rnaclassification from Other Long Non-Coding RNAS Using Multi-modal Deep Learning (2018)
17. Yang, P.: Computational Approaches for Disease Gene Identification (2013)
18. Eriksson, M., Brown, W.T., Gordon, L.B., Glynn, M.W., Singer, J., Scott, L., Erdos, M.R., Robbins, C.M., Moses, T.Y., Berglund, P., Dutra, A., Pak, E., Durkin, S., Csoka, A.B., Boehnke, M., Glover, T.W., Collins, F.S., Recurrent de novo point mutations in lamin a cause hutchinson-gilford progeria syndrome. Nature **423**(6937), 293–298 (2003)
19. Conn, S.J., Pillman, K.A., Toubia, J., Conn, V.M., Salmanidis, M., Phillips, C.A., Roslan, S., Schreiber, A.W., Gregory, P.A., Goodall, G.J.: The RNA binding protein quaking regulates formation of circRNAs. Cell **160**, 1125–1134 (2015)
20. Memczak, S., Jens, M., Elefsinioti, A., Torti, F., Krueger, J., Rybak, A., Maier, L., Mackowiak, S.D., Gregersen, L.H., Munschauer, M., Loewer, A., Ziebold, U., Landthaler, M., Kocks, C., le Noble, F., Rajewsky, N.: Circular RNAs are a large class of animal RNAs with regulatory potency. Nature **495**, 333–338 (2013)
21. Hansen, T.B., Jensen, T.I., Clausen, B.H., Bramsen, J.B., Finsen, B., Damgaard, C.K., Kjems, J.: Natural RNA circles function as efficient microRNA sponges. Nature **495**, 384–388 (2013)
22. Hansen, T.B., Jensen, T.I., Clausen, B.H., Bramsen, J.B., Finsen, B., Damgaard, C.K., Kjems, J. R.: Nature **495**, 384–8 (2013)

23. Guo, Y., Sheng, Q., Li, J.: Large scale comparison of gene expression levels by microarrays and RNAseq using TCGA data. PLoS ONE **8**, 1–10 (2013)
24. Minoche, A.E., Dohm, J.C., Himmelbauer, H.: Evaluation of genomic high-throughput sequencing data generated on Illumina HiSeq and Genome Analyzer systems. Genome Biol. **12**, R112 (2011)
25. Omberg, L., Ellrott, K., Yuan, Y.: Enabling transparent and collaborative computational analysis of 12 tumor types within the cancer genome atlas. Nat. Genet. **45**, 1121–1126 (2013)
26. Zhang, B., He, X., Ouyang, F.: Radiomic machine-learning classifiers for prognostic biomarkers of advanced nasopharyngeal carcinoma. Cancer Lett. **403**, 21–27 (2017)

Anomaly-Based Intrusion Detection System Using Support Vector Machine

S. Krishnaveni, Palani Vigneshwar, S. Kishore, B. Jothi and S. Sivamohan

Abstract In recent years, there has been an increase in digitization of records, from small databases like student details in a school and product inventory in a shop to large databases like Social Security Number in a country. This digitization even though takes lesser space than its analogy counterparts is susceptible to attacks even from remote locations. Nowadays, there has been a substantial increase in the number of cases of anomalous activities in the network which threatens network safety. So, it is important to not only store the data but also collect the session details so as to distinguish between a normal session and an abnormal session. In this paper, we propose an effective anomaly detection system for cloud computing. The support vector machine is used for profile training and intrusion detection. Experimental results show that IDS with an optimized NSL-KDD dataset using the best feature set algorithm based on Information Gain Ratio increases the accuracy of 96.24% and minimizes the false alarm rate. The machine learning-based approach such as support vector machine has significant potential benefits for the evolution of IDS programs for challenging complex environments such as cloud computing.

Keywords Intrusion detection system · Machine learning · Information gain ratio · SVM · Radial basis function

1 Introduction

Computers and networks have been under threat from viruses, worms, and attacks from hackers since they were first used. In 2008, the number of devices connected to the Internet exceeded the number of human beings and this increasing trend will see about 50 billion devices by 2020 [1], Securing these devices and the data

S. Krishnaveni (✉) · P. Vigneshwar · S. Kishore · B. Jothi · S. Sivamohan
Department of Software Engineering, SRMIST, Kattankulathur, Chennai, India
e-mail: krishnas4@srmist.edu.in; vanimithila@gmail.com

© Springer Nature Singapore Pte Ltd. 2020
S. S. Dash et al. (eds.), *Artificial Intelligence and Evolutionary Computations
in Engineering Systems*, Advances in Intelligent Systems and Computing 1056,
https://doi.org/10.1007/978-981-15-0199-9_62

passing between them is a challenging task because the number of intrusions is also increasing sharply year by year. To address this issue, a large number of defenses against network attacks have been proposed in the literature. Despite all the efforts made by researchers in the community over the last two decades, the network security problem is not completely solved. A large number of defense approaches have been proposed in the literature to provide different functions in various environments. The core element of a good defense system is intrusion detection system (IDS), which provides proper attack detection before any reaction [1]. An IDS aims to detect intrusions before they seriously damage the network. The term intrusion refers to any unauthorized attempt to access the elements of a network with the aim of making the system unreliable. Nowadays, due to the exponential growth of technologies and the increased number of available hacking tools, in both approaches, adaptability should be considered as a key requirement. To be able to react to network intrusions as fast as possible, an automatic or at least semiautomatic detection phase is required. This will decrease the amount of damage to legitimate users because an early detection system supplies more time for a proper reaction.

In this paper, we focus on anomaly-based intrusion detection using machine learning algorithm and classify them into normal or anomaly based on a certain set of features; for example, source byte, destination byte, duration flag, etc. with the help of support vector machine algorithms. We train the system with the help of the NSL-KDD dataset. We divide the dataset into 80–20 for training and testing. Furthermore, since a large amount of data is used, the time taken should be used, so we would compare the current method with other methods and also try to improve its parameters. The structure of this paper is organized as follows: the related work on network intrusion detection based on machine learning discussed in Sect. 2. Then the methodology for the proposed system presented in Sect. 3. Section 4 presented the experiment results and discussion. Finally, we summarized the paper and discussed the future work in Sect. 5.

2 Related Works

Network security has been one of the most prime concerns of the software industry as unlike the attacks on a normal server, these attacks have no geographical limitations and also the types of attacks are difficult to trace back. So, detection and prevention are used instead of basic correction (system restore). Aggarwal et al. [2], proposed to use support vector machine (SVM), a machine learning algorithm to determine the type of session. In this paper, he proposed a binary classifier to detect the type of session be it 'normal' or 'anomaly,' Jha and Ragha [3]. They proposed a method to select the features based on their information gain ratio (IGR) ranking. This can be calculated by knowing their entropy. In their paper, they proposed the value of entropy helps to determine the amount of change in the values of the session. This means that more the entropy value more will be the difference in values. Chen et al. [4] proposed a methodology to further increase the accuracy of the system. They proposed a particle swarm optimization to improve the optimization of the system.

3 Methodology

This paper proposes a comprehensive structure to choose the unique set of features from NSL-KDD datasets which would help in efficiently identifying normal traffic and differentiate it from abnormal traffic with the help of support vector machine (SVM) classifier. This proposed method implemented with a filter-based feature selection approach such as Information Gain Ratio (IGR) method; in this approach, feature has been ordered using an impartial measure: Information Gain Ratio. This is done by calculating the IGR which requires the initial entropy values. The proposed algorithm use to exclusive the significant feature set from the NSL-KDD dataset [5]. The condensed feature NSL-KDD dataset is then used for training and testing the detection model on the SVM classifier. Figure 1 shows the proposed model for anomaly-based intrusion detection; it consists of the following components:

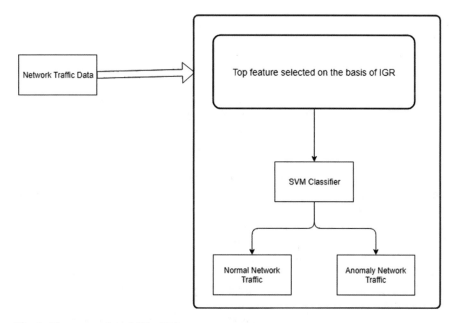

Fig. 1 The proposed model for IDS

1. Preprocessing the intrusion dataset,
2. Feature ranking based on Info gain,
3. Feature subset formation,
4. Detection model using SVM.

3.1 The Dataset

this experiment, the NSL-KDD dataset was used for evaluation. This is a comparatively newer dataset which consists of selective data of KDD 99 [8]. In total 41 features, 34 are numeric and seven are symbolic. It also has 22 training attack types and 17 testing attack types.

3.2 Dataset Preprocessing

We need dataset preprocessing as an SVM classifier cannot use the dataset features in its base form. So, we need to convert symbolic data into the numeric form and put it into a fixed range $(-1, 1)$ with the help of min–max scalar (in python) and finally, map it as normal or anomaly.

3.3 Feature Ranking

The feature ranking is done by using a filter Information Gain Ratio which uses the entropy of a given feature of a session. Entropy is a derived word from physics which basically means the amount of disorder in the values [6]. The value of entropy is less when the features correspond to a particular class. The entropy value is more when the features correspond to more than one class. Information gain helps in quantifying the usefulness of each feature in classifying the session. It is calculated using the entropy value of each feature. Information gain calculates the decrease of the weighted average entropy of the attributes to that of impurity of the complete set of data items. This means that features with the high information gain values are considered as the most important for classifying the session data. The Information Gain Ratio (IGR) is a quantitative value that is used to prioritize the feature on the basis of their values of such features in the dataset. In this research, IGR is selected as a value and not information gain because as information gain is biased toward the features with a large number of different values. Information Gain Ratio $(Ea, f) = $ Gain $(Ea, f)/$Split info (Ea, f) where E stands for dataset feature information and f denotes a particular feature. Equation (1) and (2) represent the formula of information gain.

Table 1 Top Selected Features

S. No.	Features	IG R
1	SrvSerror rate	0.508
2	Serror rate	0.464
3	flag	0.462
4	Logged in	0.456
5	dst host srv serror rate	0.442
6	diff srv rate	0.377
7	dst host serror rate	0.374
8	dst bytes	0.292
9	src bytes	0.292
10	same srv rate	0.276

$$\text{Gain } (Ea, f) = \text{Entropy } (Ea) - \sum_{ve \text{ Values } (f)} |cEa, v|/|Ea|^* \text{Entropy } (Ea)$$

$$E, v = \{a \in E/\text{value}(a, f)\} = v \tag{1}$$

We all know formula of entropy based on Shannon's entropy (Table 1):

$$\begin{aligned} Ex &= -\sum P_i \log_2(P_i) \\ \text{Split Info}(Ea, f) &= -|Ea, f||Ex|v \in \text{values}(f) \log 2|Ea, f||Ea| \end{aligned} \tag{2}$$

3.4 Radial Basis Function

In basic classification, we use linear SVM which has a straight line as its hyperplane but it is not always the case. Sometimes to reduce the number of false positives, we need a nonlinear kernel [7]. Hence, RBF is used where the hyperplane depends upon the closest point and also aims at reducing the number of false positives. Its value depends on two main factors, C—also known as the penalty factor higher C values have high complexity and lower C values have less complexity. Gamma is the kernel coefficient. High gamma values have given more influence to the farther points on the hyperplane, whereas low gamma values give more influence to closer points.

$$F(x) = \sum_{j=1}^{m} w_i h_i(x) \tag{3}$$

$$H(x) = \exp\left(-\frac{(x-c)2}{r2}\right) \tag{4}$$

where r is the standard deviation and x–c is the distance from the given point to center of the existing point.

4   Results and Discussion

In this experiment, filter Information Gain Ratio uses the entropy of a given feature of a session and optimizes the parameters of SVM (RBF) using SVM and also reduces the features of the training set. It reduces the noisy feature from the training set. The training set contains 25,149 records and the testing set contains 11,850 records. The algorithms used in the experiment are given below. During this experiment, a comparison of different kernel functions of SVM with feature selection with accuracy. The kernel function used here is linear, Gaussian, RBF, polynomial. The results show that the RBF kernel function with optimized features gives the highest accuracy. The training and testing dataset were randomly extracted from the original dataset. The attribute values intended for intrusion detection were measured through the classified results. The ROC curve for the proposed method obtained through the values can be expressed as in Fig. 2.

The proposed system was then checked for its accuracy by using the different classifiers linear SVM, RBF SVM, logistic regression and K nearest neighbor. The comparison of accuracy rate obtained through various techniques was denoted in Fig. 3. Table 2 represents the comparison of the accuracy rate. RBF SVM was achieved 96.34% higher performance than another method.

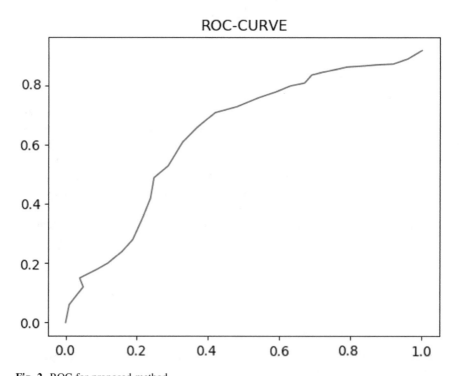

Fig. 2 ROC for proposed method

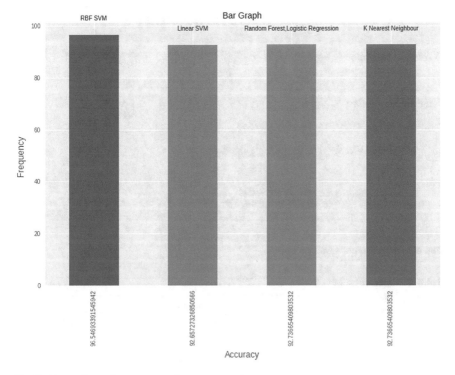

Fig. 3 Comparison of accuracy performance

Table 2 Comparison of accuracy rate

S. No	Classifier name	Accuracy (%)
1	Linear SVM	92.65
2	RBF SVM	96.34
3	Logistic regression	92.41
4	K nearest neighbor	92.87

In a similar case, the response time achieved through the proposed technique was checked with the standard classifiers, i.e., linear SVM, RBF SVM, logistic regression, and K nearest neighbor. The comparison of response time for various classifiers could be shown in Fig. 4. Table 3 represents the comparison response time of the linear SVM, RBF SVM, logistic regression, and K nearest neighbor. Hence, it is proved that RBF SVM achieved better than other methods.

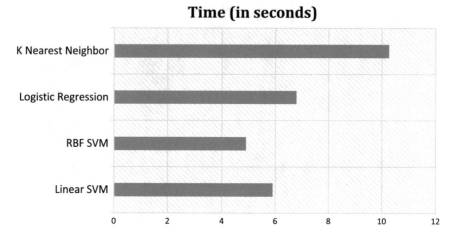

Fig. 4 Comparison of response time(S)

Table 3 Comparison of response time

S. No	Classifier name	Time (s)
1	Linear SVM	5.90
2	RBF SVM	4.90
3	Logistic regression	6.79
4	*K* nearest neighbor	10.26

5 Conclusion

In this paper, an effective anomaly detection system is implemented which trains a normal profile from features collected from publicly available NSL-KDD datasets. The support vector machine is used for profile training and intrusion detection. The efficiency is analyzed with varying network conditions by simulating four attacks. Experimental results show that IDS with an optimized NSL-KDD dataset using the best feature set algorithm based on Information Gain Ratio increases the accuracy of 96.24% and minimizes the false alarm rate. The machine learning-based approach such as support vector machine has significant potential benefits for the evolution of IDS programs for challenging complex environments such as cloud computing.

References

1. Mell, P., Grance, T.: The NIST definition of cloud computing. Tech. Rep. Natl. Inst. Stan. Technol. **5**, 1–16 (2011)
2. Sharma, S., Agrawal, J., Agarwal, S., Sharma, S.: Machine learning techniques for data mining: a survey. In: IEEE International Conference on Computational Intelligence and Computing Research, IEEE ICCIC, pp. 1–6 (2013)

3. Jha, J., Ragha, L.: Intrusion detection system using support vector machine. Int. J. Appl. Inf. Syst. (IJAIS) Foundation Comput. Sci. FCS New York USA **2**, 25–30 (2013)
4. Chen, Q., Yang, J., Gou, J.: Image compression method using improved PSO vector quantization. **2**, 490–495 (2005)
5. Li, L., Sun, H., Zhang, Z.: The research and design of honeypot system applied in the LAN security. In: IEEE 2nd International Conference on Software Engineering and Service Science (ICSESS), pp. 360–363 (2011)
6. Majidi, F., Mirzaei, H., Iranpour, T., Foroughi, F.: A diversity creation method for ensemble based classification: Application in intrusion detection. In: 7th IEEE International Conference Cybernetic Intelligent Systems, pp. 1–5 (2008)
7. Rene, C.I., Abdullah, J.: Malicious code intrusion detection using machine learning and indicators of compromise. IJCSIS **10**, 234–241 (2017)
8. Krishnaveni, S., Prabakaran, S., Sivamohan, S.A.: Survey on honeypot and honeynet systems for intrusion detection in cloud environment. J. Comput. Theor. Nanosci. **15**, 2949–2953 (2018). https://doi.org/10.1166/jctn.2018.7572

An Approach Towards the Development of Scalable Data Masking for Preserving Privacy of Sensitive Business Data

Ruby Bhuvan Jain and Manimala Puri

Abstract Large amount of digital data is generated rapidly all around the globe. Providing security to digital data is the crucial issue in almost all types of organizations. According to the Identity Theft Resource Center, there were 8069 data breaches between January 2005 and November 2017 [1]. In the year 2018, 477 cases registered about data breach [2]. In just three months of 2019, 145 such cases are already noticed [GK et al. in A study on dynamic data masking with its trends and implications. 38(2), 0975–8887, 3], and it continues to grow. Protecting the digital sensitive data from data breaches is the need of the hour. The main objective is to protect the privacy of individuals and society which is becoming crucial for effective functioning across businesses. Privacy enforcement today is being handled primarily through government monitored regulations and compliances. To overcome the limitations of existing masking methods, researcher proposed a non-zero random replacement masking method. Researcher has successfully developed a scalable data masking model which can be used for various data types—CSV, JSON, XML, and relational databases. To evaluate the proposed method, researcher used an internationally recognized UCI repository which is an open source of secondary data, out of 436 datasets available on the site; researcher selected five different datasets of various business domains. The selected business data is under five different categories—healthcare, social media, bank marketing, bank finance, and stock market. The researcher also contemplated about volume of datasets. Researcher applied three types of masking—substitution, shuffling, and proposed method on the selected datasets. The original dataset and masked datasets are classified by classification metric. Performance parameters measured on four different classifiers delivered sizeable variations. With respect to data samples used for analysis, results strongly augmented that the proposed data masking method can be used across the business-critical domains. The results strongly emphasize that the

R. B. Jain (✉)
JSPM's Abacus Institute of Computer Applications, Pune, Maharashtra, India
e-mail: ruby.jain81@gmail.com

M. Puri
Dr K. N. Modi University, Newai, Tonk, Pali, Rajasthan, India
e-mail: manimalap@yahoo.com

© Springer Nature Singapore Pte Ltd. 2020
S. S. Dash et al. (eds.), *Artificial Intelligence and Evolutionary Computations in Engineering Systems*, Advances in Intelligent Systems and Computing 1056, https://doi.org/10.1007/978-981-15-0199-9_63

proposed model is the solution which not only protects the sensitive data but also maintains the usability, accuracy, and sensitivity.

Keywords Production environment · Non-production environment · Masking · Shuffling · Substitution · Compliances

1 Introduction

Privacy of data should be maintained, as data misuse leads to threats. Due to its importance, various tools and technologies are emerging to serve the purpose. Normally, data privacy is concerned with confidentiality. Privacy is mainly related to use of personal data. It is further concerned with collection of data, use of data, and the disclosure of the data—to whom, you are giving the data. Often, people do not like to disclose private data to others for security reasons. To hide sensitive information, data masking can be used.

Data masking is the process of hiding specific data elements inside dataset. It ensures that sensitive data is replaced with realistic but not real data. The goal is that sensitive customer information is not available in unauthorized environment.

The term "unauthorized access" is very familiar to us, but who authorizes access? [4] We as consumers and customers play a role in defining that has access to our "non-public personal information" and under what conditions they can use this information. We sign and click on agreements every day that define the terms and conditions of granting access to and use of our non-public personal information. However, the service providers we grant these privileges to be required to follow security practices to ensure that "need-to-know" concepts are followed, and all unauthorized parties are denied access to our non-public personal information. You must protect, Protected Financial Information (PFI) in both production and non-production environments; the regulations do not differentiate among these environments. Remember, masking is the safest way to protect sensitive information in non-production environments.

Special Regulatory and Industry Requirements, depending on the countries that your organization does business in, are there for the protection of sensitive data. Nearly, every country in the world has a data privacy law, so you should always do a comprehensive review of each of the pertinent laws and their supporting requirement [5].

The Health Insurance Portability and Accountability Act (HIPAA) Security and Privacy Standard defines administrative, physical, and technical safeguards to protect the confidentiality, integrity, and availability of electronic Protected Health Information (PHI), sometimes referred to as personal health information [6]. The Gramm-Leach-Bliley Act (GLBA) applies to financial institutions that offer financial products or services such as loans, financial or investment advice, or insurance to individuals.

Regenerating production like data is a cumbersome lengthy process, and due to dynamic nature of production data, it is not cost effective, for data regenerating script. Hence sensitive production data are viable solution to be shared in non-production environments for various reasons. Due to data security awareness and compliances, protecting sensitive data is required in all types of businesses. When non-production environment demands the production data for various purposes, the sensitive production data is masked and the masked copy of data is provided, to protect the sensitive data and to avoid data breach. Masking methods are designed in such away so the referential integrity of data is always maintained.

Some data masking techniques like pulling out and substitution, suppress important attributes in data sets that can affect knowledge discovery. Use of such data masking techniques will not be able to match the quality of data mining results. There are various data protection mechanisms available, some of them focus on specific type of data, or specific type of data formats, or specific type of domain.

Static and dynamic are the two types of masking. Static data masking is done on the Golden copy of the database. Production data is loaded as a copy in a separate environment, as per requirement of data in non-production environment subset of the original data is sliced, and apply data masking rules while data is in static state, until requirement is freeze slices from the golden copy is pushed to the desired environment to fulfill the requirement [1]. Dynamic data masking happens at runtime, dynamically, and on-demand so that there need not to create additional copy. Shuffling, encryption, swapping, randomization, and pulling out are the various types of data masking proposed by researchers.

2　Related Work

Several methods, meeting different requirements, have been proposed to improve data confidentiality. Among them, a handful of significant researches are presented in this segment;

As per 2018 Data Breach Investigations Report [7], "Data breaches aren't just a problem for security professionals. In financial sector, 79% external and 19% internal data breaches were noticed. Education sector noticed 81% external, 19% internal data breaches. Health care noticed 43% external and 56% internal data breaches.

Ravikumar et al. [8–10] proposed random replacement techniques and also represented a comparative study of various data masking techniques with proposed technique. The study is conducted in many domains like finance, banking, and security. The results showed in the proposed method are better in terms of performance and data security. The research is applied only on statistical parameters; it can also be implemented on analytical parameters.

Ahmed and Athreya [11] focused on challenges of masking. Authors also briefed on issues like reusability, transparency, and maintainability. The model proposed in this paper is named as FAST. This is a four-step comprehensive approach, FAST starts with discovering with sensitive data, accessing the masking algorithm, executing the high-performance mask algorithm, and integration testing of solution for quantity. Hossin and Sulaiman [12], in the work studied about the specially designed evaluation metrics for optimizing generative classifier. As per the researchers, accuracy is the most commonly used evaluation metric. Authors discussed the alternative evaluation metrics—error rate, sensitivity, specificity, precision, recall, *F*-measure, and many more. The paper had figured out important factors that should be considered while designing and constructing a new suitable PS classification algorithm.

In white paper by informatica [13], focus is on sound practices to secure data. To protect sensitive data (trade secrets, intellectual property, critical business information, business partners' information, or its customers' information) company policy, regulatory requirements, and industry standards need to be followed. Data classification policies are used to store sensitive data. The data is mainly classified into three categories—Public, Internal use only, and Confidential. Sensitive data occurs in two forms: structured and unstructured. The requirements for protecting this knowledge, information, and data must be clearly defined and reflect the specific requirements within the appropriate regulatory and industry rules and standards. Definite data elements must be specified as sensitive and should never be used within their actual values in development, QA, or other non-production environments. The data classification policy should clearly identify the requirements for data masking. Note that knowledge and information can be broken down into specific data elements. Organizations must define, implement, and enforce their data classification policy and provide procedures and standards to protect both structured and unstructured sensitive data.

Various methods to protect sensitive data are already available. There are various data masking algorithms are also present obviously each and every algorithm has some pros and cons.

3   Research Methodology

In this study, researcher proposed a model named as Non-Zero Random Replacement model, which is capable of handling various types of data like numeric, alphanumeric, and unstructured data (date, email address, and so on.). The model also covers various data formats like CSV, JSON, XML, and relational databases. The proposed data masking model is compared with existing masking techniques-Shuffling and substitution. The proposed model is tested on various secondary dataset from UCI machine repository of different domains (bank, financial, stock, social networking, and healthcare dataset). The experiments are developed in Python 2.7.14 environment.

Table 1 Sample size of dataset

S. No.	Database name	Type of dataset	No. of attributes	Sample size
1	Pima Indians diabetes data	Life	08	768
2	Dow Jones index data	Business	16	750
3	Facebook performance metrics data	Business	19	500
4	Credit card clients data	Business	24	30,000
5	Bank marketing data	Business	17	45,211

3.1 Sample Design

In this work, five data sets are considered (UCI repository) for testing and evaluating the work (UC Irvine machine learning repository). This research work considers real open-source data. UCI states that machine learning community donates data sets to UCI, for promoting research and development activities. The structure promoted by UCI to collect the data sets is suitable for machine learning applications. Till date, UCI has collected 468 data sets, belonging to various domains. Being data will be kept open for others to view, hence Donors have their own policy for sharing data and Donors decides the type of data to be shared and granularity of the data [14]. The datasets used in this research available under life and business category in UCI repository. Table 1, given below, represents the design of real datasets.

3.2 Research Flow Design

Figure 1 below represents the research flow design of the system. The system starts with reading the dataset as per the user choice. Associated input files will be read and based on the inputs masking rules will be applied. To check the quality and usability of the masked data, the original data and masked data will be classified. The classification results will be compared to various evaluation metrics.

The proposed process is divided into four phases of masking life cycle–*Profiling, Environment Sizing, Sensitive field discovery, and Non-Zero random Replacement masking*. The main objective of this work is to design an effective data masking tool to protect sensitive data from unauthorized access in production and non-production environment [15]. Each phase is described below:

- **Profiling**:

It is a process recognizes the details, that data can be derived, inferred, and predicted from other data. Profiling is executed in a variety of contexts and for a range of purposes. Through the profile, sensitive data can be inferred from other non-sensitive data.

Fig. 1 Research flow design

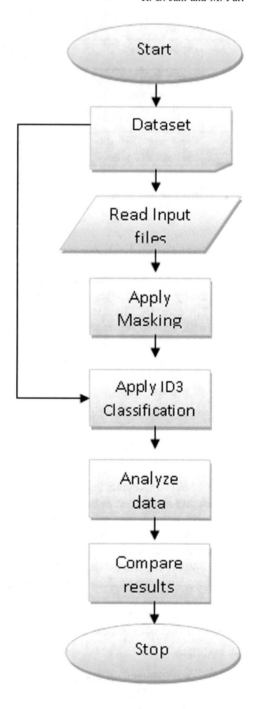

Fig. 2 Template of master masking file

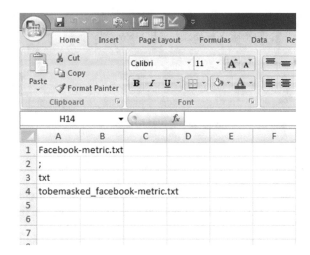

- **Extract into Hardened Environment**:

The masking environment should be hardened. The hardened environment is where access is protected by a workflow similar to production flow, software is licenses as in production, antiviruses are updated like in production, log files are generated so that in case if somebody log as well as audit trail, action can be a monitored. For masking process, we need size of $2\times$ of storage.

- **Sensitive Field Discovery**:

This step is the preliminary step of masking. In this process, two files will be prepared that will be used as input in masking process named as sensitive diligence file and master masking file. Format of master masking file (Refer Fig. 2):

First row: file name, second row: file separator, third row: type of file, and fourth row: output file name.

The template of **Sensitivity Diligence** files of CSV given in Fig. 3, below:

First row: all column names of file, second row: either "Y" to mask the column data of "N" to not mask the column data, third row: special masking condition, and fourth row: masking rule.

- **Non-Zero Random Replacement Masking**:

The proposed algorithm provides a uniform masking solution as it gives four options to the user to select the input file type [15]. Based on the input file type, the algorithm proceeds further for the next dataset and associated input files.

Start

Step1: Ask for type of data file: Enter the Data file to be Masked JSON[0]/XML [1]/ Database [2]/CSV [3] (Default set to JSON[0]).

Step 2: Read respective master masking file.

Step 3: Read Sensitive diligence file and then read the profiled data file.

Fig. 3 Sensitivity diligence files template (CSV)

Step 4: Execute Sensitivity Diligence algorithm.
Step 5: Apply non-zero masking algorithm.
Step 6: Exact cloned masked file is generated in output directory for CSV, JSON and XML. In case of relational database exact cloned masked table is created.
End

4 Result Analysis and Findings

As per the proposed research methodology in the previous section, this research considers five datasets of various sizes and domains. One dataset belongs to healthcare; second belongs to social media, and rests are of financial business domain. Validation happens sequentially on these datasets one by one, starting from chronological order. For each data set, four different files are evaluated-one original data, shuffled data, substitute data, and data masked using proposed model. The analysis is done using ID3 classification algorithm. Workflow model of the data analysis is given in Fig. 4.

The dataset is initially divided into training and test which is 80:20. Apply the ID 3 classification algorithm on the training data. This algorithm will generate ten different trees as it follows tenfold *K* approach. Out of the ten generated trees, best tree is selected using Classification Parameters (CP). The best tree is then applied to

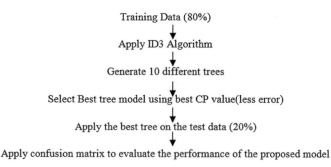

Fig. 4 Analysis workflow

20% test data; generated confusion matrix is used to evaluate the performance of proposed model in comparison of shuffling and substitution data masking techniques.

Comparing the resulting performance metric values: classification accuracy, precision, recall, *F*-measure, and specificity [5] delivered by the original data and masked data produced by three algorithms, it was found that they all figured within a percentage point of each other when measured within the control across all groups. The researcher has developed a uniform data masking architecture for various data types and comparison of non-zero random replacement method with other methods like Substitution and shuffling algorithm. The results strongly emphasize the proposed method as one of the typical methods for data masking with highest order of accuracy and usability. Observations also showed that original data and non-zero random replacement data masking algorithm, on an average, delivered matching classification performance metrics. This was a surprising finding.

Below Fig. 5, plots weighted classification performance metrics achieved by each of the masking algorithms used to predict the outcome variable within the data set. Values are weighted because they are based on the set of attributes determined to carry highest influence on the outcome class variable that they were used to predict. The proposed model and data substitution model give almost similar results when the data is small in size and contains only numeric values, but when the data is huge in size and also contains alphanumeric values then non-zero random replacement is showing results as close to the original data. Original data and non-zero random replacement delivered the highest values with almost matching 99% accuracy and precision. This was followed by substitution, and shuffling delivered the lowest performance metrics.

5   Conclusion and Future Work

There are various masking techniques available described earlier in this paper. Fact remains fact that each of them has associated pros and cons. As appeals for preserving privacy and confidentiality rises, need for masking techniques speed up quickly [16]. In this study, two most popular masking techniques substitution and shuffling compared with the proposed non-zero random replacement technique. The result showed by classification evaluation metric says that the accuracy produce by original financial data is 0.8213 with 95% CI, whereas accuracy produced by non-zero random replacement masking model is 0.8209 with 95% CI, accuracy produce by shuffled data is 0.820 with 95% CI, accuracy produce by substituted data is 0.8187 with 95% CI. Almost same pattern is followed in rest four datasets. This represents that the proposed model is giving 99% accuracy as of original data. Precision, recall, specificity, and *F*1 values also represent similar results as per original data compared to other masking algorithm like shuffling and substitution. The same set of result is showed when the model is applied on bank marketing

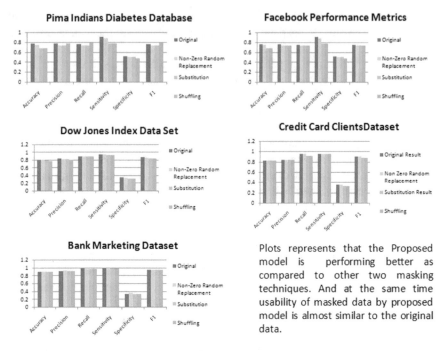

Fig. 5 Graphical analysis of various domains for masking methods

dataset. The results show that proposed model is a more efficient solution, which could be a valid alternative for protecting sensitive data. The proposed model is applied on structured (numeric and alphanumeric fields) and unstructured data (e-mail Id, date) as well. In both situations, the model provided usable data. The evaluation metrics used for the study also represented the proposed model is given better results, close to the original results as compared to other masking methods.

The scope of the study is limited to four popularly used datasets—CSV, JSON, XML, and relational database. The model is designed to mask text (numeric, alphanumeric) files only; image, video, and audio files are not covered in this study. The scope of the study is limited to classification evaluation metric. Space and time complexity were not considered which could be extended in future work.

Acknowledgements Thanks to the Research Center, Abacus Institute of Computer Applications, SavitribaiPhule Pune University for supporting me for pursuing research in the defensibly protect sensitive data topic. Thanks to, UCI center for Machine Learning Repository, for timely response to queries regarding data sets.

References

1. https://www.insurancejournal.com/search/?q=data+breach+2018. Accessed on 22 March 2019
2. https://www.insurancejournal.com/search/?q=databreach2019. Accessed on 22 March 2019
3. GK, R.K., Rabi, B.J., Manjunath, T.N.: A study on dynamic data masking with its trends and implications. Int. J. Comput. Appl. **38**(6), 0975–8887 (2012)
4. Gartner, Inc.: Market Guide for Data Masking. Marc-Antoine Meunier, AyalTirosh (2017)
5. Asenjo, J.C.: Data masking, encryption, and their effect on classification performance: trade-offs between data security and utility. Doctoral dissertation, Nova Southeastern University, Retrieved from NSUWorks, College of Engineering and Computing. (1010) http://nsuworks.nova.edu/gscis_etd/1010 (2017)
6. The keys to data protection, a guide for policy engagement on data protection, August 2018. Available at www.privacyinternational.org. Accessed on 02 Jan 2019
7. Data Breach Investigations Report. Available at: www.verizonenterprise.com/DBIR2018
8. Ravikumar, G.K., Justus Rabi, B., Manjunath ,T.N., Ravindra, S.H., Archana, R.A.: Design of data masking architecture and analysis of data masking techniques for testing. Int. J. Eng. Sci. Technol. (IJEST) **3**(6) (2011). ISSN: 0975-5462
9. Ravikumar, G.K., Manjunath, T.N., Ravindra, S.H., Umesh, I.M.: A survey on recent trends, process and development in data masking for testing. IJCSI Int. J. Comput. Sci. Iss. **8**(2) (2011). ISSN (Online): 1694-0814. www.IJCSI.org
10. Ravikumar, G.K., Manjunath, T.N.,Hegadi, R. (2011). Design of data masking architecture and analysis of data masking techniques for testing. Int. J. Eng. Sci. **3**
11. Oracle White Paper—Data Masking Best Practices (2013)
12. Hossin, M., Sulaiman, M.N.: A review on evaluation metrics for data classification evaluations. Int. J. Data Mining Knowl. Manag. Process (IJDKP) **5**(2), (2015)
13. Data privacy best practices for data protection in nonproduction environments. Informatica White Paper (2009)
14. http://archive.ics.uci.edu/ml/datasets.html
15. Jain, R.B., Puri, M., Jain, U.: A robust dynamic data masking transformation approach to safeguard sensitive data. Int. J. Future Revolution Comput. Sci. Commun. Eng. **4**(2). ISSN: 2454-4248
16. Ravikumar, GK., et al.: Experimental study of various data masking techniques with random replacement using data volume. Int. J. Comput. Sci. Info. Sec. **9**(8) (2011)
17. Compare IBM data masking solutions: InfoSphereOptim and DataStage. Options for depersonalizing sensitive production data for use in your test environments. John Haldeman (john.haldeman@infoinsightsllc.com) Information Management Consultant Information Insights LLC
18. Muralidhar, K., Sarathy, R.: Data shuffling-a new masking approach for numerical data. Manage. Sci. **52**(5), 58–670 (2006)
19. Howard, D., Howard, P.: The Sensitive Data Lifecycle: IBM vsInformaticavs MENTIS. In: Comparison Paper by Bloor, Publish date June 2018
20. Agarwal, R., Shandilya, S., Amandeep, Amit, S.: Survey on cloud computing and data masking techniques. Int. Res. J. Eng. Technol. (IRJET) **05**(04), 3697–3702 (2018). e-ISSN: 2395-0056, p-ISSN: 2395-0072
21. Muralidhar, K., Sarathy, R.: Masking numerical data: past, present, and future. In: Confidentiality and Data Access Committee of the Federal Committee on Statistical Methodology. Washington DC (2003)

Class Imbalance in Software Fault Prediction Data Set

C. Arun and C. Lakshmi

Abstract Classification has been the prominent technique in machine learning domain, due to its ability of forecasting and predicts capabilities it is widely used in various domains such as health care, networking, social network, and software engineering with enhancement of different algorithm. The performance of the classifier majorly depends on the quality and amount of data present in the training sample. In real-world scenario, the majority of training samples suffered from class imbalance problem, that is, most of the data samples belong to one particular category, i.e., majority class while very few represent the minority class. In this case, classification techniques tend to be overwhelmed by the majority class and ignore the minority class. To solve class imbalance problem people relay on the different kind of sampling techniques either by generating synthetic data or by concentrating on minority class samples, but those approaches have introduced adverse effect in the learnability. In this paper, we attempt to study different techniques proposed to solve the class imbalance problem.

Keywords Classification · Class imbalance · Machine learning · Majority · Minority · Sampling · Training

1 Introduction

In real world with the expansion of technology, applications generate vast amount of data that gets stored in data mart or data warehouse. Majority of the real-world data set are suffered due to class imbalance problem [1, 2], where sample of one particular class overrun the sample of other class, i.e., the number of instance present in one class outnumber the number of instance present in other class.

C. Arun (✉) · C. Lakshmi
School of Computing, SRM Institute of Science and Technology, Kattankulathur, India
e-mail: arunc@srmist.edu.in

C. Lakshmi
e-mail: lakshmic@srmist.edu.in

© Springer Nature Singapore Pte Ltd. 2020
S. S. Dash et al. (eds.), *Artificial Intelligence and Evolutionary Computations in Engineering Systems*, Advances in Intelligent Systems and Computing 1056,
https://doi.org/10.1007/978-981-15-0199-9_64

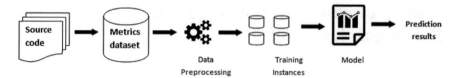

Fig. 1 Fault prediction in software component using machine learning model

Example rare instance data set such credit card fraud, medical diagnosis, network intrusion, oil spill detection [3–7], and fault information where present very few in number, because these events are very rarely occur during the life cycle. Software fault prediction is one such data set where the number of faulty information is very minimal compare to the non-faulty instances. In defect prediction, a sizeable number of software metrics, such as static code metrics, code change history, process metrics, and network metrics [8, 3, 9–13], have been used to construct different predictors for defect prediction. As the size and complexity of the system increases proportionally the number of defect prone modules get increased which makes the job of the tester is difficult to provide quality modules. The initial data set is constructed by collecting different categories of metrics from the software module. Collected data instance were pre-processed and given to class imbalance techniques which balance the sample ratio. Balanced data set is given to construct the learning model which builds the model to predict the future instance by splitting the data set into multiple subset of training samples. Learning algorithm will encounter huge difficulties while assigning correct label for the test data set when learning happens from imbalance class distribution. Researchers attempt various techniques to solve class imbalance problems such as sampling, synthetic data generation, algorithms, and ensemble techniques but these techniques solve the issue to particular extend only (Fig. 1).

2 Data Set and the Feature Set

The data set used for the class imbalance issue from the NASA [14] metric download program download from the promise software engineering repository. The data set contain 498 instance over 22 different attributes which broadly classified into five different major classes such as lines of code measure, McCabe metrics, Halstead measures, branch-count, and goal filed which either classify the instance as defective one or non-defective. The data set contain 498 instance out of which 449 instance belongs to the major category non-defective instances and 49 minority defective instances. The initial distribution of data instance where represented in Fig. 2.

Fig. 2 Initial distribution data samples from NASA metric program

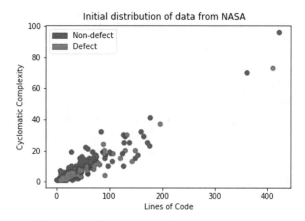

3 Performance Metrics for Class Imbalance

Class imbalance problem in the software fault prediction is two-class problem the class label of the minority class, i.e., faulty samples are positive label and the majority one, i.e., the non-faulty instances are the negative one. The performance is illustrated using confusion matrix where the first tuple represents the numbers of positive (TP) and negative (TN) samples that are correctly classified and the second tuple represents the number of misclassified positive (FP) and the negative (TN) instances (Table 1).

The common metric adopted to evaluate the performance of a learner for imbalanced data set is accuracy, which predicts the instance of majority class, i.e., negative sample compare to the prediction ratio of minority sample. Consider the scenario that the data set is extremely imbalanced and classifier classifies all the majority instances correctly and misclassifies the minority instances, the accuracy of the learner is very high, since the ratio of majority to minority samples in the data set. In the case of imbalance data set, accuracy cannot reflect the reliable behavior of the learner model.

The opted metric for the imbalanced data set is probability of false alarm (pf), which values ranges from zero to one closer to the zone better the performance (pf = 0). The pf measure is the ratio of false positive instance to the sum of false positive and the true negative instances, which indicate the performance of learning model by predicting the minority class samples accurately so as the majority class samples.

Table 1 Confusion matrix

	Predicted positive	Predicted negative
Positive	TP	FN
Negative	FP	TN

4   Class Imbalance Techniques

4.1   Sampling Techniques

Sampling techniques address the class imbalance issue by selecting a random set of samples from the original training set. Balance the training set by under sample the majority class, over sample the minority class by synthesize new minority samples and combine under and over-sampling techniques.

4.1.1   Under Sampling

Under sampling [15–17] is the simplest sampling techniques where number of instance in the majority class is reduced in-order to balance the set by either randomly or informatively removing some of the instance present in the majority class without making any change in the minority class. The NASA metric program data set was under sampled to 98 instances where each class contains 49 samples which are shown in Fig. 3.

4.1.2   Over Sampling

Over sampling [16, 18] techniques tend to balance the data set by increasing the numbers of samples in the minority class by generating new samples using repetition, bootstrapping, etc. Over sampling done either random over sampling where samples introduced by random repetition or focused over sampling in which the sample generated by samples based on the sample appears over the borderline. The distribution of sample present in both classes after over sampling is represented in Fig. 4.

Fig. 3 Distribution of data after using under-sampling technique

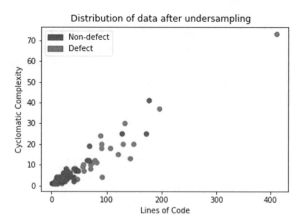

Fig. 4 Distribution of data after using over-sampling technique

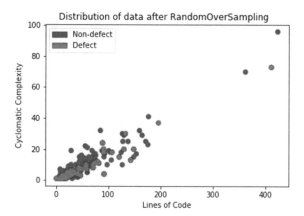

4.2 Synthetic Sampling

Balance the data set by artificial synthesize new samples belonging to the minority class till the data set becomes suitable to learning process.

4.2.1 SMOTE (Synthetic Minority Over-Sampling Technique)

SMOTE [19, 10] synthesis is a new samples by taking each minority sample and joining with nearest neighbors which falls on line segment. The minority class is over-sampled by taking each minority class sample and introducing synthetic samples along the line segments joining any or all of the k-minority class nearest neighbors. Synthetic samples are generated by computing the difference between the sample and its nearest neighbor and multiply the same by random number in the range of 0–1, and add this difference to the feature value of the original feature vector, thus creating a new feature vector for continuous features. For the nominal, features take majority vote between the feature vectors under consideration and its k-nearest neighbors for the nominal feature value. In the case of a tie, choose at random and assign that value to the new synthetic minority class sample. Figure 5 represents the sample distribution after SMOTE.

4.2.2 Borderline-SMOTE

Borderline-SMOTE [9] is an amended version where the minority samples near the borderline are over-sampled. For every sample in the minority class, calculate its m-nearest neighbors from the training set. If all the nearest neighbor belongs to the majority class then the minority sample is classified as danger sample and process is repeated for all the samples in minority class. Each sample from the danger set is picked to synthesize new data by computing the difference between samples and its

Fig. 5 Distribution of data
after using SMOTE

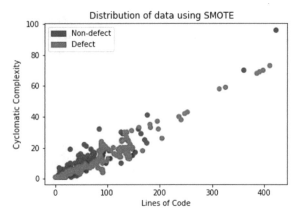

Fig. 6 Distribution of data
after using
borderline-SMOTE

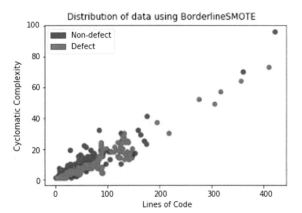

nearest neighbors and multiply the result by a random number range from 0–1. Figure 6 represents the sample distribution after borderline-SMOTE.

4.2.3 SMOTEBoost

SMOTEBoost [20] combines the techniques of synthetic minority over-sampling techniques and the standard boosting procedure to increase the distribution of minority sample to address the class imbalance. SMOTE [10] is introduced in each round of boosting will enable each learner to sample more of the minority class, and also learns better. Let D be the initial distribution of the training samples, modify it by generating N synthetic samples using SMOTE [10] algorithm. Train the data using weak learner for the obtained distribution and altered by emphasizing particular training examples. The distribution is updated to give wrong classifications higher weights than correct classifications. The entire weighted training set is given to the weak learner to compute the weak hypothesis. Figure 7 represents the sample distribution after SMOTEBoost. At the end, the different hypotheses are combined into a final hypothesis.

Fig. 7 Distribution of data after using SMOTEBoost

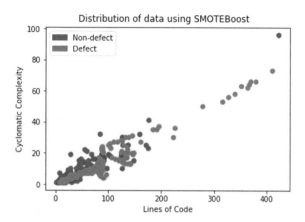

4.2.4 Cluster-Based Over Sampling

Cluster-based over sampling [21] is a two-step approach; in the first phase, minority class clustering is constructed and the second phase cluster connection is made. It is an iterative clustering approach where initial entire data structure is constructed as a single cluster then the cluster is split into number of small clusters based on TRIM criterion. Trim criterion is the product of precision, i.e., ratio of minority instance in a cluster and the generalization (minority samples). During the process majority instance outside the clusters are getting pruned in each iteration which results only minority instances. Second phase connect two samples of minority class from different cluster using proximity criterion to generate the synthetic sample using SMOTE algorithm, process repetitively continued till the user-specified percentage of sampling is satisfied. Figure 8 depicts the distribution of samples after clustered.

4.2.5 ADASYN (Adaptive Synthetic Sampling)

ADASYN [22] is based on the idea of adaptively generating minority data samples according to their distributions. Calculate the degree of imbalance in the training data set, if it is less than the pre-defined threshold compute the numbers of synthetic samples to be generated for the minority class. For each sample in minority set, find its k-nearest neighbors based on the Euclidean distance and calculate the number of synthetic samples needs to be generated for this minority sample. For each chosen minority sample identify its k-nearest neighbor and generate the synthetic sample by computing the difference in vector and multiply the same by λ is a random number: $\lambda \in [0, 1]$. Figure 9 represents the sample distribution after ADASYN.

Fig. 8 Distribution of
data after using clustered
technique

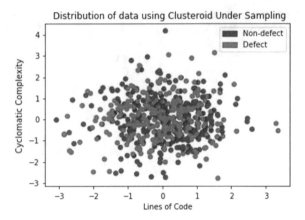

Fig. 9 Distribution data
samples after using ADASYN

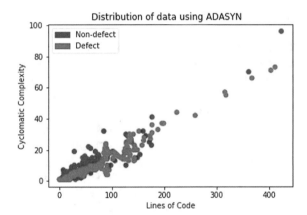

5   Algorithms

Researchers try to solve the data imbalance property on algorithmic level by
developing or extending the classification algorithms, without making any alteration on the original training set. One-class classifier model the system by considering the sample in the minority class and ignoring the majority class sample.
Biased random forest algorithm finds the nearest neighbor to find the minority
sample and its neighbor, and generates random forest and fed to random trees to
ends in result biased toward minority samples.

5.1 Ensemble

Ensembles [23, 24] are resolve the imbalance issue by increasing the classifier accuracy by making model to train using different classifiers and then combining the classifier output to a single label. Bagging is an ensemble technique where the new subsets of training samples are generated by randomly choosing the samples with replacement and use each set separately for training. Boosting is two-step approaches where results are boosted by applying cost function on the result obtained by selecting subset of samples from the original training data which might results in average performance.

5.2 Bagging (Bootstrap Aggregating)

Bagging [25, 16] is an ensemble method to reduce the high-variance in learning models. A sample is randomly selected from the training sample with replacement using bootstrapping. Bootstrapping create many subset of samples from the training set with replacement and calculate the mean of each subset. Calculate the estimated mean by computing the average of all subset means computed earlier. Model is created by selecting a subset of features and the samples and the feature which gives best split is selected and the process is repeated to create multiple models and each models training parallel and the label is computed as an aggregate function of all prediction results.

5.3 Adaptive Boosting

Ada-Boosting [16] adapts the boosting principle to generate highly accurate rule by combing many weak and inaccurate rules. In every round, each classifier is serially trained with the objective of appropriately classifying instances that were incorrectly classified in their previous round, so that the focus is primarily lies on the instances that were hard to classify in the previous case. The magnitude of focus is estimated by weight component, which is initially same for all the instances. After each iteration, the weight of the misclassified instances is getting increases on the other hand weight of the correctly classified instances is decreased. Gradient boosting builds the first learner on the training data set to predict the instances and estimate the loss and use the same to build the learner in the forthcoming stages. At every iteration, the residual of the loss function is computed using gradient descent method and used as target variable for subsequent iteration.

5.4   EasyEnsemble

The objective of the EasyEnsemble [15, 26], is to keep high efficiency of under-sampling but eliminating the possibility of ignoring the potential information contained the majority class samples. EasyEnsemble is a hybrid ensemble algorithm which adopts both boosting and bagging principle to predict the future values. It randomly generates multiple subset of samples (N1, N2, N3…) from the majority class samples (N). The size of samples present in each subset is the same as the number of minority samples present in the original data set. The new subset formed from the union of majority and minority class refers to bagging process and the resultant set is used to train using AdaBoost ensemble. The final ensemble is obtained by aggregating the results of all base learners in the AdaBoost ensembles.

5.5   BalanceCascade

BalanceCascade [15, 17] works in cascading style which initially adopted to improve the efficiency of face detection. BalanceCascade works in a supervised sequential manner, which tries to delete the samples from the majority class in an informative ways rather than randomized approach. The approach is to shrink the sample size in the majority class step by step. In each iteration a subset of sample from the majority class and the samples of minority class where the majority sample size is same as the size of the minority class which refers to the bagging process. The bagged result is used for learning using the AdaBoost classifier, in which the majority class samples that are correctly classified with higher confidences by the current trained classifiers are eliminated from the data set and will not consider for further iterations. The same ensemble approach is adopted for all the bagged set parallel and the final ensemble result is obtained by combining all the base learners in AdaBoost ensembles.

5.6   Random Forest Classifier

Random forest classifier [12, 24] is an ensemble classifier which creates a set of decision trees from a randomly selected subset of training data set and then aggregates the result of all the decision trees using voting mechanism in-order to predict the final class label. Say we have 1000 instances in the data set with as many as 10 features; it takes random sample of 100 instances and 5 randomly chosen features and build the tree. The tree construction process is repeated for about 10 times by randomly selecting different subset of instances and different subset of features. The final prediction can be the aggregate of each tree results, i.e., the number of occurrence of each leaf is counted as vote and the leaf which has the maximum number vote against its name is considered to be the final prediction result.

5.7 GASEN (Genetic Algorithm-Based Selective Ensemble)

Unlike usual neural network which trained either parallel or sequentially, GASEN [27] approach forms a set of training networks from which as subset of networks is chooses for consideration. GASEN assigns a random weight to each of the available neural networks at first. By applying genetic algorithm to evolve the weights, so they can characterize the fitness factor of joining the ensemble. Finally, it selects the networks whose weights are greater than the pre-defined threshold value that makes up the ensemble.

5.8 Improved SDA-Based Defect Prediction

Subclass discriminant analysis (SDA) [28] divides each class into set of subclasses, the optimal number of subclass for both defective and defect-free is determined using LOOT [29] (leave-one-out-test) criteria and skewed F-measure measure after computation the division is performed using nearest neighbor clustering (NNC) [28]. After sub-classing it computes the projective transformation using Eigen decomposing for non-zero values and covariance matrix. Obtain features with favorable discriminant ability from the original set of metrics. Finally, perform the prediction function using the random forest classifier.

6 Conclusion

Imbalanced learning becomes an exciting and active area of research in the field machine learning with respect fault prediction in the software components. Objective of the paper is to study the nature of class imbalance problem that exists in the fault prediction data set and to compare the different approaches that have been proposed to address the class imbalance issue. This paper provides a brief overview on various methods proposed to handle imbalanced problems based on sampling as under-sampling, over sampling, and synthetic sampling techniques such as SMOTE, Borderline-SMOTE, SMOTEBoost, Cluster-based over sampling, ADASYN, and ensemble techniques such as bagging, boosting, and GASEN. All the above-mention techniques attempt to solve the class imbalance learning problem by sampling approaches either by eliminating the instances from the major class and increasing instance count of minority class, whereas other sampling

C. Arun and C. Lakshmi

techniques increase the numbers of samples in minority class by generating synthetic samples and hence balance the data set. Ensemble techniques tend to solve the imbalance issue by using boosting, bagging approaches, thereby enhancing the performance of the weak learner to predict the label by combing the results of all the ensembles or its ensemble of ensembles.

References

1. Wang, S., Member, and Yao, X.: Multiclass imbalance problems: analysis and potential solutions. IEEE Trans. Syst. Man Cybern. B Cybern. **42**(4), August, (2012)
2. He, H., Ma, Y.: Imbalanced learning: foundations, algorithms, and applications, 1st edn. Wiley-IEEE Press, New York (2013)
3. Kubat, M., Holte, R.C., Matwin, S.: Machine learning for the detection of oil spills in satellite radar images. Mach. Learn. **30**(2–3), 195–215 (1998)
4. Rao, R.B., Krishnan, S., Niculescu, R.S.: Data mining for improved cardiac care. ACM SIGKDD Explor. Newsl **8**(1), 3–10 (2006)
5. Chan, P.K., Fan, W., Prodromidis, A.L., Stolfo, S.J.: Distributed data mining in credit card fraud detection. IEEE Intell. Syst. **14**(6), 67–74 (1999)
6. Clifton, P., Damminda, A., Vincent, L.: Minority report in fraud detection: classification of skewed Data. ACM SIGKDD Explor. Newsl **6**(1), 50–59 (2004)
7. Chan, P., Stolfo, S.: Toward scalable learning with nonuniform class and cost distributions. In: Proc. Int'l Conf. knowledge discovery and data mining, pp. 164–168 (1998)
8. He, H., Garcia, E.A.: Learning from imbalanced data. IEEE Trans. Knowl. Data Eng. **21**(9), 1263–1284 (2009)
9. Han, H., Wang, WY., Mao, BH.: Borderline-SMOTE: a new over-sampling method in imbalanced data sets learning. In: Advances in Intelligent Computing, pp. 878–887, Springer (2005)
10. Chawla, NV., Bowyer, KW., Hall, LO., Kegelmeyer, WP.: SMOTE: synthetic minority over-sampling technique. J. Artif. Intell. Res. **16**, 321–357 (2002)
11. Bhowan, U., Johnston, M., Zhang, M.: Developing new fitness functions in genetic programming for classification with unbalanced data. IEEE Trans. Syst. Man Cybern. B Cybern. **42**(2), 406–421 (2012)
12. Liaw, A., Wiener, M.: Classification and regression by randomForest. R News, **2**(3), 18–22 (2002)
13. He, P., Li, B., Liu, X., Chen, J., Mab, Y.: An empirical study on software defect prediction with a simplified metric set. J. Inf. Softw. Technol. **59**(C), March, 170–190 (2015)
14. Fault Prediction Data Set Repository. [online], available: http://promise.site.uottawa.ca/SERepository/datasets-page.html
15. Liu, XY., Wu, J., Zhou, ZH.: Exploratory undersampling for class-imbalance learning. IEEE Transactions on System Man Cybernetics-PARTB (2008)
16. Yap, BW., Rani, KA., Rahman, HAA., Fong, S., Khairudin, Z., Abdullah, NN.: An application of oversampling, undersampling, bagging and boosting in handling imbalanced datasets. In: Proceedings of the first international conference on advanced data and information engineering (DaEng-2013), pp. 13–22
17. Liu, X.Y., Wu, J., Zhou, Z.H.: Exploratory undersampling for class-imbalance learning. IEEE Trans. Syst. Man Cybern. B Cybern. **39**(2), 539–550 (2009)
18. Estabrooks, Jo, T., Japkowicz, N.: A multiple resampling method for learning from imbalanced data sets. Comput. Intell. **20**(1), 18–36 (2004)

19. Barua, S., Islam, M., Murase, K.: A novel synthetic minority oversampling technique for imbalanced data set learning. Neural Information Processing, vol. 7063. pp. 735–744. Springer, LNCS (2011)
20. Chawla, NV., Lazarevic, A., Hall, LO., Bowyer, KW.: Smoteboost: improving prediction of the minority class in boosting. In: Proc. European Conf. Principles and Practice of Knowledge Discovery in Databases, pp. 107–119, Dubrovnik, Croatia (2003)
21. Yen, S.J., Lee, Y.S.: Cluster-based under-sampling approaches for imbalanced data distributions. Expert Syst. Appl. Int. J. $36(3)$, 5718–5727 (2009)
22. He, H., Bai, Y., Garcia, EA., Li, S.: ADASYN: adaptive synthetic sampling approach for imbalanced learning. In: Proc. Int'l. J. Conf. Neural Networks, pp. 1322–1328 (2008)
23. Zhou, ZH.: Ensemble methods: foundations and algorithms. CRC Press (2012)
24. Sun, Z., Song, Q., Zhu, X., Sun, H., Xu, B., Zhou, Y.: A novel ensemble method for classifying imbalanced data. Pattern Recogn. $48(5)$, 1623–1637 (2015)
25. Breiman, L.: Bagging predictors. Mach. learn. $24(2)$, 123–140 (1996)
26. Galar, M., Fernandez, A., Barrenechea, E., Bustince, H., Herrera, F.: A review on ensembles for the class imbalance problem: bagging-, boosting-, and hybrid-based approaches. IEEE Trans. Syst. Man Cybern. C Appl. Rev. $42(4)$, July, 463–484 (2012)
27. Junfei, C., Qingfeng, W., Huailin, D.: An empirical study on ensemble selection for class-imbalance data sets. In: Proc. the 5th international conference on computer science & education, pp. 24–27 August, Hefei, China (2010)
28. Zhu, M., Martinez, A.M.: Subclass discriminant analysis. IEEE Trans. Pattern Anal. Mach. Intell. $28(8)$, 1274–1286 (2006)
29. Jing, X.Y., Wu, F., Dong, X., Xu, B.: An improved SDA based defect prediction framework for both within-project and cross-project class-imbalance problems. IEEE Trans. Softw. Eng. 43, 321–339 (2017)

A Study: Multiple-Label Image Classification Using Deep Convolutional Neural Network Architectures

S. Joseph James and C. Lakshmi

Abstract Multiple-label image classification is a kind of supervised learning approach, in which each image can belong to the set of multiple classes. The area of multiple-label image classification increases in recent years, in the field of machine learning. Multiple-label image classification has created significant scope from research people and it has been employed in an image labeling and object prediction tasks. Many conventional machine learning algorithms used for multiple-label image prediction by using relationship between objects in the image, co-occurrence objects, and rank between multiple instances in an image. Deep ConvNet architectures have proved increasing classification performance in image object classification in recent time. This study presents the task of multiple-label classification, the various literatures in area of multiple-label image classification using deep ConvNet architecture, also provides evaluation measures and performs a comparative analysis of the architecture models on PASCAL VOC dataset and various future challenges in the multiple-label image area.

Keywords Multiple-label image · Network architecture · Image annotation · Computer vision · ConvNet · VOC dataset

1 Introduction

Multiple-label image classification is a task of great significance in the field of computer vision and machine learning. Nowadays, a large amount of digital images present in the world to represent much information. Images are an important way for gaining information, appearing, and transmitting because of its visual

S. J. James (✉) · C. Lakshmi
Department of Software Engineering, SRM Institute of Science and Technology,
Chennai, India
e-mail: josephjs@srmist.edu.in

C. Lakshmi
e-mail: lakshmic@srmist.edu.in

© Springer Nature Singapore Pte Ltd. 2020 759
S. S. Dash et al. (eds.), *Artificial Intelligence and Evolutionary Computations
in Engineering Systems*, Advances in Intelligent Systems and Computing 1056,
https://doi.org/10.1007/978-981-15-0199-9_65

representation for recognizing the world. Image classification offers great support for image information retrieval and indexing because both of them require accurately label images. There are two ways for labeled classifications of images; they are single- and multiple-label classification. In single-label image classification, the input images are a collection of individual object labels. The single-object classification problem is divided into two types, binary- and multiple-class classification. In binary classification, the input image samples are categorized into one of two classes. In multiple-class classification, the input image samples belong to one class among a group of target class labels. Compared to binary-class and multiple-class problems, the case of multiple-label problem is more difficult.

Unlike single-label image classification problems, multiple-label image classification problems permit the input image to contain many class objects. The objective in multiple-label classification is to train from a collection of input images where each image belongs to many classes. Multiple-label classification was developed for automatic text classification and medical image diagnosis assistance. In recent times, the presence of multiple-label class classification tasks in real-world images attracted much research scope to this multiple-label image field. Multiple-label image classification is a much general and practical problem, due to the existence of majority of real-world images have objects from multiple categories. Multiple-label images and different type of objects are available at scattered positions with various sizes and angles. Further, the complexity of problem increases due to different compositions and relationships between objects in multiple-label images. Hence, it requires more labeled data to cover a variety of situations [1] (Figs. 1 and 2).

Fig. 1 Single label image from imagenet

Fig. 2 Multiple-label image from PASCAL VOC

2   Convolutional Neural Network (CNN)

ConvNet (CNNs) are similar to ANNs, they both are constructed with the use of neurons which does self-optimization during training process. Every neuron will get an input then computes a product and a nonlinear function computation. The network finally provides a single probability score which is weight value from the input image vector to final output class score. The last layer made up of softmax loss functions associated with the classes. The main difference between ConvNet and Artificial Neural Network is the ConvNet used in the image pattern recognition process. This is useful for extracting image dependent features from the layered architecture and makes the network applicable for image pattern recognition task. The computational complexity needed to compute image data is the biggest limitation of traditional ANN.

A. Convolutional Neural Network Architecture

The basic CNN architecture can be categorized into four areas. The input layer takes the pixel values of the raw input image samples. The convolution layer is used for calculating the output value of neurons that are connected to the input using their weights and the region connected to the input. The pooling layer will be used for downsampling along the spatial domain dimension of the input image, by minimizing the number of hyperparameters in the activation function. The fully connected layers produce object class probability scores used for classification. It is recommended that Rectifier Liner unit activation function may be used in hidden layers to increase classification performance.

B. Convolution Layer

The convolution layer is the core layer of a ConvNet which does most of the computations with use of learnable kernels. These kernels or filters are usually small in size and spreads all over the depth of input image. When the input data reaches a convolution layer, it convolves each filter across the spatial dimension space of the image input and produces a two-dimensional feature map [2].

C. Max Pooling Layer

Max pooling layer is used to reduce the size of the extracted feature image representation, thereby reduces the number of hyperparameter values which indirectly reduces the computation complexity of the architecture model. The max pooling layer operates over feature map and reduces its size using the max operation. Most of the CNN architecture uses max pooling with filter size of 2×2 and a stride value of 2 over the spatial domain of the image input. This operation reduces the feature map size to 1/4 of its original input image size [2].

D. Fully Connected Layer

Fully connected layers connect every neuron in one layer to every neuron in another layer. It is in principle the same as the traditional multi-layer perceptron neural network. The flattened matrix goes through a fully connected layer to classify the images. There is no pre-defined way for constructing ConvNet architecture. In general, CNN architecture contains convolution layers followed by max pooling layers repeatedly many times before providing the extracted feature maps to fully connected layers.

Based on the number of layers used and the order of stacking layers, we can categorize CNN model into two. The classic network architectures (LeNet-5, AlexNet, and VGG-16) were comprised simply of stacked convolution layers; modern architectures (Inception, ResNet, ResNext, DenseNet, and GoogleNet) explore new and innovative ways for constructing convolution layers in a way which allows for more efficient learning. CNN is efficient for image classification due to its automatic feature extraction and prediction using fully connected layer than softmax layer.

3 Multiple-label Images Classification

Multiple-label image classification [3] is a significant problem in the field of computer vision because most of the real-world images contain multiple objects that require multiple visual concepts for classification. Multiple-label image classification is challenging because finding the occurrences of multiple object classes requires full detailed knowledge of the image like associated classes with related regions and dependencies between classes [4]. Conventional models fine-tune multiple-label network architecture pre-trained on the single object label classification dataset images. But this may not be suitable for images which contain objects of multiple classes that are scattered at various locations, size. To overcome the issue, the multiple-label classification network contains many individual two-class classification tasks. So that one classifier will focus on one class. Though this method is very effective, the dependencies of multiple object classes significant for multiple-label object classification [3] are omitted.

Some earlier works [3, 5, 6] presented models to overcome the issue by extracting the class dependencies using recurrent neural network, and long short term memory structure attached followed by ConvNet-based architecture. Few image object detection methods [1, 7] are introduced a model by converting multiple-label image classification problem into multiple-object detection task. In these models, the individual object labels extracted using low-level image features [8]. The network architecture is trained to predict class scores of individual object labels, and the individual scores are combined to achieve multiple-label image classification task. These methods suffer repeated computation cost during testing phase. Their approach is practically not applicable for large size applications. In order to overcome these drawbacks, an effective multiple-label classification model requires three advantages that are locating meaningful object regions for individual class feature extraction and extracting dependable features from multiple-class objects, reduced computation, and annotation cost for the practical issue [9].

4 Pascal Voc Dataset

PASCAL VOC dataset is publically available data source of multiple-label image that contains 9963 images containing 24,640 annotated objects with 20 classes. The different classes are person, five animal classes, seven vehicle classes, seven object classes like bottle, chair, and table, etc. It provides image dataset for object class detection and classification thereby enables performance evaluation of various methodologies. Dataset can be downloaded from the following site http://host. robots.ox.ac.uk/pascal/VOC/.

5 Contributions

This paper will provide knowledge on important problems that come during the process of developing a multiple-label image classification system using deep learning and also will discuss possible solutions for them. The study will also provide how identified challenges and various network architectures can impact classification results of the system. This paper will also provide possible directions for future research and the architecture to follow for multiple-label object classification. This paper aims to suggest different multiple-label classification architecture approaches using deep learning and short description of benchmark multiple-label datasets and accuracy comparisons of different multiple-label image classification using deep learning algorithms architectures.

6 Models Based on Transfer Learning

Sermanet et al. [10] developed an integrated framework for object localization, region detection, and label classification using convolutional neural network. The model uses a multiple scales, sliding window method with substantial modification in convolutional neural network design that can be applied for classification. This model employed a novel deep learning method to localization combined with bounding boxes to improve detection accuracy. They showed that various object labels can be learned in parallel through single common network.

Wang et al. [11] presented a strong baseline model of deep ConvNet architecture for multiple-label image classification. Their model uses VGG16 and Resnet101 deep CNN architectures for classification. This model adapted adaptive pooling layer to the final convolutional layer feature maps so that different input sizes can be handled by same architecture. The final layer of single-label classification of ConvNet model is replaced with fully connected layer which contains N neurons equivalent to number of class labels.

Wan-Jin Yu et al. presented novel dual-path network architecture for multiple-label classification system. Their model consists of two paths, one is the multiple-object label network to extract the multiple-scale individual label-related object features through mapping the problem in a multiple-instance training framework. The multiple-instance architecture network converts the image into multiple object instances, and the subconcept layer [12] decomposes the class label into many subconcepts and computes the matching scores between the object instances and the subconcepts [12]. The multiple-instance max pooling layer integrates the subconcepts and selects the appropriate object regions. Secondly, the global prior's network used to extract global features from input images. This model experimented with PASCAL VOC 2007 and found it outperforms the ConvNet-SVM, ConvNet–RNN, and VGG-16 + SVM methods.

Hua et al. [13] presented a novel CA-BiLSTM architecture for multiple-label classification of aerial image objects. Their model is constructed with three building block elements. First one is feature extraction phase which captures high level features from input image data. Second element is the class label specific learning layer used to extract different class label specific features. Third element is the bidirectional long short term memory used for modeling class dependency and classifies multiple objects in a structured manner. Their model tested with UCM multiple-label data and DFC15 dataset and achieves F score increased by 0.044 compared with other methods.

Han et al. proposed a two-phase framework combining ConvNet transfer learning with web data augmentation [14]. In their model, the important feature maps of pre-trained model network transferred to classification task then the given dataset is augmented with images from internet for classification purpose. This model reduces the need of large dataset and also reduces the overfitting of CNN. Their mode uses Bayesian optimization for hyperparameter fine tuning. This model uses AlexNet, VGG, and ResNet which outperforms state of the art CNN.

Zhuang et al. [15] developed a deep ConvNet framework based on multiple-label learning for facial attribute classification. Their framework contains three subnets. First, the face identification model used to detect faces using fast region-based ConvNet. Second, multiple-label learning network is used for multiple attributes prediction using relation between facial attributes through attribute grouping. Third, transfer learning phase for unlabeled facial feature classification based on multiple-label transfer learning. Their framework outperforms many existing methods.

Niu et al. [16] proposed a novel framework for multiple-label image annotation. Their model contains a new two-branch deep ConvNet and a label prediction task. The two-branch deep network made up of very deep network branch for computing features and a feature integration network to integrate features with main network. The auxiliary prediction task is to estimate the appropriate label of an image (Fig. 3).

Oquab et al. [17] proposed a model that uses bounding box method information to generate training image labels. They proposed a model that generated approximately 500 square boundaries from the input images and make a single label on

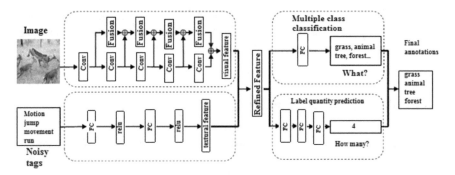

Fig. 3 An example of multiple-label image annotation framework architecture. Component used for feature extraction from input images. Component 2 extracting textual features from in out. Component 3 multiple-label classification which output the predicted multiple labels. Component 4 used to count the number of different classes and their individual class count

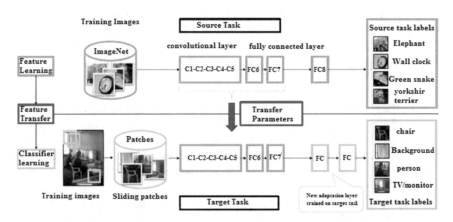

Fig. 4 An example of proposed [17] architecture. Source task network is a multiple-class classification network that provides single label as output out of multiple classes. Target network gets the tuned hyperparameters from source network recognize multiple labels and output multiple classes

individual boundary dependently to overlap with the ground truth bounding box [17] labels. Square window boundaries that intersect with more than one object are filtered out and once which are intersecting with any object labeled as a background (Fig. 4).

Sharif Razavian et al. [18] proposed a model called OverFeat that is pre-trained on ImageNet dataset, to get ConvNet activations features. Their method reuses the hidden layers pre-trained on dataset ImageNet for computing the intermediate level image representation of images in the dataset PASCAL VOC. Linear support vector machine classifier used over PASCAL VOC2007 dataset and 4096 feature representation extracted from the 22nd layer of OverFeat which achieved very good performance compared to other ConvNet-based models.

7   Model Based on Segmentation of Multiple-Label Image

Wei et al. [1] presented a flexible method called hypothesis CNN pooling model for multiple-label image classification task. The model uses ConvNet pre-trained on single-object label datasets and successfully used for multiple-label image classification. This model does not require any bounding box annotations for training multiple-label images. It can be easily adapted to multiple-label images. This model provides accuracy of 93.12% for VOC 2012 dataset.

Lingyun et al. [19] proposed a deep multiple-model for multiple instances multiple-label image classification architecture. This model utilizes the representation ability of CNN to find object instances automatically, whereas the existing methods relying on external object generators. The model incorporated images and textual context information for obtaining multiple-model object instance. Their model outperforms the baseline models like KISAR, VGG16, and Resnet152.

Zhu et al. were developed a multiple-label convolutional neural network [20] model to classify multiple attributes in an image. In their mode, first, a pedestrian object image is divided into multiple overlapping subparts that are given to multiple-label ConvNet independently for processing. Secondly, the output of multiple-label ConvNet [20] is combined in the fully connected layer followed by classification layer which is a collection of multiple-binary attribute classifier functions. This model significantly outperforms the support vector machine method for PETA dataset.

Wang et al. proposed image annotation or individual image object segmentation free framework called random crop pooling [21] for multiple-label object recognition. This framework contains three parts: a stochastic scaling component [21], a stochastic cropping component [21], and a weighted loss component [21]. The resizing ratios and crop segment locations are continuously changes while training the model because of this the network can see numerous changes within a single object then generate subimages of every image with random scale and location. This model used shared network which functions across different scaling and cropping instances. This approach dynamically generates the individual object regions from the multiple-label images during network training process instead of using fixed candidate images before training. This RCP framework experimented on PASCAL VOC 2007 and resulted in better performance (Fig. 5).

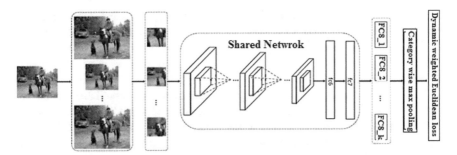

Fig. 5 An example of proposed random crop pooling ConvNet architecture. The input raw image is segmented with different size then the segmented regions are given to share ConvNet. The feature maps of different segments are integrated in the FC layer through max pooling. Euclidian loss function used for network training and classification. The different size segmentation approach provides different objects in the input image

8 Models Based on Correlated Feature Extraction and Learning

Yao et al. introduced a model for multiple-class object identification based on discriminative features available between various class objects. In this model, segmentation-oriented [22] annotations used as label individual differentiating features for each object class. The classification performance of this model improved due to combining the individual class feature and relative features available between different class objects. This model experimented on PASCAL VOC dataset using ConvNet VGG-16 architecture and achieved better performance than other models.

Yu et al. [23] developed a deep ConvNet architecture called Local–Global ConvNet (LGC). Their model contains two sublevels, one is local level [23] multiple-label classification which takes the object label co-occurrence for consideration in order to increase accuracy with graphs in the object label. Another one is global level [23] multiple-label classification which processes the global features of objects in the image. The final prediction is calculated by integrating local and global level classifier results. Their model achieves promising performance on PASCAL VOC 2007 and PASCAL VOC 2012.

Zhang et al. proposed a framework for multiple-label classification using ConvNet [24] with hidden semantics between different labels of same image. This model uses correlation between fine labels and the coarse labels to improve the performance. This model achieves 2.7% increase in accuracy compared to ConvNet model for single-label image.

Wang et al. [3] presented a framework which uses convolutional neural network combined with recurrent neural network for multiple-label image classification. This framework learns image label attachment to characterize the meaningful class dependency and the object–label relation. The ConvNet portion extracts semantic representation of images and the RNN portion models the label dependency and relationship and also used for computing probability of prediction path. The prediction path will be used for calculating prior probability of image labels. This framework applied to VOC PASCAL 2007 dataset and achieves better performance in terms of F score.

Vallet and Sakamoto [25] developed a framework multiple-label image prediction with multiple-label cost function for CNN and a prediction model which requires a single pass to produce prediction result. Multiple-nomial logistic regression (MLR) cost function used for generalizing error to multiple labels. The final cost function is the mean of error of all training samples in a batch. For prediction, this model uses global average pooling to produce confidence maps and convolution network applied to full images instead of crops to get the features map of each label in single forward.

Zhang and Wu [26] presented a strategy called multiple-label classification with label dependent features. Their model effectively captures the characteristics that are specific to each label to provide differentiating information for discrimination process. This was achieved by investigating characteristics of training samples relative to each class label like positive samples and negative samples using k-means clustering technique. The extracted specific label features are given to a collection of binary classification models for learning and prediction.

Li et al. [27] proposed a model to improve performance of pairwise ranking-based multiple-label classification task. Their model contains a loss function to compute pairwise ranking that is smooth and easier to optimize the rank. A label decision phase used to estimate optimal confidence threshold for visual object. The label decision used to decide which label should be included in the output. This model achieves better results on VOC 2007 dataset.

Jiang et al. proposed a calibrated RankSVM for multiple-label classification. The problems of classifying multiple-label image and ranking labels are solved using this technique. Using RankSVM, a virtual lookup label [28] used as a calibrated scale. The optimal coefficient of the virtual label (threshold) is implanted in learning process to obtain optimal position of the virtual labels adaptively. This makes difference between conventional RankSVM and calibrated RankSVM.

Ye et al. [29] proposed a model for multiple-label active learning classification based on cosine similarity. The correlations between all labels accurately determined and evaluated using cosine similarity. It reduces the labeling cost in multiple-label images. But only positive correlation between all labels is considered not negative correlations.

9 Models Based on Ensembling Technique

Wang et al. [30] presented a multiple-label image classification framework for surgical tool existence identification through laparoscopic device videos using VGGNet and GoogleNet. Their framework uses ensembling technique to get final prediction by computing mean of all the trained models. Their model achieves better performance than other methods and won first prize in the M2CAI challenge.

Dong et al. [31] proposed a model to transfer pre-trained deep neural network for small multiple instance multiple-label task. In their model, features extracted from each of the layers in the network architecture and add many binary classifiers [31] unit for multiple-label classification. This mode uses L1 norm regularized logistic regression to find important features for learning multiple-label classification.

Lee et al. introduced a framework based on boost ConvNet for image classification. This model uses ConvNet for extracting features from input images then applies AdaBoost algorithm on softmax [32] classifier layer to recognize image class labels. The ConvNet is pre-trained on image dataset, thereby it reduces lot of time to tune hyperparameters and extract feature maps from input data. At classification layer, ensembling technique of AdaBoost method used on softmax layer which makes improvement on probability score thereby increases the classification performance.

Wu et al. presented a new multiple-label forest algorithm [33] to train an ensembling-based hierarchical multiple-label classification trees to extract label dependency. This model develops hierarchical trees for the purpose of label transfer to find multiple-related labels. Discriminable labels are captured at higher levels of trees and they are moved to lower level nodes to find label which is harder to discriminate. The label dependency computation and final predictions are carried out by combining the labels in the hierarchy. This model achieves more label discriminating capability compared to other first-order multiple-label classifier models that treat a multiple-label classification problem as multiple separate and independent problems.

Zhang et al. [34] proposed a hierarchical ensembling-based model for classification of hierarchical multiple labels. This model computes the relationship between each class node and local prediction of all class nodes. This framework introduces the binary constraint model for optimal weight matrix learning. Their model uses fully associative label weight matrix [34] which requires large training samples and this may affect both computation and performance of the model.

Many of the models presented in this paper use transfer learning technique in different ways to increase their model performance results and used the work from the other studies to their benefit. Few conventional models use ensembling of base learning algorithm to improve the performance. Many convolutional neural network models employed pre-training, automated individual hypothesis detection, transfer learning, pre-defined object annotations, and image segmentation, etc. for improving classification performance. From the literature, we propose network architecture for multiple-label image classification based on hypothesis CNN pooling [1] and boosted CNN [35–37]. The hypothesis CNN pooling will be used for learning multiple object classes in single common network and boosted CNN will be used for improving classification performance.

Table 1 Comparision of classification performance results (accuracy in %) on PASCALVOC 2007		Bird	Car	Dog	Bike	Tv
	HPC-VGG [1]	97.1	95.8	97.3	97.1	91.5
	MIML-CNN [19]	75.1	81.4	79.3	–	–
	DELTA [12]	95.8	94.5	96.7	95.1	87.9
	PRE-1512 [17]	88.2	90.7	92.1	82.9	79.8
	LGC [23]	90.8	89.5	88.1	90.8	77.8
	CNN-RNN [3]	94.2	89.1	92.4	83.1	78.6

10 Performance Analysis of Various Models

Performance comparison of various deep convolutional neural network architectures that are applied on PASCAL VOC dataset listed in Table 1. We have taken five objects for performance comparison and found HPC-VGG [1] and DELTA [12] architectures provide better accuracy performance comparatively than other architectures.

11 Challenges in Multiple-Label Classification

Most of the convolutional neural network-based model and conventional models require lots of labeled multiple-label images which are not available in real-world scenario. Hence, a framework that learns from unlabeled data or not pre-segmented images needed to apply it real-world problems. Deep CNN architecture has many hidden layers for processing large volume of data, thereby generates very large number of hyperparameters which are optimized to make learning efficient. In this place, evolutionary algorithms like genetic algorithm particle swarm optimization can be used to find optimal hyperparameters. Convolution is the core process in CNN which extracts features map from raw images. Hence, variable size filter can be used to convolve input images to get effective feature maps that are used for prediction.

12 Conclusion and Future Study

In this transcript, a study on various multiple-label classification using deep learning algorithms, their working methodology, working principle, and performance evaluation measures have been presented. A collection of recently developed CNN architecture models has been presented based on their working methodology and a comparative accuracy performance analysis has been reported. This study provides information on the relationships between different algorithm architectures and directs path for future research work. A few future challenges in the multiple-label image classification using deep CNN architecture have also been identified: (i) The computational resource requirements and complexities of many of the algorithms architecture ask for more efficient architecture to achieve scale independence. (ii) A collection of image classification problems which need to reuse a pre-trained model or to train a new model with small changes in the architecture to find solution to many image classification problems. (iii) Having modern network architecture system may not necessarily increase end classification accuracy performance. (iv) What kind of network architecture to use is entirely depends on the particular task and hardware resource availability. (v) There is no empirical study on how and why the performance differs over different data and network architecture properties, and such a study would be useful for multiple-label image classification algorithms for any domain, (vi) The deep learning algorithm architectures require high configured GPU-based system to train large dataset; in general, a lightweight algorithm which does not require much hardware and parallel processing power may be helpful in future.

References

1. Wei, Y., Xia, W., Lin, M., Huang, J., Ni, B., Dong, J., Zhao, Y., Yan, S.: HCP: A flexible CNN framework for multi-label image classification. IEEE Trans. Pattern Anal. Mach. Intell. (2015)
2. O'Shea, K., Nash, R.: An introduction to convolutional neural networks. arXiv:1511.08458v2. Dec (2015)
3. Wang, J., Yang Y et al.: CNN-RNN: a unified framework for multiple-label image classification. arxiv:1604.04573. (2016)
4. Liu, Y., Sheng, L. et al. Multiple-label image classification via knowledge distillation fromweakly-supervised detection. arXiv:1809.05884v2. Feb (2019)
5. Devkar, R., Shiravale, S.: A survey on multiple-label classification for images. Int. J. Comput. Appl. **162**(8), 39–42 (2017)
6. Yang, H., Zhou, J.T., Cai, J.: Improving multiple-label learning with missing labels by structured semantic correlations. In: European conference on computer vision, 835–851 (2016)
7. Zhang, J., Wu, Q., Shen, C., Zhang, J., Lu, J.: Multi-label image classification with regional latent semantic dependencies. IEEE Trans. Multimedia 99 1–11 (2018)
8. van de Sande Koen, E.A., Uijlings, J.R.R., Gevers, T., Smeulders, A.W.M.: 2011. Segmentation as selective search for object recognition. In IEEE International Conference on Computer Vision. 1879–1886
9. Liu, F., Xiang, T., Hospedales, T.M., Yang, W., Sun, C.: Semantic regularisation for recurrent image annotation. In: IEEE conference on computer vision and pattern recognition. 2872–2880 (2017)
10. Sermanet, P., Eigen, D., Zhang, X. et al.: Integrated recognition, localization and detection using convolutional networks. arXiv:1312.6229v4. Feb (2014)
11. Wang, Q., Jia, N. et al.: A baseline for multiple-label image classification using an ensemble of deep convolutional neural networks. arXiv:1811.08412v2 [cs.CV] 16 Feb (2019)
12. Yu, W.J., Chen, Z.D., Luo, X., Liu, W., Xu, X.S.: DELTA: a deep dual-stream network for multi-label image classification. Pattern Recognition (2019)

13. Hua, Y., Mou, L., Zhu, X.X.: Recurrently exploring class-wise attention in a hybrid convolutional and bidirectional LSTM network for multi-label aerial image classification. ISPRS J. Photogrammetry Remote Sens. (2019)

14. Han, D., Liu, Q., Fan. W.: A new image classification method using CNN transfer learning and web data augmentation. Expert Syst. Appl. (2018)

15. Zhuang, N., Yan, Y., Chen, S., Wang, H., Shen, C.: Multi-label learning based deep transfer neural network for facial attribute classification. Pattern Recogn. (2018)

16. Niu, Y., Lu, Z., Wen, J.R., Xiang, T., Chang, S.F.: Multi-modal multi-scale deep learning for large-scale image annotation. IEEE Trans. Image Process. (2018)

17. Oquab, M., Bottou, L., Laptev, I., Sivic, J.: Learning and transferring mid-level image representations using convolutional neural networks. In: IEEE conference on computer vision and pattern recognition, Jun 2014, Columbus, OH, United States. (2014)

18. Sharif Razavian, A., Azizpour H. et al.: CNN features off-the-shelf: an astounding baseline for recognition. IEEE Explore (2014)

19. Song, L., Liu, J., et al.: A deep multiple-modal CNN for multiple-instance multiple-label image classification. IEEE Trans. Image Process. **27**(12) (December 2018)

20. Zhu, J., Liao, S., Lei, Z., Li, S.Z.: Multi-label convolutional neural network based pedestrian attribute classification. Image Vis. Comput. (2017)

21. Wang, M., Luo, C., Hong, R., Tang, J., Feng, J.: Beyond object proposals: random crop pooling for multi-label image recognition. IEEE Trans. Image Process. (2016)

22. Yao, C., Sun, P., Zhi, R., Shen, Y.: Learning coexistence discriminative features for multi-class object detection. IEEE Explore (2018)

23. Yu, Q., Wang, J., Zhang, S., Gong, Y., Zhao, J.: Combining local and global hypotheses in deep neural network for multi-label image classification. Neurocomputing (2017)

24. Zhang, G., Chen, L., Ding, Y.: A multi-label classification model using convolutional netural networks. In: 2017 29th Chinese control and decision conference (CCDC) (2017)

25. Vallet, A., Sakamoto, H.: A multiple-label convolutional neural network for automatic image annotation. J. Inf. Process. **23**(6), 767–775 (2015)

26. Zhang, M.L., Wu, L.: Lift: multi-label learning with label-specific features. IEEE Trans. Pattern Anal. Mach. Intell. (2015)

27. Li, Y., Song, Y. et al.: Improving pairwise ranking for multiple-label image classification. IEEE Explore (2017)

28. Jiang, A., Wang, C., Zhu, Y.: Calibrated rank-SVM for multi-label image categorization. In: 2008 IEEE international joint conference on neural networks (IEEE world Congress on Computational Intelligence), (2008)

29. Ye, C., Wu, J., Sheng, V.S., Zhao, P., Cui, Z.: Multi-label active learning with label correlation for image classification. In: 2015 IEEE international conference on image processing (ICIP), (2015)

30. Wang, S., Raju, A. et al.: Deep learning based multiple-label classification for surgical tool presence detection in laparoscopic videos. IEEE Explore (2017)

31. Dong, M., Pang, K., Wu, Y., Xue, J.H., Hospedales, T., Ogasawara, T.: Transferring CNNS to multi-instance multi-label classification on small datasets. In: 2017 IEEE international conference on image processing (ICIP) (2017)

32. Lee, S.J., Chen, T., Yu, L., Lai, C.H.: Image classification based on the boost convolutional neural network. IEEE Explore (2018)

33. Wu, Q., Tan, M., Song, H., Chen, J., Ng, M.K.: ML-forest: a multilabel tree ensemble method for multi-label classification. IEEE Trans. Knowl. Data Eng. (2016)

34. Zhang, L., Shah, S.K., Kakadiaris, I.A.: Hierarchical multi-label classification using fully associative ensemble learning. Pattern Recogn. (2017)

35. Lee, S.J., Chen, T. et al.: Image classification based on the boost convolutional neural network. IEEE Access, December (2017)

36. Shaikh Akib Shahriyar, Kazi Md. Rokibul Alam, Sudipta Singha Roy, Yasuhiko Morimoto.: An approach for multi label image classification using single label convolutional neural network. In: 2018 21st international conference of computer and information technology (ICCIT) (2018)
37. Oquab, M., Bottou, L., Laptev, I., Sivic, J.: 6 Learning and transferring mid-level image representations using convolutional neural networks. In: IEEE conference on computer vision and pattern recognition, 1717–1724 (2014)

Video Prediction for Automated Control of Traffic Signals Through Predictive Neural Network

C. Lakshmi and Vipanchi Chacham

Abstract In today's world of automated things, safety plays a very important role especially when it is concerned with the lives of people. Deep-learning algorithms are being used to help solve various tasks. But, using deep learning for automating traffic signals by using predictive neural network is a new challenge. In this project, the proposed model uses two predictive neural network algorithms. It is trained for next-frame prediction which learns to predict the future frames in a video sequence. This prediction can be used for improvising driving through automated control of traffic signals. A traffic simulator is made using MATLAB software. A four-way junction exists where the number of days can be provided depending on the dataset needed. An instance image of each hour shows the number of cars which appear in each direction of the road every hour. The learning algorithm uses this dataset to predict the future number of vehicles and hence, control the traffic. Using this model, the aim is to solve the traffic controlling system and make it very efficient by properly controlling and saving time through automation.

Keywords Machine learning · Deep learning · Neural network · Gradient boost · Random forest

1 Introduction

Artificial neural networks are generally unrefined electronic models in light of the neural structure of the brain [1]. The brain essentially learns from practice and experience. Neural networks process data like the human mind does. The system is made out of an extensive number of exceedingly interconnected processing components (neurons) working in parallel to resolve a particular issue. A convolutional neural system (CNN) is a feed-forward artificial neural network that has effectively

C. Lakshmi (✉) · V. Chacham
Department of Software Engineering, SRM Institute of Science and Technology,
Chennai, India
e-mail: lakshmic@srmist.edu.in

© Springer Nature Singapore Pte Ltd. 2020 775
S. S. Dash et al. (eds.), *Artificial Intelligence and Evolutionary Computations
in Engineering Systems*, Advances in Intelligent Systems and Computing 1056,
https://doi.org/10.1007/978-981-15-0199-9_66

been applied to analyze visual imagery. Convolutional systems were inspired by natural processes [2] in which the network design between neurons is propelled by the association of the animal visual cortex.

CNN has been used in wide variety of applications such as image recognition, video analysis, NLP, checkers, etc. The ImageNet Large-Scale Visual Recognition Challenge is a benchmark in object classification and detection, with millions of images and hundreds of object classes. In the ILSVRC 2014, [3] a large-scale visual recognition challenge, almost every highly ranked team used CNN as their basic framework. The winner GoogLeNet [4] (the foundation of DeepDream) increased the mean average precision of object detection to 0.439329, and reduced classification error to 0.06656, the best result to date.

Road traffic safety alludes to the techniques and measures used to keep road users from being seriously harmed. As indicated by the World Health Organization, road traffic injuries caused an expected 1.26 million demises worldwide in the year 2000. The normal rate was 20.8 for every 100,000 individuals, 30.8 for males, and 11.0 for females [5]. Measures can be taken so as to lessen the quantity of injuries. Traffic signals not only make car traffic a lot safer but also pedestrian traffic. They help decrease the quantity of accidents and make impacts at crossing points significantly less frequent. They play an essential and fundamental part with regards to security in our regular day to day life.

To address the above traffic control issues and problems, safe automation is required which can be acquired through deep-learning techniques.

2 Survey Details

In paper [6], a traffic-signal control system referred to as RHODES is discussed. The system utilizes a control architecture that predicts traffic flows, at appropriate resolution levels (individual vehicles and platoons) to enable proactive control and utilizes a data structure and computer approaches that allow for solution of the sub problems.

In paper [7], unlabeled data is given huge importance since not everything that is visible to the camera can be labeled and it is impractical to do so. To learn about physical object motion without labels, development of an action-conditional video prediction model is made that explicitly models pixel motion, by predicting a distribution over pixel motion from previous frames. By learning to transform pixels in the initial frame, this model can produce video sequences more than 10-time steps into the future, which corresponds to about 1 s.

There exist lot of difficulties in the automation of the traffic control system such as non-communication amongst vehicles and traffic lights: There will be optimal functioning of vehicles if traffic systems and vehicles could directly communicate with each other. Right now, most vehicles can not interact with the systems that experts use to control traffic lights. The cost plays a noteworthy part. Due to the high cost, the adoption of new technology becomes a hindrance. As the number of junctions in a traffic system varies, the mechanism of how it is being controlled has to vary. And since this costs more, due to the unavailability of funds, these projects might not be carried on.

3 Proposed Methodology

The main method here is done throughout a virtual traffic simulator which is made using MATLAB. The environment consists of a four-way junction. It is intended that a 360° camera is placed on top of the traffic signal. This camera is used to capture everything that happens on all the four roads in all four directions. Through MATLAB, this simulator generates different number of cars on each side of the road every hour. An instance image of each hour shows the number of cars which appear in each direction of the road. Each road is run on different time-based functions. After running the code for 1000 days, this data is saved in a file. The simulator is designed to collect data amounting to a specified number of days (Fig. 1).

Two algorithms are mainly compared using the data generated by the simulation. These are gradient boost and random forest.

Gradient boosting is an ML technique used for problems such as regression and classification, which produces a prediction model.

Random forests are a group-learning strategy for classification and regression problems too, that develops a lot of decision trees at training time and yielding the mean prediction of the individual trees. These correct the propensity of overfitting to their training set.

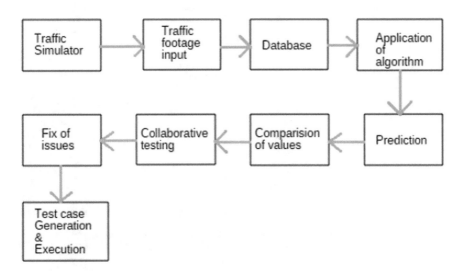

Fig. 1 Data flow diagram—level 2

The dataset has five columns. The first column represents the current day. Second column represents the number of cars in first to 24th hours on the east side of the road. The third column represents cars in the north direction, and fourth represents the cars on the west side and 5th column represents cars in the south direction (Fig. 2).

Each road is set with four different equations. The equation for the east side road is

$$(4 \times H + 2 + \text{rand}(5)) + (3 \times \text{sind}(7 \times D_{\text{current}}))$$

The dataset is then run and based on the equations, the number of cars is generated. Now, this dataset is generated for around 1000 days. The learning algorithm is trained with this dataset. The prediction of the number of cars depending on the hour is done when the two algorithms, gradient boosting and random forest are used to run on the dataset.

Fig. 2 Day 1 dataset

	A	B	C	D	E
1	1	9	9	7	16
2	1	11	13	12	17
3	1	18	14	18	20
4	1	20	18	22	24
5	1	26	20	27	27
6	1	30	25	32	30
7	1	33	28	37	32
8	1	39	29	42	37
9	1	39	33	47	39
10	1	43	36	53	40
11	1	51	39	58	43
12	1	55	43	62	49
13	1	56	44	67	51
14	1	62	49	72	52
15	1	65	50	77	56
16	1	68	53	83	60
17	1	75	57	87	64
18	1	76	61	92	66
19	1	81	64	98	69
20	1	87	67	103	72
21	1	90	69	107	75
22	1	91	72	112	76
23	1	97	75	118	82
24	1	101	78	123	84
25					

Fig. 3 Background of simulator

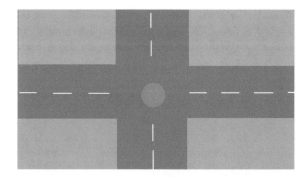

Fig. 4 Instance of 6th hour in a day

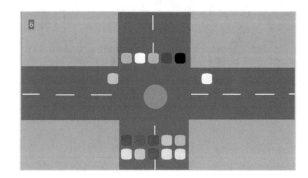

4 Experimental Setup

The setup consists of a 360° camera which is placed on the traffic signal. This helps in capturing all the roads in all the directions (Fig. 3).

Each car is represented randomly by different colors. On the top left corner of the simulator, there is a small box which shows the hour of the day. So when the simulator runs, the hour keeps changing from 1 to 24 and the number of cars in each hour in each direction also keeps changing.

If the number of days is given as 2 or more, the traffic instance again starts from the 1st hour after the 1st day's 24th hour to start the second day and so on. All of this data is saved in separate files named after each day such as "Day 1," "Day 2," etc. (Fig. 4).

5 Experimental Results and Analysis

There were mainly two algorithms that were used. One is gradient boosting algorithm which is one of the most popular and widely used linear regression algorithms. All the four directions of the junction's 100-h dataset were collected by running the code. The east road's dataset was run using the gradient descent algorithm first. This was plotted with the actual dataset. This was done in a similar manner for 1000 days and plotted again (Figs. 5 and 6).

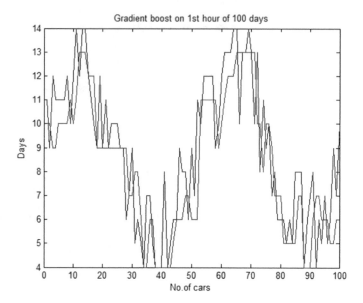

Fig. 5 Gradient boosting plot for 100 days

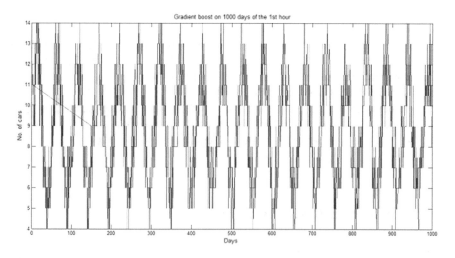

Fig. 6 Gradient boosting plot for 1000 days

Using the gradient boosting algorithm, the training dataset was run. The two graphs of the actual and predicted data were plotted for the east road. The graphs seemed to have pretty close values and gave only 25% error. The blue line represents the actual dataset and the red represents predicted gradient descent values.

When the dataset was run on Weka tool for 1000 days and it was found that it was 77% accurate (Fig. 7).

The second algorithm that was used was random forest. In a similar manner as gradient boosting, the 100 and 1000 days dataset were run using random forest method and then the graph was plotted (Fig. 8).

When the dataset was run on Weka tool for 1000 days and it was found that it was 82% accurate (Fig. 9).

From the above predictions, it was found that Random Forest has better accuracy and prediction capability than that of Gradient Boost (Table 1).

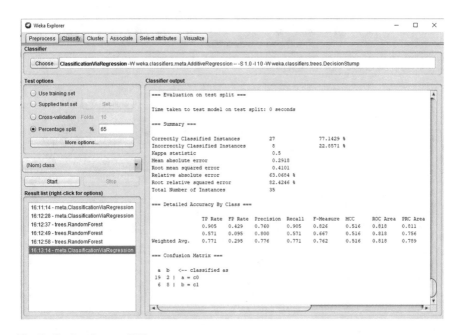

Fig. 7 Gradient boost—77% accuracy

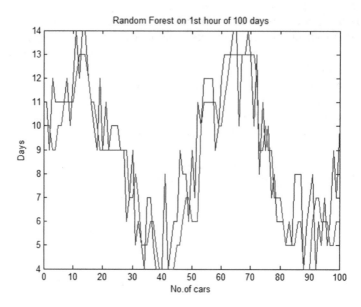

Fig. 8 Random forest plot for 100 days

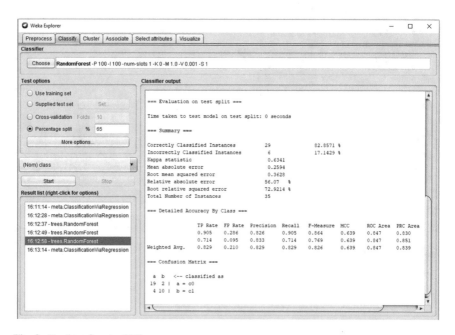

Fig. 9 Random forest—82% accuracy

Table 1 Performance of the models for different evaluation metrics

Classification model and evaluation scores	Gradient boost (%)	Random forest (%)
Precision	77.6	82.9
Recall	77.1	82.9
F measure	76.2	82.6
Error	29	25

6 Conclusion and Future Scope

In this paper, we have considered methods of automation of the traffic signals. We have used two algorithms, gradient boosting and random forest. A traffic simulator was made which was used to create dataset. These two algorithms predict the number of cars on any side of the road provided that an hour of the day is given. When this is predicted, the traffic signal can be easily controlled and made automated. After running both the algorithms, it was found that random forest is more accurate in the prediction of the cars.

This method is very beneficial than the other existing once as all the others use costly movement detectors, sensors, etc. But this method only needs one 360° camera and nothing else. So this method is very cost effective.

The future work would be focusing on analyzing the performances of each algorithm based on the type of functions used. By comparing the results to the practically observed traffic flow models, optimal algorithms can be chosen to train systems for traffic prediction and control.

References

1. Maind, S.B., Priyanka: Research paper on basic of artificial neural network. Int. J. Recent Innov. Trends Comput. Commun. 2(1), 96–100 (2014). ISSN: 2321-8169
2. Matusugu, M., Mori. K., Mitari, Y., Kaneda, Y.: Subject independent facial expression recognition with robust face detection using a convolutional neural network (PDF). Neural Netw. 16(5), 555–559 (2003). https://doi.org/10.1016/s0893-6080(03)00115-1. Accessed 17 Nov 2013
3. ImageNet Large Scale Visual Recognition Competition 2014 (ILSVRC, 2014). Accessed 30 Jan 2016
4. Szegedy, C., Liu, W., Jia, Y., Sermanet, P., Reed, S., Anguelov, D., Erhan, D., Vanhoucke, V., Rabinovich, A.: Going deeper with convolutions. Comput. Res. Repository (2014). arXiv:1409.4842
5. WHO: Global status report on road safety 2015 (PDF) (official report). Geneva: World Health Organisation (WHO), pp. vii, 1–14, 75ff (countries), 264–271 (table A2), 316–332 (table A10) (2015). ISBN 978 92 4 156506 6. Accessed 27 Jan 2016. Tables A2 & A10, data from 2013
6. Mirchandani, P., Head, L.: Rhodes: a real-time traffic signal control system: architecture, algorithms, and analysis
7. Finn, C., Goodfellow, I., Levine, S.: Unsupervised learning for physical interaction through video prediction

On (*R*, *S*)-Norm Entropy of Intuitionistic Fuzzy Sets

Tanuj Kumar, Vipin Kumar Verma and Sohan Tyagi

Abstract Fuzzy sets have been widely used to study vague phenomena. More general aspects of fuzzy sets lead the concept of intuitionistic fuzzy sets. The intuitionistic fuzzy entropy measures play a crucial role in quantitative measurement of vagueness of phenomena. In this manuscript, we propose a (*R*, *S*)-norm-based intuitionistic fuzzy entropy that quantifies the extent of fuzziness of an intuitionistic fuzzy set. Then, the correctness of the proposed entropy measure is verified by checking its axiomatic requirements.

Keywords Fuzzy sets · Intuitionistic fuzzy sets · Intuitionistic fuzzy entropy · *R*-norm fuzzy entropy · *R*-norm entropy · (*R*, *S*)-norm

1 Introduction

Fuzzy information measures are used as a computational measures of the uncertainty inherited in fuzzy variables. In the literature, it is well observed that the non-probabilistic vague or imprecise information can be deal using the fuzzy sets. The fuzziness is a kind of uncertainty inherited in the object of concern in the absence of knowledge. To measure the fuzziness in fuzzy set, the notion of fuzzy entropy is initially proposed by the Zadeh [21] in a probabilistic environment. He also explained the entropy of a fuzzy set or fuzzy event as a weighted form of Shannon's entropy [14]. Further, many definition of fuzzy entropy in view of different applications has been introduced by the various researchers, e.g., De Luca

T. Kumar (✉) · V. K. Verma · S. Tyagi
Department of Mathematics, SRM Institute of Science and Technology, NCR Campus, Ghaziabad, Uttar Pradesh, India
e-mail: tanujkhutail@gmail.com

V. K. Verma
e-mail: 1979.vipin@gmail.com

S. Tyagi
e-mail: sohanridhi@gmail.com

© Springer Nature Singapore Pte Ltd. 2020
S. S. Dash et al. (eds.), *Artificial Intelligence and Evolutionary Computations in Engineering Systems*, Advances in Intelligent Systems and Computing 1056, https://doi.org/10.1007/978-981-15-0199-9_67

and Termini [6], not used probabilistic approach to explain the fuzzy entropy of a fuzzy set. Yager [16], Kaufmann [9], Kosko [10], Pal and Pal [12], Chen et al. [5], Yao et al. [19], Ning et al. [11] and other authors also identified that uncertainty in the language is inherited from the linguistic vagueness and it is disappeared, when the fuzzy set is 'equally likely'. Nowadays, the study of new information measures based on fuzzy set theory leads lot of applications in communication and artificial intelligence systems, machine-learning process, decision support systems, medical image information processing, data analysis, etc. Hence, these applications encourage the scientist or researchers to establish new entropy and divergence measures including possible parametric generalizations of existing measures.

Further, extensions of fuzzy sets have been investigated by the Atanassov [1, 2] and he proposed the notion of intuitionistic fuzzy sets. Further, Szmidt and Kacprzyk [15] introduced the distance and entropy measures for IFSs with the help of geometric view of IFS. Hung and Yang [8] proposed entropy for IFSs by utilizing the idea of probability theory. Vlachos and Sergiadis [18] defined some new intuitionistic fuzzy entropy and established a relationship between fuzzy entropy and intuitionistic fuzzy set. Bajaj et al. [3] proposed a parametric R-norm entropy in intuitionistic fuzzy environment and discussed its applications in image thresholding and pattern recognition. Verma and Sharma [17] also proposed R-norm intuitionistic fuzzy entropy using the Szmidt and Kacprzyk's method for transforming intuitionistic fuzzy sets into fuzzy sets [15]. Presently, Joshi and Kumar [13] defined a more general (R, S)-norm intuitionistic fuzzy entropy and applied it in decision-making problems. Sometimes, it is common practice to look other possible measures, parametric generalizations and extensions of existing measures.

In this communication, we propose a new entropy called '(R, S)'-norm intuitionistic fuzzy entropy in the setting of intuitionistic fuzzy set. This manuscript is arranged as follows. In Sect. 2, we introduced some intuitive definitions of fuzzy sets, intuitionistic fuzzy sets and their corresponding entropies which are well known in literature. In Sect. 3, a new (R, S)-norm intuitionistic fuzzy entropy is proposed and its axiomatic requirements are proved. The paper is finally concluded in Sect. 4.

2 Preliminaries

2.1 Shannon's Entropy

Let $V = \{v_1, v_2, \ldots, v_n\}$ is a finite discrete random variable with the probability distribution $P = (p_1, p_2, \ldots, p_n)$, then, Shannon's entropy [14] of probability distribution P is given by

$$E(A) = -\frac{1}{n} \sum_{k=1}^{n} p_k \ln p_k \qquad (1)$$

2.2 Entropy of Fuzzy Set

Let $V = \{v_1, v_2, \ldots, v_n\}$ be a finite universal set, then a fuzzy set $A = \{<v, \phi_A(v_k)> \mid v_k \in V\}$, on V is defined by the membership function $\phi_A : V \rightarrow [0, 1]$ and the number $\phi_A(v_k)$ defines the extent of membership of an object $v \in V$ in A [20]. In fuzzy set theory, the entropy is a quantitative measurement of fuzziness in fuzzy set which gives the expected quantity of stress in taking inferences about an object weather it belongs to a set or not. De Luca and Termini [6] proposed the fuzzy entropy of a fuzzy set A as follows:

$$E = -\frac{1}{n} \sum_{k=1}^{n} [\phi_A(v_k) \log \phi_A(v_k) + (1 - \phi_A(v_k)) \log(1 - \phi_A(v_k))] \qquad (2)$$

2.3 R-norm Entropy

Let $\Theta = \{P = (p_1, p_2, \ldots, p_n) : p_k \geq 0 \ \& \ \sum_{k=1}^{n} p_k = 1\}$, $n \geq 2$ be the set of all probability distributions. Boekee and Lubbe [4] defined R-norm information measures of the distribution P for $R \in \Re^+$ as follows:

$$E_R(P) = \frac{R}{R-1} \left[1 - \sum_{k=1}^{n} p_k^R \right]; R > 0, \quad R \neq 1 \qquad (3)$$

The essential characteristics of the above R-norm information measure (3) are that when $R \rightarrow 1$, it tends to Shannon's entropy and when $R \rightarrow \infty, E_R(P) \rightarrow 1 - \max p_k; k = 1, 2, \ldots, n$. Equivalent to (3), Hooda [7] introduced R-norm fuzzy entropy of a fuzzy set A as

$$E_R(A) = \frac{R}{n(R-1)} \left[\sum_{i=1}^{n} \left(1 - \left(\phi_A^R(v_i) + (1 - \phi_{\tilde{A}}(v_i))^R \right)^{\frac{1}{R}} \right) \right]; R > 0, R \neq 1 \qquad (4)$$

2.4 Intuitionistic Fuzzy Set [1, 2]

Let $V = \{v_1, v_2, \ldots, v_n\}$ be a finite universal set, an intuitionistic fuzzy set (IFS) $\tilde{A}$

$$\tilde{A} = \{ < v, \phi_{\tilde{A}}(v_k), \psi_{\tilde{A}}(v_k) > \mid v_k \in V \}, \tag{5}$$

On V is given by is defined by the membership function $\phi_{\tilde{A}}, \psi_{\tilde{A}} : V \to [0, 1]$ and non-membership function of $\tilde{A}$, respectively. Furthermore, $\phi_{\tilde{A}}(v), \psi_{\tilde{A}}(v)$ indicate the grade of belonging and non-belonging of an object v in $\tilde{A}$, such that $0 \le \varphi_{\tilde{A}}(v) + \psi_{\tilde{A}}(v) \le 1$. The quantity $0 \le \varphi_{\tilde{A}}(v) + \psi_{\tilde{A}}(v) \le 1$. specify the grade of indefiniteness whether an element $v \in V$ belongs to $\tilde{A}$ or not.

2.5 Operations on Intuitionistic Fuzzy Set [1, 2]

Let IFS(V) denote the set of all intuitionistic fuzzy sets over the universal set V. Let $\tilde{A}, \tilde{B} \in$ IFS(V) be two IFSs. Then certain important operations may be given as follows:

- $\tilde{A} \subseteq \tilde{B}$ if $\phi_{\tilde{A}}(v) \ge \phi_{\tilde{B}}(v)\psi_{\tilde{A}}(v) \le \psi_{\tilde{B}}(v), \forall v \in V$;
- $\tilde{A} \supseteq \tilde{B}$ if $\phi_{\tilde{A}}(v) \le \phi_{\tilde{B}}(v)\psi_{\tilde{A}}(v) \ge \psi_{\tilde{B}}(v), \forall v \in V$;
- $\tilde{A} = \tilde{B}$ if $\phi_{\tilde{A}}(v) = \phi_{\tilde{B}}(v)\psi_{\tilde{A}}(v) = \psi_{\tilde{B}}(v), \forall v \in V$;
- $\tilde{A} \cup \tilde{B} = \{\max(\phi_{\tilde{A}}(v), \phi_{\tilde{B}}(v)), \min(\psi_{\tilde{A}}(v), \psi_{\tilde{B}}(v)) : v \in V\}$;
- $\tilde{A} \cap \tilde{B} = \{\min(\phi_{\tilde{A}}(v), \phi_{\tilde{B}}(v)), \max(\psi_{\tilde{A}}(v), \psi_{\tilde{B}}(v)) : v \in V\}$;,
- $\tilde{A} \cap \tilde{B} = \{\min(\phi_{\tilde{A}}(v), \phi_{\tilde{B}}(v)), \max(\psi_{\tilde{A}}(v), \psi_{\tilde{B}}(v)) : v \in V\}$;
- $\tilde{A}^c = \{v, \psi_{\tilde{A}}(v), \phi_{\tilde{A}}(v) : v \in V\}$.

2.6 R-norm Entropy of Intuitionistic Fuzzy Set

For an intuitionists fuzzy $\tilde{A}$, Bajaj et al. [3] defined a R-norm fuzzy entropy measure as follows:

$$E_R(A) = \frac{R}{n(R-1)} \left[\sum_{i=1}^{n} \left(1 - (\phi_{\tilde{A}}^R(v_i) + \psi_{\tilde{A}}^R(v_i) + \theta_{\tilde{A}}^R(v_i))^{\frac{1}{R}} \right) \right]; R > 0, R \ne 1 \tag{6}$$

Verma and Sharma [17] also proposed normalized R-norm intuitionistic fuzzy entropy measure for intuitionistic fuzzy sets A.

$$E_R(A) = \frac{1}{n\left(1 - 2^{\left(\frac{1-R}{R}\right)}\right)} \left[\sum_{i=1}^{n} \left(1 - \left(\frac{\phi_{\tilde{A}}(v_i) + 1 - \psi_{\tilde{A}}(v_i)}{2}\right)^R + \left(\frac{\psi_{\tilde{A}}(v_i) + 1 - \phi_{\tilde{A}}(v_i)}{2}\right)^R\right)\right];$$

$$R > 0, R \neq 1$$

$$(7)$$

Further, Joshi and Kumar [13] proposed a two parametric (*R*, *S*)-norm intuitionistic fuzzy entropy and used it to solve decision-making problems.

$$E_S^R(\tilde{A}) = \begin{cases} \frac{R \times S}{n(R-S)} \left[\sum_{i=1}^{n} \left(\left(\left(\phi_{\tilde{A}}^S(v_i) + (1 - \phi_{\tilde{A}}(v_i))^S\right)^{\frac{1}{S}}\right) - \left(\left(\phi_{\tilde{A}}^R(v_i) + (1 - \phi_{\tilde{A}}(v_i))^R\right)^{\frac{1}{R}}\right)\right)\right], \\ \text{either} = 0 < S < 1, R > 1 \text{ or } 0 < R < 1, S > 1 \\ \frac{R}{n(R-1)} \left[\sum_{i=1}^{n} \left(1 - \left(\phi_{\tilde{A}}^R(v_i) + (1 - \phi_{\tilde{A}}(v_i))^R\right)^{\frac{1}{R}}\right)\right], \\ \text{when } S = 1; \\ -\frac{1}{n} \sum_{k=1}^{n} \left[\phi_{\tilde{A}}(v_k) \log \phi_{\tilde{A}}(v_k) + (1 - \phi_{\tilde{A}}(v_k)) \log(1 - \phi_{\tilde{A}}(v_k))\right], \\ \text{when } S = R = 1. \end{cases}$$

$$(8)$$

3 Proposed (*R*, *S*)-Norm Intuitionistic Fuzzy Entropy Measure

In this section, we introduce two parametric intuitionistic fuzzy entropy measure-based (*R*, *S*)-norm.

Definition 1 Let $\tilde{A} = \{<v, \phi_{\tilde{A}}(v_k), \psi_{\tilde{A}}(v_k) > \mid v_k \in V\}$, belongs to collection of all IFSs Ω. Then, we define an information measure $E_S^R(A) : \Omega^n \to \Re, n \geq 2$ as follows:

1. when either $0 < S < 1 \& R \to 1$ or $0 < R < 1 \& S \to 1$

$$E_S^R(\tilde{A}) = -\frac{2^{\frac{(R-1)}{R}}}{n\left(1 - 2^{\frac{(R-S)}{SR}}\right)}$$

$$\left[\sum_{k=1}^{n} \left(\left(\left(\frac{\phi_{\tilde{A}}(v_k) + 1 - \psi_{\tilde{A}}(v_k)}{2}\right)^R + \left(\frac{\phi_{\tilde{A}}(v_k) + 1 - \psi_{\tilde{A}}(v_k)}{2}\right)^R\right)^{\frac{1}{R}}\right) - \left(\left(\frac{\phi_{\tilde{A}}(v_k) + 1 - \psi_{\tilde{A}}(v_k)}{2}\right)^S + \left(\frac{\phi_{\tilde{A}}(v_k) + 1 - \psi_{\tilde{A}}(v_k)}{2}\right)^S\right)^{\frac{1}{S}}\right)\right].$$

2. when $S = 1; R > 0, R \neq 1$

$$E_1^R(\tilde{A}) = -\frac{1}{n\left(1 - 2^{\frac{(1-R)}{R}}\right)} \sum_{k=1}^{n}\left[1 - \left(\left(\frac{\phi_{\tilde{A}}(v_k) + 1 - \psi_{\tilde{A}}(v_k)}{2}\right)^R + \left(\frac{\phi_{\tilde{A}}(v_k) + 1 - \psi_{\tilde{A}}(v_k)}{2}\right)^R\right)^{\frac{1}{R}}\right].$$

3. when $R = 1; S > 0, S \neq 1$

$$E_S^1(\tilde{A}) = -\frac{1}{n\left(1 - 2^{\frac{(1-S)}{S}}\right)} \sum_{k=1}^{n}\left[1 - \left(\left(\frac{\phi_{\tilde{A}}(v_k) + 1 - \psi_{\tilde{A}}(v_k)}{2}\right)^S + \left(\frac{\phi_{\tilde{A}}(v_k) + 1 - \psi_{\tilde{A}}(v_k)}{2}\right)^S\right)^{\frac{1}{S}}\right].$$

4. when $S = 1$ & $R \to 1$ or $R = 1$ & $S \to 1$

$$E_S^R(\tilde{A}) = -\frac{1}{n\ln 2} \sum_{k=1}^{n}\left[\left(\frac{\phi_{\tilde{A}}(v_k) + 1 - \psi_{\tilde{A}}(v_k)}{2}\right)\ln\left(\frac{\phi_{\tilde{A}}(v_k) + 1 - \psi_{\tilde{A}}(v_k)}{2}\right) + \left(\frac{\psi_{\tilde{A}}(v_k) + 1 - \phi_{\tilde{A}}(v_k)}{2}\right)\ln\left(\frac{\psi_{\tilde{A}}(v_k) + 1 - \phi_{\tilde{A}}(v_k)}{2}\right)\right]$$

Theorem 1 *An entropy of an IFS $\tilde{A}$ denoted by $E_R^S(\tilde{A})$ and defined in Definition 1 is a reliable entropy measure, if satisfies the following axiomatic properties:*

(i) **Sharpness**: $E_S^R(\tilde{A}) = 0$ *iff $\tilde{A}$ is well defined, i.e., either $\phi_{\tilde{A}}(v_k) = 1, \psi_{\tilde{A}}(v_k) = 0$ or $\phi_{\tilde{A}}(v_k) = 0, \psi_{\tilde{A}}(v_k) = 1, \forall v_k \in V$.*

(ii) **Maximality**: $E_S^R(\tilde{A}) = 1$ *iff $\phi_{\tilde{A}}(v_k) = \psi_{\tilde{A}}(v_k), \forall v_k \in V$.*

(iii) **Resolution**: $E_S^R(\tilde{A}) \leq E_S^R(\tilde{B})$ *if $\tilde{A}$ is sharper than $\tilde{B}$, i.e. if $\phi_{\tilde{A}}(v_k) \leq \phi_{\tilde{B}}(v_k)$ and $\psi_{\tilde{A}}(v_k) \geq \psi_{\tilde{B}}(v_k)$ for $\phi_{\tilde{B}}(v_k) \leq \psi_{\tilde{B}}(v_k)$ & if $\phi_{\tilde{A}}(v_k) \geq \phi_{\tilde{B}}(v_k)$ and $\psi_{\tilde{A}}(v_k) \leq \psi_{\tilde{B}}(v_k)$ for $\phi_{\tilde{B}}(v_k) \geq \psi_{\tilde{B}}(v_k)$ for any $v_k \in V$.*

(iv) **Symmetry**: $E_S^R(\tilde{A}) = E_S^R(\tilde{A}^c)$ *for all $\tilde{A} \in \text{IFS}(V)$.*

Proof In order to show that the proposed entropy defined in Definition 1 is a reliable information measure, it is sufficient to prove that it satisfies the four axioms (i)–(iv).

Sharpness: To prove Sharpness property, we require that $E_S^R(\tilde{A}) = 0$ iff $\tilde{A}$ is a crisp set, i.e. either $\phi_{\tilde{A}}(v_k) = 1, \psi_{\tilde{A}}(v_k) = 0$ or $\phi_{\tilde{A}}(v_k) = 0, \psi_{\tilde{A}}(v_k) = 1, \forall v_k \in V$ Suppose that $E_S^R(\tilde{A}) = 0$ for $R \neq S$ and $R, S > 0$.

Thus, from Definition 1, we have

$$
E_S^R(\tilde{A}) = -\frac{2^{\frac{(R-1)}{R}}}{n\left(1 - 2^{\frac{(R-S)}{SR}}\right)}
$$

$$
\left[\sum_{k=1}^{n}\left(\begin{array}{l}\left(\left(\frac{\phi_{\tilde{A}}(v_k)+1-\psi_{\tilde{A}}(v_k)}{2}\right)^R + \left(\frac{\phi_{\tilde{A}}(v_k)+1-\psi_{\tilde{A}}(v_k)}{2}\right)^R\right)^{\frac{1}{R}} \\ -\left(\left(\frac{\phi_{\tilde{A}}(v_k)+1-\psi_{\tilde{A}}(v_k)}{2}\right)^S + \left(\frac{\phi_{\tilde{A}}(v_k)+1-\psi_{\tilde{A}}(v_k)}{2}\right)^S\right)^{\frac{1}{S}}\end{array}\right)\right] = 0
$$

$$
\Rightarrow \left[\begin{array}{l}\left(\left(\frac{\phi_{\tilde{A}}(v_k)+1-\psi_{\tilde{A}}(v_k)}{2}\right)^R + \left(\frac{\phi_{\tilde{A}}(v_k)+1-\psi_{\tilde{A}}(v_k)}{2}\right)^R\right)^{\frac{1}{R}} \\ -\left(\left(\frac{\phi_{\tilde{A}}(v_k)+1-\psi_{\tilde{A}}(v_k)}{2}\right)^S + \left(\frac{\phi_{\tilde{A}}(v_k)+1-\psi_{\tilde{A}}(v_k)}{2}\right)^S\right)^{\frac{1}{S}}\end{array}\right]
$$

$$
= 0, \forall k = 1, 2, \ldots, n.
$$

i.e.,

$$
\left(\left(\frac{\phi_{\tilde{A}}(v_k)+1-\psi_{\tilde{A}}(v_k)}{2}\right)^R + \left(\frac{\phi_{\tilde{A}}(v_k)+1-\psi_{\tilde{A}}(v_k)}{2}\right)^R\right)^{\frac{1}{R}}
$$

$$
= \left(\left(\frac{\phi_{\tilde{A}}(v_k)+1-\psi_{\tilde{A}}(v_k)}{2}\right)^S + \left(\frac{\phi_{\tilde{A}}(v_k)+1-\psi_{\tilde{A}}(v_k)}{2}\right)^S\right)^{\frac{1}{S}}, \forall k = 1, 2, \ldots, n.
$$

Since $R \neq S$ and $R, S > 0$, therefore, the above equation is hold iff either $\phi_{\tilde{A}}(v_k) = 1, \psi_{\tilde{A}}(v_k) = 0$ or $\phi_{\tilde{A}}(v_k) = 0, \psi_{\tilde{A}}(v_k) = 1, \forall k = 1, 2, \ldots, n$. Conversely, we assume that the set $\tilde{A}$ is a crisp set, i.e., either $\phi_{\tilde{A}}(v_k) = 1, \psi_{\tilde{A}}(v_k) = 0$ or $\phi_{\tilde{A}}(v_k) = 0, \psi_{\tilde{A}}(v_k) = 1, \forall k = 1, 2, \ldots, n$. Then for $R \neq S$ and $R, S > 0$, we have

$$
\left(\left(\frac{\phi_{\tilde{A}}(v_k)+1-\psi_{\tilde{A}}(v_k)}{2}\right)^R + \left(\frac{\phi_{\tilde{A}}(v_k)+1-\psi_{\tilde{A}}(v_k)}{2}\right)^R\right)^{\frac{1}{R}}
$$

$$
- \left(\left(\frac{\phi_{\tilde{A}}(v_k)+1-\psi_{\tilde{A}}(v_k)}{2}\right)^S + \left(\frac{\phi_{\tilde{A}}(v_k)+1-\psi_{\tilde{A}}(v_k)}{2}\right)^S\right)^{\frac{1}{S}} = 0,
$$

$\forall k = 1, 2, \ldots, n$. This implies that $E_S^R(\tilde{A}) = 0$. Thus, $E_S^R(\tilde{A}) = 0$ iff $\tilde{A}$ is a crisp set.

Maximality: In order to prove that $E_S^R(\tilde{A}) = 1$ iff $\phi_{\tilde{A}}(v_k) = \psi_{\tilde{A}}(v_k), \forall v_k \in V$. Then it is required that the function

$$
f(\tau, v) = \frac{2^{\frac{(R-1)}{R}}}{n\left(1 - 2^{\frac{(R-S)}{SR}}\right)} \left(\left(\frac{\tau+1-v}{2}\right)^R + \left(\frac{v+1-\tau}{2}\right)^R\right)^{\frac{1}{R}}
$$
$$
- \left(\left(\frac{\tau+1-v}{2}\right)^S + \left(\frac{v+1-\tau}{2}\right)^S\right)^{\frac{1}{S}},
$$
(9)

where $\tau, v \in [0, 1]$ is a concave and it has global maxima at $\tau = v$. Differentiating f partially with respect to τ and v, respectively, we get

$$
\frac{\partial f}{\partial \tau}(\tau, v) = \frac{2^{\frac{-1}{R}}}{\left(1 - 2^{\frac{(R-S)}{SR}}\right)}
$$
$$
\left[\left(\left(\frac{\tau+1-v}{2}\right)^R + \left(\frac{v+1-\tau}{2}\right)^R\right)^{\frac{1}{R}-1}\left(\left(\frac{\tau+1-v}{2}\right)^{R-1} - \left(\frac{v+1-\tau}{2}\right)^{R-1}\right)\right]
$$
$$
\left[-\left(\left(\frac{\tau+1-v}{2}\right)^S + \left(\frac{v+1-\tau}{2}\right)^S\right)^{\frac{1}{S}-1}\left(\left(\frac{\tau+1-v}{2}\right)^{S-1} - \left(\frac{v+1-\tau}{2}\right)^{S-1}\right)\right]
$$

$$
\frac{\partial f}{\partial \tau}(\tau, v) = -\frac{2^{\frac{-1}{R}}}{\left(1 - 2^{\frac{(R-S)}{SR}}\right)}
$$
$$
\left[\left(\left(\frac{\tau+1-v}{2}\right)^R + \left(\frac{v+1-\tau}{2}\right)^R\right)^{\frac{1}{R}-1}\left(\left(\frac{\tau+1-v}{2}\right)^{R-1} - \left(\frac{v+1-\tau}{2}\right)^{R-1}\right)\right]
$$
$$
\left[-\left(\left(\frac{\tau+1-v}{2}\right)^S + \left(\frac{v+1-\tau}{2}\right)^S\right)^{\frac{1}{S}-1}\left(\left(\frac{\tau+1-v}{2}\right)^{S-1} - \left(\frac{v+1-\tau}{2}\right)^{S-1}\right)\right]
$$

The critical point of f can be obtain by setting

$$
\frac{\partial f}{\partial \tau}(\tau, v) = 0 \,\&\, \frac{\partial f}{\partial v}(\tau, v) = 0
$$
(10)

Since $R \neq S$ and $R, S > 0$, therefore from (10), we obtain $\tau = v$.

To examine the convexity of the function f, we evaluate the second-order partial derivative of f:

$$\frac{\partial^2 f}{\partial \tau^2}(\tau, v) = \frac{2^{-\frac{(R+1)}{R}}}{\left(1 - 2^{\frac{(R-S)}{SR}}\right)}$$

$$\left[\begin{array}{l} (R-1)\left(\left(\frac{\tau+1-v}{2}\right)^R + \left(\frac{v+1-\tau}{2}\right)^R\right)^{\frac{1}{R}-1}\left(\left(\frac{\tau+1-v}{2}\right)^{R-2} - \left(\frac{v+1-\tau}{2}\right)^{R-2}\right) \\ + (1-R)\left(\left(\frac{\tau+1-v}{2}\right)^R + \left(\frac{v+1-\tau}{2}\right)^R\right)^{\frac{1}{R}-2}\left(\left(\frac{\tau+1-v}{2}\right)^{R-1} - \left(\frac{v+1-\tau}{2}\right)^{R-1}\right)^2 \\ - (S-1)\left(\left(\frac{\tau+1-v}{2}\right)^S + \left(\frac{v+1-\tau}{2}\right)^S\right)^{\frac{1}{S}-1}\left(\left(\frac{\tau+1-v}{2}\right)^{S-2} + \left(\frac{v+1-\tau}{2}\right)^{S-2}\right) \\ - (1-S)\left(\left(\frac{\tau+1-v}{2}\right)^S + \left(\frac{v+1-\tau}{2}\right)^S\right)^{\frac{1}{S}-2}\left(\left(\frac{\tau+1-v}{2}\right)^{S-1} + \left(\frac{v+1-\tau}{2}\right)^{S-1}\right)^2 \end{array} \right]$$

$$\frac{\partial^2 f}{\partial v^2}(\tau, v) = \frac{2^{-\frac{(R+1)}{R}}}{\left(1 - 2^{\frac{(R-S)}{SR}}\right)}$$

$$\left[\begin{array}{l} (R-1)\left(\left(\frac{\tau+1-v}{2}\right)^R + \left(\frac{v+1-\tau}{2}\right)^R\right)^{\frac{1}{R}-1}\left(\left(\frac{\tau+1-v}{2}\right)^{R-2} - \left(\frac{v+1-\tau}{2}\right)^{R-2}\right) \\ + (1-R)\left(\left(\frac{\tau+1-v}{2}\right)^R + \left(\frac{v+1-\tau}{2}\right)^R\right)^{\frac{1}{R}-2}\left(\left(\frac{\tau+1-v}{2}\right)^{R-1} - \left(\frac{v+1-\tau}{2}\right)^{R-1}\right)^2 \\ - (S-1)\left(\left(\frac{\tau+1-v}{2}\right)^S + \left(\frac{v+1-\tau}{2}\right)^S\right)^{\frac{1}{S}-1}\left(\left(\frac{\tau+1-v}{2}\right)^{S-2} + \left(\frac{v+1-\tau}{2}\right)^{S-2}\right) \\ - (1-S)\left(\left(\frac{\tau+1-v}{2}\right)^S + \left(\frac{v+1-\tau}{2}\right)^S\right)^{\frac{1}{S}-2}\left(\left(\frac{\tau+1-v}{2}\right)^{S-1} + \left(\frac{v+1-\tau}{2}\right)^{S-1}\right)^2 \end{array} \right]$$

Now, we consider two cases:

Case 1: $0 < S < 1$, $R > 1$: In this case, at the critical point $\tau = v$, we get

$$\frac{\partial^2 f}{\partial \tau^2}(\tau, v) < 0 \text{ and } \frac{\partial^2 f}{\partial \tau^2}(\tau, v)\frac{\partial^2 f}{\partial v^2}(\tau, v) - \left(\frac{\partial^2 f}{\partial \tau \partial v}(\tau, v)\right) > 0.$$

Case 2: $0 < S < 1$, $R > 1$: In this case, at the critical point $\tau = v$, also we get

$$\frac{\partial^2 f}{\partial \tau^2}(\tau, v) < 0 \text{ and } \frac{\partial^2 f}{\partial \tau^2}(\tau, v)\frac{\partial^2 f}{\partial v^2}(\tau, v) - \left(\frac{\partial^2 f}{\partial \tau \partial v}(\tau, v)\right)^2 > 0.$$

Therefore, in both cases, we conclude that f is a concave function.

Hence, by the concavity, f has a relative maxima at the critical point $\tau = v$ and its maximum value is 1. Consequently, $E_S^R(\tilde{A})$ has a relative maxima at the critical point $\phi_{\tilde{A}}(\tau_k) = \psi_{\tilde{A}}(\tau_k)$ and its maximum value is 1.

Resolution: In order to show that entropy defined in Definition 1 satisfies (iii), we need to show that the function f defined in Eq. (9) is an increasing function with respect to τ, when $\tau \leq \upsilon$ and decreasing with respect to τ, when $\tau \geq \upsilon$.

The critical point of f can be obtained by setting

$$\frac{\partial f}{\partial \tau}(\tau, v) = 0 \,\&\, \frac{\partial f}{\partial v}(\tau, v) = 0. \tag{11}$$

Since $R \neq S$ and R, $S > 0$, therefore, from (11), we obtain $\tau = \upsilon$.

Now, first, we let $\tau \leq \upsilon$, and then following two cases arise:

Case 1: $0 < R < 1, S > 1$, then $\dfrac{2^{-\frac{1}{R}}}{\left(1 - 2^{\frac{(R-S)}{SR}}\right)} > 0$

$$\left(\left(\frac{\tau+1-v}{2}\right)^{R-1} - \left(\frac{v+1-\tau}{2}\right)^{R-1}\right) \geq 0 \,\&\, \left(\left(\frac{\tau+1-v}{2}\right)^{S-1} - \left(\frac{v+1-\tau}{2}\right)^{S-1}\right) \leq 0.$$

Case 2: If $R > 1, 0 < S < 1$, then, we have $\dfrac{2^{-\frac{1}{R}}}{\left(1 - 2^{\frac{(R-S)}{SR}}\right)} < 0$ and

$$\left(\left(\frac{\tau+1-v}{2}\right)^{R-1} - \left(\frac{v+1-\tau}{2}\right)^{R-1}\right) \leq 0 \,\&\, \left(\left(\frac{\tau+1-v}{2}\right)^{S-1} - \left(\frac{v+1-\tau}{2}\right)^{S-1}\right) \geq 0$$

which implies $\frac{\partial f}{\partial \tau}(\tau, v) \geq 0$.

Next, we let $\tau \geq v$, then the following two cases arise:

Case 1: If $0 < R < 1, S > 1$, then we have $\dfrac{2^{-\frac{1}{R}}}{\left(1 - 2^{\frac{(R-S)}{SR}}\right)} > 0$ and

$$\left(\left(\frac{\tau+1-v}{2}\right)^{R-1} - \left(\frac{v+1-\tau}{2}\right)^{R-1}\right) \leq 0 \,\&\, \left(\left(\frac{\tau+1-v}{2}\right)^{S-1} - \left(\frac{v+1-\tau}{2}\right)^{S-1}\right) \geq 0$$

which implies $\frac{\partial f}{\partial \tau}(\tau, v) \leq 0$.

Case 2: If $R > 1, 0 < S < 1$, then we have $\dfrac{2^{-\frac{1}{R}}}{\left(1 - 2^{\frac{(R-S)}{SR}}\right)} < 0$ and

$$\left(\left(\frac{\tau+1-v}{2}\right)^{R-1} - \left(\frac{v+1-\tau}{2}\right)^{R-1}\right) \geq 0 \,\&\, \left(\left(\frac{\tau+1-v}{2}\right)^{S-1} - \left(\frac{v+1-\tau}{2}\right)^{S-1}\right) \leq 0$$

which implies $\frac{\partial f}{\partial \tau}(\tau, v) \leq 0$.

Thus, for any $\tau, v \in [0,1]\tau$, and $R, S > 0$, $R \neq S$, f is an increasing function with respect to τ for $\tau \leq v$ and f is a decreasing function with respect to τ, when $\tau \geq v$. On the other hand, we obtain that $\frac{\partial f}{\partial \tau}(\tau, v) \leq 0$, when $\tau \leq v$ and $R, S > 0$, $R \neq S$, $f \frac{\partial f}{\partial \tau}(\tau, v) \geq 0$, when $\tau \geq v$ and $R, S > 0$, $R \neq S$.

Now, we consider two sets $\tilde{A}, \tilde{B} \in \text{IFS}(V)$ such that $\tilde{A} \subseteq \tilde{B}$. Also, let universal set be $V = \{v_1, v_2, \ldots, v_n\}$ be divided into two disjoint sets V_1 and V_2 with $V_1 \cap V_2 = V$, such that for all $v_k \in V_1$, we have $\phi_{\tilde{A}}(v_k) \leq \phi_{\tilde{B}}(v_k) \leq \psi_{\tilde{A}}(v_k) \leq \phi_{\tilde{B}}(v_k)$, and for all $v_k \in V_2$, we have $\phi_{\tilde{A}}(v_k) \geq \phi_{\tilde{B}}(v_k) \geq \psi_{\tilde{A}}(v_k) \geq \phi_{\tilde{B}}(v_k)$. Then by the monotonicity of the function f and proposed entropy defined in Definition 1, we obtain that $E_S^R(\tilde{A}) \leq E_S^R(\tilde{B})$ when $\tilde{A} \subseteq \tilde{B}$.

Symmetry: From Definition 1 of $E_S^R(\tilde{A})$, we achieve that $E_S^R(\tilde{A}) = E_S^R(\tilde{A}^c)$. Hence, $E_S^R(\tilde{A})$ satisfies all the axiomatic requirement of the intuitionistic fuzzy entropy measure and Hence, it is a reliable entropy measure of intuitionistic fuzzy set.

4 Conclusion

In order measure the quantitative uncertainty or vagueness of intuitionistic fuzzy sets, we have successfully proposed new parametric (R, S)-norm-based intuitionistic fuzzy entropy. Then, we have proved that the proposed intuitionistic fuzzy entropy measure is a reliable information measure. The proposed (R, S)-norm-based intuitionistic fuzzy entropy provides flexibility and wider applications of intuitionistic fuzzy entropy in the field of image processing and data classification problems.

References

1. Atanassov, K.T.: Intuitionistic fuzzy sets. Fuzzy Sets Syst. **20**, 87–96 (1986)
2. Atanassov, K.T.: Intuitionistic Fuzzy Sets. Springer Physica-Verlag, Heidelberg (1999)
3. Bajaj, R., Kumar, T., Gupta, N.: R-norm intuitionistic fuzzy information measures and its computational applications. In: Communications in Computer and Information Science, vol. 305, no. 8, pp. 372–380. Springer (2012)
4. Boekee, D.E., Van der Lubbe, J.C.A.: The R-norm information measures. Inf. Control **45**, 136–155 (1980)
5. Chen, X., Kar, S., Ralescu, D.A.: Cross-entropy measure of uncertain variables. Inf. Sci. **201**, 53–60 (2012)
6. De Luca, A., Termini, S.: Definition of a non-probabilistic entropy in the setting of fuzzy sets theory. Inf. Control **20**, 301–312 (1972)
7. Hooda, D.S.: On generalized measure of fuzzy entropy. Mathematica Slovaca **54**, 315–325 (2004)
8. Hung, W.L., Yang, M.S.: Fuzzy entropy on intuitionistic fuzzy sets. Int. J. Intell. Syst. **21**, 443–451 (2006)

13. Joshi, R.J., Kumar, S.: Parametric (R, S)-norm entropy on intuitionistic fuzzy sets with a new approach in multiple atttribute decision making. Fuzzy Inf. Eng. **9**, 181–203 (2017)
 9. Kauffmann, A.: Introduction to the Theory of Fuzzy Subsets. New York Academic (1975)
10. Kosko, B.: Fuzzy entropy and conditioning. Inf. Sci. **40**, 165–174 (1986)
11. Ning, Y., Ke, H., Fu, Z.: Triangular entropy of uncertain variables with application to portfolio selection. Soft. Comput. **19**, 2203–2209 (2015)
12. Pal, N.R., Pal, S.K.: Object background segmentation using new definitions of entropy. In: Institute of Electrical Engineering, Proceedings Computers and Digital Techniques, vol. 136, pp. 284–295 (1989)
14. Shannon, C.E.: A mathematical theory of communication. Bell Syst. Tech. J. **27**, 379–423 (1948)
15. Szmidt, E., Kacprzyk, J.: Entropy for intuitionistic fuzzy sets. Fuzzy Sets Syst. **118**, 467–477 (2001)
16. Yager, R.R.: On measures of fuzziness and negation, part I: membership in the unit interval. Int. J. Gen Syst. **5**, 221–229 (1979)
17. Verma, R., Sharma, B.: R-norm entropy on intuitionistic fuzzy sets. J. Intell. Fuzzy Syst. **28**, 327–335 (2015)
18. Vlachos, I.K., Sergiadis, G.D.: Intuitionistic fuzzy information-applications to pattern recognition. Pattern Recogn. Lett. **28**, 197–206 (2007)
19. Yao, K., Gao, J., Dai, W.: Sine entropy for uncertain variables. Int. J. Uncertain. Fuzziness Knowl. Syst. **21**, 743–753 (2013)
20. Zadeh, L.A.: Fuzzy sets. Inf. Control **8**, 338–353 (1965)
21. Zadeh, L.A.: Probability measure of fuzzy events. J. Math. Anal. Appl. **23**, 421–427 (1968)

Author Index

© Springer Nature Singapore Pte Ltd. 2020
S. S. Dash et al. (eds.), *Artificial Intelligence and Evolutionary Computations in Engineering Systems*, Advances in Intelligent Systems and Computing 1056,
https://doi.org/10.1007/978-981-15-0199-9

Printed in the United States
By Bookmasters